IET ELECTROMAGNETIC WAVES SERIES 26

Series Editors: Professor P.J.B. Clarricoats
Professor J.R. Wait

Spherical Near-Field Antenna Measurements

Other volumes in this series:

Volume 1 **Geometrical theory of diffraction for electromagnetic waves, 3rd edition** G.L. James
Volume 10 **Aperture antennas and diffraction theory** E.V. Jull
Volume 11 **Adaptive array principles** J.E. Hudson
Volume 12 **Microstrip antenna theory and design** J.R. James, P.S. Hall and C. Wood
Volume 15 **The handbook of antenna design, volume 1** A.W. Rudge, K. Milne, A.D. Oliver and P. Knight (Editors)
Volume 16 **The handbook of antenna design, volume 2** A.W. Rudge, K. Milne, A.D. Oliver and P. Knight (Editors)
Volume 18 **Corrugated horns for microwave antennas** P.J.B. Clarricoats and A.D. Oliver
Volume 19 **Microwave antenna theory and design** S. Silver (Editor)
Volume 21 **Waveguide handbook** N. Marcuvitz
Volume 23 **Ferrites at microwave frequencies** A.J. Baden Fuller
Volume 24 **Propagation of short radio waves** D.E. Kerr (Editor)
Volume 25 **Principles of microwave circuits** C.G. Montgomery, R.H. Dicke and E.M. Purcell (Editors)
Volume 26 **Spherical near-field antenna measurements** J.E. Hansen (Editor)
Volume 28 **Handbook of microstrip antennas, 2 volumes** J.R. James and P.S. Hall (Editors)
Volume 31 **Ionospheric radio** K. Davies
Volume 32 **Electromagnetic waveguides: theory and applications** S.F. Mahmoud
Volume 33 **Radio direction finding and superresolution, 2nd edition** P.J.D. Gething
Volume 34 **Electrodynamic theory of superconductors** S.A. Zhou
Volume 35 **VHF and UHF antennas** R.A. Burberry
Volume 36 **Propagation, scattering and diffraction of electromagnetic waves** A.S. Ilyinski, G. Ya.Slepyan and A. Ya.Slepyan
Volume 37 **Geometrical theory of diffraction** V.A. Borovikov and B.Ye. Kinber
Volume 38 **Analysis of metallic antenna and scatterers** B.D. Popovic and B.M. Kolundzija
Volume 39 **Microwave horns and feeds** A.D. Olver, P.J.B. Clarricoats, A.A. Kishk and L. Shafai
Volume 41 **Approximate boundary conditions in electromagnetics** T.B.A. Senior and J.L. Volakis
Volume 42 **Spectral theory and excitation of open structures** V.P. Shestopalov and Y. Shestopalov
Volume 43 **Open electromagnetic waveguides** T. Rozzi and M. Mongiardo
Volume 44 **Theory of nonuniform waveguides: the cross-section method** B.Z. Katsenelenbaum, L. Mercader Del Rio, M. Pereyaslavets, M. Sorella Ayza and M.K.A. Thumm
Volume 45 **Parabolic equation methods for electromagnetic wave propagation** M. Levy
Volume 46 **Advanced electromagnetic analysis of passive and active planar structures** T. Rozzi and M. Farinai
Volume 47 **Electromagnetic mixing formulae and applications** A. Sihvola
Volume 48 **Theory and design of microwave filters** I.C. Hunter
Volume 49 **Handbook of ridge waveguides and passive components** J. Helszajn
Volume 50 **Channels, propagation and antennas for mobile communications** R. Vaughan and J. Bach-Anderson
Volume 51 **Asymptotic and hybrid methods in electromagnetics** F. Molinet, I. Andronov and D. Bouche
Volume 52 **Thermal microwave radiation: applications for remote sensing** C. Matzler (Editor)
Volume 53 **Principals of planar near-field antenna measurements** S. Gregson, J. McCormick and C. Parini
Volume 502 **Propagation of radiowaves, 2nd edition** L.W. Barclay (Editor)

Spherical Near-Field Antenna Measurements

Edited by J.E. Hansen

The Institution of Engineering and Technology

Published by The Institution of Engineering and Technology, London, United Kingdom

First published 1988
Reprinted 2008

The Institution of Engineering and Technology
Michael Faraday House
Six Hills Way, Stevenage
Herts, SG1 2AY, United Kingdom

www.theiet.org

British Library Cataloguing in Publication Data
Spherical near-field antenna measurements theory and practice. –
(Electromagnetic waves series; 26).
1. Antennas (Electronics) – Measurement
I. Hansen, J.E. II. Hald J.
III. Institute of Electrical Engineers
IV. Series
621.38'028'3 TK7871.6

ISBN (10 digit) 0 86341 110 X
ISBN (13 digit) 978-0-86341-110-6

Reprinted in the UK by Lightning Source UK Ltd, Milton Keynes

Contributing authors

J. Hald, Brüel & Kjær, Nærum, Denmark
J. E. Hansen, Technical University of Denmark
F. Jensen, TICRA, Copenhagen, Denmark
F. Holm Larsen, Technical University of Denmark

Contents

Preface

Spherical near-field antenna testing is an interesting and powerful method within the general area of antenna measurements. The present book aims at presenting theoretical and practical aspects of the spherical test technique in a comprehensive manner. The book is intended for use primarily by the research and development engineer engaged in antenna measurements. It is hoped, however, that it may also be found useful to others by providing reference material on related topics, notably spherical electromagnetic waves.

While some of the results described in the book have not been published before, most of its contents is based on research documented in doctoral dissertations, research reports and papers from the Electromagnetics Institute at the Technical University of Denmark (TUD) over the last twenty years.

The original inspiration to the work done at TUD in this area is owed to E. V. Jull of the University of British Columbia, Canada, who did some of the early work on near-field antenna theory and measurements in the late fifties and who was a visiting scientist at TUD for two periods in 1963–65. With the inauguration in 1967 of the large Radio Anechoic Chamber at TUD and the acquiring, little by little, of automated equipment and computing facilities, the idea of embarking into a research programme on near-field antenna measurements became obvious. The first result in this connection was achieved by F. Jensen whose thesis (1970) contains a basic theoretical element of spherical near-field testing, the transmission formula.

A. C. Ludwig, formerly at the Jet Propulsion Laboratory, Pasadena, USA, who was a visiting professor at TUD in 1972 and 1979, developed the first useful algorithm for utilizing Jensen's transmission formula. While at TUD, he also presented his now well-known definitions of cross polarization. In 1975, F. Holm Larsen spent a short period of time at the National Bureau of Standards (NBS) in Boulder, USA, where he learnt of the new, efficient algorithm published by P. Wacker for inverting Jensen's transmission formula and acquainted himself with a computer code developed by R. L. Lewis for the case without probe correction. As a part of his thesis work, Larsen presented in 1977 the first complete spherical near-field measurement with transformation to the far field based on his own algorithm in which probe correction was included.

Already in 1976 an agreement between the European Space Agency (ESA) and TUD for a joint effort regarding further development of spherical near-field testing was signed. In a series of theoretical and experimental contracts, in which TICRA of Copenhagen took part, an experimental spherical near-field facility at TUD was constructed which enables in-depth studies of the technique. A number of TUD staff members were involved in the project. Among these H. Bach, E. Lintz Christensen and F. Conrad were prominent. The TUD–ESA facility is in extensive use for research, measurement of antennas for space applications and calibration of standard-gain horns. The spherical test technique has become commercially available and is now implemented at several large test ranges in Europe and elsewhere.

The authors wish to acknowledge the support received from ESA during many years of co-operation. Special thanks are due to J. Aasted and N. E. Jensen, former and present head of the ESTEC antenna section, and J. Hammer for supervising the major part of our contracts. On many occasions over the years, the authors have had the opportunity to discuss antenna measurements with staff members of NBS. This fact has markedly influenced our work within spherical near-field testing and the presentations in this book. In this connection, the authors would like to thank A. Newell, P. Wacker, A. D. Yaghjian, R. L. Lewis, R. Wittmann and C. Stubenrauch. Thanks are also due to D. Hess of Scientific-Atlanta Inc., Atlanta, USA for many discussions.

The valuable help and suggestions from colleagues at TUD and TICRA in the preparation of the manuscript are gratefully acknowledged. J. Lemanczyk undertook the time-consuming task of reading the manuscript and making linguistic improvements and corrections. He also did the measurements presented in Chapter 5. S. K. Lynggaard computed and prepared the figures illustrating special functions in Chapter 2 and Appendix A2. He also produced a large number of the other figures in the book. A Frandsen made the computations for Section 4.4.6 and proofread some of the chapters. S. Boel Pedersen read Appendix A4 and suggested improvements.

As mentioned above, all chapters and appendices reflect previous work although generally in a considerably reworked and revised form. F. Holm Larsen wrote Sections 5.2 and 5.4 as well as Appendices A2 and A3. F. Jensen wrote Chapter 6. J. Hald wrote Chapter 7 and Section A1.5. The remainder of the book was written by J. E. Hansen.

November, 1986 J.E.H.

Chapter 1

Introduction

1.1 Background

1.1.1 Introduction

A need for precise measurement of the radiation from microwave antennas has arisen in connection with the development of advanced antenna design concepts and improved theoretical approaches for antenna analysis. The need has been felt in particular within space applications where antennas are being constructed to tight specifications, i.e. down to the order of one per cent in gain. Antenna testing has therefore attracted considerable interest in recent years where a large number of studies have been made and new developments effected.

The oldest type of test range, the outdoor far-field range, has seen a few modern implementations. However, it is to the indoor test techniques that most of the efforts have been devoted. The compact range as well as near-field scanning ranges in planar, cylindrical and spherical geometries have all more or less reached mature states. They are now being installed at many places and seem to be the natural choice for contemporary antenna testing.

The present book concerns itself with one of the newer measurement schemes: spherical near-field scanning.

1.1.2 Brief outline of antenna testing

The angular dependence of the electromagnetic field radiated from a transmitting antenna is not invariant but changes gradually with the distance. At large distances from the antenna it approaches a definitive shape, the far-field radiation pattern. For most antennas, the far-field pattern is the characteristic of primary interest. To obtain the far field experimentally for a given antenna is the objective of antenna testing.

For any fixed direction in space relative to the antenna under test, the radiated far field is a plane wave propagating in that particular direction. As a consequence of reciprocity, the amplitude and phase of the far field as a function of angle may therefore be recorded experimentally as the received signal when the test antenna is immersed and rotated in a plane wave field.

The ability to provide plane wave illumination of the antenna under test is a basic requirement for any test range. This is accomplished with an auxiliary antenna: the probe. The creation of a uniform plane wave zone of sufficient quality without parasitic interference fields can be achieved in a number of different ways.

In far-field test ranges, ample separation of the test antenna and the probe is used as a means of obtaining the plane wave condition. At the Rayleigh distance R, given by $R = 2D^2/\lambda$, where λ is the wavelength, the curvature of the spherical wave radiated from a small probe is small enough for the deviation from a plane wave over a volume containing a test antenna of diameter D to be less than $\pi/8$ in phase. Although the Rayleigh distance is commonly taken as sufficient when a true far-field recording is required, errors in the recorded pattern may not be negligible at this distance. Far-field ranges can be impractical due to the required measurement distance, typically several hundred metres, which in itself makes the suppression of reflections from the ground and nearby objects difficult. Nevertheless, accurate far-field ranges exist, although they are few in number, at least in Europe.

In general, solutions to the problem of constructing precision test ranges for antennas have been found in the protected indoor environment of radio anechoic chambers. The basic principle of the compact range is to use a probe in the shape of a large collimating device to generate a plane wave zone without requiring the large separation of a far-field range.

Systems for near-field scanning employ a single, small probe which is scanned in a regular grid across a measurement surface in the near field of the test antenna. The measurement surface may be plane (planar scanning) or it may be cylindrical (cylindrical scanning). It may also have the form of a spherical surface enclosing the test antenna. In this case the near-field scanning method is spherical scanning, the subject of this book. (See Fig. 1.1.) In all scanning methods, the far-field characteristics of the test antenna are derived by applying suitable processing schemes to the near-field data obtained from the scanning process. This is equivalent to viewing the plane wave zones of the near-field scanning ranges as being synthesized in a computer.

It is not the intention here to discuss the relative merits and drawbacks of the various types of test ranges. Such comparisons can probably only lead to meaningful results when particular applications are in mind. Instead, some of the features of spherical near-field testing are given below.

1.1.3 Spherical near-field testing

Some of the features of spherical near-field testing (SNFT) of antennas are:

1. SNFT employs the fact that antenna radiation may always be expressed in terms of a truncated expansion in electromagnetic spherical waves satisfying Maxwell's equations. The truncation error is small compared to other errors, i.e. measurement errors and computational round off errors. Due to a cut-off property of spherical waves the number of terms in such truncated spherical

Fig. 1.1 *Photograph of a scaled model of the DFS–Kopernikus communication satellite mounted for TM–TC antenna coverage testing on the spherical near-field test range at the Technical University of Denmark. The measuring probe is seen at the far right. (By permission of MBB/ERNO, Munich, FRG.)*

wave expansions is manageable and only increases quadratically with the antenna diameter.

2. SNFT is exact for spatially band limited radiation fields, i.e. fields for which the spherical wave expansion has a finite number of terms.
3. SNFT is based on a description of the complete antenna field in the same spherical coordinate system as is always used for far-field characteristics. The SNFT measurement set-up is similar to a far-field set-up collapsed to a short distance. The probe is always pointing towards the antenna under test. Probe calibration can take place using the same computer code and the same set-up as used for measurement of the antenna under test.

4. SNFT inherently provides antenna directivity. This makes it ideally suited to precision gain calibration of (for example) standard gain antennas of known ohmic losses.
5. In SNFT the measurement surface (a sphere) has a finite area. Truncations, which might introduce errors, are not required.
6. SNFT is suitable for the measurement of narrow beam antennas as well as broad beam or omnidirectional antennas.
7. Antenna gain can be determined in SNFT by a substitution method.
8. SNFT provides easy means of non-trivial self checking by change of the measurement distance or by translation of the antenna under test within the measurement sphere. The influence of reflections and multiple reflections can be assessed in this way.
9. Disturbing reflections from the environment may be suppressed in SNFT by using a directive probe.
10. SNFT can be configured in a number of different mechanical set-ups. Movement of the antenna under test can be completely avoided, if required. The physical dimensions of a set-up are only slightly larger than the size of the largest antenna that can be measured.

1.2 Organization of (this) book

Several review papers on antenna testing including spherical near-field methods have been published in recent years [1–8]. It is not the intention, therefore, to provide another comprehensive survey of the literature on test methods in this book. Instead, brief references to papers relevant to the material presented are included in the chapters.

Chapter 2 provides the necessary theoretical background for the later chapters by describing electromagnetic spherical wave expansions, their definitions and properties. Power normalized spherical waves in a new, simple notation are used throughout. An account of the antenna scattering matrix, a concept from general antenna theory, is given using the new notation. It is shown how the scattering matrix elements—i.e. the reflection coefficient, the receiving and transmitting coefficients as well as scattering coefficients—describe all of the antenna's properties in a concise manner.

In Chapter 3, a spherical near-field transmission formula is derived using the scattering matrix formulation of Chapter 2. As all other scanning techniques for antenna near-field testing, spherical scanning utilizes a transmission formula as its theoretical basis. A transmission formula expresses the complex signal received by an antenna when another antenna, usually within the near field, is transmitting. The receiving coefficient of the first antenna as well as the transmitting coefficients of the second antenna and various geometrical factors enter the spherical near-field transmission formula. It is shown that for large separations between the two antennas the spherical near-field transmission formula reduces to the well-known Friis's transmission formula.

In Chapter 4 the application to spherical near-field testing of the transmission formula derived in Chapter 3 is demonstrated. It is shown in detail how the transmitting coefficients of a given antenna, and thereby its directivity and pattern at all distances, can be computed numerically from data obtained experimentally in a spherical near-field range. An efficient algorithm which takes as its input the measured near-field data and provides the antenna transmitting coefficients, directivity and radiation pattern as output is outlined. The computations essentially amount to solving, or inverting, the transmission formula numerically.

Chapter 5 is devoted to experimental aspects of spherical near-field antenna testing. A test range basically contains two antennas, the antenna to be measured (the test antenna) and the probe. In ordinary far-field antenna measurements, the measured data is a function of probe polarization which must therefore be known in order that meaningful results can be obtained. In near-field measurements, however, the measured data is a function of probe pattern as well. Probe calibration for spherical near-field measurements therefore comprises polarization and pattern. Procedures for these calibrations are described in the chapter.

The probe is usually chosen as a medium-gain horn antenna. The chapter begins by studying an important class of rotationally symmetric probes, introduced in Chapter 4, which is particularly useful. It is shown how dual polarized probes may be constructed for enhanced measurement efficiency. The technique of correcting for non-perfect polarization of probes is also presented.

Design of spherical near-field test ranges is covered in some detail. Various mechanical set-ups are mentioned together with aspects of RF-circuit design and data collection and control systems.

Measurement procedures are covered in one of the sections. Emphasis here is on mechanical alignment, which is a critical step in near-field measurements. The complete alignment procedure applied at The Technical University of Denmark (TUD) is described as an example.

The chapter concludes with some results of specific measurements and calibrations carried out at the TUD spherical near-field test range.

Chapter 6 presents results from computer simulations of errors in spherical near-field measurements. It is shown that as far as the measurement of directivity is concerned, the accuracy is mainly affected by alignment errors which may reduce (or increase) the apparent area of the test antenna aperture. The accuracy of side lobe levels and cross-polar levels is mainly affected by such errors as receiver nonlinearity and drift. The influence of omitting probe correction is also discussed. These results are compared to the accuracy evaluations of the measurements presented in Chapter 5.

Truncation of the measurements by omitting all areas outside the main lobe and the first few side lobes can sometimes be exploited in order to save measurement time. The error associated with such truncations are studied in the chapter.

In Chapter 7 the alternative viewpoint on antenna measurements known as plane wave synthesis is presented. All types of antenna test ranges are in fact based upon the creation of a plane wave. The plane wave may exist physically as in compact ranges or as in far-field ranges. In these, the test antenna (if receiving) is rotated in a test zone where a field exists that is close to being a plane wave. The plane wave may also be synthesized in a computer, as a suitable weighted sum of the contributions from a transmitting probe which is moved stepwise (or from several probes present simultaneously) on a surface surrounding the test zone. The test antenna far field is then simply the sum of its responses to each transmitting probe, weighted in the same way.

For spherical near-field measurements, the plane-wave synthesis viewpoint has been introduced at the University of Sheffield in England. The theory is outlined in the chapter together with alternative approaches. The problem of numerical optimization of plane wave zones from spherical or planar arrays is addressed.

The book is concluded by five appendices of which the first three provide a systematic listing of equations and definitions pertaining to spherical waves. The fourth appendix summarizes elements of data processing theory pertinent to antenna measurements. Appendix 5 is a list of principle symbols and their uses.

Although a coherent exposition of spherical near-field measurements is attempted and the presentation therefore contains a number of references from chapter to chapter, much of the material can be read independently. This is particularly true as far as Chapter 2 and the appendices A1, A2 and A3 are concerned. These cover a large part of the classical theory on electromagnetic spherical waves and include a number of formulas that have not been published before. Some of the experimental techniques described in Chapter 5 are also of interest in connections other than spherical near-field measurements. As an example, the section on dual polarized probes is also applicable to plane polar near-field techniques. Chapter 7 on plane-wave synthesis can also be read independently of the previous chapters. Appendix A4 deals with the discrete Fourier transform (DFT) and applications within general antenna theory and measurements.

References

1. JOHNSON, R. C., ECKER, H. A. and HOLLIS, J. S., 1973, 'Determination of far-field antenna patterns from near-field measurements', *Proc. IEEE*, Vol. **61**, No. 12, pp. 1668–1694, Dec.
2. WACKER, P. F. and NEWELL, A. C., 1977, 'Advantages and disadvantages of planar, circular cylindrical, and spherical scanning and description of the NBS antenna scanning facilities', ESA Workshop on Antenna Testing Techniques, ESTEC, Holland, **ESA SP 127**, pp. 115–121, June.
3. KUMMER, W. H. and GILLESPIE, F. S., 1978, 'Antenna measurements—1978', *Proc. IEEE*, Vol. **66**, No. 4, pp. 483–507, Apr.
4. PARIS, D. T., LEACH, W. M., Jr. and JOY, E. B., 1978, 'Basic theory of probe-compensated near-field measurements', *IEEE Trans.*, Vol. **AP-26**, No. 3, pp. 373–379, May.

5. TURCHIN, V. I. and TSEYTLIN, N. M., 1979, 'Antenna testing based on near-field measurements (Review)', *Radio Engineering & Electronic Physics*, Vol. **24**, pp. 1–26, Dec.
6. APPEL-HANSEN, J. et al., 1982, 'Antenna Measurements', Chapter 8 in *The Handbook of Antenna Design*, Vol. **1**, Peter Peregrinus, London.
7. VOKURKA, V. J., 1984, 'Advanced antenna measurements', 14th European Microwave Conference, pp. 60–70, Liege, Belgium, Sept.
8. YAGHJIAN, A. D., 1986, 'An overview of near-field antenna measurements', *IEEE Trans.*, Vol. **AP-34**, pp. 30–45, Jan.

Chapter 2

Scattering matrix description of an antenna

2.1 Introduction

The presentation in later chapters of spherical near-field testing of antennas rests upon the characterization of each of the two antennas involved in the measurement situation, the antenna under test and the probe, in terms of a scattering matrix.

The scattering matrix description of an antenna was introduced by Dicke [1]. Since then the approach has been used in a number of studies [2, 3, 4, 5] of the radiating, scattering and coupling properties of arbitrary, structurally unspecified, lossless antennas. The scattering matrix contains the properties of an antenna system with respect to a set of antenna terminals or local port. It also describes the behaviour with respect to the electromagnetic field outside a 'minimum sphere' enclosing the system. Here, an infinite set (denumerable) of radiation ports are assumed one for each set of two spherical vector waves, usually an incoming and an outgoing wave. Similar to the waves in the transmission lines, the spherical waves must be carefully defined and should preferably be normalized with respect to power. The scattering matrix thus relates the complex amplitudes of all waves. Since an infinity of spherical waves is needed for the general field it must contain an infinite number of rows and columns. However, as we shall touch upon several times, the infinite matrix can be truncated. This accounts for an important physical principle, band limitation, which applies to all practical antenna structures. Such ordinary antenna properties as power gain and directivity are functionals on the scattering matrix elements.

Scattering matrix theory plays an important role in connection with spherical near-field testing. Firstly, the theory is convenient for the derivation of a 'spherical' transmission formula. Here, it is not required that the test antenna and the probe are both reciprocal antennas. Reciprocity, if present, is expressed conveniently as a relationship between elements of the scattering matrix. Secondly, scattering matrix theory provides the link between quantities measurable in transmission lines and external fields without reference to particular structural types. The theory constitutes a means for *absolute measurements* of electromagnetic fields without unknown or unspecified constants.

The present chapter begins (Section 2.2) with a description of spherical wave theory in sufficient detail for the purpose of a comprehensive treatment in later chapters of the theoretical sides of spherical near-field antenna measurements without probe correction (Section 4.2) and with probe correction (Section 4.3 and later chapters). The presentation of spherical waves is traditional, however with a new and simplified notation for the wave functions.

In Section 2.3 the main elements of power-normalized scattering matrix theory for antennas are introduced. The chapter concludes (Section 2.4) by giving formulas for important antenna properties and parameters in terms of spherical waves and scattering matrix elements. The theory of maximum directivity (Section 2.4.4) is a classical subject of general interest which is sometimes useful in computer simulation of near-field measurements.

2.2 Spherical waves

2.2.1 Introduction

Most of the material in this book depends on the particular solutions to Maxwell's equations known as spherical wave functions or spherical waves. These are well-defined and convenient building blocks for general electromagnetic fields. Their usefulness in antenna theory and spherical near-field measurements rests on many interesting properties of which several will be discussed here (and in further detail in Appendices A1, A2 and A3).

The applications of spherical waves within electromagnetic theory and antennas are numerous. A survey of the extensive literature on applications falls outside the scope of the present book. Spherical wave functions were introduced by W. W. Hansen [6] who proposed a method of generating solutions to the vector wave equation in any coordinate system in which the scalar wave equation is separable. A detailed formulation and derivation is given by Stratton [7]. Later presentations have been given in several books. Some of these are Morse and Feshbach [8], Harrington [9], Tai [10], Collin and Zucker [11], to mention a few. A more recent exposition is given by Ludwig [12].

With an assumed time dependence of $\exp(-i\omega t)$, Maxwell's equations for the electromagnetic field in a region of a linear, isotropic and homogeneous medium are

$$\nabla \times \vec{H} = -i\omega\varepsilon\vec{E} + \vec{J} \tag{2.1}$$

$$\nabla \times \vec{E} = i\omega\mu\vec{H} - \vec{M} \tag{2.2}$$

where $\vec{H}$ and $\vec{E}$ are the magnetic and electric field vectors, ω is the angular frequency, and ε and μ permittivity and permeability, respectively, of the medium. The assumed sources are $\vec{J}$ (electric current density) and $\vec{M}$ (magnetic current density).

In a source-free region, Maxwell's equations imply that both $\vec{E}$ and $\vec{H}$ satisfy the vector wave equation

$$\nabla \times (\nabla \times \vec{C}) - k^2\vec{C} = 0 \tag{2.3}$$

where k, the propagation constant, is given by $k = \omega\sqrt{\mu\varepsilon} = 2\pi/\lambda$, where λ is the wavelength. Only the loss-free case is considered. However, lossy media can be accounted for by making k complex in eqn (2.3). In fact, most of the formulas presented later can be applied with k and the specific admittance $\eta = \sqrt{(\varepsilon/\mu)}$ of the medium assumed complex. Exceptions are the formulas for power flow in Section 2.2.4 and the expressions for directivity and gain in Section 2.4.3.

The spherical wave functions are defined in the spherical coordinates (r, θ, ϕ) with unit vectors $(\hat{r}, \hat{\theta}, \hat{\phi})$ illustrated in Fig. 2.1 together with a rectangular coordinate system (x, y, z) with unit vectors $(\hat{x}, \hat{y}, \hat{z})$. With $\vec{r} = r\hat{r}$, and $\mathrm{f} = \mathrm{f}(r, \theta, \phi)$ being a generating function obtained by solving the scalar wave equation

$$(\nabla^2 + k^2)\mathrm{f} = 0 \tag{2.4}$$

it may be shown that the vector functions $\vec{m}$ and $\vec{n}$ defined through

$$\vec{m} = \nabla \mathrm{f} \times \vec{r} \tag{2.5}$$

$$\vec{n} = k^{-1}\nabla \times \vec{m} \tag{2.6}$$

satisfy eqn (2.3).

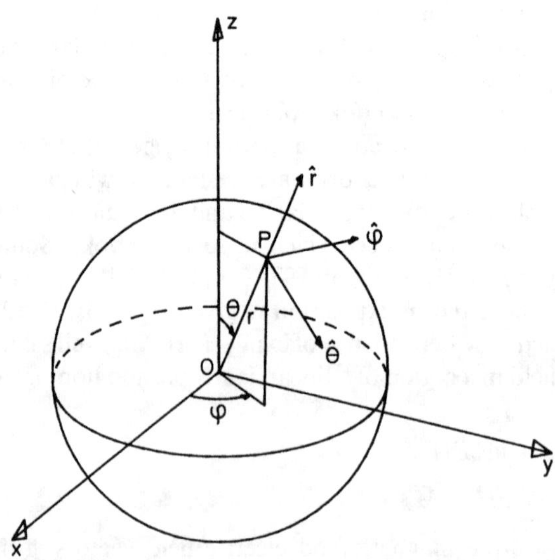

Fig. 2.1 *Rectangular coordinate system (x, y, z) and spherical coordinate system (r, θ, ϕ) with unit vectors $\hat{r}$, $\hat{\theta}$, $\hat{\phi}$. The spherical coordinates r, θ and ϕ are confined to the intervals $0 \leqslant r < \infty$, $0 \leqslant \theta \leqslant \pi$ and $0 \leqslant \phi < 2\pi$, respectively*

From eqns (2.3) and (2.6) we have

$$\vec{m} = k^{-2}\nabla \times (\nabla \times \vec{m}) \tag{2.7}$$

$$= k^{-1}\nabla \times \vec{n}. \tag{2.8}$$

Hence, $\vec{m}$ and $\vec{n}$ are related by curl operations and are ideally suited to represent an electromagnetic field $(\vec{E}, \vec{H})$ in a homogeneous medium.†

In Stratton [7], a generating function, obtained by separation of variables

$$\mathrm{f}^{(c)}_{\substack{e\\o}mn}(r, \theta, \phi) = z_n^{(c)}(kr)\, P_n^m(\cos\theta) \begin{matrix}\cos\\ \sin\end{matrix}\, m\phi \tag{2.9}$$

with $n = 1, 2, 3, \ldots$ and $m = 0, 1, 2, \ldots, n$ is used to generate $\vec{m}$ and $\vec{n}$. Here, e and o signify even and odd, respectively, referring to the choice of the trigonometric function in the ϕ-dependence. The function $P_n^m(\cos\theta)$ describing the θ-dependence of f is the associated Legendre function of nth degree and mth order. The radial function $z_n^{(c)}(kr)$ is specified by an upper index (c) as one of the functions

$$z_n^{(1)} = j_n(kr) \qquad \text{(spherical Bessel function)} \tag{2.10a}$$

$$z_n^{(2)} = n_n(kr) \qquad \text{(spherical Neumann function)} \tag{2.10b}$$

$$z_n^{(3)} = h_n^{(1)}(kr) = j_n(kr) + in_n(kr) \qquad \text{(spherical Hankel function of the first kind)} \tag{2.10c}$$

$$z_n^{(4)} = h_n^{(2)}(kr) = j_n(kr) - in_n(kr) \qquad \text{(spherical Hankel function of the second kind)} \tag{2.10d}$$

where $c = 1$ and $c = 2$ indicate standing waves, while $c = 3$ represents an outward travelling wave, and $c = 4$ an inward travelling wave.

In passing, it may be noted that the right-hand side of eqn (2.9) has the surprisingly simple property of being expressible by a finite number of trigonometric functions and inverse powers of kr.

With the proper choice of one or two values of c in eqn (2.9), a complete and orthogonal set of functions is obtained which may be used as a basis for all finite, continuous and single-valued solutions to (2.4) in any of the regions 1, 2 and 3 in Fig. 2.2.

By employing $\mathrm{f}^{(c)}_{\substack{e\\o}mn}(r, \theta, \phi)$ in eqns (2.5) and (2.6), we obtain the well-known solutions to the vector wave equation [7]

$$\vec{m}^{(c)}_{\substack{e\\o}mn}(r, \theta, \phi) = \mp z_n^{(c)}(kr)\frac{mP_n^m(\cos\theta)}{\sin\theta}\begin{matrix}\sin\\ \cos\end{matrix}\, m\phi\hat{\theta}$$

$$- z_n^{(c)}(kr)\frac{\mathrm{d}P_n^m(\cos\theta)}{\mathrm{d}\theta}\begin{matrix}\cos\\ \sin\end{matrix}\, m\phi\hat{\phi} \tag{2.11}$$

† The third (irrotational) W. W. Hansen vector wave function $\vec{l} = \nabla \mathrm{f}$ is not needed in the present book. For applications of $\vec{l}$, see ref. [33].

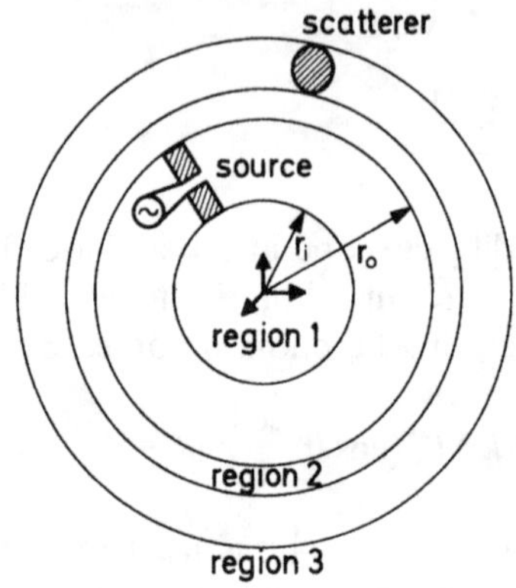

Fig. 2.2 *Source-free regions 1, 2 and 3 limited by spherical surfaces centred at the origin of the coordinate system*

$$\vec{n}^{(c)}_{\substack{e\\o}mn}(r,\theta,\phi) = \frac{n(n+1)}{kr} z_n^{(c)}(kr) P_n^m(\cos\theta) \begin{matrix}\cos\\ \sin\end{matrix} m\phi\hat{r}$$
$$+ \frac{1}{kr}\frac{\mathrm{d}}{\mathrm{d}(kr)}\{krz_n^{(c)}(kr)\} \frac{\mathrm{d}P_n^m(\cos\theta)}{\mathrm{d}\theta} \begin{matrix}\cos\\ \sin\end{matrix} m\phi\hat{\theta}$$
$$\mp \frac{1}{kr}\frac{\mathrm{d}}{\mathrm{d}(kr)}\{krz_n^{(c)}(kr)\} \frac{mP_n^m(\cos\theta)}{\sin\theta} \begin{matrix}\sin\\ \cos\end{matrix} m\phi\hat{\phi}. \tag{2.12}$$

For use in connection with the definitions of far fields, we shall also need the asymptotic form of the $\vec{m}$ and $\vec{n}$ functions with $c = 3$ and $c = 4$ as kr tends to infinity. This is obtained by introducing the large argument approximations valid for $kr \gg n$

$$z_n^{(3)}(kr) \to (-i)^{n+1}\frac{\mathrm{e}^{ikr}}{kr} \qquad kr \to \infty \tag{2.13}$$

$$z_n^{(4)}(kr) \to i^{n+1}\frac{\mathrm{e}^{-ikr}}{kr} \qquad kr \to \infty \tag{2.14}$$

$$\frac{1}{kr}\frac{\mathrm{d}}{\mathrm{d}(kr)}\{krz_n^{(3)}(kr)\} \to (-i)^n\frac{\mathrm{e}^{ikr}}{kr} \qquad kr \to \infty \tag{2.15}$$

$$\frac{1}{kr}\frac{\mathrm{d}}{\mathrm{d}(kr)}\{krz_n^{(4)}(kr)\} \to i^n\frac{\mathrm{e}^{-ikr}}{kr} \qquad kr \to \infty \tag{2.16}$$

and replace the radial functions in (2.11) and (2.12) accordingly.

2.2.2 *Power-normalized spherical waves*

In order to obtain a more compact notation, Jensen [13] introduced one symbol $\vec{f}^{(c)}_{smn}$ to designate $\vec{m}$ and $\vec{n}$ with the convention that $s = 1$ denotes the $\vec{m}$-function while $s = 2$ gives the $\vec{n}$-function. At the same time, a slightly different generating function than that of eqn (2.9) was used

$$g_{mn}^{(c)}(r, \theta, \phi) = z_n^{(c)}(kr) P_n^{|m|}(\cos\theta)\, \mathrm{e}^{im\phi} \tag{2.17}$$

and where $n = 1, 2, 3, \ldots$ as before, but now $m = -n, -n+1, \ldots, 0, \ldots, n-1, n$. The factor of $\exp(im\phi)$, instead of the cosine and sine in eqn (2.9), is more convenient in connection with the rotation of spherical waves discussed in Appendix A2. Also, the use of the exponential is more adapted to the Fast Fourier Transformation algorithm used later.

For these reasons we shall use Jensen's notation, however in a power-normalized form, such that any single $c = 3$ spherical wave with amplitude 1 will radiate a power of $\frac{1}{2}$ watt. The power normalization introduced in this way turns out to be important in connection with the application to spherical near-field measurements in later chapters. The form of the generating function needed for this purpose is slightly more complicated:

$$F_{mn}^{(c)}(r, \theta, \phi) = \frac{1}{\sqrt{2\pi}} \frac{1}{\sqrt{n(n+1)}} \left(-\frac{m}{|m|}\right)^m z_n^{(c)}(kr)\, \bar{P}_n^{|m|}(\cos\theta)\, \mathrm{e}^{im\phi} \tag{2.18}$$

where $\bar{P}_n^m(\cos\theta)$ is the normalized associated Legendre function as defined by Belousov [14] (see Appendix A1). The factor $(-m/|m|)^m$ ensures that the phase of the modes follows the phase of the spherical harmonics as defined by Edmonds [15]. Coefficients for coordinate rotation, to be defined later, can then be applied directly to the vector modes.

We define

$$\left(-\frac{m}{|m|}\right)^m = 1 \qquad \text{for } m = 0. \tag{2.19}$$

Insertion of the generating function (2.18) into the right-hand side of (2.5) defines the spherical wave function

$$\begin{aligned}\vec{F}_{1mn}^{(c)}(r, \theta, \phi) &= \nabla F_{mn}^{(c)}(r, \theta, \phi) \times \vec{r} \\ &= \frac{1}{\sqrt{2\pi}} \frac{1}{\sqrt{n(n+1)}} \left(-\frac{m}{|m|}\right)^m \left\{ z_n^{(c)}(kr) \frac{im\bar{P}_n^{|m|}(\cos\theta)}{\sin\theta} \mathrm{e}^{im\phi}\hat{\theta} \right. \\ &\quad \left. - z_n^{(c)}(kr) \frac{\mathrm{d}\bar{P}_n^{|m|}(\cos\theta)}{\mathrm{d}\theta} \mathrm{e}^{im\phi}\hat{\phi} \right\}.\end{aligned} \tag{2.20}$$

Subsequent use of $\vec{F}_{1mn}^{(c)}$ in the right-hand side of (2.6) defines

$$\begin{aligned}\vec{F}_{2mn}^{(c)}(r, \theta, \phi) &= k^{-1}\nabla \times \vec{F}_{1mn}^{(c)}(r, \theta, \phi) \\ &= \frac{1}{\sqrt{2\pi}} \frac{1}{\sqrt{n(n+1)}} \left(-\frac{m}{|m|}\right)^m \left\{ \frac{n(n+1)}{kr} z_n^{(c)}(kr) \bar{P}_n^{|m|}(\cos\theta)\, \mathrm{e}^{im\phi}\hat{r} \right. \\ &\quad + \frac{1}{kr} \frac{\mathrm{d}}{\mathrm{d}(kr)} \{kr z_n^{(c)}(kr)\} \frac{\mathrm{d}\bar{P}_n^{|m|}(\cos\theta)}{\mathrm{d}\theta} \mathrm{e}^{im\phi}\hat{\theta} \\ &\quad \left. + \frac{1}{kr} \frac{\mathrm{d}}{\mathrm{d}(kr)} \{kr z_n^{(c)}(kr)\} \frac{im\bar{P}_n^{|m|}(\cos\theta)}{\sin\theta} \mathrm{e}^{im\phi}\hat{\phi} \right\}.\end{aligned} \tag{2.21}$$

Equations (2.20) and (2.21) are the (dimensionless) power-normalized spherical wave functions used throughout this book. Asymptotic forms of the radial functions (2.13–2.16) may be inserted in a similar manner into $\vec{F}^{(c)}_{1mn}(r, \theta, \phi)$ and $\vec{F}^{(c)}_{2mn}(r, \theta, \phi)$ as was done in connection with the $\vec{m}$ and $\vec{n}$ functions.

The electric field in a source-free region of space such as region 2 of Fig. 2.2 may now be written as a weighted sum of the spherical wave functions defined above

$$\vec{E}(r, \theta, \phi) = \frac{k}{\sqrt{\eta}} \sum_{csmn} Q^{(c)}_{smn} \vec{F}^{(c)}_{smn}(r, \theta, \phi). \tag{2.22}$$

From eqn (2.2), we have for the associated magnetic field

$$\begin{aligned}\vec{H}(r, \theta, \phi) &= (i\omega\mu)^{-1} \nabla \times \vec{E}(r, \theta, \phi) \\ &= -ik\sqrt{\eta} \sum_{csmn} Q^{(c)}_{smn} \vec{F}^{(c)}_{3-s,m,n}(r, \theta, \phi).\end{aligned} \tag{2.23}$$

The wave functions $\vec{F}^{(c)}_{smn}$ are dimensionless and the additional factors in front of the summation signs in (2.22) and (2.23) are needed in order to obtain the dimension of watts$^{1/2}$ for the wave coefficients $Q^{(c)}_{smn}$. With this convention, the expression for the power P radiated by outward travelling modes is as simple as possible:

$$P = \frac{1}{2} \sum_{smn} |Q^{(3)}_{smn}|^2 \text{ watts} \tag{2.24}$$

as will be shown in Section 2.2.4.

In (2.24), the summation is understood in this way

$$\sum_{smn} = \sum_{s=1}^{2} \sum_{n=1}^{\infty} \sum_{m=-n}^{n} \tag{2.25}$$

while for the summation in (2.22) and (2.23)

$$\sum_{csmn} = \sum_{c=3}^{4} \sum_{s=1}^{2} \sum_{n=1}^{\infty} \sum_{m=-n}^{n}. \tag{2.26}$$

The choice of $c = 3$ and 4 in (2.26) indicates that both inward and outward propagating waves are present in the particular summation. In the general case, the proper choice of one or two values of c warrants a complete and orthogonal representation of all finite, continuous and single-valued solutions to Maxwell's equations in any of the regions 1, 2 or 3 in Fig. 2.2. In later chapters, the upper index c in expansions such as (2.22) and (2.23) is sometimes omitted. This is done when only one value of c is present in a particular sum and, furthermore, it is clear from the context which type of wave is needed.

In representations such as (2.22) and (2.23), the employment of the index s in the spherical wave functions $\vec{F}^{(c)}_{smn}$ serves the purpose of distinguishing between two different functions. One of the two wave functions, i.e. $\vec{F}^{(c)}_{1mn}$, has no radial component. It will therefore always represent that part of a (magnetic or electric) field which is transverse with respect to the radial coordinate r. The magnetic field of a transverse magnetic (TM) wave is proportional to $\vec{F}^{(c)}_{1mn}$ whereas the associated electric field is proportional to $\vec{F}^{(c)}_{2mn}$. For transverse electric (TE) waves, the roles of the wave functions are reversed. Here, the electric field is proportional to $\vec{F}^{(c)}_{1mn}$, while the associated magnetic field is represented by $\vec{F}^{(c)}_{2mn}$. Note that the sum of the indices s in the wave functions describing the electric and magnetic field vectors of a given electromagnetic field is always 3, as evident from (2.22) and (2.23).

When attached to the wave coefficient $Q^{(c)}_{smn}$, however, the index s has a slightly different meaning. Here, $s = 1$ always indicates a coefficient to a TE-wave while $s = 2$ indicates a TM-wave coefficient.

In practice the summations will always be truncated. The maximum value of n and m retained in the sums will be denoted by N and M, respectively. Since $\bar{P}^{|m|}_n(\cos\theta)$ vanishes for $|m| > n$, we must have $M \leqslant N$. In most cases, $M = N$ because a full mode set in m is usually needed. For this latter case, we shall define for later use an index transformation which converts the index triplet (s, m, n) to the single index j. The transformation is

$$j = 2\{n(n+1) + m - 1\} + s \tag{2.27}$$

whereby the triple summation is reduced to a single summation

$$\sum_{s=1}^{2}\sum_{n=1}^{N}\sum_{m=-n}^{n} = \sum_{j=1}^{J} \tag{2.28}$$

with

$$J = 2N(N+2). \tag{2.29}$$

The index transformation implies that we may use substitutions like, for example, $Q^{(c)}_{smn} = Q^{(c)}_j$ and $\vec{F}^{(c)}_{smn} = \vec{F}^{(c)}_j$ whenever a condensed notation is convenient.

The summation over s, m and n may be depicted in an (n, m) coordinate system (the nm plane) where all values of n and m present in the summation can be indicated. To each of these points correspond two terms of the summation, i.e. the terms with $s = 1$ and $s = 2$. The index transformation (2.27) and the nm plane is illustrated in Fig. 2.3.

2.2.3 The spherical waveguide

Spherical wave theory can be given a physical interpretation which turns out to be particularly useful in connection with spherical near-field measurements. As will be discussed in this section and in later chapters, spherical wave radiation may be visualized as taking place in a spherical waveguide [9, 16, 17]. Many of the well-known properties of conventional cylindrical guides are found to have their counterparts in the spherical waveguide. Concepts such as orthogonal modes, cut-off, propagation and evanescence are common features.

We consider a transmitting antenna which is not necessarily centred in a

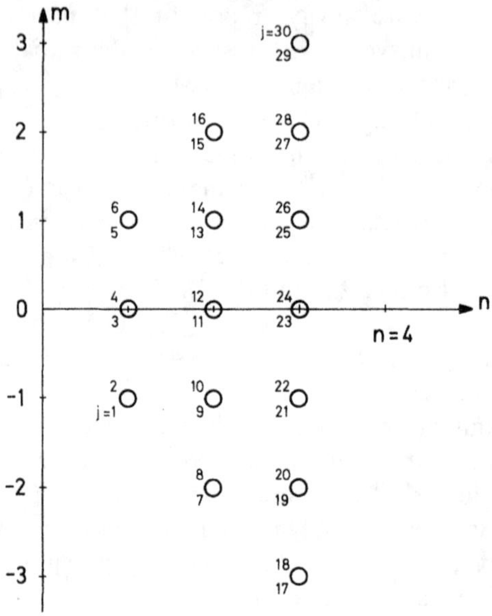

Fig. 2.3 *Illustrating the transformation $j = 2[n(n+1) + m - 1] + s$ between the index triplet (s, m, n) and the single index j. Since s takes the values 1 and 2, each indicated point in the nm-plane is associated with two values of the single index j*

spherical coordinate system. Let r_0 denote the radius of the smallest possible spherical surface (the *minimum sphere*) with centre at the origin and circumscribing the antenna. The space outside the sphere may now be viewed as a waveguide where the propagation takes place in the radial direction. The spherical guide is a waveguide extending from $r = r_0$ to infinity. Waveguide modes in the shape of spherical waves may propagate in this guide. If a scatterer is present as in Fig. 2.2, each of the regions 1, 2 and 3 may be treated as a section of a spherical waveguide. In the absence of any antenna or scatterer, the whole space may be viewed as a spherical waveguide where a wave coming in from infinity will converge towards the origin, pass through it, and diverge back to infinity.

All cross-sections of the spherical waveguide have a finite area and the mode spectrum is discrete. The spherical wave functions introduced in the previous sections are the modes of the spherical waveguide. They are divided into two classes, TE and TM modes as mentioned in Section 2.2.2. In TE-modes, the radial component of $\vec{E}$ is equal to zero and for TM-modes $H_r = 0$. At large values of kr ($kr \gg n$), the radial component vanishes as $(kr)^{-2}$ whereas the θ- and ϕ-components of the spherical waves decay as $(kr)^{-1}$ (see eqns (2.13)–(2.16)). In the far field all modes are purely transverse.

The r-variation of a mode is linked to the polar index n and is m-independent. The field will in general consist of both incoming and outgoing modes. With the chosen time dependence of $\exp(-i\omega t)$, the type $c = 3$ radial functions, $h_n^{(1)}(kr)$ and $(1/kr)(\mathrm{d}/\mathrm{d}(kr))\{krh_n^{(1)}(kr)\}$, belong to outward propagating modes satisfying the radiation condition at infinity. The functions of type $c = 4$, namely $h_n^{(2)}(kr)$ and $(1/kr)(\mathrm{d}/\mathrm{d}(kr))\{krh_n^{(2)}(kr)\}$, correspond to inward propagation. If a source-free region contains the centre of the spherical waveguide (region 1 of Fig. 2.2), the type $c = 1$ functions, $j_n(kr)$ and $(1/kr)(\mathrm{d}/\mathrm{d}(kr))\{krj_n(kr)\}$, must be chosen since only these functions are finite at the origin. Alternatively, equal amounts of ingoing and outgoing waves may be used since

$$j_n(kr) = \tfrac{1}{2}(h_n^{(1)}(kr) + h_n^{(2)}(kr)) \tag{2.30}$$

In a conventional (loss-free) cylindrical waveguide with a constant cross-section, some modes may be above cut-off and propagate along the guide without decay while others are below cut-off and are decayed exponentially. In a spherical waveguide, the cross-section of the guide is increasing with r. This means that more and more spherical modes may propagate as r grows. Similarly, when r is diminished, fewer modes may propagate. The transition between propagation and evanescence for a mode occurs around a radial distance of $r_t = n/k$. With $r \ll r_t$, the decay is extremely rapid but diminishes as r approaches r_t. Over the transition region around r_t, propagation begins gradually and for $r \gg r_t$, a $c = 3$ mode of index n propagates in the outward direction with the simple spherical divergence rate of r^{-1} in amplitude.

As mentioned above, the minimum sphere for a given antenna is defined as the sphere of smallest radius centred at the origin to completely circumscribe the antenna. For an antenna with a minimum sphere of radius r_0, the radiated field may in principle contain an infinity of modes. However, if it is assumed that the various modes all have amplitudes of the same order of magnitude at $r = r_0$, those with $n > kr_0$ will be heavily attenuated outside r_0. Only modes with $n < kr_0$ are of importance in the far field or at some close distance where a near-field measurement may be taken. This is a consequence of the cut-off property of the radial functions for the outward propagating modes. The radial functions are very large (and dominantly imaginary) for r smaller than the cut-off distance n/k. For r larger than n/k, the amplitude of the radial functions behaves as r^{-1} to a good approximation. The features of the amplitude variation with r of spherical modes are illustrated in Fig. 2.4.

The important conclusion to be drawn from the above considerations is that representations such as (2.22–2.23) may be truncated at some $n = N$. A sufficient number of terms for convergence of a series representation of spherical wave functions is for all practical purposes given by the empirical rule

$$N = [kr_0] + n_1 \tag{2.31}$$

where r_0 is the radius of the minimum sphere and where the square brackets indicate the largest integer smaller than or equal to kr_0. The integer n_1 depends on the locations in the coordinate system of the source and the field point as well as on the accuracy required. Note that if the source is centred so that r_0 is as small as possible, N is also minimized according to eqn (2.31).

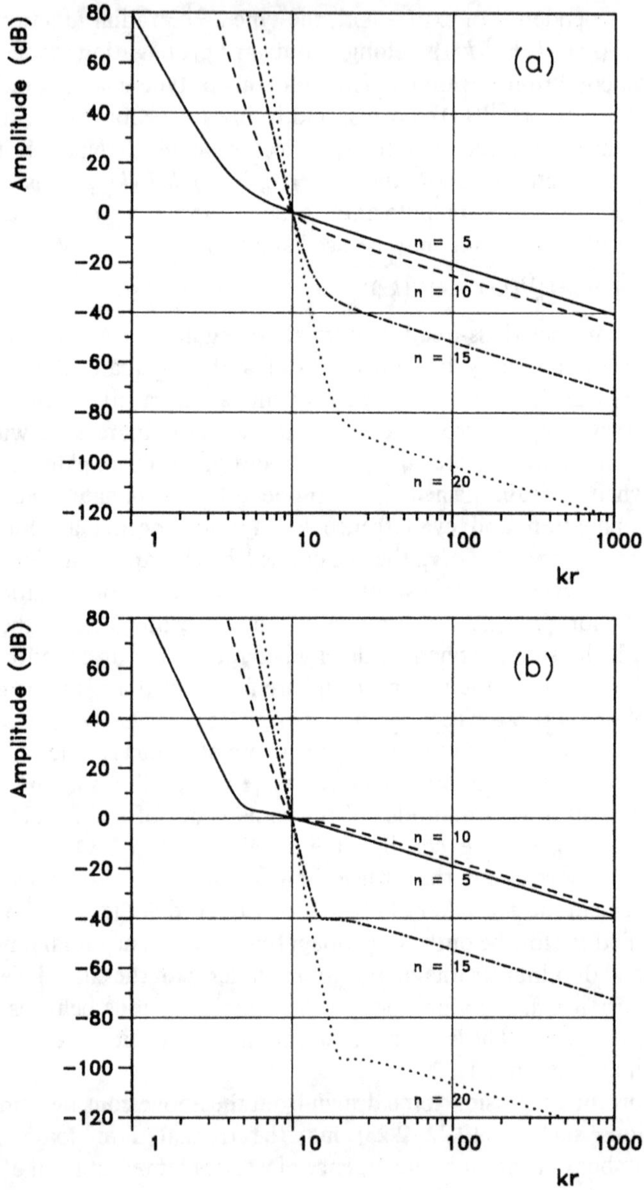

Fig. 2.4 *Relative amplitude variations in decibels (i.e.* $20\log_{10}|f|$*) of (a) the function* $f = h_n^{(1)}(kr)$ *and (b) the function* $f = \frac{1}{kr}\frac{d}{d(kr)}\{kr\,h_n^{(1)}(kr)\}$ *for* $n = 5, 10, 15$ *and* 20

For comparison the amplitudes are normalized to 0 dB at the surface of an assumed minimum sphere with $kr_0 = 10$. As can be seen, already at the distance $kr = 15$ a function with $n = 20$ is damped more than 50 dB relative to the $n = 10$ function. In the far field the difference in amplitude is about 75 dB

As an example, let us investigate the field at a point P from the disc-shaped, x-polarized current distribution with a 10 dB edge taper shown in Fig. 2.5. This distribution typifies the equivalent currents of an aperture field from a small reflector antenna. The symmetry of the distribution implies that only modes with $m = \pm 1$ will be present in the spherical wave series of the field.

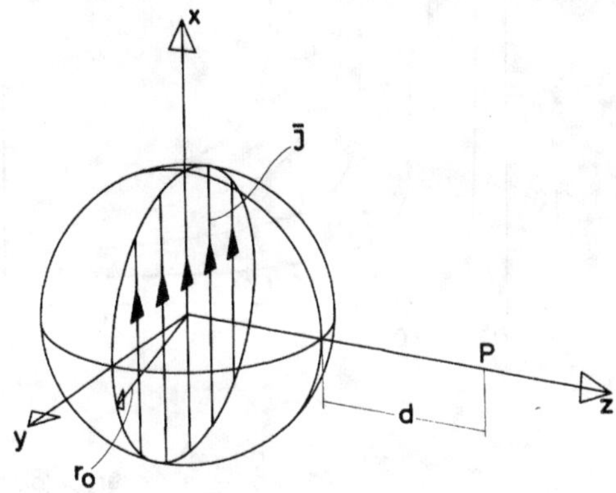

Fig. 2.5 *Geometry of the current distribution used for illustrating the convergence properties of a spherical wave series representation*

Using the spherical wave expansion of the current distribution the x-component of the electric field at the point P has been computed for various kd (see Fig. 2.5) and various values of n_1 (see eqn 2.31). Smoothed parametric curves for the error due to the truncation of the series are shown in Figs 2.6(a, b) for $kr_0 = 22$ and 44, respectively.†

It can be seen that for d larger than 5 wavelengths ($kd \gtrsim 30$), the truncation error is almost independent of d. Close to the minimum sphere ($d < \lambda$), there is a strong dependence upon d. Here, a relatively large n_1 is needed in order to obtain more than a few correct digits in the result.

The two sets of curves shown in Fig. 2.6 appear visually alike. The major difference is a spread to the right of the curves in Fig. 2.6b relative to the curves in Fig. 2.6a. The number n_1 required for a given fixed error and a fixed distance kd (radians) is approximately proportional to $(kr_0)^{1/3}$, (see Turchin and Tseytlin [18]).

If the field point is more than a few wavelengths from the minimum sphere and 4 correct digits in the field computation are sufficient, an obvious choice is $n_1 = 10$.

The spherical wave expansions of other current distributions show similar behaviour.

† Figures 2.5 and 2.6, and the discussion hereof, are contributed by S. K. Lynggaard.

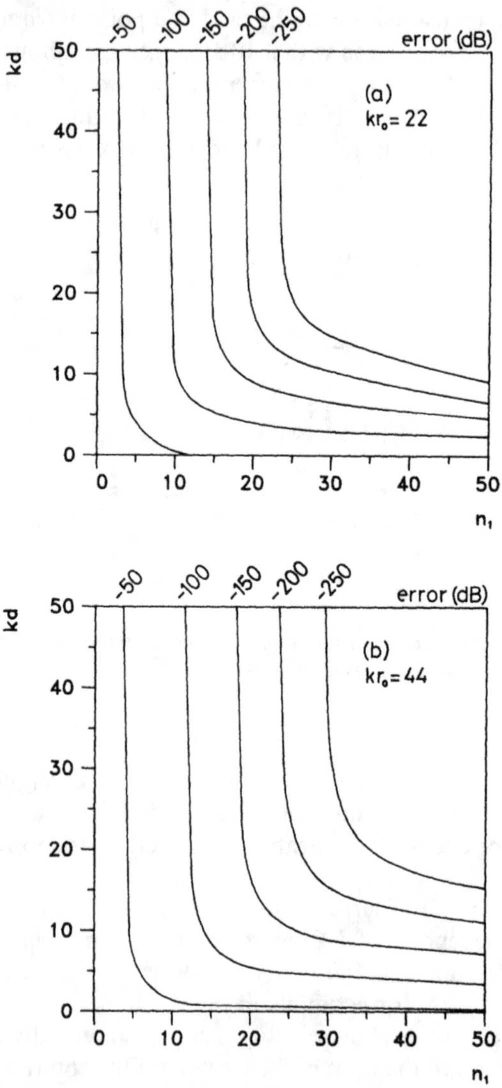

Fig. 2.6 *Truncation error in dB for the spherical wave expansion of the current distribution depicted in Fig. 2.5*
The interpolated error curves are shown versus the distance kd and the number n_1; see eqn. (2.31)
(a): $kr_0 = 22$ (b): $kr_0 = 44$

When the source is so far from the origin that a source-free inner spherical volume of radius r_i exists (see Fig. 2.2) spherical Bessel functions ($c = 1$ type functions) must be chosen in the spherical wave expansion for this region. We have here

$$N = [kr] + n_1 \tag{2.32}$$

valid for points of evaluation with $r < r_i$. The upper limit N is thus associated with the position of the point of evaluation and arises from the decay of the spherical Bessel function for $r < n/k$. Typically, $n_1 = 10$ may be used as in eqn (2.31).

The related question of whether field representations (2.22–2.23) can be truncated at some $|m| = M$ (where $M < N$) will be touched upon later.

The phase of the radial functions approximates the simple $\exp(ikr)$ dependence for $kr \gg n$. For smaller values of kr, the phase begins to deviate and at a certain distance given by $r = 4n^2/\pi k$, the phase deviation is close to $\pi/8$ for all modes [19]. The phase variation with r of spherical modes is illustrated in Fig. 2.7.

The amplitude and phase behaviour of the spherical modes divide, for each mode, the space into the regions

$0 \lesssim r \lesssim n/k$	evanescent region
$n/k \lesssim r \lesssim 4n^2/\pi k$	Fresnel region (near field)
$4n^2/\pi k \lesssim r < \infty$	Fraunhofer region (far field).

The three regions of space are also defined for an antenna radiating several modes. In this case, the limits are as follows

$r_0 \lesssim r \lesssim N/k$	evanescent region
$N/k \lesssim r \lesssim 4N^2/\pi k$	Fresnel region (near field)
$4N^2/\pi k \lesssim r < \infty$	Fraunhofer region (far field)

where as before, r_0 is the radius of the minimum sphere and N given by eqn (2.31).

For a source with a minimum sphere of diameter $D = 2r_0$, the Rayleigh distance R is conventionally defined as

$$R = \frac{2D^2}{\lambda} = \frac{2}{\lambda}(2r_0)^2 \tag{2.33}$$

$$\cong \frac{2}{\lambda}\left(2\frac{N}{k}\right)^2 = \frac{4}{\pi}\frac{N^2}{k} \tag{2.34}$$

which coincides with the above limit between the Fresnel and Fraunhofer regions.

The phi-variation $\exp(im\phi)$ of a mode depends on the azimuthal index m which indicates the number of oscillations in the interval $0 \leqslant \phi < 2\pi$. For $m \neq 0$, the field of a mode rotates around the z-axis with time. The rotation is in the positive sense from the x-axis towards the y-axis with m positive, and in

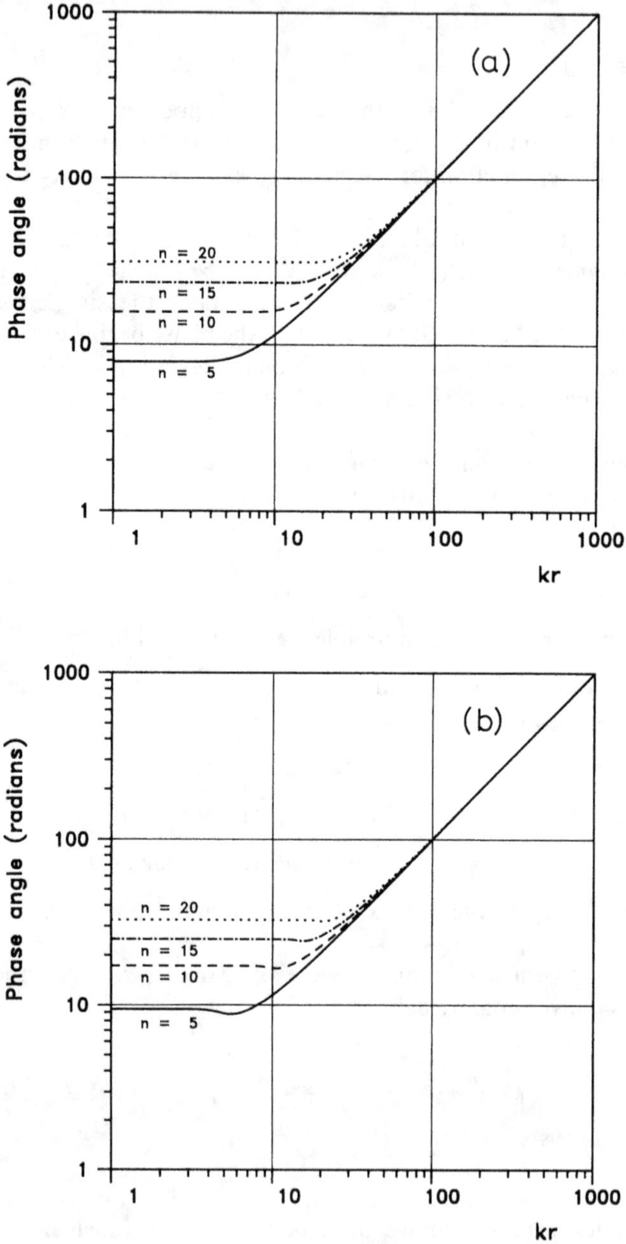

Fig. 2.7 *Relative phase variations of (a) the function* $f = h_n^{(1)}(kr)$ *and (b) the function* $f = \dfrac{1}{kr}\dfrac{d}{d(kr)}\{kr\,h_n^{(1)}(kr)\}$ *for* $n = 5, 10, 15$ *and 20. The phases are normalized so that all phase angles tend towards* kr *as* $kr \to \infty$

the negative sense if m is negative. The mode functions $\vec{F}^{(c)}_{smn}(r, \theta, \phi)$ are single-valued and continuous functions. For this reason, only modes with $m = 0$ can have a radial component on the z-axis, and only modes with $m = \pm 1$ will have θ- and ϕ-components here. Thus

$$[\vec{F}^{(c)}_{smn}(r, 0, \phi)]_r = [\vec{F}^{(c)}_{smn}(r, \pi, \phi)]_r = 0 \qquad \text{for } m \neq 0 \tag{2.35}$$

and

$$[\vec{F}^{(c)}_{smn}(r, 0, \phi]_{\substack{\theta\\ \phi}} = [\vec{F}^{(c)}_{smn}(r, \pi, \phi)]_{\substack{\theta\\ \phi}} = 0 \qquad \text{for } m \neq \pm 1. \tag{2.36}$$

For $m = \pm 1$, the field of each mode is circularly polarized on the z-axis.

The theta-variation of a mode depends on both n and m. It has the form of a standing wave with zeros spaced non-equidistantly between 0 and π. The zeros are more closely spaced in the neighbourhood of $\theta = \pi/2$ than around the poles. In certain regions close to the poles, i.e. for $0 < \sin\theta < \sin\theta_c = |m|/n$, the associated Legendre functions are below cut-off. Here, the mode amplitude tends rapidly towards zero as $\sin\theta \to 0$. From the formulas of Appendix A2 on coordinate rotation of spherical waves, one can obtain a Fourier expansion in theta of the components of a spherical mode. In this expansion the fastest oscillating terms have a theta-dependence of $\exp(in\theta)$ or $\exp(-in\theta)$. This property will be used later to derive a sampling criterion.

The field configurations showing phi- and theta-variations for individual components of certain modes are illustrated in Fig. 2.8.

2.2.4 *Power flow*

In this section we shall consider the power flow associated with an outgoing spherical wave field in a lossless media.

The power flow in a time-harmonic electromagnetic field is described by the complex Poynting vector $\vec{S}$ defined by†

$$\vec{S} = \tfrac{1}{2}\vec{E} \times \vec{H}^*. \tag{2.37}$$

The outward flux of the real part of $\vec{S}$ through a closed surface expresses the total power P radiated by the sources inside the surface. For a spherical surface of radius r

$$P = \int_{\phi=0}^{2\pi} \int_{\theta=0}^{\pi} \mathrm{Re}(\hat{r} \cdot \vec{S}) r^2 \sin\theta \, d\theta \, d\phi. \tag{2.38}$$

The outward flux of the imaginary part of $\vec{S}$ equals 2ω multiplied by the time average W_e of the electric energy minus the time average W_m of the magnetic energy stored outside the integration surface [20].

$$2\omega(W_e - W_m) = \int_{\phi=0}^{2\pi} \int_{\theta=0}^{\pi} \mathrm{Im}(\hat{r} \cdot \vec{S}) r^2 \sin\theta \, d\theta \, d\phi. \tag{2.39}$$

† The superscript * is used to denote complex conjugate.

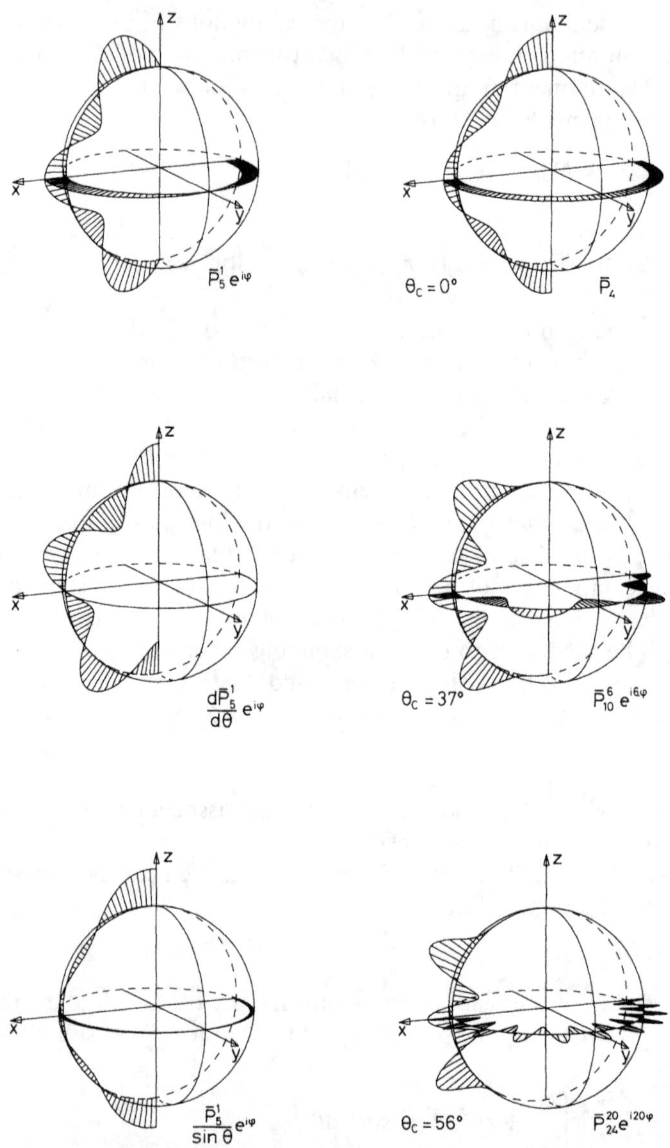

Fig. 2.8 *Angular dependence of six different spherical wave components at a fixed radius and at a given instant of time. The (θ, ϕ)-variation of the modes are shown with positive values outside the spheres and negative values inside*
In the left column of the Figure are shown (from the top) the angular dependence of the r, θ, and ϕ components, respectively, of $\vec{F}_{215}(r, \theta, \phi)$. The right column shows (from the top) the angular dependence of the r-component of $\vec{F}_{204}$, $\vec{F}_{2,6,10}$ and $\vec{F}_{2,20,24}$, respectively

In the far field of any given source, the electric and magnetic field vectors are related as in a plane wave

$$\vec{H} = \eta \hat{r} \times \vec{E} \tag{2.40}$$

and the complex Poynting vector is real

$$\begin{aligned}\vec{S} &= \tfrac{1}{2}\vec{E} \times \vec{H}^* \\ &= \tfrac{1}{2}\vec{E} \times (\eta \hat{r} \times \vec{E}^*) \\ &= \tfrac{1}{2}\eta |\vec{E}|^2 \hat{r} \quad \text{watts per square meter.}\end{aligned} \tag{2.41}$$

The power P_1 radiated per unit solid angle in the direction (θ, ϕ) is thus expressed by

$$P_1(\theta, \phi) = \tfrac{1}{2}\eta |\vec{E}|^2 r^2 \quad \text{watts per steradian.} \tag{2.42}$$

Let us now compute the outward flux of the complex Poynting vector through a sphere of a finite radius r

$$\frac{1}{2}\int_{\phi=0}^{2\pi}\int_{\theta=0}^{\pi} (\vec{E} \times \vec{H}^*) \cdot \hat{r} r^2 \sin\theta \, \mathrm{d}\theta \, \mathrm{d}\phi \tag{2.43}$$

for the case of an outgoing wave field, i.e. a field which can be represented by spherical wave functions of type $c = 3$

$$\vec{E}(r, \theta, \phi) = \frac{k}{\sqrt{\eta}} \sum_{smn} Q_{smn}^{(3)} \vec{F}_{smn}^{(3)}(r, \theta, \phi) \tag{2.44a}$$

$$\vec{H}(r, \theta, \phi) = -ik\sqrt{\eta} \sum_{\sigma\mu\nu} Q_{\sigma\mu\nu}^{(3)} \vec{F}_{3-\sigma,\mu,\nu}^{(3)}(r, \theta, \phi). \tag{2.44b}$$

In order to be able to identify contributions from products of terms with different indices when the product $\vec{E} \times \vec{H}^*$ is computed, we have introduced the Greek letters σ, μ, ν to denote the summation indices in (2.44b). The flux may readily be determined by using the relation

$$\vec{F}_{smn}^{(3)*}(r, \theta, \phi) = (-1)^m \vec{F}_{s,-m,n}^{(4)}(r, \theta, \phi) \tag{2.45}$$

and the orthogonality integral for the vector product of two spherical wave functions (see Appendix A1):

$$\begin{aligned}&\int_{\phi=0}^{2\pi}\int_{\theta=0}^{\pi} \{\vec{F}_{smn}^{(c)}(r, \theta, \phi) \times \vec{F}_{\sigma\mu\nu}^{(\gamma)}(r, \theta, \phi)\} \cdot \hat{r} \sin\theta \, \mathrm{d}\theta \, \mathrm{d}\phi \\ &\qquad = \delta_{3-s,\sigma}\delta_{m,-\mu}\delta_{n\nu}(-1)^{3-s}(-1)^m R_{sn}^{(c)}(kr) R_{3-s,n}^{(\gamma)}(kr)\end{aligned} \tag{2.46}$$

where we have used the abbreviation

$$R_{sn}^{(c)}(kr) = \begin{cases} z_n^{(c)}(kr) & \text{for } s = 1 \\ \dfrac{1}{kr}\dfrac{\mathrm{d}}{\mathrm{d}(kr)}\{kr z_n^{(c)}(kr)\} & \text{for } s = 2 \end{cases} \tag{2.47}$$

and the Kronecker delta

$$\delta_{ij} = \begin{cases} 0 & \text{for } i \neq j \\ 1 & \text{for } i = j. \end{cases} \tag{2.48}$$

The result is

$$\frac{1}{2}\int_{\phi=0}^{2\pi}\int_{\theta=0}^{\pi} (\vec{E} \times \vec{H}^*) \cdot \hat{r} r^2 \sin\theta \, d\theta \, d\phi = \sum_{smn} \tfrac{1}{2} i(-1)^{3-s}(kr)^2 R_{sn}^{(3)}(kr) R_{3-s,n}^{(4)}(kr) |Q_{smn}^{(3)}|^2. \tag{2.49}$$

The only contributions to (2.49) arise when the index triplet (σ, μ, ν) used in (2.44b) coincides with the triplet (s, m, n) of eqn (2.44a). This means that the complex power flow of each mode is independent of all other modes. Separation of (2.49) into real and imaginary parts can be accomplished as follows. Using

$$R_{sn}^{(3)}(kr) = R_{sn}^{(1)}(kr) + iR_{sn}^{(2)}(kr), \tag{2.50}$$

$$R_{sn}^{(4)}(kr) = R_{sn}^{(1)}(kr) - iR_{sn}^{(2)}(kr) \tag{2.51}$$

and the Wronskian

$$R_{1n}^{(1)}(kr) R_{2n}^{(2)}(kr) - R_{2n}^{(1)}(kr) R_{1n}^{(2)}(kr) = (kr)^{-2} \tag{2.52}$$

gives

$$\frac{1}{2}\int_{\phi=0}^{2\pi}\int_{\theta=0}^{\pi} (\vec{E} \times \vec{H}^*) \cdot \hat{r} r^2 \sin\theta \, d\theta \, d\phi = \sum_{smn} \{\tfrac{1}{2} + \tfrac{1}{2} i(-1)^{3-s}(kr)^2 V_n(kr)\} |Q_{smn}^{(3)}|^2 \tag{2.53}$$

where the cross product

$$V_n(kr) = R_{1n}^{(1)}(kr) R_{2n}^{(1)}(kr) + R_{1n}^{(2)}(kr) R_{2n}^{(2)}(kr) = \left(\frac{1}{kr} + \frac{1}{2}\frac{d}{d(kr)}\right) |h_n^{(1)}(kr)|^2 \tag{2.54}$$

is a real quantity which may be shown to be always negative (see formula 10.1.27 in Abramowitz and Stegun [21]). For a TE-mode ($s = 1$) the stored magnetic energy exceeds the stored electric energy while for a TM-mode ($s = 2$) the opposite is true; see eqn (2.39).

The real power flow associated with the field (2.44a, b) is thus given by

$$P = \frac{1}{2}\sum_{smn} |Q_{smn}^{(3)}|^2. \tag{2.55}$$

Note that P is independent of the radius r at which the flux is evaluated.

The complex power flow of an incoming field, i.e. a field that can be represented by spherical wave functions of type $c = 4$, can be analysed similarly and shows similar behaviour.

2.3 The antenna scattering matrix

2.3.1 Definitions

The general antenna with an associated spherical coordinate system may be considered as a waveguide junction with several ports. One port, the 'local port', is connected to a generator or a load. The remaining ports are 'radiation ports' connected to equivalent modal transmission lines, one for each spherical mode in the field outside the antenna's minimum sphere. The modal transmission lines together form the spherical waveguide discussed in Section 2.2.3. All of the properties of the antenna as a transmitting, receiving or scattering device are then contained in linear relationships between the complex amplitude coefficients of incoming and outgoing modes on a set of transmission lines each carrying a single mode. A linear relationship of this kind is the antenna scattering matrix formulation introduced by Dicke [1]. Obviously, such a formulation must be based upon careful definitions and normalizations of the modes involved.

Figure 2.9 illustrates a horn antenna from the scattering point of view. The complex amplitudes of incoming and outgoing waves on the guide to the local port (in this case a circular waveguide) are referred to a reference plane and denoted v and w, respectively. The normalization is chosen such that the power carried by these waves are $\frac{1}{2}|v|^2$ and $\frac{1}{2}|w|^2$, respectively.† The complex amplitudes of the spherical waves outside $r = r_0$ (radius of the minimum sphere) are $Q_{smn}^{(3)}$ for the outgoing waves and $Q_{smn}^{(4)}$ for the incoming waves. With the single index convention of Section 2.2.2 we have $Q_{smn}^{(3)} = Q_j^{(3)}$ and $Q_{smn}^{(4)} = Q_j^{(4)}$. In accordance with common practice in connection with scattering matrices, however, we shall here use the notation $Q_j^{(4)} = a_j$ and $Q_j^{(3)} = b_j$. Also, following the single index convention, we have $a_j = a_{smn}$ and $b_j = b_{smn}$, which will be used later when convenient.

The total electric field outside the minimum sphere is thus given by

$$\vec{E}(r, \theta, \phi) = \frac{k}{\sqrt{\eta}} \sum_{j=1}^{J} \{a_j \vec{F}_j^{(4)}(r, \theta, \phi) + b_j \vec{F}_j^{(3)}(r, \theta, \phi)\} \qquad r > r_0. \quad (2.56)$$

The number of spherical modes can be assumed to be finite as explained in Section 2.2.3. Hence, the number of modal transmission lines is also finite and the linear relationship between incoming and outgoing waves may be written as a matrix equation of order $J + 1$

$$\begin{bmatrix} \Gamma & \mathbf{R} \\ \mathbf{T} & \mathbf{S} \end{bmatrix} \begin{bmatrix} v \\ \mathbf{a} \end{bmatrix} = \begin{bmatrix} w \\ \mathbf{b} \end{bmatrix} \quad (2.57)$$

where Γ is a complex number (the antenna reflection coefficient), $\mathbf{R}$ is a $1 \times J$ row matrix with elements R_j, $j = 1, 2, \ldots, J$ (the antenna receiving coefficients),

† A further discussion of the waveguide field is given in Section 2.3.2.

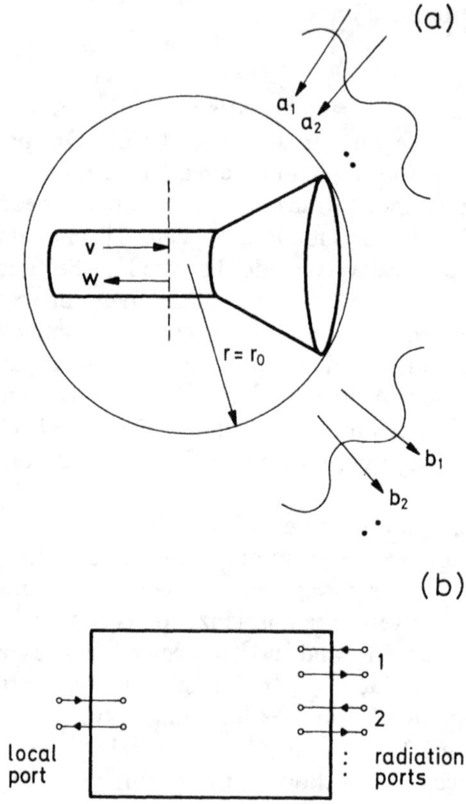

Fig. 2.9 *Illustrating a horn antenna from the scattering point of view*
(*a*) Antenna with waveguide modes and spherical modes
(*b*) Network representation

T is a $J \times 1$ column matrix with elements T_i, $i = 1, 2, ..., J$ (the antenna transmitting coefficients), and **S** is $J \times J$ square matrix with elements S_{ij} (the antenna scattering coefficients). The $J \times 1$ column matrices **a** and **b** contain as their elements $a_1, a_2, ..., a_J$, and $b_1, b_2, ..., b_J$. The three matrices **R**, **T** and **S** describe in an absolute and concise manner the antenna's receiving, transmitting and scattering properties, respectively. In expanded form, eqn (2.57) becomes

$$\Gamma v + \sum_{j=1}^{J} R_j a_j = w \tag{2.58}$$

$$T_i v + \sum_{j=1}^{J} S_{ij} a_j = b_i, \quad i = 1, 2, ..., J. \tag{2.59}$$

In a condensed form, eqn (2.57) may be written

$$\hat{\mathbf{S}}\hat{\mathbf{a}} = \hat{\mathbf{b}} \tag{2.60}$$

where $\hat{\mathbf{S}}$ is called the total scattering matrix for the antenna. The elements of the total scattering matrix are dimensionless.

For a lossless antenna, the total scattering matrix is unitary, i.e. $\hat{\mathbf{S}}^{+}\hat{\mathbf{S}} = \hat{\mathbf{I}}$, [1] where $\hat{\mathbf{I}}$ denotes the unit matrix of $(J + 1)$th order†. The unitarity condition implies that the norm, the sum of the elements absolute squared, of every column and every row in $\hat{\mathbf{S}}$ equals unity. In particular, for the first column

$$|\Gamma|^2 + |\mathrm{T}|^2 = 1. \tag{2.61}$$

For a lossy antenna in the purely transmitting case, the total radiated power is equal to the power incident in the waveguide minus the sum of the reflected power and the power dissipated in the antenna

$$\frac{1}{2}\sum_{i=1}^{J} |b_i|^2 = \tfrac{1}{2}|v|^2 - (\tfrac{1}{2}|w|^2 + P_{\text{loss}}). \tag{2.62}$$

Dividing through by the incident power $P_{\text{inc}} = \frac{1}{2}|v|^2$, and making use of eqn (2.57) with $\mathbf{a} = \mathbf{0}$, we arrive at the relationship

$$|\Gamma|^2 + |\mathrm{T}|^2 = 1 - \frac{P_{\text{loss}}}{P_{\text{inc}}} \tag{2.63}$$

analogous to eqn (2.61). It may be shown that in the lossy case, all other columns and all rows in $\hat{\mathbf{S}}$ have norms equal to or less than one.

The total scattering matrix for empty space is not defined. However, the scattering matrix $\mathbf{S}$ has a meaning in this case. For empty space, in which a spherical coordinate system is defined, any incident spherical wave will converge towards the origin as a focal point, pass through it, and emerge as a diverging spherical wave. In other words, for empty space, $\mathbf{b} = \mathbf{a}$, so we must have $\mathbf{S} = \mathbf{I}$, where $\mathbf{I}$ is the unit matrix of infinite order.

The symmetry properties of $\hat{\mathbf{S}}$ will be discussed in Section 2.3.2.

A convenient way to visualize the properties of the scattering matrix formulation outlined above is to utilize scattering flow graphs [22]. The complete flow graph corresponding to eqn (2.57) contains $2J + 2$ nodes and $(J + 1)^2$ branches. The incoming waves define nodes v and a_j, $j = 1, 2, \ldots, J$, while the outgoing waves determine the nodes w and b_i, $i = 1, 2, \ldots, J$. Each of the incoming wave nodes is connected to each of the outgoing wave nodes, and the wave transformations that take place in the antenna are indicated by designations on the oriented branches. The complete flow graph becomes very large even for a relatively small number of waves. However, a representation of sufficient generality need not contain more than the four nodes shown in Fig. 2.10.

An incoming wave v at the local port is partly reflected via branch Γ and

† The superscript $^{+}$ is used to denote the Hermitian conjugate of a matrix. For later use we shall also define the superscripts ^{T} and * for the transposed matrix and the complex conjugate matrix, respectively, so that $S^{+} = (S^{*})^{\mathrm{T}}$.

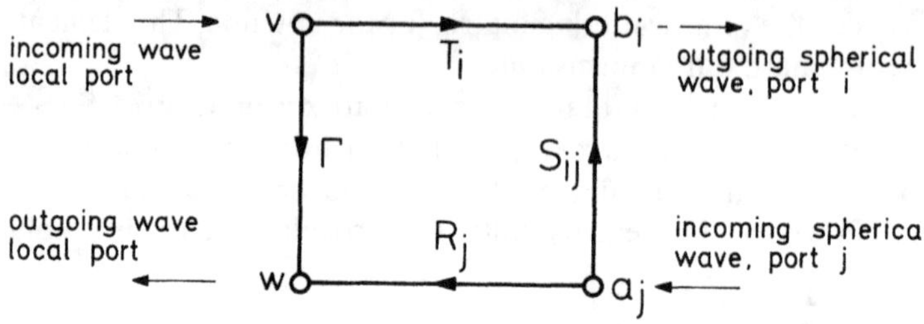

Fig. 2.10 *Representative part of complete scattering flow graph for eqn (2.57)*
Nodes represent waves while branches indicate transformations between waves

contributes to an outgoing wave w at the same port. The remaining part is transmitted through branch T_i and contributes to an outgoing spherical wave b_i. An incoming spherical wave a_j from the surrounding space is partly scattered by way of branch S_{ij} and contributes to the outgoing wave b_i. The remaining part of a_j is received due to branch R_j and contributes to the wave w.

In the remainder of this section, we shall consider two applications of the scattering matrix formulation. First, we determine the transmitted wave amplitude when an antenna is coupled to a known generator.

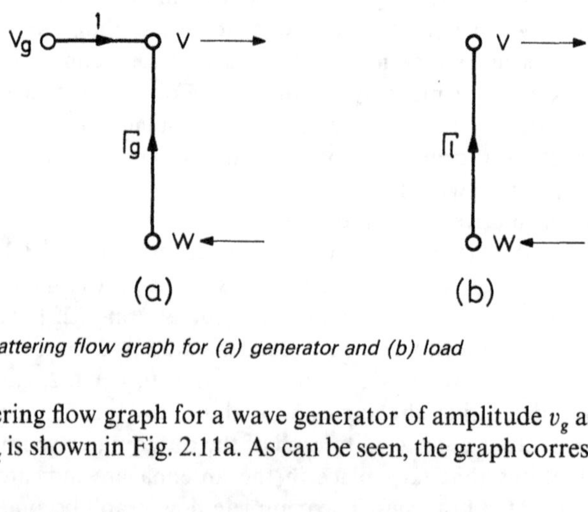

Fig. 2.11 *Scattering flow graph for (a) generator and (b) load*

The scattering flow graph for a wave generator of amplitude v_g and reflection coefficient Γ_g is shown in Fig. 2.11a. As can be seen, the graph corresponds to the equation

$$v = v_g + \Gamma_g w \tag{2.64}$$

which expresses the output amplitude v as the sum of the generator amplitude v_g and a contribution from any wave w that may be present as a reflection from the antenna to which the generator is connected. Thus, v_g is the amplitude of the signal when the generator is connected to a load without reflection ($w = 0$). The generator reflection coefficient is denoted by Γ_g. For a matched generator Γ_g is equal to zero. Assuming free-space environments of the antenna, we may let $a_j = 0$ for $j = 1, 2, \ldots, J$. Then, from eqn (2.57)

$$\Gamma v = w \tag{2.65}$$

$$\mathbf{T}v = \mathbf{b}. \tag{2.66}$$

By solving (2.64) and (2.65) for v and inserting into (2.66), we obtain for the amplitude of the radiated waves

$$\mathbf{b} = \frac{v_g}{1 - \Gamma_g \Gamma} \mathbf{T}. \tag{2.67}$$

The radiated electric field is

$$\vec{E}(r, \theta, \phi) = \frac{k}{\sqrt{\eta}} \sum_{i=1}^{J} b_i \vec{F}_i^{(3)}(r, \theta, \phi). \tag{2.68}$$

Next, let us find the received signal and the scattered wave amplitude in the receiving case when the antenna is coupled to a known load (receiver). The scattering flow graph for a receiving load of reflection coefficient Γ_l is shown in Fig. 2.11b. The graph corresponds to the equation for the load

$$v = \Gamma_l w. \tag{2.69}$$

We now assume the load to be coupled to an antenna. The incident waves are of amplitude $\mathbf{a}$ and we have from (2.57)

$$\Gamma v + \mathbf{Ra} = w \tag{2.70}$$

which by means of (2.69) gives the amplitude of the wave towards the load

$$w = \frac{1}{1 - \Gamma_l \Gamma} \mathbf{Ra}. \tag{2.71}$$

The power accepted by the load is

$$\begin{aligned} P &= \tfrac{1}{2}(1 - |\Gamma_l|^2)\,|w|^2 \\ &= \tfrac{1}{2}(1 - |\Gamma_l|^2)\frac{|\mathbf{Ra}|^2}{|1 - \Gamma_l \Gamma|^2}. \end{aligned} \tag{2.72}$$

For a matched load, i.e. $\Gamma_l = 0$, the accepted power becomes

$$\begin{aligned} P &= \tfrac{1}{2}|\mathbf{Ra}|^2 \\ &= P'. \end{aligned} \tag{2.73}$$

Maximum accepted power is obtained for $\Gamma_l = \Gamma^*$ as may be readily shown by considering equation (2.72). In this case

$$\begin{aligned} P &= \frac{1}{2}\frac{|\mathbf{Ra}|^2}{1 - |\Gamma|^2} \\ &= P_a \end{aligned} \tag{2.74}$$

is called the available power. As may be seen from the equations, $P' \leqslant P_a$.

The amplitude of the waves scattered by the antenna is obtained as follows. From equation (2.57) we have

$$\mathbf{T}v + \mathbf{S}\mathbf{a} = \mathbf{b}. \tag{2.75}$$

Insertion for v from eqn (2.69) and subsequently for w from eqn (2.71) gives the outgoing wave amplitude

$$\mathbf{b} = \{\mathbf{T}\Gamma_l(1 - \Gamma\Gamma_l)^{-1}\mathbf{R} + \mathbf{S}\}\mathbf{a}. \tag{2.76}$$

This is, however, not the wave amplitude of the field scattered by the antenna itself. We must remember that in our formulation, the scattering matrix for empty space is not equal to zero but equal to the unit matrix **I**. We must therefore subtract the outgoing wave which, in our formulation, is inherently present whenever a spherical wave is incident towards the origin in empty space.

The wave amplitude $\mathbf{b}'$ for the field scattered by the antenna becomes

$$\mathbf{b}' = \{\mathbf{T}\Gamma_l(1 - \Gamma\Gamma_l)^{-1}\mathbf{R} + (\mathbf{S} - \mathbf{I})\}\mathbf{a}. \tag{2.77}$$

The first term in the parenthesis on the right-hand side describes the part of the received waves that is reflected by the load and therefore is reradiated with the directional characteristics of transmitted waves. The second term relates to the part of the received waves that is scattered directly off from the antenna structure without entering the antenna waveguide.

When the antenna is terminated in a matched load, Γ_l is equal to zero, and

$$\mathbf{b}' = (\mathbf{S} - \mathbf{I})\mathbf{a}. \tag{2.78}$$

The scattered electric field is

$$\vec{E}'(r, \theta, \phi) = \frac{k}{\sqrt{\eta}} \sum_{i=1}^{J} b_i' \vec{F}_i^{(3)}(r, \theta, \phi) \tag{2.79}$$

where b_i', $i = 1, 2, \ldots, J$ are the elements of the column matrix $\mathbf{b}'$.

2.3.2 Reciprocity

A (linear) non-reciprocal antenna or scatterer is characterized by the fact that part of or all of the material media from which they have been fabricated is described by the general complex unsymmetrical permeability and permittivity tensors $\boldsymbol{\mu}$ and $\boldsymbol{\varepsilon}$. In practice, non-reciprocal antennas or scatterers occur whenever a magnetized ferrite (e.g. in a microwave isolator) is a part of the structure. By reversing the direction of the static biasing magnetic field, $\boldsymbol{\mu}$ and $\boldsymbol{\varepsilon}$ are in effect replaced by the transposed tensors $\boldsymbol{\mu}^{\mathrm{T}}$ and $\boldsymbol{\varepsilon}^{\mathrm{T}}$, and a new device evolves, called the adjoint antenna or adjoint scatterer [23]. The adjoint of the adjoint antenna or scatterer is the original device itself.

In this section we shall derive the specific symmetry relations which exist between the elements of the receiving matrix for a given antenna and the elements of the transmitting matrix for the adjoint antenna. We shall also, without proof, state a relationship between the elements of the scattering matrices of adjoint antennas.

A reciprocal antenna has symmetric tensors $\boldsymbol{\mu}$ and $\boldsymbol{\varepsilon}$ and is its own adjoint antenna. Hence, the symmetry relations derived for two adjoint antennas contain the symmetry relations for a reciprocal antenna as a special case.

Reference is made to Fig. 2.12. Here, S is a surface which encloses a volume V containing an antenna. The antenna is connected to some feed system shown as

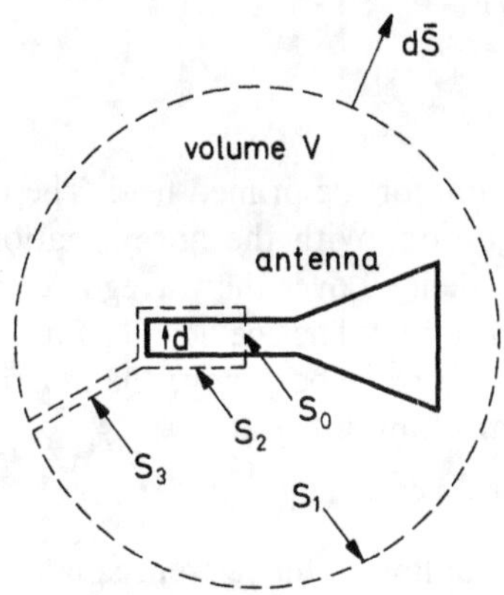

Fig. 2.12 *For the derivation of the reciprocity relations. The Figure shows the integration surface $S = S_0 + S_1 + S_2 + S_3$ and the outward oriented vector area element $d\vec{S}$*

a piece of waveguide with a short dipole d for excitation or load. The surface S is composed of four parts, $S = S_0 + S_1 + S_2 + S_3$. Of these, S_0 is the reference plane where waveguide and antenna meet. S_1 is a sphere of radius r completely enclosing the antenna and feed system such that a spherical wave expansion can be used on and outside S_1. The surface S_2 follows the outer surface of the feed system, and S_3 is a narrow tube connecting S_1 and S_2. All media inside S is assumed characterized by symmetrical or unsymmetrical (complex) permeability and permittivity tensors $\boldsymbol{\mu}$ and $\boldsymbol{\varepsilon}$.

We consider two different situations. In the first situation, a set of sources $(\vec{J}, \vec{M})$ outside S is producing the field $(\vec{E}, \vec{H})$ in V and on S. In the second situation, the media in V are replaced by media characterized by $\boldsymbol{\mu}^{\mathrm{T}}$ and $\boldsymbol{\varepsilon}^{\mathrm{T}}$, and a new set of sources $(\vec{J}', \vec{M}')$ outside V is producing the field $(\vec{E}', \vec{H}')$ in V and on S. In other words, in the second situation the adjoint of the antenna in the first situation is considered. The surface S encloses no sources and the generalization of the Lorentz form of the reciprocity theorem may be applied [11]

$$\int_S (\vec{E} \times \vec{H}' - \vec{E}' \times \vec{H}) \cdot \mathrm{d}\vec{S} = 0 \tag{2.80}$$

where $\mathrm{d}\vec{S}$ is the outward oriented area element on S. We shall specify the field vectors on the left-hand side in terms of incoming and outgoing modes. In the waveguide, only a single mode is assumed while on S_1, a set of spherical modes is present. The coefficients of the incoming modes are v and a_{smn}, respectively, for the unprimed field, and v' and $a'_{s'm'n'}$, respectively, for the primed field. Similarly, the outgoing modes have the coefficients w, b_{smn} and w', $b'_{s'm'n'}$.

For the integral over the reference plane in the waveguide, a suitable field representation is as follows [16]. For the unprimed field we shall use for the tangential components

$$\vec{E}_t = \frac{k}{\sqrt{\eta}}(v + w)\vec{e} \tag{2.81}$$

$$\vec{H}_t = k\sqrt{\eta}(v - w)\vec{h} \tag{2.82}$$

with a similar representation for the primed field. The normal components are not needed for the integration. With the normalization indicated, the vector modal functions $\vec{e}$ and $\vec{h}$ defined over the waveguide cross-section are dimensionless and may be assumed real. The coefficient v for the incident wave and the coefficient w for the reflected wave both have the dimension watts$^{1/2}$. We shall also introduce the normalization integral

$$\text{Re}\,\{k^2 \textstyle\int_{S_0} \vec{e} \times \vec{h}^* \cdot d\vec{S}\} = -1 \tag{2.83}$$

where the total *inward* power flow at the reference plane is given by $\frac{1}{2}|v|^2 - \frac{1}{2}|w|^2$. We obtain

$$\begin{aligned}\textstyle\int_{S_0}(\vec{E} \times \vec{H}' - \vec{E}' \times \vec{H}) \cdot d\vec{S} &= k^2 \textstyle\int_{S_0}(v + w)(v' - w')(\vec{e} \times \vec{h}) \cdot d\vec{S} \\ &\quad - k^2 \textstyle\int_{S_0}(v' + w')(v - w)(\vec{e} \times \vec{h}) \cdot d\vec{S} \\ &= 2(vw' - v'w)\end{aligned} \tag{2.84}$$

The integration over S_1 can be performed using the orthogonality integral (2.46). For the unprimed field we shall use the representation

$$\vec{E} = \frac{k}{\sqrt{\eta}} \sum_{smn} (a_{smn}\vec{F}^{(4)}_{smn} + b_{smn}\vec{F}^{(3)}_{smn}) \tag{2.85}$$

$$\vec{H} = -ik\sqrt{\eta} \sum_{smn} (a_{smn}\vec{F}^{(4)}_{3-s,m,n} + b_{smn}\vec{F}^{(3)}_{3-s,m,n}) \tag{2.86}$$

with a similar representation for the primed field. For the first term in the generalized reciprocity theorem we have

$$\begin{aligned}&\int_{S_1} (\vec{E} \times \vec{H}') \cdot d\vec{S} \\ &= \int_{S_1} \left\{\frac{k}{\sqrt{\eta}} \sum_{smn} (a_{smn}\vec{F}^{(4)}_{smn} + b_{smn}\vec{F}^{(3)}_{smn})\right\} \\ &\quad \times \left\{-ik\sqrt{\eta} \sum_{s'm'n'} (a'_{s'm'n'}\vec{F}^{(4)}_{3-s',m',n'} + b'_{s'm'n'}\vec{F}^{(3)}_{3-s',m',n'})\right\} \cdot d\vec{S}.\end{aligned} \tag{2.87}$$

Use of the orthogonality integral (A1.71) gives

$$\int_{S_1} (\vec{E} \times \vec{H}') \cdot d\vec{S} = -i(kr)^2 \sum_{smn} (-1)^{3-s}(-1)^m$$
$$(a_{smn} a'_{s,-m,n} R^{(4)}_{sn} R^{(4)}_{3-s,n} + b_{smn} a'_{s,-m,n} R^{(3)}_{sn} R^{(4)}_{3-s,n}$$
$$+ a_{smn} b'_{s,-m,n} R^{(4)}_{sn} R^{(3)}_{3-s,n} + b_{smn} b'_{s,-m,n} R^{(3)}_{sn} R^{(3)}_{3-s,n}). \quad (2.88)$$

The second term of the integral is calculated in a similar way. After changes in the summation indices it can be subtracted from (2.88). The following expression results

$$\int_{S_1} (\vec{E} \times \vec{H}' - \vec{E}' \times \vec{H}) \cdot d\vec{S} = -i(kr)^2 \sum_{smn} (-1)^{3-s}(-1)^m$$
$$\{b_{smn} a'_{s,-m,n} - a_{smn} b'_{s,-m,n}\} \left\{-(-1)^{3-s} \frac{2i}{(kr)^2}\right\} \quad (2.89)$$
$$= -2 \sum_{smn} (-1)^m (b_{smn} a'_{s,-m,n} - a_{smn} b'_{s,-m,n}) \quad (2.90)$$

where we have used the Wronskian

$$R^{(3)}_{sn} R^{(4)}_{3-s,n} - R^{(4)}_{sn} R^{(3)}_{3-s,n} = -(-1)^{3-s} \frac{2i}{(kr)^2}. \quad (2.91)$$

In the integral over S_2, the integrand vanishes identically since on the conducting surface we have $\zeta_m \vec{H}_t = \hat{n} \times \vec{E}$ and $\zeta_m \vec{H}'_t = \hat{n} \times \vec{E}'$ where t indicates tangential component and ζ_m denotes the surface impedance of the material [11]. The contribution from S_3 also vanishes because the integrand is finite and we may let the radius of the narrow tube S_3 tend to zero. Therefore,

$$\int_{S_2+S_3} (\vec{E} \times \vec{H}' - \vec{E}' \times \vec{H}) \cdot d\vec{S} = 0. \quad (2.92)$$

As a result of eqns (2.80), (2.84), (2.90) and (2.92) we finally obtain

$$vw' - v'w = \sum_{smn} (-1)^m (b_{smn} a'_{s,-m,n} - a_{smn} b'_{s,-m,n}) \quad (2.93)$$

which is valid for an arbitrary choice of the coefficients v, v', a and a'. Equation (2.93) expresses the generalized Lorentz form of the reciprocity theorem for the fields in two different situations as described above.

In order to arrive at relations between transmitting and receiving properties of antennas, we now consider a special case of eqn (2.93).

We shall assume that the unprimed quantities refer to a situation where the antenna of Fig. 2.12 is acting as a transmitting antenna with unity excitation (1 watt$^{1/2}$) and no spherical waves are coming in from the surrounding space, i.e.

$$v = 1 \quad (2.94)$$

$$a_{smn} = 0 \qquad \text{for all } s, m \text{ and } n. \quad (2.95)$$

The primed quantities correspond to a second situation where we shall assume the (adjoint) antenna to be purely receiving with a single spherical mode (of indices σ, μ, ν and unity amplitude) as the only incident field, i.e.

$$v' = 0 \tag{2.96}$$

$$a'_{s,-m,n} = \begin{cases} 1 & \text{for } (s, -m, n) = (\sigma, \mu, \nu) \\ 0 & \text{otherwise.} \end{cases} \tag{2.97}$$

With these wave amplitudes, we obtain from equation (2.93)

$$w' = (-1)^{\mu} b_{\sigma,-\mu,\nu}. \tag{2.98}$$

This equation expresses a relationship between quantities belonging to the two situations. In each of these situations, we also have the dependencies following from the general scattering matrix equation (2.57), i.e. in the first situation

$$\Gamma = w \tag{2.99}$$

$$\mathbf{T} = \mathbf{b} \tag{2.100}$$

and in the second situation

$$\mathbf{R}'\mathbf{a}' = w' \tag{2.101}$$

$$\mathbf{S}'\mathbf{a}' = \mathbf{b}'. \tag{2.102}$$

Use of (2.100) and (2.101) on the right- and left-hand side of eqn (2.98), respectively, yields

$$R'_{\sigma\mu\nu} = (-1)^{\mu} T_{\sigma,-\mu,\nu} \tag{2.103}$$

or, since the indices $\sigma\mu\nu$ may be chosen arbitrarily

$$R'_{smn} = (-1)^{m} T_{s,-m,n}. \tag{2.104}$$

Equation (2.104) is the desired relationship between elements of the receiving and transmitting matrices for two adjoint antennas.

For the scattering properties of adjoint antennas, the following relations may be shown to hold

$$S'^{\sigma\mu\nu}_{smn} = (-1)^{m+\mu} S^{s,-m,n}_{\sigma,-\mu,\nu} \tag{2.105}$$

and

$$\Gamma' = \Gamma. \tag{2.106}$$

As important special cases of eqns (2.104) and (2.105) we have for a reciprocal antenna

$$R_{smn} = (-1)^{m} T_{s,-m,n} \tag{2.107}$$

and

$$S^{\sigma\mu\nu}_{smn} = (-1)^{m+\mu} S^{s,-m,n}_{\sigma,-\mu,\nu} \tag{2.108}$$

respectively.

2.3.3 The field from electric and magnetic dipoles

For use in the next section we shall establish the fields of electric and magnetic dipoles in terms of spherical wave functions. A short z-directed electric dipole located at the origin produces the electromagnetic field [11]

$$\vec{E}_e^z(r,\theta,\phi) = -\frac{\zeta k^2}{2\pi} d_e \frac{h_1^{(1)}(kr)}{kr}\cos\theta\,\hat{r} + \frac{\zeta k^2}{4\pi} d_e \frac{1}{kr}\frac{\mathrm{d}}{\mathrm{d}(kr)}\{krh_1^{(1)}(kr)\}\sin\theta\,\hat{\theta} \tag{2.109}$$

$$\vec{H}_e^z(r,\theta,\phi) = \frac{ik^2}{4\pi} d_e h_1^{(1)}(kr)\sin\theta\,\hat{\phi} \tag{2.110}$$

where $d_e = Il$ is the dipole moment (product of current I and dipole length l) and where $h_1^{(1)}(kr)$ is the Hankel function of first kind and first order

$$h_1^{(1)}(kr) = -\frac{e^{ikr}}{kr}\left(1+\frac{i}{kr}\right) \tag{2.111}$$

and

$$\frac{1}{kr}\frac{\mathrm{d}}{\mathrm{d}(kr)}\{krh_1^{(1)}(kr)\} = \frac{e^{ikr}}{kr}\left\{-i+\frac{1}{kr}+\frac{i}{(kr)^2}\right\}. \tag{2.112}$$

As may be seen by comparing (2.109) and (2.110) to the expressions given in Section A1.3.1 for the wave functions with $n = 1$, the field of the z-directed dipole can be written

$$\vec{E}_e^z(r,\theta,\phi) = \frac{k}{\sqrt{\eta}}\sum_{smn} Q_{smn}\vec{F}_{smn}^{(3)}(r,\theta,\phi) \tag{2.113}$$

$$\vec{H}_e^z(r,\theta,\phi) = -ik\sqrt{\eta}\sum_{smn} Q_{smn}\vec{F}_{3-s,m,n}^{(3)}(r,\theta,\phi). \tag{2.114}$$

With only one term in the summations, i.e. $s = 2$, $m = 0$ and $n = 1$, we have

$$\vec{E}_e^z(r,\theta,\phi) = \frac{k}{\sqrt{\eta}} Q_{201}\vec{F}_{201}^{(3)}(r,\theta,\phi) \tag{2.115}$$

$$\vec{H}_e^z(r,\theta,\phi) = -ik\sqrt{\eta}Q_{201}\vec{F}_{101}^{(3)}(r,\theta,\phi) \tag{2.116}$$

with

$$Q_{201} = -\frac{1}{\sqrt{6\pi}}\frac{k}{\sqrt{\eta}}d_e. \tag{2.117}$$

Having thus expressed the field of the z-directed dipole in terms of spherical waves, we shall now turn to the x- and y-directed dipoles. The radial behaviour of the field components of the z-oriented dipole is independent of the angular coordinates θ and ϕ. A reorientation of the dipole will therefore not give rise to spherical modes with values of n other than $n = 1$. There are only six spherical modes altogether with $n = 1$, and it would not be difficult from inspection and symmetry considerations to obtain the correct combination needed for expressing the fields of the x-oriented and y-oriented dipoles. However, we shall here prefer to use a more systematic approach with rotation of the coordinate system.

Let us consider again the z-oriented dipole at the origin of an (x, y, z) coordinate system. Let us assume another coordinate system (x', y', z') to be initially coincident with the (x, y, z) system. We shall now keep the dipole fixed in space but rotate the primed coordinate system through the Euler angles

$$(\chi_o, \theta_o, \phi_o) = \left(0, -\frac{\pi}{2}, 0\right) \tag{2.118}$$

thereby obtaining a new system (x', y', z') which is rotated with respect to the (x, y, z) system. As may be easily seen using the definitions of Euler angles (see Section A2.1), the dipole will then be oriented along the (positive) x'-axis. The following formula (see Section A2.2) gives the expansion of a spherical mode $\vec{F}^{(c)}_{smn}(r, \theta, \phi)$ defined in the unrotated, unprimed coordinate system in terms of a finite series of spherical modes $\vec{F}^{(c)}_{s\mu n}(r', \theta', \phi')$ defined in the rotated, primed coordinate system.

$$\vec{F}^{(c)}_{smn}(r, \theta, \phi) = \sum_{\mu=-n}^{n} e^{im\varphi_o} d^n_{\mu m}(\theta_o)\, e^{i\mu\chi_o} \vec{F}^{(c)}_{s\mu n}(r', \theta', \phi') \tag{2.119}$$

where $d^n_{\mu m}(\theta_o)$ is a rotation coefficient. Using

$$d^1_{-10}\left(-\frac{\pi}{2}\right) = \frac{\sqrt{2}}{2} \tag{2.120}$$

$$d^1_{00}\left(-\frac{\pi}{2}\right) = 0 \tag{2.121}$$

$$d^1_{10}\left(-\frac{\pi}{2}\right) = -\frac{\sqrt{2}}{2} \tag{2.122}$$

readily obtained from formulas (A2.13), (A2.7) and the tables A2.6 we find that

$$\vec{F}^{(3)}_{201}(r, \theta, \phi) = \frac{\sqrt{2}}{2} \vec{F}^{(3)}_{2,-1,1}(r', \theta', \phi') - \frac{\sqrt{2}}{2} \vec{F}^{(3)}_{211}(r', \theta', \phi'). \tag{2.123}$$

Insertion of (2.123) into eqns (2.115) and (2.116) and omitting the primes gives the field produced by a short x-directed dipole located at the origin:

$$\vec{E}^x_e(r, \theta, \phi) = \frac{k}{\sqrt{\eta}} Q_{201} \frac{\sqrt{2}}{2} \{\vec{F}^{(3)}_{2,-1,1}(r, \theta, \phi) - \vec{F}^{(3)}_{211}(r, \theta, \phi)\} \tag{2.124}$$

$$\vec{H}_e^x(r,\theta,\phi) = -ik\sqrt{\eta}Q_{201}\frac{\sqrt{2}}{2}\{\vec{F}^{(3)}_{1,-1,1}(r,\theta,\phi) - \vec{F}^{(3)}_{111}(r,\theta,\phi)\}. \tag{2.125}$$

The field from the y-directed dipole may be found from eqn (2.124) and (2.125) by rotation of the coordinate system through the Euler angles

$$(\chi_o, \theta_o, \phi_o) = \left(0, 0, -\frac{\pi}{2}\right) \tag{2.126}$$

or from eqn (2.115) and (2.116) by rotation of the coordinate system through the Euler angles

$$(\chi_o, \theta_o, \phi_o) = \left(-\frac{\pi}{2}, -\frac{\pi}{2}, 0\right). \tag{2.127}$$

The result is

$$\vec{E}_e^y(r,\theta,\phi) = \frac{k}{\sqrt{\eta}}Q_{201}\frac{i\sqrt{2}}{2}\{\vec{F}^{(3)}_{2,-1,1}(r,\theta,\phi) + \vec{F}^{(3)}_{211}(r,\theta,\phi)\} \tag{2.128}$$

$$\vec{H}_e^y(r,\theta,\phi) = -ik\sqrt{\eta}Q_{201}\frac{i\sqrt{2}}{2}\{\vec{F}^{(3)}_{1,-1,1}(r,\theta,\phi) + \vec{F}^{(3)}_{111}(r,\theta,\phi)\} \tag{2.129}$$

in which the Q_{201} is given by eqn (2.117).

For a short magnetic dipole located at the origin, a similar analysis may be carried out. A magnetic dipole may be described either as a short element of magnetic current or as a small electric current loop. In the latter case, the orientation of the dipole is defined by using the direction of the current in the loop and the right-hand rule. The magnetic dipole moment d_m is given by

$$d_m = I_m l \tag{2.130}$$

$$= -i\omega\mu SI' \tag{2.131}$$

where I_m is the magnetic current element, and l its length, and where S is the area of the loop and I' the loop current.

The fields of a magnetic dipole oriented along the z-axis, x-axis and y-axis are given by, respectively

$$\vec{E}_m^z(r,\theta,\phi) = \frac{k}{\sqrt{\eta}}Q_{101}\vec{F}^{(3)}_{101}(r,\theta,\phi) \tag{2.132}$$

$$\vec{H}_m^z(r,\theta,\phi) = -ik\sqrt{\eta}Q_{101}\vec{F}^{(3)}_{201}(r,\theta,\phi) \tag{2.133}$$

$$\vec{E}_m^x(r,\theta,\phi) = \frac{k}{\sqrt{\eta}}Q_{101}\frac{\sqrt{2}}{2}\{\vec{F}^{(3)}_{1,-1,1}(r,\theta,\phi) - \vec{F}^{(3)}_{111}(r,\theta,\phi)\} \tag{2.134}$$

$$\vec{H}_m^x(r,\theta,\phi) = -ik\sqrt{\eta}Q_{101}\frac{\sqrt{2}}{2}\{\vec{F}^{(3)}_{2,-1,1}(r,\theta,\phi) - \vec{F}^{(3)}_{211}(r,\theta,\phi)\} \tag{2.135}$$

$$\vec{E}_m^y(r,\theta,\phi) = \frac{k}{\sqrt{\eta}}Q_{101}\frac{i\sqrt{2}}{2}\{\vec{F}^{(3)}_{1,-1,1}(r,\theta,\phi) + \vec{F}^{(3)}_{111}(r,\theta,\phi)\} \tag{2.136}$$

$$\vec{H}_m^y(r,\theta,\phi) = -ik\sqrt{\eta}Q_{101}\frac{i\sqrt{2}}{2}\{\vec{F}^{(3)}_{2,-1,1}(r,\theta,\phi) + \vec{F}^{(3)}_{211}(r,\theta,\phi)\}. \tag{2.137}$$

In these expansions Q_{101} is given by

$$Q_{101} = -\frac{i}{\sqrt{6\pi}}k\sqrt{\eta}d_m \tag{2.138}$$

If d_e and d_m are chosen such that

$$d_m = -\zeta d_e \tag{2.139}$$

the fields of the electric dipole and a magnetic dipole with the same orientation are interrelated by

$$\vec{E}_m = \zeta\vec{H}_e \tag{2.140}$$

$$\vec{H}_m = -\eta\vec{E}_e. \tag{2.141}$$

This may be verified by insertion of eqn (2.139) in the field expansions. The magnetic dipole is in this case the *dual source* of the electric dipole and the field $(\vec{E}_m, \vec{H}_m)$ is the *dual field* of $(\vec{E}_e, \vec{H}_e)$. As may be easily verified from eqns (2.117), (2.138) and (2.139), the requirement that a z-directed magnetic dipole is the dual source of a given similarly oriented electric dipole leads to the relation

$$Q_{101} = -iQ_{201} \tag{2.142}$$

between the expansion coefficients in the spherical wave expansion of the two fields.

2.3.4 Scattering matrices for electric and magnetic dipoles

As examples of scattering matrices, we shall now consider the scattering matrices of the short electric and magnetic dipoles. The scattering matrix formulation is essentially a tool for the bookkeeping of modes. We shall therefore in this section visualize the short electric and magnetic dipoles as connected to a generator (transmitting case) or to a load (receiving/scattering case) through a transmission line with only one mode propagating. The minimum sphere for the dipoles has vanishing radius, and the field is described in terms of spherical modes (see Section 2.3.3) using the single index convention of eqn (2.27).

We intend to determine the elements of the total scattering matrix $\mathbf{\hat{S}}$ as given by eqns (2.57) and (2.60). We assume the dipoles to be lossless and matched to the transmission line as well so that $\Gamma = 0$. Then, from eqn (2.63), $|\mathbf{T}|^2 = 1$.

Let us first consider the z-directed electric dipole. In the transmitting case, the outgoing wave amplitudes $\mathbf{b}$ are obtained by equating the right-hand sides of eqns (2.68) and (2.115)

$$\frac{k}{\sqrt{\eta}} \sum_{i=1}^{J} b_i \vec{F}_i^{(3)}(r, \theta, \phi) = \frac{k}{\sqrt{\eta}} Q_{201} \vec{F}_{201}^{(3)}(r, \theta, \phi) \tag{2.143}$$

which shows that the power is radiated in a single mode. From the single index convention, this mode has the index $i = 2\{n(n+1) + m - 1\} + s = 4$ and

$$b_i = \begin{cases} 0 & i \neq 4 \\ Q_{201} & i = 4. \end{cases} \tag{2.144}$$

Since $\mathbf{T}v = \mathbf{b}$, and $|v|^2 = |Q_{201}|^2$ due to power conservation, we can write

$$T_i = \begin{cases} 0 & i \neq 4 \\ 1 & i = 4 \end{cases} \tag{2.145}$$

by properly choosing the reference plane in the transmission line.

For the receiving/scattering case we have by the reciprocity equation (2.107)

$$R_j = \begin{cases} 0 & j \neq 4 \\ 1 & j = 4 \end{cases} \tag{2.146}$$

which shows that only one mode of the incoming field will transfer power to the load.

There now remain the elements S_{ij} of $\mathbf{S}$. The total scattering matrix is unitary and the sum of the squares of the absolute values of the elements is equal to one for each row and column in $\mathbf{\hat{S}}$. Equations (2.145) and (2.146) then yield $|S_{i4}|^2 = 0$ for all i, and $|S_{4j}|^2 = 0$ for all j. The remaining elements are determined by making the observation that the dipole is *invisible* to all incoming modes with indices different from 4 and, therefore, will act as empty space in relation to these modes. This is in accordance with the fact that the mode $\vec{F}_4^{(1)}$ is the only spherical mode which possesses a non-vanishing $\hat{z}$-component at the origin (see eqns (A1.51–1.53)). In other words

$$S_{ij} = \begin{cases} 1 & (i = j) \wedge (i \neq 4) \\ 0 & \text{otherwise.} \end{cases} \tag{2.147}$$

The total scattering matrix for the $\hat{z}$-directed short electric dipole has now been found, and is given by

$$\hat{\mathbf{S}}_e^z = \begin{bmatrix} 0 & 0 & 0 & 0 & 1 & 0 & 0 & \cdots \\ 0 & 1 & 0 & 0 & 0 & 0 & 0 & \\ 0 & 0 & 1 & 0 & 0 & 0 & 0 & \\ 0 & 0 & 0 & 1 & 0 & 0 & 0 & \\ 1 & 0 & 0 & 0 & 0 & 0 & 0 & \\ 0 & 0 & 0 & 0 & 0 & 1 & 0 & \\ 0 & 0 & 0 & 0 & 0 & 0 & 1 & \\ \vdots & & & & & & & \ddots \end{bmatrix}. \tag{2.148}$$

As an example of the use of (2.148), we consider the receiving case with an arbitrary incident field specified by a set of wave coefficients **a**.

In the absence of any dipole, the incoming modes undergo scattering by empty space. The scattering matrix for empty space is the unit matrix as explained in Section 2.3.1. We therefore have in this case **b** = **a** which shows that equal amounts of incoming and outgoing modes must be present in the total field. With no dipole present, the total field is

$$\begin{aligned} \vec{E}(r,\theta,\phi) &= \frac{k}{\sqrt{\eta}} \sum_{j=1}^{J} \{a_j \vec{F}_j^{(4)}(r,\theta,\phi) + b_j \vec{F}_j^{(3)}(r,\theta,\phi)\} \\ &= \frac{k}{\sqrt{\eta}} 2 \sum_{j=1}^{J} a_j \vec{F}_j^{(1)}(r,\theta,\phi). \end{aligned} \tag{2.149}$$

In the presence of a $\hat{z}$-directed dipole, we have the scattering matrix equation

$$[\hat{\mathbf{S}}_e^z] \begin{bmatrix} v \\ \mathbf{a} \end{bmatrix} = \begin{bmatrix} w \\ \mathbf{b} \end{bmatrix}. \tag{2.150}$$

Assuming the dipole transmission line to be connected to a matched load, i.e. $v = 0$, and, since eqn (2.146) holds, we have $w = a_4$. Hence, all the energy of the incoming $\vec{F}_4^{(4)}$ mode is received by the dipole and absorbed in the load. From (A1.51–1.53) $\vec{F}_4^{(1)} = \vec{F}_{201}^{(1)}$ is the only mode to have a $\hat{z}$-component at the origin. Setting $r = 0$ in eqn (2.149) therefore yields

$$\vec{E}(0,\theta,\phi) = \frac{k}{\sqrt{\eta}} \frac{\sqrt{6}}{3\sqrt{\pi}} a_4 \hat{z} \tag{2.151}$$

whereby

$$w = a_4 = \frac{\sqrt{6\pi}}{2} \frac{\sqrt{\eta}}{k} E_z(0,\theta,\phi). \tag{2.152}$$

Equation (2.152) expresses the received signal in a dipole in terms of the electric field component parallel to the dipole.

For the scattered field we have from (2.150)

$$b_i = \begin{cases} 0 & i = 4 \\ a_i & i \neq 4. \end{cases} \tag{2.153}$$

The total scattering matrices for $\hat{x}$- and $\hat{y}$-directed dipoles are given below without proof.

$$\hat{\mathbf{S}}_e^x = \begin{bmatrix} 0 & 0 & \frac{\sqrt{2}}{2} & 0 & 0 & 0 & -\frac{\sqrt{2}}{2} & 0 & \cdots \\ 0 & 1 & 0 & 0 & 0 & 0 & 0 & 0 & \\ \frac{\sqrt{2}}{2} & 0 & \frac{1}{2} & 0 & 0 & 0 & \frac{1}{2} & 0 & \\ 0 & 0 & 0 & 1 & 0 & 0 & 0 & 0 & \\ 0 & 0 & 0 & 0 & 1 & 0 & 0 & 0 & \\ 0 & 0 & 0 & 0 & 0 & 1 & 0 & 0 & \\ -\frac{\sqrt{2}}{2} & 0 & \frac{1}{2} & 0 & 0 & 0 & \frac{1}{2} & 0 & \\ 0 & 0 & 0 & 0 & 0 & 0 & 0 & 1 & \\ \vdots & & & & & & & & \ddots \end{bmatrix} \tag{2.154}$$

$$\hat{\mathbf{S}}_e^y = \begin{bmatrix} 0 & 0 & -\frac{i\sqrt{2}}{2} & 0 & 0 & 0 & -\frac{i\sqrt{2}}{2} & 0 & \cdots \\ 0 & 1 & 0 & 0 & 0 & 0 & 0 & 0 & \\ \frac{i\sqrt{2}}{2} & 0 & \frac{1}{2} & 0 & 0 & 0 & -\frac{1}{2} & 0 & \\ 0 & 0 & 0 & 1 & 0 & 0 & 0 & 0 & \\ 0 & 0 & 0 & 0 & 1 & 0 & 0 & 0 & \\ 0 & 0 & 0 & 0 & 0 & 1 & 0 & 0 & \\ \frac{i\sqrt{2}}{2} & 0 & -\frac{1}{2} & 0 & 0 & 0 & \frac{1}{2} & 0 & \\ 0 & 0 & 0 & 0 & 0 & 0 & 0 & 1 & \\ \vdots & & & & & & & & \ddots \end{bmatrix} \tag{2.155}$$

For the dual magnetic dipoles, the scattering matrices are as follows:

$$\hat{\mathbf{S}}_m^z = \begin{bmatrix} 0 & 0 & 0 & -i & 0 & 0 & \cdots \\ 0 & 1 & 0 & 0 & 0 & 0 & \\ 0 & 0 & 1 & 0 & 0 & 0 & \\ -i & 0 & 0 & 0 & 0 & 0 & \\ 0 & 0 & 0 & 0 & 1 & 0 & \\ 0 & 0 & 0 & 0 & 0 & 1 & \\ \vdots & & & & & & \ddots \end{bmatrix} \tag{2.156}$$

$$\hat{\mathbf{S}}_m^x = \begin{bmatrix} 0 & -i\frac{\sqrt{2}}{2} & 0 & 0 & 0 & i\frac{\sqrt{2}}{2} & 0 & \cdots \\ -i\frac{\sqrt{2}}{2} & \frac{1}{2} & 0 & 0 & 0 & \frac{1}{2} & 0 & \\ 0 & 0 & 1 & 0 & 0 & 0 & 0 & \\ 0 & 0 & 0 & 1 & 0 & 0 & 0 & \\ 0 & 0 & 0 & 0 & 1 & 0 & 0 & \\ i\frac{\sqrt{2}}{2} & \frac{1}{2} & 0 & 0 & 0 & \frac{1}{2} & 0 & \\ 0 & 0 & 0 & 0 & 0 & 0 & 1 & \\ \vdots & & & & & & & \ddots \end{bmatrix} \tag{2.157}$$

$$\hat{\mathbf{S}}_m^y = \begin{bmatrix} 0 & -\frac{\sqrt{2}}{2} & 0 & 0 & 0 & -\frac{\sqrt{2}}{2} & 0 & \cdots \\ \frac{\sqrt{2}}{2} & \frac{1}{2} & 0 & 0 & 0 & -\frac{1}{2} & 0 & \\ 0 & 0 & 1 & 0 & 0 & 0 & 0 & \\ 0 & 0 & 0 & 1 & 0 & 0 & 0 & \\ 0 & 0 & 0 & 0 & 1 & 0 & 0 & \\ \frac{\sqrt{2}}{2} & -\frac{1}{2} & 0 & 0 & 0 & \frac{1}{2} & 0 & \\ 0 & 0 & 0 & 0 & 0 & 0 & 1 & \\ \vdots & & & & & & & \ddots \end{bmatrix} \tag{2.158}$$

For antennas consisting of more than one dipole, receiving and transmitting elements can be obtained by adding the corresponding elements for the individual dipoles multiplied by their excitation weights. The scattering elements, however, must be found by considerations involving the scattering of the antenna for various incident fields.

A z-directed *Huygens source* is defined as the combination of an $\hat{x}$-directed electric dipole and a dual but $\hat{y}$-directed magnetic dipole. The total scattering matrix for the Huygens source is

$$\hat{\mathbf{S}}_H^z = \begin{bmatrix} 0 & -\frac{1}{2} & \frac{1}{2} & 0 & 0 & -\frac{1}{2} & -\frac{1}{2} & 0 & \cdots \\ \frac{1}{2} & \frac{3}{4} & \frac{1}{4} & 0 & 0 & -\frac{1}{4} & -\frac{1}{4} & 0 & \\ \frac{1}{2} & -\frac{1}{4} & \frac{1}{4} & 0 & 0 & -\frac{1}{4} & \frac{3}{4} & 0 & \\ 0 & 0 & 0 & 1 & 0 & 0 & 0 & 0 & \\ 0 & 0 & 0 & 0 & 1 & 0 & 0 & 0 & \\ \frac{1}{2} & -\frac{1}{4} & \frac{1}{4} & 0 & 0 & \frac{3}{4} & -\frac{1}{4} & 0 & \\ -\frac{1}{2} & \frac{1}{4} & \frac{3}{4} & 0 & 0 & \frac{1}{4} & \frac{1}{4} & 0 & \\ 0 & 0 & 0 & 0 & 0 & 0 & 0 & 1 & \\ \vdots & & & & & & & & \ddots \end{bmatrix}. \tag{2.159}$$

As another example, the combination of an $\hat{x}$-directed electric dipole and a $\hat{y}$-directed electric dipole excited in phase quadrature forms a $\hat{z}$-directed *turnstile* with the total scattering matrix

$$\hat{\mathbf{S}}_T^z = \begin{bmatrix} 0 & 0 & 1 & 0 & 0 & 0 & 0 & 0 & \cdots \\ 0 & 1 & 0 & 0 & 0 & 0 & 0 & 0 & \\ 0 & 0 & 0 & 0 & 0 & 0 & 1 & 0 & \\ 0 & 0 & 0 & 1 & 0 & 0 & 0 & 0 & \\ 0 & 0 & 0 & 0 & 1 & 0 & 0 & 0 & \\ 0 & 0 & 0 & 0 & 0 & 1 & 0 & 0 & \\ -1 & 0 & 0 & 0 & 0 & 0 & 0 & 0 & \\ 0 & 0 & 0 & 0 & 0 & 0 & 0 & 1 & \\ \vdots & & & & & & & & \ddots \end{bmatrix} \tag{2.160}$$

Note that $\hat{\mathbf{S}}_T^z$ is rather simple with the only non-vanishing elements in the first column and the first row being T_6 and R_2, respectively. This reflects the fact that the given turnstile has an electric field which is represented by the spherical wave $\vec{F}_6 = \vec{F}_{211}^{(3)}$ only. The far field is right-hand circularly polarized at $\theta = 0$ and left-hand circularly polarized at $\theta = \pi$.

The electric and magnetic short dipoles discussed in this section are members of the class of antennas called *minimum scattering antennas* [1]. The Huygens source and the turnstile antenna must include interconnecting transmission lines in order that they act as antennas with one port only. These antennas are not minimum-scattering.

2.3.5 Remarks on the definition of antenna scattering matrices

The scattering matrices considered in the present book are based on incoming and outgoing spherical waves. For a source-free region $r_0 \leqslant r < \infty$ *external* to an antenna with a minimum sphere of radius r_0 we have defined (see Section 2.3.1) the antenna scattering matrix equation

$$\begin{bmatrix} \Gamma & \mathbf{R} \\ \mathbf{T} & \mathbf{S} \end{bmatrix} \begin{bmatrix} v \\ \mathbf{a} \end{bmatrix} = \begin{bmatrix} w \\ \mathbf{b} \end{bmatrix} \tag{2.161}$$

where $\mathbf{a}$ and $\mathbf{b}$ contain as elements the wave amplitudes of incoming and outgoing spherical modes

$$\vec{E}(r, \theta, \phi) = \frac{k}{\sqrt{\eta}} \sum_{j=1}^{J} \{a_j \vec{F}_j^{(4)}(r, \theta, \phi) + b_j \vec{F}_j^{(3)}(r, \theta, \phi)\} \qquad r_0 \leqslant r < \infty \tag{2.162}$$

and where v and w are the amplitudes of suitably normalized waveguide modes incident and reflected at the waveguide port of the antenna.

An alternative formulation has been introduced by Yaghjian [24]. Based upon the same waveguide modes as above, Yaghjian's *source scattering matrix* equation is of the form

$$\begin{bmatrix} \Gamma' & \mathbf{R}' \\ \mathbf{T}' & \mathbf{S}' \end{bmatrix} \begin{bmatrix} v \\ \mathbf{a}' \end{bmatrix} = \begin{bmatrix} w \\ \mathbf{b}' \end{bmatrix}. \tag{2.163}$$

For the source-free external region $r_0 \leqslant r < \infty$, $\mathbf{a}'$ contains the amplitudes of standing waves of type $\vec{F}_j^{(1)}(r, \theta, \phi)$ while $\mathbf{b}'$ represents outgoing waves of type $\vec{F}_j^{(3)}(r, \theta, \phi)$. Hence,

$$\vec{E}(r, \theta, \phi) = \frac{k}{\sqrt{\eta}} \sum_{j=1}^{J} \{a'_j \vec{F}_j^{(1)}(r, \theta, \phi) + b'_j \vec{F}_j^{(3)}(r, \theta, \phi)\} \qquad r_0 \leqslant r < \infty. \tag{2.164}$$

Since a standing spherical wave may be expressed by travelling waves as follows

$$\vec{F}_j^{(1)}(r, \theta, \phi) = \tfrac{1}{2}\{\vec{F}_j^{(3)}(r, \theta, \phi) + \vec{F}_j^{(4)}(r, \theta, \phi)\} \tag{2.165}$$

it is easy to verify that the source scattering matrix becomes

$$\begin{bmatrix} \Gamma' & \mathbf{R}' \\ \mathbf{T}' & \mathbf{S}' \end{bmatrix} = \begin{bmatrix} \Gamma & \tfrac{1}{2}\mathbf{R} \\ \mathbf{T} & \tfrac{1}{2}(\mathbf{S} - \mathbf{I}) \end{bmatrix} \tag{2.166}$$

when expressed by the elements of the classical scattering matrix of eqn (2.161). For empty space, as seen previously, the classical theory has $\mathbf{S} = \mathbf{I}$, where $\mathbf{I}$ is the unit matrix. Therefore, in the source scattering formulation we must have $\mathbf{S}' = \mathbf{0}$. Note that the antenna reflection coefficients and the transmitting coefficients are the same in the two formulations, i.e. $\Gamma' = \Gamma$ and $\mathbf{T}' = \mathbf{T}$.

Hitherto we have dealt with external regions only. Although we shall in general avoid internal regions, let us briefly touch upon the situation illustrated in Fig. 2.13. Here, for a source-free region $0 \leqslant r \leqslant r_i$ *internal* to a given antenna,

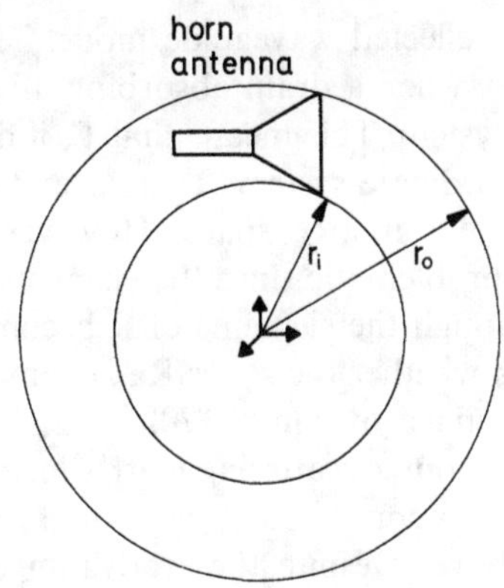

Fig. 2.13 *Horn antenna with source free internal region of radius r_i and minimum sphere of radius r_0*

we may define a scattering matrix equation identical in form to (2.161). However, since we prefer to retain the symbol $\mathbf{a}$ to designate incident modes, and the symbol $\mathbf{b}$ for reflected modes, the electric field is in this case given by

$$\vec{E}(r, \theta, \phi) = \frac{k}{\sqrt{\eta}} \sum_{j=1}^{J} \{a_j \vec{F}_j^{(3)}(r, \theta, \phi) + b_j \vec{F}_j^{(4)}(r, \theta, \phi)\} \qquad 0 \leqslant r \leqslant r_i. \tag{2.167}$$

For the internal region, a source scattering matrix equation similar to (2.163) still applies. However, the wave amplitudes $\mathbf{a}'$ and $\mathbf{b}'$ are in this case defined through

$$\vec{E}(r, \theta, \phi) = \frac{k}{\sqrt{\eta}} \sum_{j=1}^{J} \{a'_j \vec{F}_j^{(3)}(r, \theta, \phi) + b'_j \vec{F}_j^{(1)}(r, \theta, \phi)\} \qquad 0 \leqslant r \leqslant r_i. \tag{2.168}$$

Having thus defined the classical and source scattering partitioning of the fields for interior regions, we shall consider a simple example. Let us assume that the antenna in Fig. 2.13 is excited with a waveguide mode of amplitude v and that no other sources are present. The source scattering matrix equation then yields

$$\Gamma' v + \mathbf{R}'\mathbf{0} = w \tag{2.169}$$

as the field, ref. eqn (2.168), must be only represented by $\mathbf{b}'$-modes which are finite at the origin. As can be seen from this equation, Γ' represents the antenna reflection coefficient. In contrast, the classical scattering matrix equation formally gives

$$\Gamma v + \mathbf{R}\mathbf{a} = w \tag{2.170}$$

for the amplitude of the reflected waveguide mode. Thus, Γ is the reflection coefficient when $\mathbf{a} = \mathbf{0}$, i.e. when a drain absorbing all modes is placed at the origin of the coordinate system. This means that Γ, if defineable, is dependent on the location of the coordinate system. Therefore, it does not represent the antenna reflection coefficient in free space. However, the interpretation of eqn (2.170) is even more problematic since the elements of $\mathbf{a}$ will start growing large for $n > kr_i$. Even though the elements of $\mathbf{R}$ become small from the same kr_i, it is an open question whether the series $\mathbf{Ra}$ is convergent and accordingly whether Γ is rigorously defined by eqn (2.170).

For interior regions, the source scattering matrix formulation therefore is the only useable formulation. For empty space, the source scattering matrix is equal to $\mathbf{0}$ as we have seen above. Despite these advantages of using the source scattering matrices, we have preferred the classical scattering matrices in this book. The reason is that the distinction between incoming and outgoing waves is very clear in this formulation and moreover, the necessary considerations of power flow and power normalization are simplest in the framework of the classical theory.

Note, that for the special case of a pure scatterer, i.e. an antenna without any local port, Yaghjian's source scattering matrix reduces to the *T*-matrix introduced by Waterman [25] in connection with the null-field method in electromagnetic scattering theory.

2.4 Antenna parameters expressed by spherical waves

2.4.1 Far-field patterns

A general outgoing wave field is

$$\vec{E}(r, \theta, \phi) = \frac{k}{\sqrt{\eta}} \sum_{smn} Q_{smn}^{(3)} \vec{F}_{smn}^{(3)}(r, \theta, \phi) \tag{2.171}$$

$$\vec{H}(r, \theta, \phi) = -ik\sqrt{\eta} \sum_{smn} Q_{smn}^{(3)} \vec{F}_{3-s,m,n}^{(3)}(r, \theta, \phi). \tag{2.172}$$

In order to obtain expressions for the far field we need the asymptotic form of the $\vec{F}^{(3)}_{smn}$ functions as kr tends to infinity. This is obtained by replacing the radial functions entering $\vec{F}^{(3)}_{smn}$ by their large argument approximations (see Section 2.2.1). For the electric and magnetic fields at large distances we obtain

$$\vec{E}(r,\theta,\phi) \rightarrow \frac{k}{\sqrt{\eta}} \sum_{smn} Q^{(3)}_{smn} \vec{F}^{(3),a}_{smn}(r,\theta,\phi) \qquad \text{as } kr \rightarrow \infty \tag{2.173}$$

$$\vec{H}(r,\theta,\phi) \rightarrow -ik\sqrt{\eta} \sum_{smn} Q^{(3)}_{smn} \vec{F}^{(3),a}_{3-s,m,n}(r,\theta,\phi) \qquad \text{as } kr \rightarrow \infty \tag{2.174}$$

where the superscript indicates that the asymptotic form of the radial functions entering the spherical wave function has been inserted. Expressions more convenient for computation can be obtained by introducing a function describing the far field in terms of the angular variables only. To this end we define the *far-field pattern functions* $\vec{K}_{smn}(\theta,\phi)$ by the equation

$$\vec{K}_{smn}(\theta,\phi) = \lim_{kr\rightarrow\infty}\left[\sqrt{4\pi}\,\frac{kr}{e^{ikr}}\,\vec{F}^{(3)}_{smn}(r,\theta,\phi)\right]. \tag{2.175}$$

Explicit expressions for the far-field pattern functions become

$$\vec{K}_{1mn}(\theta,\phi) = \sqrt{\frac{2}{n(n+1)}}\left(-\frac{m}{|m|}\right)^m e^{im\phi}(-i)^{n+1} \left\{\frac{im\bar{P}^{|m|}_n(\cos\theta)}{\sin\theta}\hat{\theta} - \frac{d\bar{P}^{|m|}_n(\cos\theta)}{d\theta}\hat{\phi}\right\} \tag{2.176}$$

$$\vec{K}_{2mn}(\theta,\phi) = \sqrt{\frac{2}{n(n+1)}}\left(-\frac{m}{|m|}\right)^m e^{im\phi}(-i)^{n} \left\{\frac{d\bar{P}^{|m|}_n(\cos\theta)}{d\theta}\hat{\theta} + \frac{im\bar{P}^{|m|}_n(\cos\theta)}{\sin\theta}\hat{\phi}\right\}. \tag{2.177}$$

Note, that

$$\vec{K}_{smn} = i\hat{r} \times \vec{K}_{3-s,m,n} \tag{2.178}$$

In terms of $\vec{K}_{smn}(\theta,\phi)$, the electric and magnetic fields at large distances are:

$$\begin{aligned}\vec{E}(r,\theta,\phi) &\rightarrow \frac{k}{\sqrt{\eta}}\frac{1}{\sqrt{4\pi}}\frac{e^{ikr}}{kr}\sum_{smn} Q^{(3)}_{smn}\vec{K}_{smn}(\theta,\phi)\\ &= \frac{k}{\sqrt{\eta}}\frac{1}{\sqrt{4\pi}}\frac{e^{ikr}}{kr}v\sum_{smn} T_{smn}\vec{K}_{smn}(\theta,\phi) &(2.179)\\ &= \frac{k}{\sqrt{\eta}}\frac{1}{\sqrt{4\pi}}\frac{e^{ikr}}{kr}v\vec{K}(\theta,\phi) &(2.180)\end{aligned}$$

$$\vec{H}(r,\theta,\phi) \to \eta\hat{r} \times \vec{E}(r,\theta,\phi)$$
$$= k\sqrt{\eta}\frac{1}{\sqrt{4\pi}}\frac{e^{ikr}}{kr}v\hat{r} \times \vec{K}(\theta,\phi) \tag{2.181}$$

where $\vec{K}(\theta,\phi)$ is the *absolute far-field pattern* defined by

$$\vec{K}(\theta,\phi) = \sum_{smn} T_{smn}\vec{K}_{smn}(\theta,\phi) \tag{2.182}$$

Both $\vec{K}_{smn}(\theta,\phi)$ and $\vec{K}(\theta,\phi)$ are dimensionless. The quantity $C\vec{K}(\theta,\phi)$, where C is a constant, is denoted a *relative far-field pattern* or *far-field pattern.*

2.4.2 Polarization

For the description of polarization characteristics of a given antenna far-field pattern $\vec{K}(\theta,\phi)$, it is convenient to use a well-defined complex unit vector distribution tangential to the far-field sphere as a reference. Normally, one will have to choose a suitable distribution $\hat{\imath}_{co} = \hat{\imath}_{co}(\theta,\phi)$ for characterizing the copolar component. The orthogonal distribution for describing the cross-polar component then follows from

$$\hat{\imath}_{cross}(\theta,\phi) = \hat{r} \times \hat{\imath}^*_{co}(\theta,\phi) \tag{2.183}$$

It is well known [26] that the orthogonal polarization relative to a given elliptical polarization is also elliptical. Its ellipticity is the same but the sense is opposite, and the major axes of the two polarization ellipses are orthogonal. Clearly, these criteria are all fulfilled by the definition of $\hat{\imath}_{cross}(\theta,\phi)$ in eqn (2.183). Furthermore, it is easy to verify from eqn (2.183) that

$$\hat{\imath}_{co}(\theta,\phi) \cdot \hat{\imath}^*_{cross}(\theta,\phi) = 0 \tag{2.184}$$

for all (θ,ϕ) as required for orthogonality.

We now consider a given far-field pattern $\vec{K}(\theta,\phi)$. The *copolar component* $K_{co}(\theta,\phi)$ is given by

$$K_{co}(\theta,\phi) = \vec{K}(\theta,\phi) \cdot \hat{\imath}^*_{co}(\theta,\phi) \tag{2.185}$$
$$= \sum_{smn} T_{smn}\vec{K}_{smn}(\theta,\phi) \cdot \hat{\imath}^*_{co}(\theta,\phi) \tag{2.186}$$

whereas the *cross-polar component* follows from

$$K_{cross}(\theta,\phi) = \vec{K}(\theta,\phi) \cdot \hat{\imath}^*_{cross}(\theta,\phi) \tag{2.187}$$
$$= \sum_{smn} T_{smn}\vec{K}_{smn}(\theta,\phi) \cdot \hat{\imath}^*_{cross}(\theta,\phi). \tag{2.188}$$

The given far-field pattern is thus decomposed into the two polarization components

$$\vec{K}(\theta,\phi) = K_{co}(\theta,\phi)\hat{\imath}_{co}(\theta,\phi) + K_{cross}(\theta,\phi)\hat{\imath}_{cross}(\theta,\phi). \tag{2.189}$$

Some specific definitions of $\hat{\imath}_{co}$ and $\hat{\imath}_{cross}$ are of particular interest and are commonly used. For linear polarization, the so-called 'definition 3' by Ludwig [27] is based on the unit vectors

$$\hat{\imath}_{co,3L}(\theta,\phi) = \hat{\theta}\cos(\phi-\phi_o) - \hat{\phi}\sin(\phi-\phi_o) \qquad 0 \leqslant \theta < \pi \tag{2.190}$$

$$\begin{aligned}\hat{\imath}_{cross,3L}(\theta,\phi) &= \hat{r} \times \hat{\imath}^*_{co,3L}(\theta,\phi)\\ &= \hat{\theta}\sin(\phi-\phi_o) + \hat{\phi}\cos(\phi-\phi_o) \qquad 0 \leqslant \theta < \pi\end{aligned} \tag{2.191}$$

Before applying eqns (2.190) and (2.191) to any given antenna pattern, the positive z-axis must be made coincident with the antenna boresight direction, and the reference angle ϕ_o shall be chosen to equal the angle from the x-axis to the major axis of the polarization ellipse at $\theta = 0$.

If ϕ_o is chosen equal to zero, the reference polarization as given by (2.190) is identical to the polarization of a linearly polarized Huygens source consisting of an $\hat{x}$-directed electric dipole combined with a $\hat{y}$-directed magnetic dipole, (see Section 2.3.4). Figure 2.14 illustrates the distribution of reference and cross-polarization unit vectors on the far-field sphere. Note, that one set of vectors is derived from the other by a rotation about the z-axis.

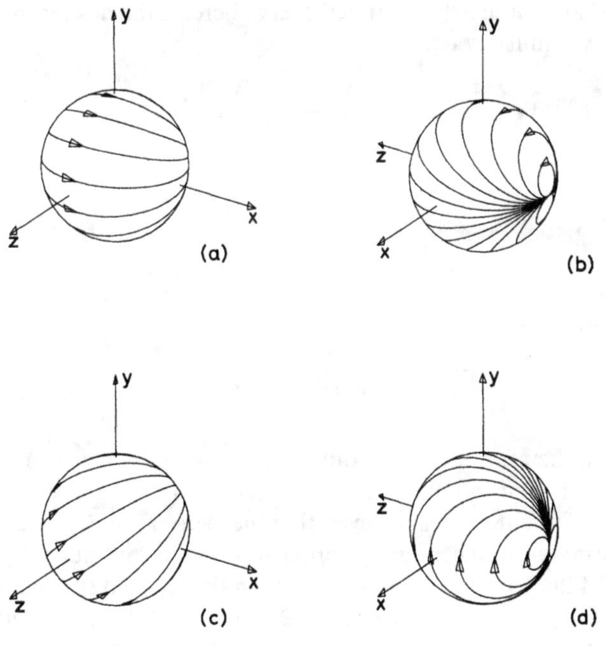

Fig. 2.14 *Co- and cross-polarization unit vector distributions on the far-field sphere*
(*a*) $\hat{\imath}_{co,3L}(\theta,\phi)$ with $\phi_o = 0°$
(*b*) same as (*a*)
(*c*) $\hat{\imath}_{co,3L}(\theta,\phi)$ with $\phi_o = 20°$
(*d*) $\hat{\imath}_{cross,3L}(\theta,\phi)$ with $\phi_o = 0°$

In the case of circular polarization, a generalization of the above definition (with $\phi_o = 0$) may be used. If the desired polarization is right-handed, we define [28]

$$\hat{\imath}_{co,RC}(\theta,\phi) = \frac{1}{\sqrt{2}}[\hat{\imath}_{co,3L}(\theta,\phi) + i\hat{\imath}_{cross,3L}(\theta,\phi)]_{\phi_o=0}$$

$$= \frac{1}{\sqrt{2}} e^{i\phi}(\hat{\theta} + i\hat{\phi}) \qquad 0 \leqslant \theta < \pi \tag{2.192}$$

which, with our time factor $\exp(-i\omega t)$, everywhere is right-hand circularly polarized [29] and identical to the polarization of a circularly polarized Huygens source, i.e. the superposition of two orthogonal linearly polarized Huygens sources fed in phase quadrature. The associated cross-polar unit vector is

$$\hat{\imath}_{cross,RC}(\theta,\phi) = \hat{r} \times \hat{\imath}^*_{co,RC}(\theta,\phi)$$

$$= \frac{i}{\sqrt{2}} e^{-i\phi}(\hat{\theta} - i\hat{\phi}) \qquad 0 \leqslant \theta < \pi \tag{2.193}$$

which is left-hand circularly polarized everywhere. If the desired polarization is left-handed, we similarly define

$$\hat{\imath}_{co,LC}(\theta,\phi) = \frac{1}{\sqrt{2}}[\hat{\imath}_{co,3L}(\theta,\phi) - i\hat{\imath}_{cross,3L}(\theta,\phi)]_{\phi_o=0}$$

$$= \frac{1}{\sqrt{2}} e^{-i\phi}(\hat{\theta} - i\hat{\phi}) \qquad 0 \leqslant \theta < \pi \tag{2.194}$$

whereby

$$\hat{\imath}_{cross,LC}(\theta,\phi) = \hat{r} \times \hat{\imath}^*_{co,LC}(\theta,\phi)$$

$$= \frac{-i}{\sqrt{2}} e^{i\phi}(\hat{\theta} + i\hat{\phi}) \qquad 0 \leqslant \theta < \pi. \tag{2.195}$$

In most cases, the absolute phase difference between two circular polarization components is not of importance and the factors i or $-i$ in eqns (2.193) and (2.195) may be left out. Note, however, that the factors $\exp(i\phi)$ and $\exp(-i\phi)$ in the above formulas are necessary in order to provide continuity at $\theta = 0$ of the distribution of the polarization unit vectors on the far-field sphere. At $\theta = \pi$, the unit vector distributions in eqns (2.190–2.195) are all discontinuous.

We shall conclude this section by providing, for a given far-field pattern, formulas for the axial ratio r and the tilt angle β of the polarization ellipse.

Let us assume, in accordance with common practice, that the far field is specified in terms of circularly polarized components without a factor i in the cross-polar component. This may be obtained by choosing instead $\hat{\imath}_{co,RC}$

and $\hat{\imath}_{co,LC}$ as the basis of the decomposition into orthogonal components. The advantage hereof is that for $\phi = 0$ the two θ-components have the phase zero.

$$\vec{K}(\theta, \phi) = K_R(\theta, \phi)\hat{\imath}_{co,RC}(\theta, \phi) + K_L(\theta, \phi)\hat{\imath}_{co,LC}(\theta, \phi). \tag{2.196}$$

Defining

$$K_R = |K_R| \exp(i\psi_R) \tag{2.197}$$

and

$$K_L = |K_L| \exp(i\psi_L) \tag{2.198}$$

and introducing the complex circular polarization ratio

$$Q = \frac{K_R}{K_L} \tag{2.199}$$

the *axial ratio* $r = tg\alpha$ of the polarization ellipse may be expressed in two ways

$$tg\alpha = \frac{|K_R| - |K_L|}{|K_R| + |K_L|} \qquad -\frac{\pi}{4} \leqslant \alpha \leqslant \frac{\pi}{4} \tag{2.200}$$

$$= \frac{|Q| - 1}{|Q| + 1}. \tag{2.201}$$

Here, $\alpha = 0$ corresponds to linear polarization. Positive and negative values of α correspond to right- and left-handed elliptical polarizations, respectively.

The tilt angle β of the polarization ellipse is given by

$$\beta = \frac{\psi_L - \psi_R}{2} \tag{2.202}$$

$$= -\tfrac{1}{2}\arg(Q) \tag{2.203}$$

with respect to the *linearly* polarized vector $[\hat{\imath}_{co,3L}(\theta, \phi)]_{\phi_o=0}$, see Fig. 2.15.

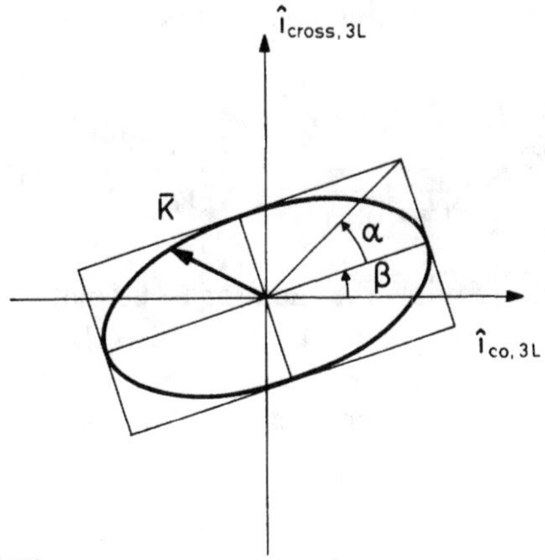

Fig. 2.15 *Polarization ellipse*

The ratio Q facilitates the discussion of polarization measurements given in Chapter 5.

2.4.3 Directivity and gain

The *directivity* $D = D(\theta, \phi)$ is defined for the far field of an antenna as the ratio between the power radiated per unit solid angle in the direction (θ, ϕ) and the power radiated per unit solid angle if the antenna radiated isotropically. According to eqn (2.42) the power radiated per unit solid angle in the direction (θ, ϕ) is given by

$$r^2 \tfrac{1}{2} \operatorname{Re}\{\vec{E}(r, \theta, \phi) \times \vec{H}^*(r, \theta, \phi)\} \cdot \hat{r} = r^2 \tfrac{1}{2}\eta\, |\vec{E}(r, \theta, \phi)|^2 \tag{2.204}$$

$$= \frac{1}{2}\frac{1}{4\pi}\left|\sum_{smn} Q^{(3)}_{smn} \vec{K}_{smn}(\theta, \phi)\right|^2 \tag{2.205}$$

in the far field.

The power radiated per unit solid angle if the antenna radiated isotropically is equal to the total radiated power P divided by 4π. From eqn (2.55)

$$\frac{P}{4\pi} = \frac{1}{4\pi}\frac{1}{2}\sum_{smn} |Q^{(3)}_{smn}|^2. \tag{2.206}$$

The directivity of the outgoing wave field becomes

$$D(\theta, \phi) = \frac{\left|\sum_{smn} Q^{(3)}_{smn} \vec{K}_{smn}(\theta, \phi)\right|^2}{\sum_{smn} |Q^{(3)}_{smn}|^2}. \tag{2.207}$$

From eqn (2.66), $Q^{(3)}_{smn} = b_{smn} = vT_{smn}$. The directivity may therefore also be expressed through the transmitting coefficients as

$$D(\theta, \phi) = \frac{\left|\sum_{smn} T_{smn} \vec{K}_{smn}(\theta, \phi)\right|^2}{\sum_{smn} |T_{smn}|^2} = \frac{|\vec{K}(\theta, \phi)|^2}{\sum_{smn} |T_{smn}|^2} = |\vec{K}(\theta, \phi)|^2 \tag{2.208}$$

where the last equality requires the radiator to be matched and lossless.

The *gain* $G = G(\theta, \phi)$ is defined in the far field as the ratio between the power radiated per unit solid angle in the direction (θ, ϕ) and the power radiated per unit solid angle if the input power accepted by the antenna

$$P_{\text{in}} = \tfrac{1}{2}|v|^2(1 - |\Gamma|^2) \tag{2.209}$$

was radiated isotropically. The gain therefore becomes

$$G(\theta, \phi) = \frac{\left|\sum\limits_{smn} Q_{smn}^{(3)} \vec{K}_{smn}(\theta, \phi)\right|^2}{|v|^2(1 - |\Gamma|^2)} \tag{2.210}$$

$$= \frac{\left|\sum\limits_{smn} T_{smn} \vec{K}_{smn}(\theta, \phi)\right|^2}{1 - |\Gamma|^2} = \frac{|\vec{K}(\theta, \phi)|^2}{1 - |\Gamma|^2} \tag{2.211}$$

$$= \frac{\left|\sum\limits_{smn} T_{smn} \vec{K}_{smn}(\theta, \phi)\right|^2}{\dfrac{P_{\text{loss}}}{P_{\text{inc}}} + \sum |T_{smn}|^2} \tag{2.212}$$

where the last equation results from using eqn (2.63).

In this section, directivity and gain has been defined without regard to polarization. However, directivity and gain may equally well be defined for a particular field component; see Section 2.4.4. The directivity or gain is in all directions equal to the sum of the directivities or gains for any two orthogonal polarizations.

2.4.4 *Maximum directivity*

It has been shown by Chu [30] for fan beam radiation and by Harrington [31] for the pencil beam case that a maximum exists for the directivity which can be obtained in spherical wave expansions like (2.171–2.172) when wave functions up to $n = N$ are included.

Expansions where the set of coefficients are adjusted for maximum directivity in some direction (θ', ϕ') are useful as theoretical models of pencil beam antenna radiation, e.g. when simulations of near-field antenna measurements are carried out. The Huygens source (ref. Section 2.3.4) is an example of a maximum directivity antenna. For this antenna, N is equal to 1 and the directivity is 3. In this section, we shall derive the coefficients in maximum directivity expansions.

The directivity is in all directions equal to the sum of the directivities of the two orthogonal polarization components

$$D(\theta, \phi) = D_{\text{co}}(\theta, \phi) + D_{\text{cross}}(\theta, \phi) \tag{2.213}$$

where

$$D_{\text{co}}(\theta, \phi) = \frac{\left|\sum\limits_{smn} Q_{smn} \vec{K}_{smn}(\theta, \phi) \cdot \hat{\imath}_{\text{co}}^*\right|^2}{\sum\limits_{smn} |Q_{smn}|^2}$$

$$= |\vec{K}(\theta, \phi) \cdot \hat{\imath}_{\text{co}}^*|^2 \tag{2.214}$$

and

$$D_{\text{cross}}(\theta, \phi) = \frac{\left| \sum_{smn} Q_{smn} \vec{K}_{smn}(\theta, \phi) \cdot \hat{\imath}^*_{\text{cross}} \right|^2}{\sum_{smn} |Q_{smn}|^2}$$

$$= |\vec{K}(\theta, \phi) \cdot \hat{\imath}^*_{\text{cross}}|^2. \tag{2.215}$$

Applying the Cauchy–Schwartz inequality [32] to equation (2.214) yields

$$D_{\text{co}}(\theta', \phi') \leqslant \frac{\sum_{smn} |Q_{smn}|^2 \sum_{smn} |\vec{K}_{smn}(\theta', \phi') \cdot \hat{\imath}^*_{\text{co}}|^2}{\sum_{smn} |Q_{smn}|^2} \tag{2.216}$$

$$= \sum_{smn} |\vec{K}_{smn}(\theta', \phi') \cdot i^*_{\text{co}}|^2$$

$$= D_{\text{co,max}}(\theta', \phi') \tag{2.217}$$

where equality in (2.216) is obtained for

$$Q_{smn} = c(\vec{K}_{smn}(\theta', \phi') \cdot \hat{\imath}^*_{\text{co}})^* \tag{2.218}$$

c being an arbitrary constant.

The maximum value of $D_{\text{co}}(\theta, \phi)$ in any given direction (θ', ϕ') is thus given by eqn (2.217). The set of wave coefficients required for actually obtaining the maximum is given by (2.218).

Let the desired polarization unit vector $\hat{\imath}_{\text{co}}$ be given by

$$\hat{\imath}_{\text{co}} = \alpha\hat{\theta} + \beta\hat{\phi} \tag{2.219}$$

where α and β are complex constants satisfying the relation

$$|\alpha|^2 + |\beta|^2 = 1. \tag{2.220}$$

We then have

$$D_{\text{co,max}}(\theta', \phi') = \sum_{mn} \{|(\vec{K}_{1mn}(\theta', \phi'))_\theta \alpha^*|^2 + |(\vec{K}_{1mn}(\theta', \phi'))_\phi \beta^*|^2$$
$$+ |(\vec{K}_{2mn}(\theta', \phi'))_\theta \alpha^*|^2 + |(\vec{K}_{2mn}(\theta', \phi'))_\phi \beta^*|^2\}. \tag{2.221}$$

From Section A1.3.2 we have

$$\{\vec{K}_{1mn}(\theta', \phi')\}_\theta = -i\{\vec{K}_{2mn}(\theta', \phi')\}_\phi \tag{2.222}$$

and

$$\{\vec{K}_{1mn}(\theta', \phi')\}_\phi = i\{\vec{K}_{2mn}(\theta', \phi')\}_\theta \tag{2.223}$$

so that

$$D_{\text{co,max}}(\theta', \phi') = \sum_{mn} \{|\vec{K}_{1mn}(\theta', \phi')|^2(|\alpha|^2 + |\beta|^2)\}$$
$$= \sum_{mn} |\vec{K}_{1mn}(\theta', \phi')|^2 \qquad (2.224)$$

independent of the choice of reference polarization.

In order to determine $D_{\text{co,max}}$ we first consider the case $(\theta', \phi') = (0, 0)$. For this choice the calculation becomes particularly simple since all other terms than those with $m = +1$ and $m = -1$ vanish (see Section A1.3.2).

$$D_{\text{co,max}}(0, 0) = \sum_{mn} |\vec{K}_{1mn}(0, 0)|^2$$
$$= \sum_{n=1}^{N} \{|\vec{K}_{11n}(0, 0)|^2 + |\vec{K}_{1,-1,n}(0, 0)|^2\}$$
$$= \sum_{n=1}^{N} (2n + 1)$$
$$= N^2 + 2N. \qquad (2.225)$$

For arbitrary values of (θ', ϕ'), the maximum directivity must equal the maximum directivity obtainable for $(\theta', \phi') = (0, 0)$, as it must be directionally independent. The reason is that N, the maximum value of n, is independent of rotation; see Section A2.2. This means that if a certain directivity is obtainable within N terms in n in one direction, it is also obtainable in any other direction within the same number of terms simply by rotating the spherical wave expansion.

The conclusion is therefore that independent of the direction (θ', ϕ') and of the chosen copolar polarization, we have

$$D_{\text{co,max}}(\theta', \phi') = N^2 + 2N. \qquad (2.226)$$

The set of expansion coefficients required for obtaining the maximum directivity in the direction $(\theta', \phi') = (0, 0)$ follows from (2.218). As an example, let us consider the case where the polarization is linear and parallel to the x-axis. We then get from Section A1.3.2:

$$Q_{11n} = Q_{1,-1,n} = Q_{21n} = -Q_{2,-1,n} = c(-i^n \tfrac{1}{2}\sqrt{2n + 1}) \qquad (2.227)$$

where c is an arbitrary constant.

All other coefficients are equal to zero. Note, that these coefficients are proportional to the coefficients for an x-polarized plane wave travelling in the positive direction of the z-axis (see Section A1.6). The expansion for maximum directivity, however, is employing the far-field pattern functions $\vec{K}_{smn}$ whereas the plane wave expansion utilizes $\vec{F}^{(1)}_{smn}$.

It may be shown that maximum directivity expansions with a prescribed polarization such as derived above have zero cross-polarization in all directions. Examples of radiation patterns are shown in Fig. 2.16.

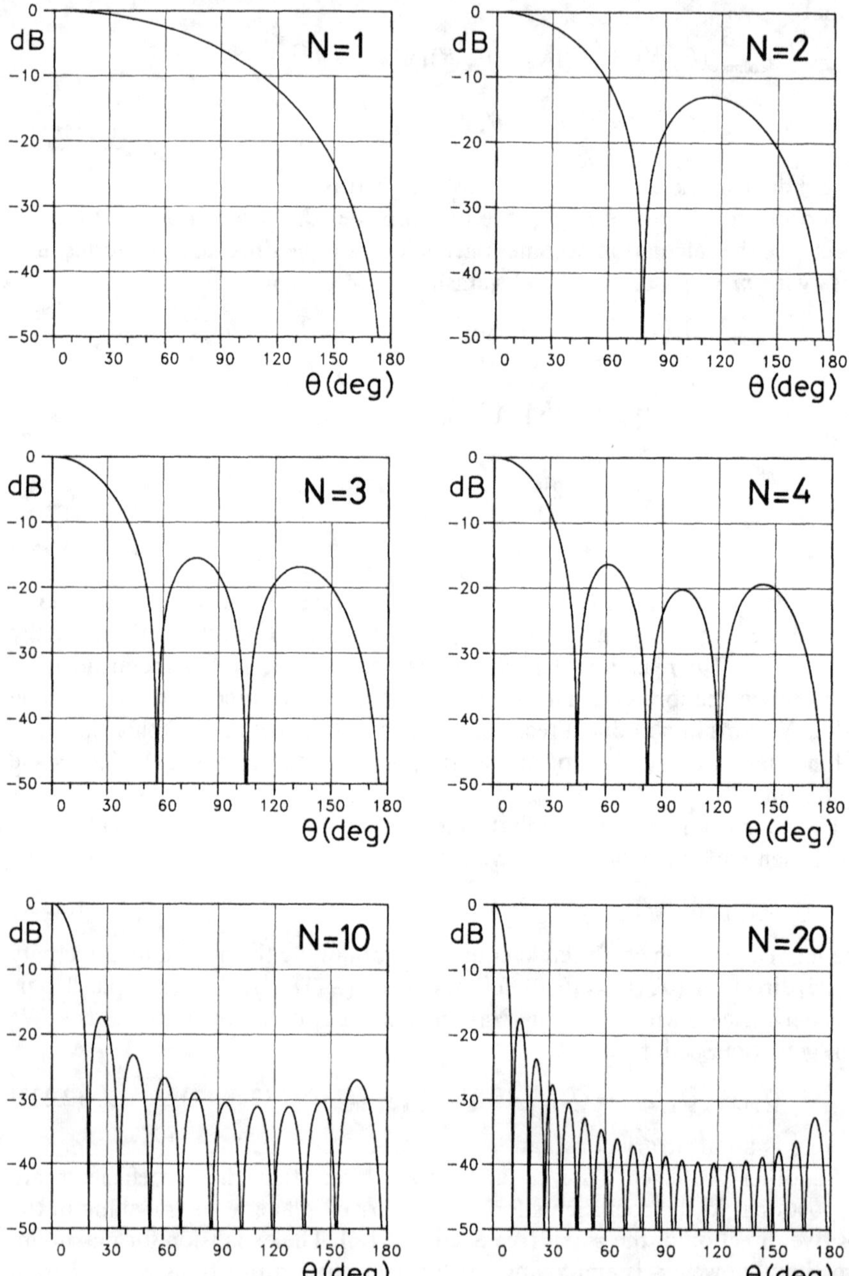

Fig. 2.16 *Maximum directivity far-field patterns* $|\vec{K}(\theta, 0)|$ *in decibels (i.e.* $20\log_{10}|\vec{K}|$*) for prescribed upper limits* $N = 1, 2, 3, 4, 10$ *and* 20 *in the spherical wave expansion. The peak maximum directivities* $N^2 + 2N$ *are normalized to 0 decibels*

References

1. DICKE, R. H., MONTGOMERY, C. G. and PURCELL, E. M., 1965, *Principles of microwave circuits*, Dover Publications, Inc., New York.
2. KAHN, W. K. and KURSS, H., 1965, 'Minimum-scattering antennas', *IEEE Trans.*, Vol. **AP-13**, No. 5, pp. 671–675, Sep.
3. WASYLKIWSKYJ, W. and KAHN, W. K., 1970, 'Theory of mutual coupling among minimum-scattering antennas', *IEEE Trans.*, Vol. **AP-18**, No. 2, pp. 204–216, Mar.
4. WASYLKIWSKYJ, W. and KAHN, W. K., 1970, 'Scattering properties and mutual coupling of antennas with prescribed radiation pattern', *IEEE Trans.*, Vol. **AP-18**, No. 6, pp. 741–752, Nov.
5. GATELY, A. C., Jr., STOCK, D. J. R. and RU-SHAO CHEO, B., 1968, 'A network description for antenna problems', *Proc. IEEE*, Vol. **56**, No. 7, pp. 1181–1193, July.
6. HANSEN, W. W., 1935, 'A new type of expansion in radiation problems', *Phys. Review*, Vol. **47**, pp. 139–143, Jan.
7. STRATTON, J. A., 1941, *Electromagnetic Theory*, Chapter VII, McGraw-Hill, Inc., New York.
8. MORSE, P. M. and FESHBACH, H., 1953, *Methods of theoretical physics*, Chapter 13, Vol. 2, McGraw-Hill, Inc., New York.
9. HARRINGTON, R. F., 1961, *Time-harmonic electromagnetic fields*, Appendix E, McGraw-Hill, Inc., New York.
10. TAI, C. T., 1971, *Dyadic Green's functions in electromagnetic theory*, Intext Educational Publishers, San Francisco.
11. COLLIN, R. E. and ZUCKER, F. J. (editors), 1969, *Antenna Theory*, McGraw-Hill, Inc., New York.
12. LUDWIG, A. C., 1982, 'Spherical wave theory', Chapter 2.3 in *The Handbook of Antenna Design*, Vol. **1**, Peter Peregrinus, London.
13. JENSEN, F., 1970, 'Electromagnetic near-field–far-field correlations', Ph.D. Thesis, Laboratory of Electromagnetic Theory, Technical University of Denmark, Report **LD 15**, July.
14. BELOUSOV, S. L., 1962, *Tables of normalized associated Legendre polynomials*, Pergamon Press, Oxford.
15. EDMONDS, A. R., 1974, *Angular momentum in quantum mechanics*, 2nd edition, Princeton University Press, Princeton, New Jersey.
16. MARCUVITZ, N., 1951, *Waveguide handbook*, Dover Publications, Inc., New York.
17. FELSEN, L. and MARCUVITZ, N., 1973, *Radiation and scattering of waves*, Prentice-Hall, Englewood Cliffs, New Jersey.
18. TURCHIN, V. I. and TSEYTLIN, N. M., 1979, 'Antenna testing based on near-field measurements (Review)', *Radio Engineering & Electronic Physics*, Vol. **24**, pp. 1–26, Dec.
19. POTTER, P. D., 1967, 'Application of spherical wave theory to Cassegrainian-fed paraboloids', *IEEE Trans.*, Vol. **AP-15**, No. 6, pp. 727–736, Nov.
20. COLLIN, R. E. and ROTHSCHILD, S., 1964, 'Evaluation of antenna Q', *IEEE Trans.*, Vol. **AP-12**, No. 1, pp. 23–27, Jan.
21. ABRAMOWITZ, M. and STEGUN, I. A. (editors), 1965, *Handbook of mathematical functions*, Dover Publications, Inc., New York.
22. KERNS, D. M. and BEATTY, R. W., 1966, *Basic theory of waveguide junctions and introductory microwave network analysis*, Pergamon Press, Oxford.
23. KERNS, D. M., 1981, 'Plane-wave scattering-matrix theory of antennas and antenna–antenna interactions', *NBS Monograph* **162**, Nat. Bur. Stds., Boulder Colorado, June.
24. YAGHJIAN, A. D., 1977, 'Near-field antenna measurements on a cylindrical surface: A source scattering-matrix formulation', Nat. Bur. Stds., *Tech. Note* **696**, Boulder, Colorado, July.
25. WATERMAN, P. C., 1971, 'Symmetry, Unitarity and Geometry in Electromagnetic Scattering', *Phys. Review D*, Vol. **3**, No. 4, pp. 825–839, Feb.
26. RUMSEY, V. H., DESCHAMPS, G. A., KALES, M. L. and BOHNERT, J. I., 1951, 'Techniques for handling elliptically polarized waves with special reference to antennas', *Proc. IRE*, Vol. **39**, pp. 533–552, May.

27. LUDWIG, A. C., 1973, 'The definition of cross polarization', *IEEE Trans.*, Vol. **AP-21**, No. 1, pp. 116–119, Jan.
28. WOOD, P. J., 1980, *Reflector antenna analysis and design*, Peter Peregrinus, London.
29. IEEE, 1969, 'IEEE Standard definitions of terms for antennas', *IEEE Trans.*, Vol. **AP-17**, No. 3, pp. 262–269, May.
30. CHU, L. J., 1948, 'Physical limitations of omni-directional antennas', *J. Appl. Phys.*, Vol. **19**, pp. 1163–1175, Dec.
31. HARRINGTON, R. F., 1960, 'Effect of antenna size on gain, bandwidth and efficiency', *J. Res. NBS*, Vol. **64D**, No. 1, pp. 1–12, Jan.–Feb.
32. RALL, L. B., 1969, *Computational solution of nonlinear operator equations*, John Wiley & Sons, Inc., New York.
33. COLLIN, R. E., 1986, 'Dyadic Green's function expansions in spherical coordinates', *Electromagnetics*, Vol. **6**, pp. 183–207.

Chapter 3

Scattering matrix description of antenna coupling

3.1 Introduction

The scattering matrix theory for antennas, (see Chapter 2), is the obvious tool for analyzing the interaction between the test antenna and the probe in near-field test ranges with spherical geometries. In particular, the theory forms a natural basis for a simple derivation of the *transmission formula* in spherical coordinates. This formula is fundamental to spherical near-field testing of antennas, and its derivation is the main subject of this chapter.

Historically, the first derivation of a transmission formula for spherical scanning did not utilize scattering matrices. Instead, the Lorentz reciprocity theorem (see Appendix A1) was used in the derivation [1]. Furthermore, the first application to near-field testing of scattering matrix theory was related to planar scanning [2] and employed an extended theory where continuous spectra of plane waves were used rather than the discrete set of spherical waves. The scattering matrix formulation of a spherical near-field transmission formula was first given by Wacker [3] and further developed and clarified by Larsen [4] who introduced a power-normalized formulation. A scattering matrix formulation for the spherical case was presented by Lewis [5] who extended the source scattering matrix concept, developed by Yaghjian [6] for the cylindrical case, to spherical coordinates. (See also Appel-Hansen [7].)

In this chapter, we shall present in Section 3.2.2 the derivation of the spherical near-field transmission formula [4]. The formula expresses the complex signal received by a probe of known receiving coefficients as a function of the probe coordinates and the probe rotation when a test antenna of unknown coefficients is transmitting. In the original derivation [1] the probe rotation was not included. A rotation of the probe must be included in the theory before the orthogonality of the mathematical factors entering the formula can be fully utilized in the calculations (ref. Chapter 4) leading to the wanted transmitting coefficients of the test antenna. This fact was recognized independently in [8] and in [9].

In Section 3.2.3 a different transmission formula, slightly more complicated and valid for the opposite direction of transmission, is derived. Both transmission formulas are deduced without assuming reciprocity for either of the two antennas involved. Of course, in case of reciprocal antennas, the two transmission formulas become identical. In Section 3.2.4 it is shown that one and the same computer algorithm (detailed in later chapters) may be used in the data processing of near-field measurements irrespective of whether or not the test antenna and probe are reciprocal and independent of which antenna transmits and which receives.

Any near-field measurement requires some kind of probe calibration, i.e. determination of the set of probe receiving coefficients. This must be done in a separate spherical near-field scan before the test antenna can be measured. A special measuring procedure may be required here and is outlined in Section 3.2.5. Probe calibration from a practical point of view, and in much more detail, is described in Chapter 5.

Some theoretical results of a general interest are reported in Section 3.3. The transmission formula takes a particularly simple form for the important class of probes with azimuthal indices ± 1 only. This form is derived in Section 3.3.1 where also the concept of the *probe response constants* is defined for later use. In Section 3.3.2 a number of results pertaining to electric and magnetic dipole probes are derived. The well known *Friis' transmission formula* [10] is shown in Section 3.3.3 to be a special case of the transmission formula.

The chapter concludes with the application of scattering matrix theory to the derivation of a transmission formula with multiple reflections included (Section 3.4), [11], [12], [13].

3.2 The transmission formula

3.2.1 Geometry

The geometry of the measurement set-up for spherical near-field measurements differs from the geometry of far-field measurements only in the measurement distance. While in far-field measurements the distance between the test antenna and the measurement probe usually is kept larger than the Rayleigh distance, the measurement distance in spherical near-field measurement is of the order of a few times the radius of the test antenna minimum sphere.

During the measurement, the test antenna is kept fixed in an unprimed (x, y, z) coordinate system while the probe, including its own primed coordinate system (x', y', z'), is moved on a spherical surface of radius A centred at the origin. The probe coordinates are thus (A, θ, ϕ) in the test antenna coordinate system. The probe is constantly pointing towards the origin of the (x, y, z) coordinate system while scanning the measurement sphere. The possibility of rotating the probe by an angle χ around its own axis is also required in spherical near-field measurements, although normally the values $\chi = 0$ and $\chi = \pi/2$ will be sufficient as we shall see later. Figure 3.1 shows the test antenna and probe minimum spheres (non-intersecting) as well as their coordinate systems.

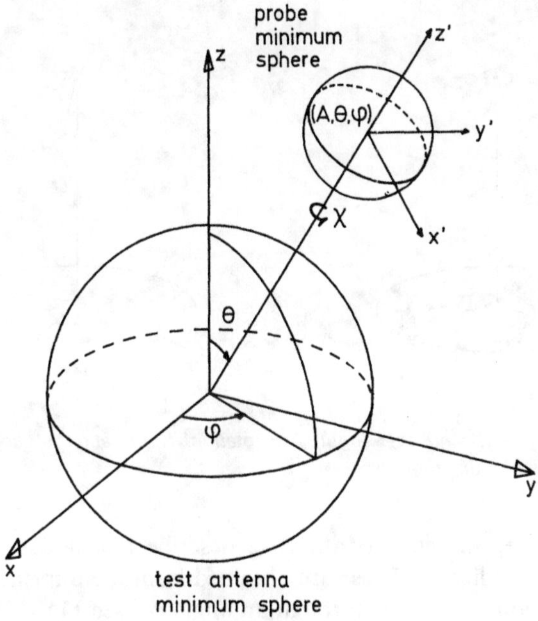

Fig. 3.1 *Test antenna and probe minimum spheres*
The probe coordinate system is shown for $\chi = 0$.

3.2.2 Test antenna transmitting, probe receiving

Let us consider the field radiated by the test antenna expressed in its own spherical system (r, θ, ϕ) with associated rectangular system (x, y, z)

$$\vec{E}_t(r, \theta, \phi) = \frac{k}{\sqrt{\eta}} \sum_{smn} Q^{(3)}_{smn} \vec{F}^{(3)}_{smn}(r, \theta, \phi)$$

$$= \frac{k}{\sqrt{\eta}} \sum_{smn} vT_{smn} \vec{F}^{(3)}_{smn}(r, \theta, \phi), \qquad r > r_0 \tag{3.1}$$

where r_0 is the radius of the test antenna minimum sphere and where v and T_{smn} is the excitation amplitude and test antenna transmitting coefficients, respectively. In (3.1) use has been made of the relationships $Q^{(3)}_{smn} = b_{smn} = vT_{smn}$; see Section 2.3.1 and eqn (2.66).

We shall first seek for this field an expansion in the primed probe coordinate system (before the probe is introduced). Such an expansion shall be obtained by successively describing each of the modes in eqn (3.1) in a sequence of rotated and translated coordinate systems as shown in Fig. 3.2. In doing so, we shall use the convention that a wave function expresses a spherical mode belonging to the same coordinate system as the coordinates in the argument of the function.

Further, we shall use mode indices *smn* for modes expressed in the unprimed coordinates, while mode indices $\sigma\mu\nu$ designate modes in the primed coordinate system.

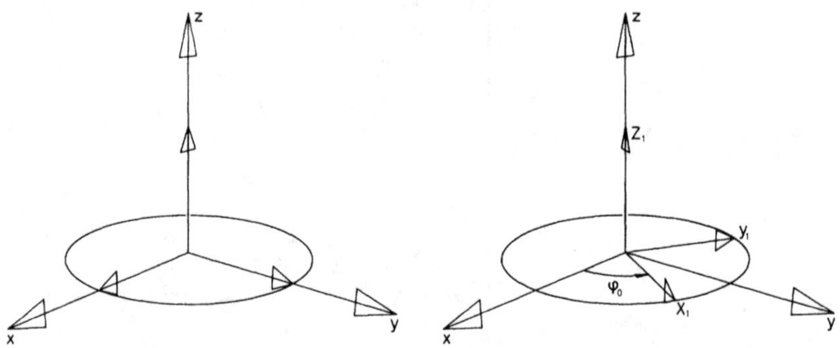

Fig. 3.2(a) *Before (left) and after (right) first rotation. The rotation is about the z-axis and defined by the angle ϕ_o*

As the first step in the derivation we describe the modes in eqn (3.1) in coordinates with indices 1. These are obtained from the unprimed coordinates by a rotation about the z-axis through an angle ϕ_o; see Fig. 3.2(a).

$$\vec{F}^{(3)}_{smn}(r, \theta, \phi) = e^{im\phi_o}\vec{F}^{(3)}_{smn}(r_1, \theta_1, \phi_1), \qquad r_1 > r_0. \tag{3.2}$$

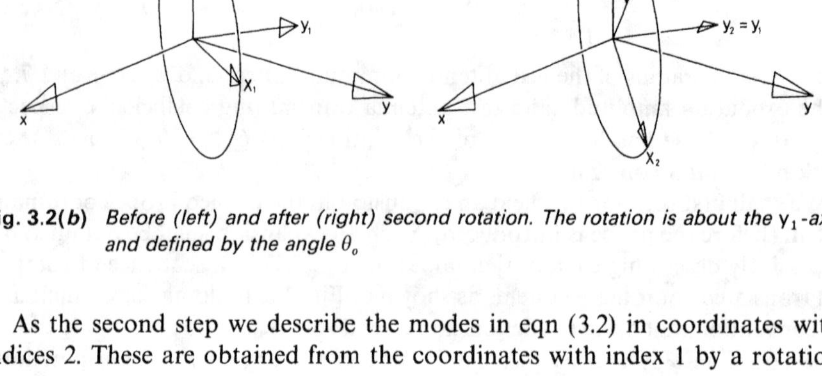

Fig. 3.2(b) *Before (left) and after (right) second rotation. The rotation is about the y_1-axis and defined by the angle θ_o*

As the second step we describe the modes in eqn (3.2) in coordinates with indices 2. These are obtained from the coordinates with index 1 by a rotation about the y_1-axis through an angle θ_o; see Fig. 3.2(b).

$$\vec{F}^{(3)}_{smn}(r_1, \theta_1, \phi_1) = \sum_{\mu=-n}^{n} d^n_{\mu m}(\theta_o) \vec{F}^{(3)}_{s\mu n}(r_2, \theta_2, \phi_2), \qquad r_2 > r_0 \tag{3.3}$$

where $d^n_{\mu m}(\theta_o)$ is a rotation coefficient (see Appendix A2).

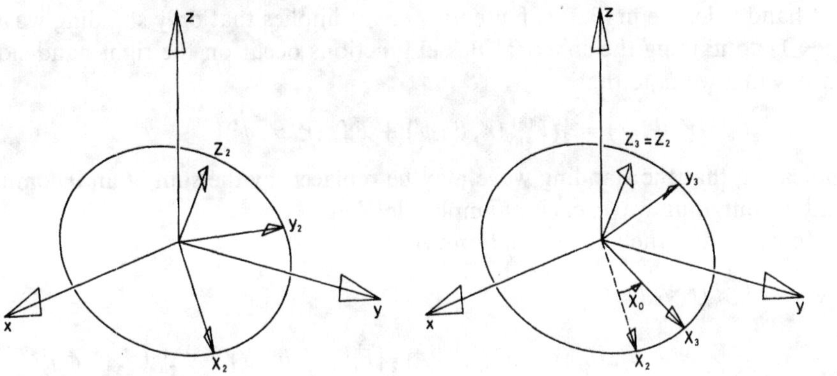

Fig. 3.2(*c*) *Before (left) and after (right) third rotation. The rotation is about the z_2-axis and defined by the angle χ_o*

As the third step we describe the modes in eqn (3.3) in coordinates with index 3. These are obtained from the coordinates with index 2 by a rotation about the z_2-axis through an angle χ_o; see Fig. 3.2(c).

$$\vec{F}^{(3)}_{s\mu n}(r_2, \theta_2, \phi_2) = e^{i\mu\chi_o} \vec{F}^{(3)}_{s\mu n}(r_3, \theta_3, \phi_3), \qquad r_3 > r_0. \tag{3.4}$$

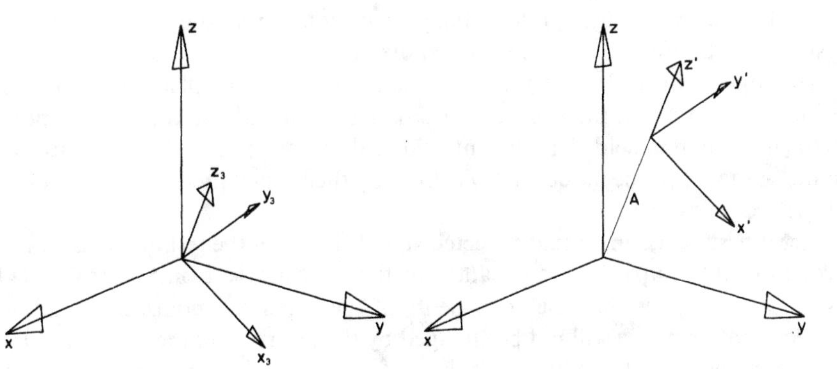

Fig. 3.2(*d*) *Before (left) and after (right) translation. The translation is in the positive direction of the z_3-axis*

Finally, we describe the modes in eqn (3.4) in the primed probe coordinates. These are obtained from the coordinates with index 3 by a translation of $A > 0$

along the z_3-axis; see Fig. 3.2(d).

$$\vec{F}^{(3)}_{s\mu n}(r_3, \theta_3, \phi_3) = \sum_{\sigma=1}^{2} \sum_{\substack{\nu=|\mu| \\ \nu \neq 0}}^{\infty} C^{sn(3)}_{\sigma\mu\nu}(kA)\vec{F}^{(1)}_{\sigma\mu\nu}(r', \theta', \phi') \qquad r' < (A - r_0) \tag{3.5}$$

where $C^{sn(3)}_{\sigma\mu\nu}(kA)$ is a translation coefficient, ref. Appendix A3. The fact that the left-hand side of eqn (3.5) is finite for $r_3 > r_0$ implies that only standing waves ($c = 1$) containing the spherical Bessel functions occur on the right-hand side. In passing, we note that

$$\vec{F}^{(1)}_{\sigma\mu\nu}(r', \theta', \phi') = \tfrac{1}{2}(\vec{F}^{(3)}_{\sigma\mu\nu}(r', \theta', \phi') + \vec{F}^{(4)}_{\sigma\mu\nu}(r', \theta', \phi'))$$

indicating that the standing wave may be replaced by the sum of an incoming and an outgoing wave, each of amplitude $\frac{1}{2}$.

In summary, the above equations yield

$$\vec{F}^{(3)}_{smn}(r, \theta, \phi) = \sum_{\sigma\mu\nu} e^{im\phi_o} d^n_{\mu m}(\theta_o) e^{i\mu\chi_o} C^{sn(3)}_{\sigma\mu\nu}(kA) \tfrac{1}{2}\{\vec{F}^{(3)}_{\sigma\mu\nu}(r', \theta', \phi') + \vec{F}^{(4)}_{\sigma\mu\nu}(r', \theta', \phi')\}. \tag{3.6}$$

The expansion sought for the test antenna field on the left-hand side of eqn (3.1) is now obtained by inserting eqn (3.6) into the right-hand side of eqn (3.1). The resulting equation is

$$\vec{E}_t = \frac{k}{\sqrt{\eta}} \sum_{\substack{smn \\ \sigma\mu\nu}} v T_{smn} e^{im\phi_o} d^n_{\mu m}(\theta_o) e^{i\mu\chi_o} C^{sn(3)}_{\sigma\mu\nu}(kA) \tfrac{1}{2}(\vec{F}^{(3)}_{\sigma\mu\nu} + \vec{F}^{(4)}_{\sigma\mu\nu}) \tag{3.7}$$

where the angles $(\chi_o, \theta_o, \phi_o)$ describing the orientation of the probe coordinate system are the Euler angles (ref. Appendix A2).

Having established the above expansion we shall now place the probe antenna at the origin of the primed coordinate system and determine its response to the test antenna field. The currents induced on the probe will only contribute to the outgoing probe modes. They cannot by themselves modify the coefficients of the incoming modes.

Let it first be assumed that reflected signals between the test antenna and the probe do not contribute significantly to the field incident on the probe. This assumption implies that the coefficients of the spherical modes $\vec{F}^{(4)}_{\sigma\mu\nu}(r', \theta', \phi')$ incident on the probe will not be changed by the insertion of the probe. In other words, the field scattered by the probe and subsequently rescattered by the test antenna is assumed to be negligible at the probe location. The field scattered by the probe may of course be determined explicitly, if needed. This is done in Section 3.4 on the influence of multiple reflections.

We shall also assume that the receiver (load) connected to the probe is perfectly matched to the transmission line. The reflection coefficient Γ_l is thus zero. This second assumption is made on the grounds that in most practical cases the signal received by the probe is not measured in absolute units (i.e. watts). A factor accounting for mismatch is not of any significance in such cases. Furthermore, $\Gamma_p = 0$ can always be assumed in theoretical considerations where abstract sources such as the Hertzian dipole are involved. The restriction on Γ_l (or Γ_p) can be relinquished, as described in Section 5.4, on the measurement of gain.

The signal w received by the probe may now be computed. To this end we make use of eqn (2.71) without the mismatch factor

$$w = \mathbf{R}^p \mathbf{a} \tag{3.8}$$

where $\mathbf{R}^p$ contains the probe receiving coefficients $R^p_{\sigma\mu\nu}$ and $\mathbf{a}$ has as its components the wave amplitudes $a_{\sigma\mu\nu}$ of those modes in eqn (3.7) which are incident on the probe. The incident modes are the $\vec{F}^{(4)}_{\sigma\mu\nu}(r', \theta', \phi')$ wave functions, hence

$$a_{\sigma\mu\nu} = \frac{v}{2} \sum_{smn} T_{smn} \, \mathrm{e}^{im\phi_o} \, d^n_{\mu m}(\theta_o) \, \mathrm{e}^{i\mu\chi_o} \, C^{sn(3)}_{\sigma\mu\nu}(kA). \tag{3.9}$$

Omitting the subscripts o on the probe angles $(\chi_o, \theta_o, \phi_o)$ and inserting eqn (3.9) into eqn (3.8) gives the transmission formula

$$w(A, \chi, \theta, \phi) = \frac{v}{2} \sum_{\substack{smn \\ \sigma\mu\nu}} T_{smn} \, \mathrm{e}^{im\phi} \, d^n_{\mu m}(\theta) \, \mathrm{e}^{i\mu\chi} \, C^{sn(3)}_{\sigma\mu\nu}(kA) R^p_{\sigma\mu\nu}. \tag{3.10}$$

The transmission formula (3.10) expresses the complex signal received by the probe as a function of probe position and orientation relative to the transmitting test antenna. The formula is central for spherical near-field antenna testing and will be used later to determine the unknown test antenna coefficients T_{smn} from samples of the received probe signal $w(A, \chi, \theta, \phi)$.

Note, that no assumption about reciprocity of the test antenna or the probe has been made in the derivation.

3.2.3 Test antenna receiving, probe transmitting

A transmission formula may also be derived for the case where the test antenna is receiving and the probe is transmitting. For this purpose we may use the same geometry (Fig. 3.1) and use a similar procedure as in Section 3.2.2. The field radiated by the matched probe is

$$\vec{E}_p(r', \theta', \phi') = \frac{k}{\sqrt{\eta}} \sum_{\sigma\mu\nu} v_p T^p_{\sigma\mu\nu} \vec{F}^{(3)}_{\sigma\mu\nu}(r', \theta', \phi') \qquad r' > r'_0 \tag{3.11}$$

where v_p is the amplitude of the signal applied to the probe, and where r'_0 is the radius of the probe minimum sphere. In order to express this field in the unprimed test antenna coordinate system, we may go through the same series of coordinate transformations as previously, however in the opposite order of succession and with the opposite arguments. Instead of using ϕ_o, θ_o, χ_o and A as in Section 3.2.2 we shall here need transformations with $-A$, $-\chi_o$, $-\theta_o$ and $-\phi_o$. The formula expressing $\vec{E}_p$ in the unprimed test antenna coordinate system then becomes

$$\vec{E}_p = \frac{k}{\sqrt{\eta}} \sum_{\substack{\sigma\mu\nu \\ smn}} v_p T^p_{\sigma\mu\nu} C^{\sigma\nu(3)}_{s\mu n}(-kA)\, e^{-i\mu\chi_o} d^n_{m\mu}(-\theta_o)\, e^{-im\phi_o} \tfrac{1}{2}(\vec{F}^{(3)}_{smn} + \vec{F}^{(4)}_{smn}) \tag{3.12}$$

by analogy with eqn (3.7).

The following two assumptions will now be made:

1. Reflected signals between the probe and the test antenna shall not contribute significantly to the field incident on the test antenna.
2. The receiver (load) or the test antenna is perfectly matched to its transmission line, thus $\Gamma_l = 0$ or $\Gamma = 0$.

The signal received by the test antenna now becomes

$$w_t = \mathbf{Ra} \tag{3.13}$$

where $\mathbf{a}$ contains the incident wave amplitudes a_{smn} as defined through eqn (3.12)

$$a_{smn} = \frac{v_p}{2} \sum_{\sigma\mu\nu} T^p_{\sigma\mu\nu} C^{\sigma\nu(3)}_{s\mu n}(-kA)\, e^{-i\mu\chi_o} d^n_{m\mu}(-\theta_o)\, e^{-im\phi_o} \tag{3.14}$$

and where $\mathbf{R}$ contains the test antenna receiving coefficients R_{smn}. Omitting again the subscripts o on the probe angles $(\chi_o, \theta_o, \phi_o)$ and inserting eqn (3.14) into eqn (3.13) we obtain

$$w_t = \frac{v_p}{2} \sum_{\substack{\sigma\mu\nu \\ smn}} T^p_{\sigma\mu\nu} C^{\sigma\nu(3)}_{s\mu n}(-kA)\, e^{-i\mu\chi} d^n_{m\mu}(-\theta)\, e^{-im\phi} R_{smn} \tag{3.15}$$

which can be further reduced by application of the symmetry relations (A3.13), (A2.7)

$$w_t = \frac{v_p}{2} \sum_{\substack{\sigma\mu\nu \\ smn}} T^p_{\sigma\mu\nu} C^{sn(3)}_{\sigma,-\mu,\nu}(kA)\, e^{-i\mu\chi} d^n_{\mu m}(\theta)\, e^{-im\phi} R_{smn}. \tag{3.16}$$

By using (A2.9) and subsequently replacing m by $-m$ and μ by $-\mu$ under the summation sign, which does not change the value of the sum, we obtain

$$w_t = \frac{v_p}{2} \sum_{\substack{\sigma\mu\nu \\ smn}} (-1)^\mu T^p_{\sigma,-\mu,\nu} C^{sn(3)}_{\sigma\mu\nu}(kA)\, e^{i\mu\chi} d^n_{\mu m}(\theta)\, e^{im\phi} (-1)^m R_{s,-m,n}. \tag{3.17}$$

The transmission formula (3.17) expresses the complex signal received by the test antenna as a function of the position and orientation of a transmitting probe. No assumption about reciprocity of the test antenna or the probe has been made in the derivation of (3.17). The formula is slightly more complicated than the formula (3.10) for the opposite direction of transmission. This is not surprising as the derivation in the two cases is not symmetrical with respect to the two antennas. Note, however, that by introducing from eqns (2.103) and (2.104) the relations

$$T^{p}_{\sigma,-\mu,\nu} = (-1)^{\mu} R^{p'}_{\sigma\mu\nu} \tag{3.18}$$

$$R_{s,-m,n} = (-1)^{m} T'_{smn} \tag{3.19}$$

between the coefficients of the original antennas and the coefficients of their adjoint antennas (indicated by the primes), we obtain the following transmission formula

$$w_t = \frac{v_p}{2} \sum_{\substack{\sigma\mu\nu \\ smn}} R^{p'}_{\sigma\mu\nu} C^{sn(3)}_{\sigma\mu\nu}(kA)\, e^{i\mu\chi} d^{n}_{\mu m}(\theta)\, e^{im\phi}\, T'_{smn}. \tag{3.20}$$

With respect to known and unknown quantities, eqn (3.20) is similar in form to eqn (3.10). The two equations, however, apply to opposite directions of transmission.

Provided the excitations are the same, i.e. $v_p = v$, the signal received in a given spherical near-field set-up is equal to the signal received in the same set-up, where the receiver and transmitter are interchanged and the adjoint antennas substituted for the original antennas. In particular, if the test antenna and the probe are both reciprocal, eqns (3.10) and (3.20) yield identical received signals, since in this case the primes disappear and

$$T^{p}_{\sigma,-\mu,\nu} = (-1)^{\mu} R^{p}_{\sigma\mu\nu} \tag{3.21}$$

$$R_{s,-m,n} = (-1)^{m} T_{smn}. \tag{3.22}$$

In the next two sections applications of the derived transmission formulas will be shown.

3.2.4 Reciprocal and non-reciprocal test antennas

In this section, application of the transmission formulas in connection with the measurement of the test antenna is considered. Later chapters will cover practical aspects.

The choice of the direction of transmission in a near-field experiment of course depends upon the test antenna to be measured. If the test antenna is of a non-reciprocal transmitting type, obviously it must be transmitting during the measurements for physical reasons. If it is of a non-reciprocal receiving type it must be receiving during the measurement. Examples of non-reciprocal transmitting and receiving antennas are shown schematically in Fig. 3.3. Should the test antenna be reciprocal, the choice of direction of transmission is unimportant. In the two preceding sections, transmission formulas for both directions of transmission have been derived.

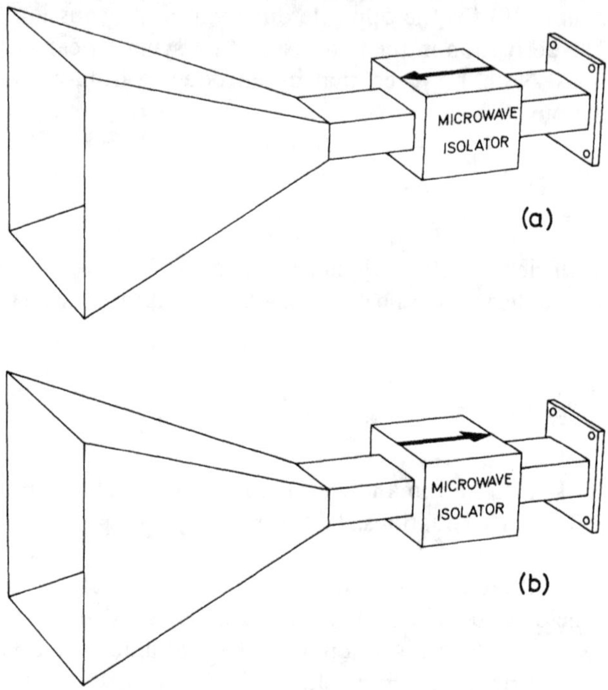

Fig. 3.3 *(a) A non-reciprocal antenna of the transmitting type; (b) A non-reciprocal antenna of the receiving type*
If the two horns are identical and the two microwave isolators differ in the direction of transmission only, the antennas (a) and (b) are the adjoints of each other.

For the test antenna transmitting case we found

$$w(A, \chi, \theta, \phi) = \frac{v}{2} \sum_{\substack{smn \\ \sigma\mu\nu}} T_{smn}\, e^{im\phi}\, d^n_{\mu m}(\theta)\, e^{i\mu\chi}\, C^{sn(3)}_{\sigma\mu\nu}(kA) R^p_{\sigma\mu\nu}. \tag{3.23}$$

This formula will be used in Chapter 4 to determine the test antenna's unknown transmitting coefficients T_{smn}. A computer algorithm, to be discussed later, for performing this task will need as its input data:

1. Samples of the received probe signal $w(A, \chi, \theta, \phi)$ as obtained in a spherical near-field scanning procedure with measuring distance A, angular variables (χ, θ, ϕ) and test antenna input signal v.
2. The receiving coefficients $R^p_{\sigma\mu\nu}$ of the probe.

If the test antenna is reciprocal, eqn (3.22) may be applied, whereby its receiving coefficients R_{smn} can also be determined. We shall discuss in the next section how the receiving coefficients $R^p_{\sigma\mu\nu}$ of the probe can be found in the probe calibration procedure.

For the test antenna receiving case we found

$$w_t(A, \chi, \theta, \phi) = \frac{v_p}{2} \sum_{\substack{smn \\ \sigma\mu\nu}} T'_{smn} \, e^{im\phi} d^n_{\mu m}(\theta) \, e^{i\mu\chi} C^{sn(3)}_{\sigma\mu\nu}(kA) R^{p'}_{\sigma\mu\nu} \tag{3.24}$$

which is of the same form with respect to known and unknown quantities as eqn (3.23). The above mentioned algorithm may therefore be used without change for determining T'_{smn}, the transmitting coefficients of the adjoint test antenna. For this application of the computer algorithm, the input data must be

1. Samples of the received test antenna signal $w_t(A, \chi, \theta, \phi)$ as obtained in a spherical near-field scanning procedure with measuring distance A, angular variables (χ, θ, ϕ), and probe input signal v_p.
2. The receiving coefficients $R^{p'}_{\sigma\mu\nu}$ of the adjoint probe.

Apart from the reciprocal case where $T'_{smn} = T_{smn}$, the coefficients T'_{smn} are not of direct interest. However, eqn (3.19) may be invoked and the physical relevant test antenna receiving coefficients R_{smn} thereby found. We shall discuss in the next section how the receiving coefficients $R^{p'}_{\sigma\mu\nu}$ of the adjoint probe can be found in the probe calibration procedure.

If the test antenna and the probe are both reciprocal, the two transmission formulas become identical. Note, that in this case a knowledge of the probe receiving coefficients is required by the computer algorithm even when the probe is transmitting. Also, the algorithm will provide test antenna transmitting coefficients even when the test antenna is receiving in the experiment.

3.2.5 Iterative scheme for probe calibration

In this section, application of the transmission formulas in connection with the calibration of the probe is considered.

Calibration of the probe antenna, i.e. determination of its set of receiving coefficients, must be done in a separate spherical near-field scan before the test antenna can be measured.

In itself, of course, the probe calibration requires another measuring probe, the auxiliary probe, which must be known, and the question then arises: How do we obtain the data for the auxiliary probe? One solution to this problem is to assume the auxiliary probe to be a Hertzian dipole or a Huygens source for which the receiving coefficients are known; see Section 2.3.4. This will be sufficient if the actual auxiliary probe is chosen, e.g. as an open waveguide or a small horn, and if the measurement distance is so large that the variation over the probe minimum sphere of the auxiliary probe field is correct (i.e. as the far field of a Hertzian dipole).

It is also possible to calibrate the auxiliary probe as well as the probe itself in one single operation. Two near-field scans carried out in the spherical near-field set-up is needed for this purpose. In the first spherical scan, the probe plays the role of an unknown test antenna and is measured with the given auxiliary probe.

Let the set of recorded data be denoted by A. In the second scan, the two probes are interchanged and the direction of transmission in the set-up reversed. Let the set of data obtained in this way be denoted by B.

Further, let the known set of theoretical coefficients for a simple source (a Hertzian dipole or Huygens source with the same polarization on axis as the auxiliary probe) be denoted by c_0. An iterative process can now be established in a succession of runs with the computer program; see Table 3.1.

Table 3.1 *Iteration scheme for determining coefficients for probe and auxiliary probe.*

Computer program run No.	Input to program: Measured data	Input to program: Set of probe coefficients	Output from program: Set of test antenna coefficients generated
1	A	c_0	c_1
2	B	c_1	c_2
3	A	c_2	c_3
4	B	c_3	c_4
etc.	etc.	etc.	etc.

In the first run the input to the program is the measured data set A and the set of coefficients c_0. The output from the program is another set of coefficients called c_1. In the second run, the measured data set B together with the set of coefficients c_1 is taken as input to the program. The output is a new set of coefficients called c_2. Provided the assumed simple source used in the first approximation has the correct on-axis polarization, the process will converge rapidly, e.g. in two steps. As can be seen, both the coefficients of the auxiliary probe and the coefficients of the probe itself will be determined by the process.

In the above discussion we have deliberately avoided specifying which coefficients—transmitting or receiving, original or adjoint—are used in the single steps. This depends on whether the actual probe is of a non-reciprocal receiving type, a non-reciprocal transmitting type, or perhaps is reciprocal. The choice of probe is dictated in the last end by the type of test antenna to be measured.

If the probe is chosen as a non-reciprocal receiving type, it must necessarily be calibrated in an experiment where it receives. The coefficients provided by the calibration will be the transmitting coefficients $T^{p\prime}_{\sigma\mu\nu}$ of the adjoint probe. Application of eqn (3.18) subsequently yields the required probe receiving coefficients $R^{p}_{\sigma\mu\nu}$.

If the probe is chosen as a non-reciprocal transmitting type, it must necessarily be calibrated in an experiment where it transmits. The coefficients provided by the calibration will be the transmitting coefficients $T^{p}_{\sigma\mu\nu}$. The required receiving coefficients of the adjoint probe $R^{p'}_{\sigma\mu\nu}$ is then easily obtained using eqn (3.18).

If the probe is reciprocal, the result of the probe calibration is the transmitting coefficients $T^{p}_{\sigma\mu\nu}$. The required receiving coefficients then follow from eqn (3.21).

The above discussion may leave the impression that probe calibration is a complicated, time-consuming task. In most cases actual practice shows the contrary; see Chapter 5.

3.3 Special cases of the transmission formula

3.3.1 Linearly polarized probe with $\mu = \pm 1$

For reference in later chapters we shall now consider an important special case of the transmission formula. In this version, the transmission formula concerns an arbitrary test antenna. The probe, however, is linearly polarized and has a pattern where only the azimuthal modes $\mu = \pm 1$ occur. Such a probe is commonly used in spherical near-field measurements. In practice it may consist of a conical horn excited from a circular waveguide with only the TE_{11} mode propagating.

Figure 3.4 illustrates the geometry. The probe is aligned with its axis coincident with the z'-axis of the primed probe coordinate system. The probe

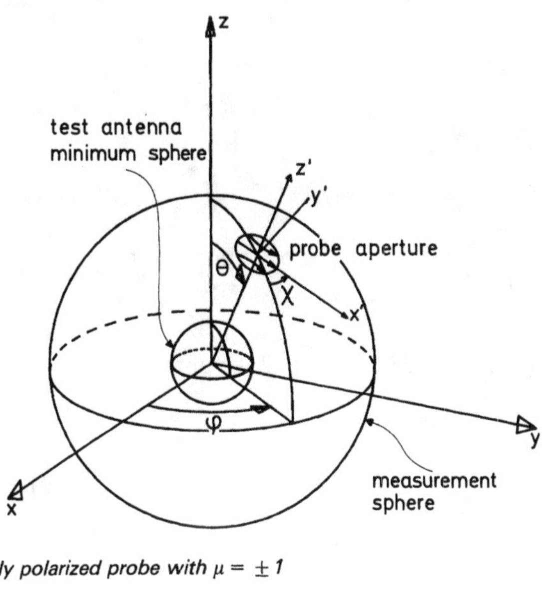

Fig. 3.4 *Linearly polarized probe with $\mu = \pm 1$*

aperture distribution is such that its radiation in the principal planes is linearly polarized. The direction of polarization is parallel to the $x'z'$-plane. For this class of probes we have $R^p_{1,-1,\nu} = R^p_{11\nu}$ and $R^p_{2,-1,\nu} = -R^p_{21\nu}$; see eqns (5.8–5.9). All other probe receiving coefficients $R^p_{\sigma\mu\nu}$ are equal to zero. The test antenna is placed in the unprimed test antenna coordinate system such that its minimum sphere is centred at the origin.

The test antenna has the transmitting coefficients T_{smn} and the applied signal v watts$^{1/2}$. The received probe signal is $w(A, \chi, \theta, \phi)$ where (A, χ, θ, ϕ) are coordinates defining the probe location relative to the test antenna as previously. From eqn (3.23) we have the transmission formula

$$w(A, \chi, \theta, \phi) = \frac{v}{2} \sum_{\substack{smn \\ \sigma\mu\nu}} T_{smn} \, e^{im\phi} d^n_{\mu m}(\theta) \, e^{i\mu\chi} C^{sn(3)}_{\sigma\mu\nu}(kA) R^p_{\sigma\mu\nu}. \tag{3.25}$$

We shall now introduce a new concept: the *probe response constants*, defined by

$$P_{s\mu n}(kA) = \frac{1}{2} \sum_{\sigma\nu} C^{sn(3)}_{\sigma\mu\nu}(kA) R^p_{\sigma\mu\nu}. \tag{3.26}$$

Using eqn (A3.12) for the translation coefficients with negative μ, we then have from (3.26)

$$P_{s,-1,n}(kA) = (-1)^{s+1} P_{s1n}(kA). \tag{3.27}$$

The probe response coefficients vanish for $\mu \neq \pm 1$. Insertion into eqn (3.25) yields

$$w = v \sum_{smn} T_{smn} \, e^{im\phi} \{ d^n_{1m}(\theta) \, e^{i\chi} + (-1)^{s+1} d^n_{-1m}(\theta) \, e^{-i\chi} \} P_{s1n}(kA). \tag{3.28}$$

The curly bracket in eqn (3.28) is now considered separately. For $s = 1$ we obtain

$$\{ d^n_{1m}(\theta) \, e^{i\chi} + d^n_{-1m}(\theta) \, e^{-i\chi} \}$$
$$= \{ d^n_{1m}(\theta) + d^n_{-1m}(\theta) \} \cos\chi + \{ d^n_{1m}(\theta) - d^n_{-1m}(\theta) \} i \sin\chi \tag{3.29}$$
$$= -\frac{2i^n}{\sqrt{2n+1}} e^{-im\phi} \{ \vec{K}_{1mn}(\theta, \phi) \cdot \hat{\theta} \cos\chi + \vec{K}_{1mn}(\theta, \phi) \cdot \hat{\phi} \sin\chi \} \tag{3.30}$$

by virtue of eqns (A2.20–2.23). For $s = 2$ we obtain in a similar way

$$\{ d^n_{1m}(\theta) \, e^{i\chi} - d^n_{-1m}(\theta) \, e^{-i\chi} \}$$
$$= -\frac{2i^n}{\sqrt{2n+1}} e^{-im\phi} \{ \vec{K}_{2mn}(\theta, \phi) \cdot \hat{\theta} \cos\chi + \vec{K}_{2mn}(\theta, \phi) \cdot \hat{\phi} \sin\chi \} \tag{3.31}$$

Equations (3.30–3.31) are now inserted into eqn (3.28). The transmission formula for the type of probe considered here becomes

$$w(A, \chi, \theta, \phi) = v \sum_{smn} T_{smn} P_{s1n}(kA) \frac{-2i^n}{\sqrt{2n+1}} \vec{K}_{smn}(\theta, \phi) \cdot \hat{x}' \tag{3.32}$$

where the χ-dependence on the right-hand side is contained in $\hat{x}'$. For $\chi = 0$ and $\chi = \pi/2$ we have $\hat{x}' = \hat{\theta}$ and $\hat{x}' = \hat{\phi}$, respectively. Let us introduce the notations $w_\theta = w(A, 0, \theta, \phi)$ and $w_\phi = w(A, \pi/2, \theta, \phi)$. Equation (3.32) can then be written in the vector form

$$w_\theta \hat{\theta} + w_\phi \hat{\phi} = v \sum_{smn} T_{smn} \left\{ P_{s1n}(kA) \frac{-2i^n}{\sqrt{2n+1}} \right\} \vec{K}_{smn}(\theta, \phi). \tag{3.33}$$

From this equation, it is seen that if a probe with response constants satisfying eqn (3.27) and

$$P_{s1n}(kA) = -\tfrac{1}{2} i^{-n} \sqrt{2n+1} \tag{3.34}$$

can be found, the far field from the test antenna is measured directly. It is observed that the required probe constants are of the same form as plane wave coefficients (ref. Appendix A1.6). This is related to the fact that the far field of a test antenna can be measured as its response to a plane wave. A detailed proof, however, is outside the scope here.

We shall conclude this section by deriving a useful alternative form of eqns (3.32–3.33).

The far-field pattern functions $\vec{K}_{smn}(\theta, \phi)$ have the same angular dependence as the component tangential to the measurement sphere of the spherical wave function

$$[\vec{F}^{(3)}_{smn}(A, \theta, \phi)]_{\text{tang}} = \frac{-i^{n-s}}{\sqrt{4\pi}} R^{(3)}_{sn}(kA) \vec{K}_{smn}(\theta, \phi) \tag{3.35}$$

which may be seen by comparing the definitions (A1.45–1.46) and (A1.59–1.60). The probe response constants $P_{s1n}(kA)$ is shown in the next section to take a particular simple form for a tangential electric x'-directed Hertzian dipole probe

$$P^e_{s1n}(kA) = \frac{\sqrt{6}}{8} i^{-s} \sqrt{2n+1}\, R^{(3)}_{sn}(kA). \tag{3.36}$$

Introducing now the electric Hertzian dipole response constant $P^e_{s1n}(kA)$ and the tangential component of the spherical wave function, ref. eqn (3.35), into eqn (3.32), yields the following version of the transmission formula

$$w(A, \chi, \theta, \phi) = \frac{\sqrt{6\pi}}{2} v \sum_{smn} \frac{P_{s1n}(kA)}{P^e_{s1n}(kA)} T_{smn} [\vec{F}^{(3)}_{smn}(A, \theta, \phi)] \cdot \hat{x}' \tag{3.37}$$

or, in vector form

$$w_\theta \hat{\theta} + w_\phi \hat{\phi} = \frac{\sqrt{6\pi}}{2} v \sum_{smn} \frac{P_{s1n}(kA)}{P^e_{s1n}(kA)} T_{smn} [\vec{F}^{(3)}_{smn}(A, \theta, \phi)]_{\text{tang}} \tag{3.38}$$

valid for a $\mu = \pm 1$ probe, linearly x'-polarized in its principal planes. The probe signal is seen to have the form of the tangential component of a spherical wave expansion. Note also that the role of the probe is only to provide weight factors $P_{s1n}(kA)/P^e_{s1n}(kA)$ to the spherical wave functions in eqns (3.37–3.38).

3.3.2 The Hertzian dipole as a probe

The x'-directed tangential electric Hertzian dipole is a special case of the linearly polarized probe with $\mu = \pm 1$ considered in the last section. In this section, we shall state some results pertaining to x'-directed electric and magnetic dipole probes; see Fig. 3.5.

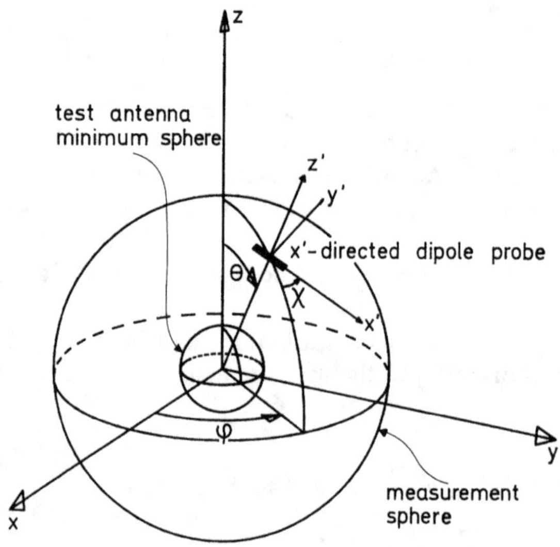

Fig. 3.5 *Hertzian dipole as a receiving probe*

Let us first consider the probe response constants. For the x'-directed electric dipole, the only receiving coefficients different from zero are $R^p_{211} = -R^p_{2,-1,1} = -\sqrt{2}/2$, ref. eqn (2.154). From eqn (3.26) we therefore have

$$\begin{aligned} P^e_{s1n}(kA) &= -\frac{\sqrt{2}}{4} C^{sn(3)}_{211}(kA) \\ &= -\frac{\sqrt{2}}{4}\frac{\sqrt{3}}{2}\sqrt{2n+1}\,\{\delta_{s2} R^{(3)}_{2n}(kA) + \delta_{3-s,2}\, i R^{(3)}_{1n}(kA)\} \\ &= \frac{\sqrt{6}}{8} i^{-s} \sqrt{2n+1}\, R^{(3)}_{sn}(kA) \end{aligned} \tag{3.39}$$

and

$$P^{e}_{s,-1,n}(kA) = -\frac{\sqrt{6}}{8} i^{s} \sqrt{2n+1}\, R^{(3)}_{sn}(kA) \tag{3.40}$$

where we have used the formulas (A3.20–3.21) for the translation coefficients. Similar results may be obtained for the x'-directed magnetic dipole. Here, the only receiving coefficients different from zero are $R^{p}_{111} = -R^{p}_{1,-1,1} = -i(\sqrt{2}/2)$, ref. eqn (2.157). For this probe it may be shown that the response constants are

$$P^{m}_{s1n}(kA) = \frac{\sqrt{6}}{8} i^{s} \sqrt{2n+1}\, R^{(3)}_{3-s,n}(kA) \tag{3.41}$$

and

$$P^{m}_{s,-1,n}(kA) = \frac{\sqrt{6}}{8} i^{-s} \sqrt{2n+1}\, R^{(3)}_{3-s,n}(kA). \tag{3.42}$$

Note that the x'-directed magnetic dipole is y'-polarized.

Next, let us examine the signal received by dipole probes. For the x'-directed electric dipole, eqn (3.37) takes the form

$$w^{e}(A,\chi,\theta,\phi) = \frac{\sqrt{6\pi\eta}}{2k}\left\{\frac{k}{\sqrt{\eta}} v \sum_{smn} T_{smn} \vec{F}^{(3)}_{smn}(A,\theta,\phi)\right\}\cdot \hat{x}'. \tag{3.43}$$

For the x'-directed magnetic dipole, the corresponding relation is

$$w^{m}(A,\chi,\theta,\phi) = \frac{\sqrt{6\pi}}{2k\sqrt{\eta}}\left\{-ik\sqrt{\eta} v \sum_{smn} T_{smn} \vec{F}^{(3)}_{3-s,m,n}(A,\theta,\phi)\right\}\cdot \hat{x}'. \tag{3.44}$$

The quantities in the curly brackets in the above equations are the electric and magnetic field vectors $\vec{E}(A,\theta,\phi)$ and $\vec{H}(A,\theta,\phi)$, respectively, of the test antenna field. In other words,

$$w^{e}(A,0,\theta,\phi) = \frac{\sqrt{6\pi\eta}}{2k} E_{\theta}(A,\theta,\phi) \tag{3.45}$$

$$w^{e}\left(A,\frac{\pi}{2},\theta,\phi\right) = \frac{\sqrt{6\pi\eta}}{2k} E_{\phi}(A,\theta,\phi) \tag{3.46}$$

$$w^{m}(A,0,\theta,\phi) = \frac{\sqrt{6\pi}}{2k\sqrt{\eta}} H_{\theta}(A,\theta,\phi) \tag{3.47}$$

$$w^{m}\left(A,\frac{\pi}{2},\theta,\phi\right) = \frac{\sqrt{6\pi}}{2k\sqrt{\eta}} H_{\phi}(A,\theta,\phi). \tag{3.48}$$

Hence, the signal received by an electric (magnetic) dipole is proportional to the incident electric (magnetic) field parallel to the dipole. The constant of proportionality for the electric dipole has been derived earlier (eqn (2.152)) and is well known [14].

In the far field, the signal in an electric dipole probe may be found from eqn (3.43) by introducing the far-field pattern $\vec{K}(\theta, \phi)$, ref. eqn (2.182).

$$w^e(A, \chi, \theta, \phi) \to \frac{\sqrt{6}}{4}\frac{e^{ikA}}{kA} v\vec{K}(\theta, \phi) \cdot \hat{x}, \qquad kA \to \infty. \tag{3.49}$$

It is convenient to define the *normalized far-field probe signal* W^e as

$$W^e(\chi, \theta, \phi) = \lim_{kA \to \infty} \left[w^e(A, \chi, \theta, \phi) \frac{kA}{e^{ikA}} \right] \tag{3.50}$$

whereby eqn (3.49) may be replaced by

$$W^e(\chi, \theta, \phi) = \frac{\sqrt{6}}{4} v\vec{K}(\theta, \phi) \cdot \hat{x}'. \tag{3.51}$$

By letting χ take the values 0 and $\pi/2$, respectively, eqn (3.51) is equivalent to

$$W^e(0, \theta, \phi) = \frac{\sqrt{6}}{4} vK_\theta(\theta, \phi) \tag{3.52}$$

and

$$W^e\left(\frac{\pi}{2}, \theta, \phi\right) = \frac{\sqrt{6}}{4} vK_\phi(\theta, \phi). \tag{3.53}$$

The test antenna far-field pattern can thus be expressed in terms of the normalized far-field probe signal

$$\vec{K}(\theta, \phi) = \frac{2\sqrt{6}}{3v}\left\{ W^e(0, \theta, \phi)\hat{\theta} + W^e\left(\frac{\pi}{2}, \theta, \phi\right)\hat{\phi} \right\}. \tag{3.54}$$

The gain $G(\theta, \phi)$ of the matched test antenna becomes (cf eqn (2.211))

$$\begin{aligned} G(\theta, \phi) &= |\vec{K}(\theta, \phi)|^2 \\ &= \frac{8}{3|v|^2}\left\{ \left| W^e(0, \theta, \phi) \right|^2 + \left| W^e\left(\frac{\pi}{2}, \theta, \phi\right) \right|^2 \right\}. \end{aligned} \tag{3.55}$$

Finally, as an application of some of the formulas derived above, let us investigate the transfer of power between an arbitrary linearly polarized test antenna and a receiving electric Hertzian dipole in the limit of large separation (far-field conditions) between the two antennas.

The power P_p received by the probe when aligned parallel to the electric field of the test antenna is given by

$$\begin{aligned} P_p &= \tfrac{1}{2}|w^e|^2 \\ &\to \frac{|v|^2}{2}\frac{3}{8(kA)^2}|\vec{K}(\theta, \phi)|^2 \qquad \text{for } kA \to \infty \end{aligned} \tag{3.56}$$

in virtue of eqn (3.49). The input power P_t accepted by the matched test antenna is given by $P_t = \frac{1}{2}|v|^2$.

Hence,

$$\frac{P_p}{P_t} = \left(\frac{\lambda}{4\pi A}\right)^2 G_p G_t \tag{3.57}$$

where $G_p = \frac{3}{2}$ is the gain of the probe and where G_t is the gain of the test antenna. Equation (3.57) is *Friis' transmission formula* for the case of an arbitrary linearly polarized transmitting test antenna and a receiving electric dipole aligned to be parallel to the incident electric field. In the next section we shall derive Friis' transmission formula for a more general case.

3.3.3 Friis' transmission formula [10]

In the present section, we shall derive the Friis' transmission formula in a more general form than considered above; see eqn (3.57). Here, the only assumptions made about the probe are that it is matched and linearly polarized in the direction towards the test antenna. Since the test antenna coefficients can always be expressed in rotated coordinates, there is no loss of generality in assuming the probe to be located on the z-axis of the test antenna coordinate system. Hence, we may let $(\theta, \phi) = (0, 0)$; see Fig. 3.6.

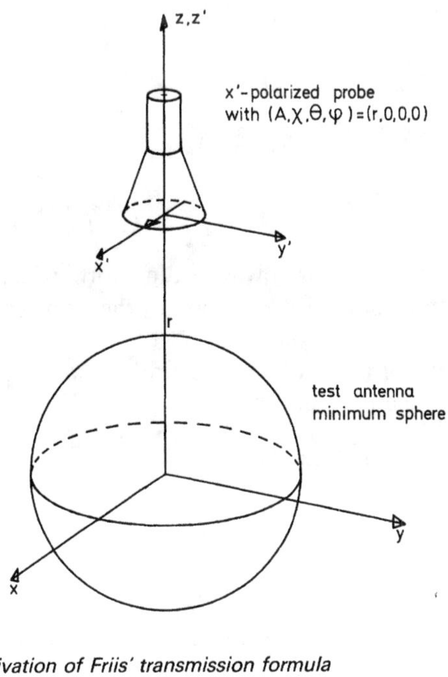

Fig. 3.6 *For the derivation of Friis' transmission formula*

For convenience, the transmission formula is restated here

$$w(A, \chi, \theta, \phi) = \frac{v}{2} \sum_{\substack{smn \\ \sigma\mu\nu}} T_{smn} \, e^{im\phi} d^n_{\mu m}(\theta) \, e^{i\mu\chi} C^{sn(3)}_{\sigma\mu\nu}(kA) R^p_{\sigma\mu\nu} \tag{3.58}$$

where the symbols have been defined previously.

Let us examine the right-hand side of (3.58) as kA tends to infinity. The asymptotic behaviour of the translation coefficients $C^{sn(3)}_{\sigma\mu\nu}(kA)$ is given by eqns (A3.22–3.24) and dictates that only terms with $\mu = 1$ and $\mu = -1$ in (3.58) are of importance for large separations. Assuming the probe to be polarized in the direction of the x'-axis, we have $\chi = 0$, as well as $R^p_{11\nu} = R^p_{1,-1,\nu}$ and $R^p_{21\nu} = -R^p_{2,-1,\nu}$. Equation (3.58) yields

$$\begin{aligned} \underset{(kA\to\infty)}{w} \to \frac{v}{2} \sum_{\substack{smn \\ \nu}} T_{smn} \{ & d^n_{1m}(0) C^{sn(3)}_{11\nu}(kA) R^p_{11\nu} + d^n_{1m}(0) C^{sn(3)}_{21\nu}(kA) R^p_{21\nu} \\ & + d^n_{-1,m}(0) C^{sn(3)}_{1,-1,\nu}(kA) R^p_{11\nu} - d^n_{-1,m}(0) C^{sn(3)}_{2,-1,\nu}(kA) R^p_{21\nu} \}. \end{aligned} \tag{3.59}$$

The rotation coefficients take the simple form $d^n_{\mu m}(0) = \delta_{\mu m}$ for all values of n, when the argument theta is equal to zero (no rotation). We thus obtain

$$\begin{aligned} \underset{(kA\to\infty)}{w} \to \frac{v}{2} \sum_{s\nu n} \{ & T_{s1n} C^{sn(3)}_{11\nu}(kA) R^p_{11\nu} + T_{s1n} C^{sn(3)}_{21\nu}(kA) R^p_{21\nu} \\ & + T_{s,-1,n} C^{sn(3)}_{1,-1,\nu}(kA) R^p_{11\nu} - T_{s,-1,n} C^{sn(3)}_{2,-1,\nu}(kA) R^p_{21\nu} \}. \end{aligned} \tag{3.60}$$

Finally, by introducing the asymptotic form of the translation coefficients

$$\begin{aligned} \underset{(kA\to\infty)}{w} \to \frac{v}{4} \frac{e^{ikA}}{kA} \sum_{\nu n} & i^{\nu-n-1} \sqrt{(2\nu+1)(2n+1)} \\ & (R^p_{11\nu} + R^p_{21\nu})(T_{11n} + T_{21n} + T_{1,-1,n} - T_{2,-1,n}). \end{aligned} \tag{3.61}$$

The transmission formula (3.61) is now recast in a form which is suitable for the derivation of the Friis' formula. Introduction of the gains of the probe and the test antenna remains.

The far-field pattern of the probe in the probe coordinate system is in the direction of the test antenna given by

$$\begin{aligned} \vec{K}^p(\pi, \phi) &= \sum_{\sigma\mu\nu} T^p_{\sigma\mu\nu} \vec{K}_{\sigma\mu\nu}(\pi, \phi) \\ &= \sum_{\nu} i^\nu \sqrt{2\nu+1} (R^p_{11\nu} + R^p_{21\nu}) \hat{x} \end{aligned} \tag{3.62}$$

by using the relations (A1.63–1.64) for the far-field pattern functions $\vec{K}_{\sigma\mu\nu}$. Similarly, for the far-field pattern of the test antenna in the direction of the probe, we find that

$$\vec{K}(0, \phi) = \sum_{smn} T_{smn} \vec{K}_{smn}(0, \phi)$$

$$= -\sum_{n} i^{-n} \tfrac{1}{2}\sqrt{2n+1}\,(T_{11n} + T_{21n} + T_{1,-1,n} - T_{2,-1,n})\hat{x}$$

$$-\sum_{n} i^{-n+1} \tfrac{1}{2}\sqrt{2n+1}\,(T_{11n} + T_{21n} - T_{1,-1,n} + T_{2,-1,n})\hat{y}. \tag{3.63}$$

It is now clear that eqn (3.61) may be written as†

$$w \to \frac{v}{2}\frac{e^{ikA}}{kA}\{i\vec{K}^p(\pi, \phi)\cdot\hat{x}\}\{\vec{K}(0, \phi)\cdot\hat{x}\} \qquad \text{for } kA \to \infty. \tag{3.64}$$

Taking the absolute squared on both sides of eqn (3.64) and introducing the gain of the probe

$$G_p = |\vec{K}^p(\pi, \phi)\cdot\hat{x}|^2 \tag{3.65}$$

as well as the gain of the x-polarized part of the test antenna field

$$G_t = |\vec{K}(0, \phi)\cdot\hat{x}|^2 \tag{3.66}$$

ref. eqn (2.211) with $\Gamma = 0$, yields

$$\frac{\frac{1}{2}|w|^2}{\frac{1}{2}|v|^2} = \frac{G_p G_t}{4(kA)^2}. \tag{3.67}$$

By introducing the normalized far-field signal

$$W = \lim_{(kA\to\infty)}\left[w\frac{kA}{e^{ikA}}\right] \tag{3.68}$$

we obtain

$$\frac{\frac{1}{2}|W|^2}{\frac{1}{2}|v|^2} = \frac{G_p G_t}{4}. \tag{3.69}$$

Equations (3.67, 3.69) are equivalent forms of Friis' transmission formula for the far-field transfer of power between two antennas. The Friis' transmission formula expresses the power received by the matched probe relative to the input power accepted by the matched test antenna in terms of the gain product for the two antennas. The probe is assumed to be linearly polarized. The test antenna is in general elliptically polarized with G_t being the gain of the field component (linearly polarized) that is parallel to the probe polarization.

† It should be noted that the 90 degrees phase shift implied by the factor of i in (3.64) is also present in the generalized reciprocity theorem by Brown [15].

If the *accepted power* $\frac{1}{2}|v|^2$ is replaced by the *radiated power*, we get

$$\frac{\frac{1}{2}|w|^2}{\frac{1}{2}\sum_{smn}|vT_{smn}|^2} = \frac{G_p D_t}{4(kA)^2} \tag{3.70}$$

or, equivalently,

$$\frac{\frac{1}{2}|W|^2}{\frac{1}{2}\sum_{smn}|vT_{smn}|^2} = \frac{G_p D_t}{4} \tag{3.71}$$

where D_t is the directivity of that test antenna field component which is parallel to the probe.

Polarization mismatch between the test antenna and the probe may easily be included in the above formulas, if needed.

3.4 The transmission formula with multiple reflections included

We shall show in this section that multiple reflections between the probe and the test antenna may be taken into account in the transmission formula. In the derivations in Section 3.2 as well as in the applications in Section 3.3, we have assumed that the field radiated by a transmitting test antenna was not changed by the insertion of the probe into the field. Of course, the currents induced in the probe create a scattered field which in turn must introduce currents on the test antenna. These extra test antenna currents will radiate a reflected field which in most cases can be considered small as compared to the direct radiated field from the test antenna. This first-order reflected field as well as higher-order multiple reflected fields, which are even smaller, may be accounted for in the transmission formula, however at the expense of simplicity. The transmission formula with multiple reflections included turns out to become nonlinear in the unknown test antenna coefficients. Furthermore, as the scattering coefficients of both test antenna and probe will become involved, full correction for multiple reflections in near-field measurements is extremely complicated. On the other hand, the transmission formula with multiple reflections included may provide some insight into the interaction mechanisms which take place in spherical near-field measurements.

With reference to Fig. 3.7 where the signals at the probe and test antenna reference planes are defined, we have the individual scattering matrix equations for the two antennas

$$\begin{bmatrix} \Gamma_p & \mathbf{R}^p \\ \mathbf{T}^p & \mathbf{S}^p \end{bmatrix} \begin{bmatrix} v_p \\ \mathbf{a}^p \end{bmatrix} = \begin{bmatrix} w \\ \mathbf{b}^p \end{bmatrix} \tag{3.72}$$

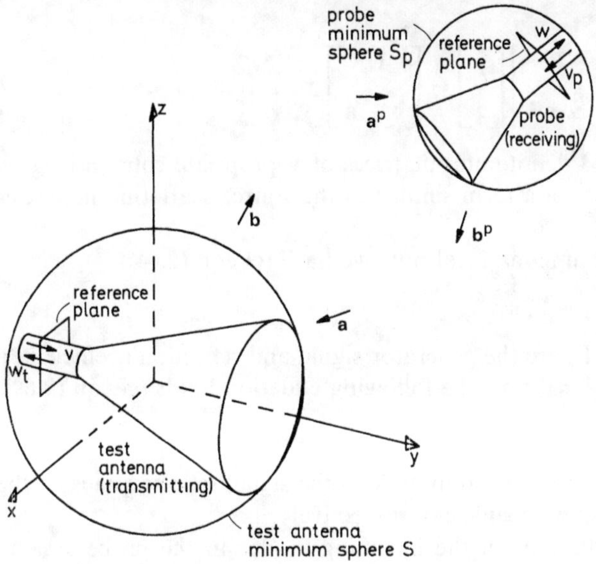

Fig. 3.7 *Test antenna and probe with non-intersecting minimum spheres and local ports defined by reference planes in waveguides. Also shown are the incoming and outgoing signals at the reference planes*

and

$$\begin{bmatrix} \Gamma & \mathbf{R} \\ \mathbf{T} & \mathbf{S} \end{bmatrix} \begin{bmatrix} v \\ \mathbf{a} \end{bmatrix} = \begin{bmatrix} w_t \\ \mathbf{b} \end{bmatrix} \tag{3.73}$$

where the indices p and t denote the probe and test antenna, respectively, and where the other symbols have their usual meaning in the single index convention; see Section 2.2.2. The column lengths in eqns (3.72) and (3.73) are $J_p + 1$ and $J + 1$, respectively, where $J_p = 2N_p(N_p + 2)$ and $J = 2N(N + 2)$. The probe and the test antenna minimum spheres have different radii so the maximum ν and n in their spherical wave expansions, viz. $\nu = N_p$ and $n = N$, respectively, are in general different.

Similarly to what was done in connection with the discussion following eqn (2.76) we subtract the column matrices

$$\begin{bmatrix} 0 \\ \mathbf{a}^p \end{bmatrix} \quad \text{and} \quad \begin{bmatrix} 0 \\ \mathbf{a} \end{bmatrix} \tag{3.74}$$

from both sides of eqns (3.72) and (3.73), respectively. This gives

$$\begin{bmatrix} \Gamma_p & \mathbf{R}^p \\ \mathbf{T}^p & \mathbf{S}^p - \mathbf{I}^p \end{bmatrix} \begin{bmatrix} v_p \\ \mathbf{a}^p \end{bmatrix} = \begin{bmatrix} w \\ \mathbf{b}^p - \mathbf{a}^p \end{bmatrix} \tag{3.75}$$

and

$$\begin{bmatrix} \Gamma & \mathbf{R} \\ \mathbf{T} & \mathbf{S}-\mathbf{I} \end{bmatrix} \begin{bmatrix} v \\ \mathbf{a} \end{bmatrix} = \begin{bmatrix} w_t \\ \mathbf{b}-\mathbf{a} \end{bmatrix} \tag{3.76}$$

where $\mathbf{I}^p$ and $\mathbf{I}$ denote unit matrices of appropriate column lengths. Equations (3.75–3.76) are of a form similar to the source scattering matrix equation; see Section 2.3.5.

At the test antenna local port we have (cf eqn (2.64))

$$v = v_g + \Gamma_g w_t \tag{3.77}$$

where v_g and Γ_g are the generator signal and reflection coefficient, respectively. At the probe local port the following equation holds (cf eqn (2.69))

$$v_p = \Gamma_l w \tag{3.78}$$

where v_p and Γ_l are the amplitude of the signal reflected back to the probe and the load reflection coefficient, respectively.

The amplitude $\mathbf{a}^p$ of the incoming modes in the probe system are purely associated with currents on the test antenna. The field from these currents is the total field with outgoing mode coefficients $\mathbf{b}$ minus the incident field with outgoing mode coefficients $\mathbf{a}$. We may therefore write

$$a_\beta^p = \sum_i G_{\beta i}^+ (b_i - a_i) \tag{3.79}$$

where

$$G_{\beta i}^+ = \tfrac{1}{2} e^{im\phi} d_{\mu m}^n(\theta) \, e^{i\mu\chi} C_{\sigma\mu\nu}^{sn(3)}(kA) \tag{3.80}$$

cf eqn (3.6). The single index i on the left-hand side is representing the index triplet (s, m, n) referring to the test antenna and should not be confused with the imaginary unit i on the right-hand side. The single index β represents (σ, μ, ν) referring to the probe. Equation (3.79) may be written in matrix form

$$\mathbf{a}^p = \mathbf{G}^+ (\mathbf{b} - \mathbf{a}). \tag{3.81}$$

Similarly, the incoming test antenna modes may be related to the probe modes through

$$a_j = \sum_\alpha G_{j\alpha}^- (b_\alpha^p - a_\alpha^p) \tag{3.82}$$

or, in matrix form

$$\mathbf{a} = \mathbf{G}^- (\mathbf{b}^p - \mathbf{a}^p) \tag{3.83}$$

where

$$G_{j\alpha}^- = \tfrac{1}{2} C_{s\mu n}^{\sigma\nu(3)}(-kA) \, e^{-im\phi} d_{m\mu}^n(-\theta) \, e^{-i\mu\chi} \tag{3.84}$$

are the matrix elements of the matrix $\mathbf{G}^-$, ref. eqn (3.12).

Having established above the equations pertaining to the waves of Fig. 3.7, we shall now introduce the corresponding scattering flow graphs (see also Section 2.3.1).

The complete flow graph covering all waves is of course very large. However, the diagram in Fig. 3.8(a) is of sufficient generality to fully illustrate all mechanisms involved in the mutual coupling between the test antenna and the probe. Wave nodes with indices or double indices are representatives for many waves. Similarly, the branches with indices represent wave transformations of many waves.

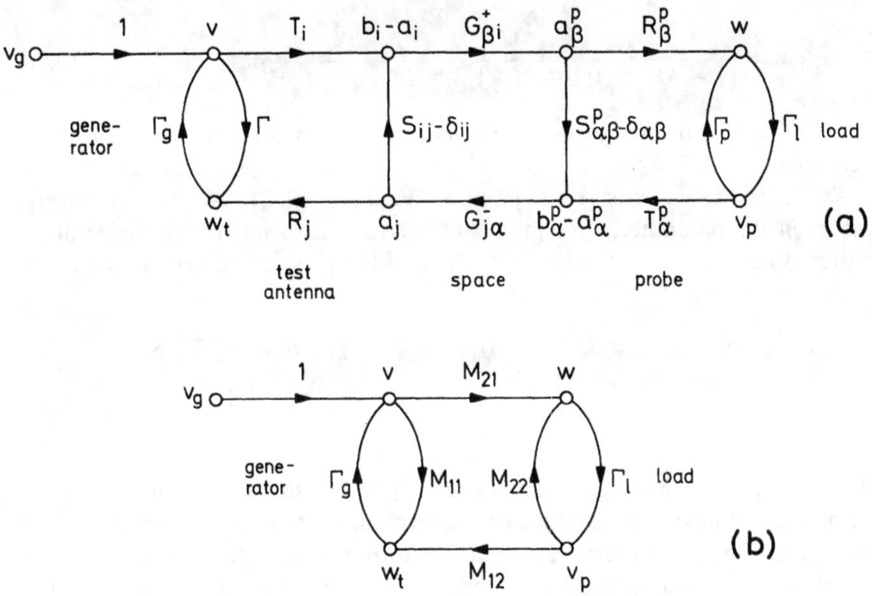

Fig. 3.8 *Scattering flow graphs with generator and load included for the two-antenna system in Fig. 3.7. (a) Full diagram. (b) Reduced diagram.*

An overall scattering matrix equation relating the waves at the two antenna reference planes can be found by eliminating the four column matrices $\mathbf{a}^p$, $(\mathbf{b}^p - \mathbf{a}^p)$, $\mathbf{a}$ and $(\mathbf{b} - \mathbf{a})$ from eqns (3.75–3.76), (3.81) and (3.83). The calculations are lengthy and shall be omitted here. The result is

$$\begin{bmatrix} M_{11} & M_{12} \\ M_{21} & M_{22} \end{bmatrix} \begin{bmatrix} v \\ v_p \end{bmatrix} = \begin{bmatrix} w_t \\ w \end{bmatrix} \tag{3.85}$$

where

$$M_{11} = \Gamma + \mathbf{R}\{\mathbf{I} - \mathbf{G}^-(\mathbf{S}^p - \mathbf{I}^p)\mathbf{G}^+(\mathbf{S} - \mathbf{I})\}^{-1}\mathbf{G}^-(\mathbf{S}^p - \mathbf{I}^p)\mathbf{G}^+\mathbf{T} \tag{3.86}$$

$$M_{12} = \mathbf{R}\{\mathbf{I} - \mathbf{G}^-(\mathbf{S}^p - \mathbf{I}^p)\mathbf{G}^+(\mathbf{S} - \mathbf{I})\}^{-1}\mathbf{G}^-\mathbf{T}^p \tag{3.87}$$

$$M_{21} = \mathbf{R}^p\{\mathbf{I}^p - \mathbf{G}^+(\mathbf{S} - \mathbf{I})\mathbf{G}^-(\mathbf{S}^p - \mathbf{I}^p)\}^{-1}\mathbf{G}^+\mathbf{T} \tag{3.88}$$

$$M_{22} = \Gamma_p + \mathbf{R}^p\{\mathbf{I}^p - \mathbf{G}^+(\mathbf{S} - \mathbf{I})\mathbf{G}^-(\mathbf{S}^p - \mathbf{I}^p)\}^{-1}\mathbf{G}^+(\mathbf{S} - \mathbf{I})\mathbf{G}^-\mathbf{T}^p \tag{3.89}$$

Figure 3.8(b) shows the scattering flow graph (with generator and load included) corresponding to eqn (3.85).

In addition to eqn (3.85) it is of interest to avail of a relationship between v_g, the generator signal amplitude, and the amplitude w of the wave received by the probe. Such an equation results when v, w_t and v_p are eliminated from eqn (3.85) by introducing the relations (3.77–3.78). We obtain

$$w = \frac{M_{21}}{(1 - M_{11}\Gamma_g)(1 - M_{22}\Gamma_l) - M_{21}M_{12}\Gamma_g\Gamma_l} v_g \tag{3.90}$$

which is a general form of the transmission formula with multiple reflections included.

As discussed by Kerns [12, 16] and by Wacker [13], the two different matrix factors in the parenthesis in eqns (3.86–3.89) are associated with the multiple scattered fields between the test antenna and the probe. The first term **I** in the Neumann series

$$\{\mathbf{I} - \mathbf{G}^-(\mathbf{S}^p - \mathbf{I}^p)\mathbf{G}^+(\mathbf{S} - \mathbf{I})\}^{-1} = \mathbf{I} + \{\mathbf{G}^-(\mathbf{S}^p - \mathbf{I}^p)\mathbf{G}^+(\mathbf{S} - \mathbf{I})\} + \{\mathbf{G}^-(\mathbf{S}^p - \mathbf{I}^p)\mathbf{G}^+(\mathbf{S} - \mathbf{I})\}^2 + \ldots \tag{3.91}$$

represents in eqn (3.86) the field scattered back to the test antenna from the probe without having reached the reference plane in the probe transmission line; see Fig. 3.7. The second term represents the three times scattered field, and so on. Equation (3.87) is associated with the field reflected from the load. From this equation, the first term of the Neumann series describes the transmission through the probe back into the test antenna. The second term represents a field which has undergone two additional scatterings, etc. The second and higher-order terms in eqn (3.91) tend to zero when the separation between the test antenna and the probe tends to infinity. The series converges at all separations larger than some finite distance.

If no probe is present, the scattering matrix $\mathbf{S}^p$ must be substituted by $\mathbf{I}^p$, which is the scattering matrix of the point at the origin of the empty probe coordinate system. In eqn (3.86), this yields $M_{11} = \Gamma$ as expected.

The back-scattering cross-section of a matched horn is often more than 10 dB below the value when the horn is short-circuited [17]. Introduction of matched conditions, e.g. through insertion of a microwave isolator between the probe and the load, will therefore reduce the effect of probe scattering and the consequent change, through multiple reflections, of the probe signal w; see Fig. 3.9. This is of particular relevance since the probe is always pointing towards the test antenna.

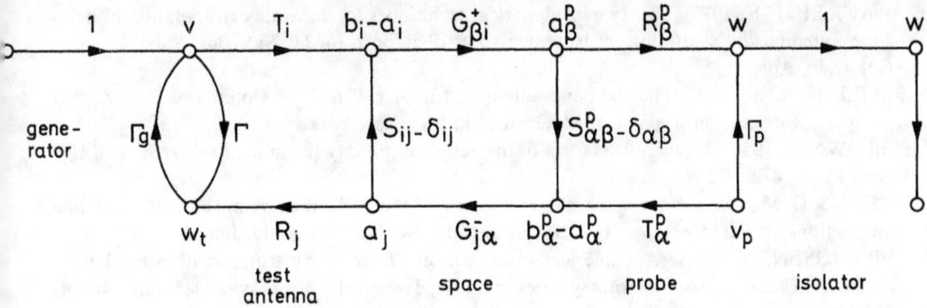

Fig. 3.9 *Scattering flow graph for measurement set-up with a microwave isolator inserted between the probe and its load*

Note, that in Fig. 3.9 no signal arrives at node v_p. This node and the two branches leading from it may therefore be omitted from the flow graph.

References

1. JENSEN, F., 1970, 'Electromagnetic near-field–far-field correlations', Ph.D. Thesis, Laboratory of Electromagnetic Theory, Technical University of Denmark, Report **LD 15**, July.
2. KERNS, D. M., 1970, 'Correction of near-field antenna measurements made with an arbitrary but known measuring antenna', *Electronics Letters*, Vol. **6**, No. 11, pp. 346–347, May.
3. WACKER, P. F., 1975, 'Non-planar near-field measurements: Spherical scanning', Report **NBSIR 75-809**, Electromagnetics Division, Inst. for Basic Standards, Nat. Bur. Stds., Boulder, Colorado, June.
4. LARSEN, FL. HOLM, 1980, 'Probe-corrected spherical near-field antenna measurements', Ph.D. Thesis, Electromagnetics Institute, Technical University of Denmark, Report **LD 36**, Dec.
5. LEWIS, R. L., 1981, 'Spherical-wave source-scattering-matrix analysis of the mutual coupling between two antennas', *IEEE AP-S Symposium Digest*, Vol. **1**, pp. 261–264, Los Angeles, USA, June.
6. YAGHJIAN, A. D., 1977, 'Near-field antenna measurements on a cylindrical surface: A source scattering-matrix formulation', Nat. Bur. Stds., Tech. Note **696**, Boulder, Colorado, July.
7. APPEL-HANSEN, J., GILLESPIE, E. S., HICKMAN, T. G. and DYSON, J. D., 1982, 'Antenna measurements', Chapter 8 in *The Handbook of Antenna Design*, Vol. **1**, Peter Peregrinus, London.
8. JENSEN, F., 1975, 'On the probe compensation for near-field measurements on a sphere', *Archiv für Elektronik und Übertragungstechnik*, Vol. **29**, No. 7/8, pp. 305–308, July–Aug.
9. WACKER, P. F., 1974, 'Near-field antenna measurements using a spherical scan: Efficient data reduction with probe correction', Conference on Precision Electromagnetic Measurements, *IEE Conference Publication No.* **113**, London.
10. FRIIS, H. T., 1946, 'A note on a simple transmission formula', *Proc. IRE*, Vol. **34**, pp. 254–256, May.
11. KINBER, B. YE. and ARSAYEV, I. YE., 1971, 'Mutual coupling of antennas in the Fresnel zone', *Radio Engineering & Electronic Physics*, Vol. **11**, pp. 1844–1851, Nov.
12. KERNS, D. M., 1976, 'Plane-wave scattering-matrix theory of antennas and antenna-antenna interactions: formulation and applications', *J. Res. NBS*, Vol. **80B**, No. 1, pp. 5–51, Jan.–Mar.

13. WACKER, P. F., 1972, 'Theory and numerical techniques for accurate extrapolation of near-zone antenna and scattering measurements', NBS Report **10733**, Nat. Bur. Stds., Boulder, Colorado, Apr.
14. COLLIN, R. E., 1969, 'The receiving antenna', Chapter 4 in R. E. Collin and F. J. Zucker (editors), *Antenna Theory*, Vol. **1**, McGraw-Hill, Inc., New York.
15. BROWN, J., 1958, 'A generalized form of the aerial reciprocity theorem', *Proc. IEE*, Vol. **105**, Sect. C, pp. 472–475.
16. KERNS, D. M., 1981, 'Plane-wave scattering-matrix theory of antennas and antenna–antenna interactions', *NBS Monograph* **162**, Nat. Bur. Stds., Boulder, Colorado, June.
17. FRANDSEN, A., 1985, 'A numerical investigation of scattering from small conical horn antennas', Ph.D. Thesis, Electromagnetics Institute, Technical University of Denmark, Report **LD 58**, Nov.

Chapter 4

Data reduction in spherical near-field measurements

4.1 Introduction

As we have seen in Chapter 2, an antenna is characterized by its total scattering matrix $\hat{\mathbf{S}}$. The elements of this matrix fall into four categories: the antenna reflection coefficient (one element), the antenna scattering coefficients (J^2 elements), the antenna receiving coefficients (J elements) and the antenna transmitting coefficients (J elements). Here, J equals $2N(N + 2)$, where N is given by the electrical radius kr_0 (in radians) of the minimum sphere plus a small number, typically about 10; see Section 2.2.3.

The determination of the reflection coefficient is relatively simple and direct, and it does not present a problem as far as data processing is concerned. The scattering coefficients on the other hand, are very difficult and cumbersome to obtain from near-field measurements. However, in comparison with the receiving and transmitting properties of an antenna, the scattering behaviour is usually of minor importance and will only be touched briefly upon in this book. The main problem in spherical near-field testing is to determine receiving and transmitting coefficients for an antenna from measurements in its near field. As discussed in Chapter 3, it is usually sufficient to deal with one of these sets of coefficients.

We shall choose here to consider a measurement set-up in which the test antenna transmits and the probe receives. The required data processing for obtaining the test antenna transmitting coefficients and subsequently the radiated field is then the subject of this chapter. Two different theoretical approaches are possible: measurement *with* and *without* probe correction.

Measurement with probe correction

In the first approach, which is also the most important, the influence of the extent of a realistic probe, e.g. a conical horn, is taken into account. Such a probe will not measure the field at its location but rather some weighted average of the field over the horn aperture. The measured quantity here is the received signal w. Since w is expressed in the transmission formula derived in Chapter 3, this formula is the natural theoretical starting point. What is required for evaluating the test antenna coefficients from the probe signal is then an inversion or solution of the transmission formula. In Section 3.2.4, the existence of a computer algorithm for this purpose has already been anticipated. This chapter deals with its theory and construction.

The transmission formula, eqn (3.10), is repeated below.

$$w(A, \chi, \theta, \phi) = \frac{v}{2} \sum_{\substack{smn \\ \sigma\mu\nu}} T_{smn} \, e^{im\phi} d^n_{\mu m}(\theta) \, e^{i\mu\chi} C^{sn(3)}_{\sigma\mu\nu}(kA) R^p_{\sigma\mu\nu} \tag{4.1}$$

where $d^n_{\mu m}(\theta)$ and $C^{sn(3)}_{\sigma\mu\nu}(kA)$ are well-defined rotation and translation coefficients, respectively (see Appendices A2 and A3), and where $R^p_{\sigma\mu\nu}$ are the probe receiving coefficients. These are known from the probe calibration procedure; see section 3.2.5. The probe signal w is assumed available for all values of (θ, ϕ) on the measurement sphere (radius A) and at all orientation angles χ of the measurement probe, or *input probe*; see Fig. 3.1.

Obviously, eqn (4.1) cannot be solved for T_{smn} on a computer using the continuous data assumed on the left-hand side. At some step in the solution process a discretization by selection of specific measurement points must be made. The number of unknown T_{smn} coefficients is $2N(N + 2)$ (see eqn (2.29)) so at least this number of discrete data samples must be recorded in the measurements.

A straightforward but inefficient method of solving the transmission formula is to form a system of linear equations which can be solved numerically, by inserting $2N(N + 2)$ discrete values of (A, χ, θ, ϕ) in eqn (4.1). The method leads to ill-conditioned systems of equations and is not practical except possibly for small N, e.g. $N = 2$ or $N = 3$. It is interesting to note, however, that the method does not depend upon equidistant sampling, or for that matter, upon the spherical shape of the measurement surface.

An efficient and accurate algorithm adapted to practical measurements and based upon equidistant sampling on a spherical surface is presented in Sections 4.3 and 4.4. The experimental results in Chapter 5 are derived from application of this algorithm to measured data.

After a brief introduction in Section 4.3.1, the transmission formula is *solved analytically* for the transmitting coefficients in Section 4.3.2 by exploitation of the orthogonality exhibited by the functions involved on the right-hand side. As a result, the coefficients are expressed explicitly in terms of three integrals, one for each of the angular variables χ, ϕ and θ. In Section 4.3.3 these integrals are calculated numerically. It is conveniently left to the numerical integration schemes to determine in which points the integrand, and thereby the measured data, is actually needed. In this way a *discrete solution* of the transmission formula is obtained and the wanted transmitting coefficients can be found by applying the specified discrete operations to discrete data. The chi and phi integrals (Sections 4.3.3.2 and 4.3.3.3) turn out to be Fourier integrals while the evaluation of the theta integral is more complex. For spatially band-limited signals, as assumed here, the numerical evaluation of all three integrals are *exact* provided certain *sampling criteria* are fulfilled.

An efficient reduction of the theta integral based on the *Wacker algorithm* [1, 2] with improvements by Larsen [3, 4] is described in detail in Sections 4.3.3.4 and 4.3.3.5.

When the unknown coefficients T_{smn} have been found as outlined above, the test antenna far field or the field at other distances may be readily evaluated by using eqn (4.1) once again, this time by insertion of known quantities on the right-hand side. If the electric far field is wanted, the receiving coefficients of an *output probe* in the shape of an electric dipole in the required orientation is inserted. If the magnetic field is wanted, the output probe must be a magnetic dipole. It is clear that as soon as T_{smn} has been determined, the transmission formula can be used to transform the field back and forth between different spheres. The concepts of input and output probes and their application in field transformations are discussed in Section 4.3.4.

Section 4.4 contains some considerations on the practical construction of a computer code for the procedures formulated in Section 4.3. It also includes a test case for such a code.

Measurement without probe correction

In the second and much simpler approach, the probe is assumed to be a good approximation to a short electric or magnetic dipole or a combination thereof. This approach is relevant for probes of low directivity applied at a not-too-close distance to the test antenna; see Section 3.2.5.

The received signal in an electric or magnetic dipole is merely proportional to the component of the electric or magnetic field, respectively, parallel to the dipole as discussed in Section 3.3.2.

For a given test antenna, the field components are contained in the spherical wave expansion

$$\vec{E}(A, \theta, \phi) = \frac{k}{\sqrt{\eta}} \sum_{s=1}^{2} \sum_{n=1}^{N} \sum_{m=-n}^{n} v T_{smn} \vec{F}^{(3)}_{smn}(A, \theta, \phi), \qquad A > r_0 \tag{4.2}$$

$$\vec{H}(A, \theta, \phi) = -ik\sqrt{\eta} \sum_{s=1}^{2} \sum_{n=1}^{N} \sum_{m=-n}^{n} v T_{smn} \vec{F}^{(3)}_{3-s,m,n}(A, \theta, \phi), \qquad A > r_0 \tag{4.3}$$

which then must be solved for the unknown coefficients T_{smn}. In eqns (4.2) and (4.3) r_0 and A are the radii of the test antenna minimum sphere and the measurement sphere, respectively. The upper limit N in the n-summation is given by eqn (2.31). The components of $\vec{E}(A, \theta, \phi)$ or $\vec{H}(A, \theta, \phi)$ are assumed available for all (θ, ϕ).

An obvious, but impractical 'big matrix' method of solution consists in insertion of a sufficient number of discrete measurement points in eqns (4.2) and

(4.3). Thereby, a system of linear equations is formed from which T_{smn} can be found numerically [5].

More efficient methods can be constructed by first carrying out an *analytical solution* by making use of the orthogonalities of the functions involved. Several different schemes are possible here and are explored in Section 4.2. In Section 4.2.2 we follow Stratton [6] and use the orthogonality, eqn (A1.70), for the vector function $\vec{F}^{(c)}_{smn}$. More fundamental orthogonalities, cf eqns (A1.68–1.69), are exhibited by the components of $\vec{F}^{(c)}_{smn}$. These are investigated in Sections 4.2.3–4.2.5.

The eventual *discrete solution* of the spherical wave expansion can be obtained by evaluating numerically the integrals of the analytical solution. The integrals are of the same form as in the probe corrected case. Numerical integration and sampling, therefore, will not be given a separate treatment in connection with measurements without probe correction.

Having found the test antenna coefficients, the far field or the field at other distances may readily be evaluated by insertion of the coefficients into the spherical wave expansion.

Short electric or magnetic dipoles have well-known receiving coefficients and are of course only special cases of the general probe assumed in eqn (4.1). The test antenna coefficients T_{smn} found by doing measurements with dipoles and subsequently solving eqns (4.2–4.3) are therefore equally well obtainable from the transmission formula (4.1). It is felt, however, that the approach in which near-field measurements of the $\vec{E}$- and $\vec{H}$-vectors themselves form the basis for prediction of the far field is conceptually the simplest.

We shall therefore derive details of this method first. By doing so, we are also following the line in the historical development over the last few decades by which spherical near-field theory has moved from being a rather academic exercise to become a practical tool in the hands of engineers.

4.2 Measurement without probe correction

4.2.1 Introduction

The problem of deriving the transmitting coefficients for a given source when some or all of the field components on a spherical surface have been measured is addressed in the present section. Only analytical expressions in terms of certain integrals of measured continuous data will be derived. The question of constructing formulas that apply to discrete sampled data will not be considered in this connection but is the subject of Section 4.3 for the case of probe-corrected measurements.

4.2.2 General case

Instead of considering a test antenna with unknown transmitting coefficients T_{smn} and input signal amplitude v, we shall choose to formulate the general case in terms of the wave coefficients $Q^{(c)}_{smn}$. This does not really make any difference, but stresses the fact that we are dealing with an academic problem which, besides its theoretical interest, may serve as a model for our investigations in the next few sections.

Let us consider an unknown electromagnetic field $(\vec{E}, \vec{H})$ existing in a source-free region $r_0 < r < r_1$ of space; see Fig. 4.1. In general, this field may contain

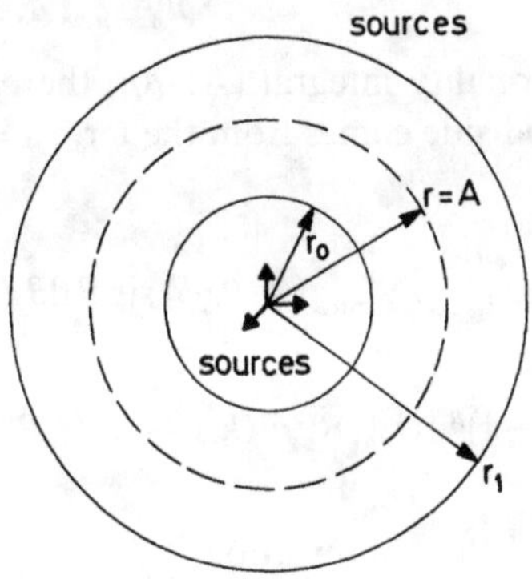

Fig. 4.1 *Measurement sphere of radius $r=A$ contained in a source-free region of space $r_0<r<r_1$*

both outgoing ($c = 3$) and incoming ($c = 4$) waves. We then have

$$\vec{E}(r, \theta, \phi) = \frac{k}{\sqrt{\eta}} \sum_{c=3}^{4} \sum_{s=1}^{2} \sum_{n=1}^{N} \sum_{m=-n}^{n} Q^{(c)}_{smn} \vec{F}^{(c)}_{smn}(r, \theta, \phi), \qquad r_0 < r < r_1 \tag{4.4}$$

$$\vec{H}(r, \theta, \phi) = -ik\sqrt{\eta} \sum_{c=3}^{4} \sum_{s=1}^{2} \sum_{n=1}^{N} \sum_{m=-n}^{n} Q^{(c)}_{smn} \vec{F}^{(c)}_{3-s,m,n}(r, \theta, \phi), \qquad r_0 < r < r_1 \tag{4.5}$$

where the upper index N is given by eqn (2.31). Assuming that we have been able to measure $\vec{E}$ and $\vec{H}$ on the complete surface of a sphere with radius A where $r_0 < A < r_1$, let us now see how the wave coefficients $Q^{(3)}_{smn}$ of the outgoing waves as well as the coefficients $Q^{(4)}_{smn}$ of the incoming waves may be determined.

We consider the electric field first. For this

$$\vec{E}(A, \theta, \phi) = \frac{k}{\sqrt{\eta}} \sum_{s=1}^{2} \sum_{n=1}^{N} \sum_{m=-n}^{n} \{Q^{(3)}_{smn} \vec{F}^{(3)}_{smn}(A, \theta, \phi) + Q^{(4)}_{smn} \vec{F}^{(4)}_{smn}(A, \theta, \phi)\} \tag{4.6}$$

where the left-hand side is the measured field distribution. Scalar multiplication of eqn (4.6) by the wave function $\vec{F}^{(\gamma)}_{\sigma,-\mu,\nu}(A, \theta, \phi)$ and integration over θ and ϕ

gives

$$\int_{\phi=0}^{2\pi}\int_{\theta=0}^{\pi} \vec{E}(A,\theta,\phi)\cdot \vec{F}^{(\gamma)}_{\sigma,-\mu,\nu}(A,\theta,\phi)\sin\theta\,d\theta\,d\phi$$

$$=\frac{k}{\sqrt{\eta}}\sum_{s=1}^{2}\sum_{n=1}^{N}\sum_{m=-n}^{n}\int_{\phi=0}^{2\pi}\int_{\theta=0}^{\pi}\{Q^{(3)}_{smn}\vec{F}^{(3)}_{smn}(A,\theta,\phi)+Q^{(4)}_{smn}\vec{F}^{(4)}_{smn}(A,\theta,\phi)\}$$

$$\cdot\vec{F}^{(\gamma)}_{\sigma,-\mu,\nu}(A,\theta,\phi)\sin\theta\,d\theta\,d\phi. \quad (4.7)$$

In view of the orthogonality integral (A1.70), the only non-vanishing contribution to the right-hand side comes from the term $s=\sigma$, $m=\mu$ and $n=\nu$ so that

$$\int_{\phi=0}^{2\pi}\int_{\theta=0}^{\pi} \vec{E}(A,\theta,\phi)\cdot \vec{F}^{(\gamma)}_{\sigma,-\mu,\nu}(A,\theta,\phi)\sin\theta\,d\theta\,d\phi$$

$$=\frac{k}{\sqrt{\eta}}(-1)^{\mu}\{Q^{(3)}_{\sigma\mu\nu}\tilde{R}^{(3,\gamma)}_{\sigma\nu}(kA)+Q^{(4)}_{\sigma\mu\nu}\tilde{R}^{(4,\gamma)}_{\sigma\nu}(kA)\} \quad (4.8)$$

where we have introduced the abbreviation

$$\tilde{R}^{(c,\gamma)}_{sn}(kr)=\left\{R^{(c)}_{sn}(kr)R^{(\gamma)}_{sn}(kr)+\delta_{s2}n(n+1)\frac{z^{(c)}_n(kr)}{kr}\frac{z^{(\gamma)}_n(kr)}{kr}\right\}. \quad (4.9)$$

Substituting for convenience in eqn (4.8) the Greek indices (σ,μ,ν) by the Latin indices (s,m,n) and rearranging, we finally obtain

$$Q^{(3)}_{smn}\tilde{R}^{(3,\gamma)}_{sn}(kA)+Q^{(4)}_{smn}\tilde{R}^{(4,\gamma)}_{sn}(kA)$$

$$=\frac{\sqrt{\eta}}{k}(-1)^{m}\int_{\phi=0}^{2\pi}\int_{\theta=0}^{\pi}\vec{E}(A,\theta,\phi)\cdot\vec{F}^{(\gamma)}_{s,-m,n}(A,\theta,\phi)\sin\theta\,d\theta\,d\phi. \quad (4.10)$$

By choosing $\gamma=3$ and $\gamma=4$ in the above equation, we obtain two linear equations in the two unknown coefficients $Q^{(3)}_{smn}$ and $Q^{(4)}_{smn}$.

However, it is not difficult to prove from eqn (A1.11) that the determinant of the system of equations is given by

$$\begin{vmatrix}\tilde{R}^{(3,3)}_{sn}(kA) & \tilde{R}^{(4,3)}_{sn}(kA)\\ \tilde{R}^{(3,4)}_{sn}(kA) & \tilde{R}^{(4,4)}_{sn}(kA)\end{vmatrix}=\begin{cases}0 & s=1\\ -\dfrac{4n(n+1)}{(kA)^6} & s=2.\end{cases} \quad (4.11)$$

In other words, from a measurement of $\vec{E}$ on the complete spherical surface $r=A$, only the TM-wave coefficients $Q^{(3)}_{2mn}$ and $Q^{(4)}_{2mn}$ may be determined. This may be explained by the fact that for $s=1$, the electric field has no radial component. The remaining tangential components are not sufficient in distinguishing between inward and outward travelling waves.

A similar analysis may be carried out for the magnetic field $\vec{H}$. The outcome of this is that from a measurement of $\vec{H}$ alone, only the TE-wave coefficients

$Q^{(3)}_{1mn}$ and $Q^{(4)}_{1mn}$ may be determined. In conclusion, therefore, by measuring both $\vec{E}$ and $\vec{H}$ on a spherical surface of radius A, where $r_0 < A < r_1$ (cf Fig. 4.1) all coefficients in the expansion (4.4) and (4.5) may be found.

In Section 4.2.5, we shall see that the full information of all coefficients may in fact be obtained from fewer measurements, namely the measurements of the four tangential components E_θ, E_ϕ, H_θ and H_ϕ only.

Let us briefly discuss the cases where either the inner region $0 \leqslant r < r_0$ or the outer region $r_1 \leqslant r < \infty$ is source free. If the inner region is source free (see Fig. 4.2) we have the following expansions

$$\vec{E}(r, \theta, \phi) = \frac{k}{\sqrt{\eta}} \sum_{s=1}^{2} \sum_{n=1}^{N} \sum_{m=-n}^{n} Q^{(1)}_{smn} \vec{F}^{(1)}_{smn}(r, \theta, \phi), \qquad 0 \leqslant r < r_1 \tag{4.12}$$

$$\vec{H}(r, \theta, \phi) = -ik\sqrt{\eta} \sum_{s=1}^{2} \sum_{n=1}^{N} \sum_{m=-n}^{n} Q^{(1)}_{smn} \vec{F}^{(1)}_{3-s,m,n}(r, \theta, \phi), \qquad 0 \leqslant r < r_1. \tag{4.13}$$

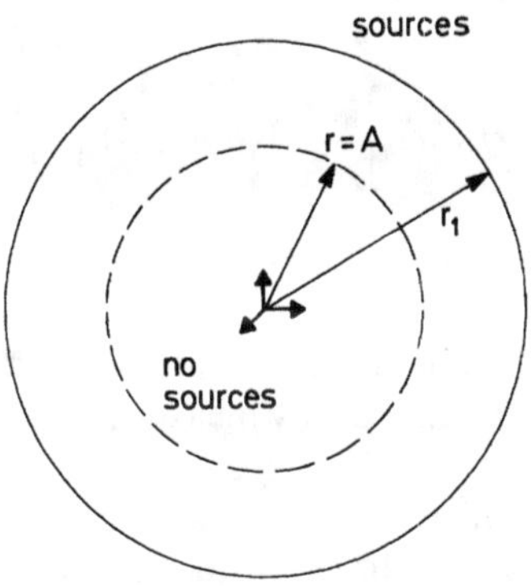

Fig. 4.2 *Measurement sphere of radius $r = A$ contained in a source-free region of space $0 \leqslant r < r_1$*

Scalar multiplication of eqn (4.12) by $\vec{F}^{(1)}_{\sigma,-\mu,\nu}(A, \theta, \phi)$ and integration over θ and ϕ gives

$$\int_{\phi=0}^{2\pi} \int_{\theta=0}^{\pi} \vec{E}(A, \theta, \phi) \cdot \vec{F}^{(1)}_{\sigma,-\mu,\nu}(A, \theta, \phi) \sin\theta \, d\theta \, d\phi$$
$$= \frac{k}{\sqrt{\eta}} \sum_{s=1}^{2} \sum_{n=1}^{N} \sum_{m=-n}^{n} \int_{\phi=0}^{2\pi} \int_{\theta=0}^{\pi} Q^{(1)}_{smn} \vec{F}^{(1)}_{smn}(A, \theta, \phi) \cdot \vec{F}^{(1)}_{\sigma,-\mu,\nu}(A, \theta, \phi) \sin\theta \, d\theta \, d\phi. \tag{4.14}$$

In view of (A1.70), the only non-vanishing contribution to the right-hand side comes from the terms $s = \sigma$, $m = \mu$ and $n = \nu$, and

$$\int_{\phi=0}^{2\pi} \int_{\theta=0}^{\pi} \vec{E}(A, \theta, \phi) \cdot \vec{F}^{(1)}_{\sigma,-\mu,\nu}(A, \theta, \phi) \sin\theta \, d\theta \, d\phi$$

$$= \frac{k}{\sqrt{\eta}} (-1)^{\mu} Q^{(1)}_{\sigma\mu\nu} \tilde{R}^{(1,1)}_{\sigma\nu}(kA) \tag{4.15}$$

where $\tilde{R}^{(1,1)}_{\sigma\nu}(kA)$ is given by eqn (4.9). Substituting for convenience in eqn (4.15) the Greek indices (σ, μ, ν) by the Latin (s, m, n) and rearranging, we finally obtain

$$Q^{(1)}_{smn} = \frac{\sqrt{\eta}}{k} (-1)^m \frac{1}{\tilde{R}^{(1,1)}_{sn}(kA)} \int_{\phi=0}^{2\pi} \int_{\theta=0}^{\pi} \vec{E}(A, \theta, \phi)$$

$$\cdot \vec{F}^{(1)}_{s,-m,n}(A, \theta, \phi) \sin\theta \, d\theta \, d\phi. \tag{4.16}$$

From this equation, $Q^{(1)}_{smn}$ may be computed provided the chosen measurement distance is such that kA is not a zero of the function $\tilde{R}^{(1,1)}_{sn}(kA)$. Since we have

$$\tilde{R}^{(1,1)}_{sn}(kA) = \begin{cases} \{j_n(kA)\}^2, & s = 1 \\ \left\{\frac{1}{kr}\frac{d}{d(kr)}(kr j_n(kr))\right\}^2_{r=A} + n(n+1)\left\{\frac{j_n(kA)}{kA}\right\}^2, & s = 2 \end{cases} \tag{4.17}$$

it is clear that zeros exist for $s = 1$. For $s = 2$, $\tilde{R}^{(1,1)}_{sn}(kA)$ is always positive, although it may become small.

With eqn (4.13) as a starting point, a similar analysis may be carried out for the magnetic field $\vec{H}$. In other words, by measuring either $\vec{E}$ or $\vec{H}$ on a spherical surface of radius A, where $0 < A < r_1$ (see Fig. 4.2), all coefficients $Q^{(1)}_{smn}$ in eqns (4.12–4.13) may be determined with the possible exception of a single coefficient in cases where an unfortunate choice of the measurement radius A has been made.

If all sources are confined to the inner region (see Fig. 4.3) we have the expansions

$$\vec{E}(r, \theta, \phi) = \frac{k}{\sqrt{\eta}} \sum_{s=1}^{2} \sum_{n=1}^{N} \sum_{m=-n}^{n} Q^{(3)}_{smn} \vec{F}^{(3)}_{smn}(r, \theta, \phi) \qquad r_0 < r < \infty \tag{4.18}$$

$$\vec{H}(r, \theta, \phi) = -ik\sqrt{\eta} \sum_{s=1}^{2} \sum_{n=1}^{N} \sum_{m=-n}^{n} Q^{(3)}_{smn} \vec{F}^{(3)}_{3-s,m,n}(r, \theta, \phi)$$

$$r_0 < r < \infty \tag{4.19}$$

In practical measurements, this is the most frequently encountered situation. An examination of the solution of these equations reveals that here all coefficients $Q^{(3)}_{smn}$ may be determined from a knowledge of either $\vec{E}$ or $\vec{H}$ on the measurement sphere. This holds true even for $kA \to \infty$, i.e. in the far field, where

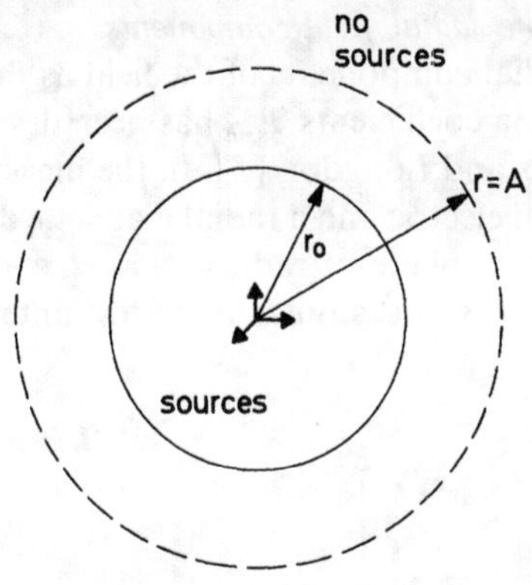

Fig. 4.3 *Measurement sphere of radius $r = A$ contained in a source-free region of space $r_0 < r < \infty$*

E_r and H_r both vanish. In Section 4.2.4, however, we shall see that the tangential components E_θ and E_ϕ, alternatively H_θ and H_ϕ, are in fact sufficient for the determination of $Q^{(3)}_{smn}$.

The findings of this section are summarized in Table 4.1.

Table 4.1 *Summary of coefficients determined by the orthogonality integral, eqn (A1.70).*

Situation illustrated in	Unknown coefficients present in expansion	Measurement for all (θ, ϕ) of	Coefficients obtainable from the orthogonality integral (A1.70)
Fig. 4.1	$Q^{(3)}_{smn}$ and $Q^{(4)}_{smn}$	$\vec{E}(A, \theta, \phi)$	$Q^{(3)}_{2mn}$ and $Q^{(4)}_{2mn}$ (all TM coefficients)
Fig. 4.1	$Q^{(3)}_{smn}$ and $Q^{(4)}_{smn}$	$\vec{H}(A, \theta, \phi)$	$Q^{(3)}_{1mn}$ and $Q^{(4)}_{1mn}$ (all TE coefficients)
Fig. 4.2	$Q^{(1)}_{smn}$	$\vec{E}(A, \theta, \phi)$	$Q^{(1)}_{smn}$ except possibly a single TE coefficient
Fig. 4.2	$Q^{(1)}_{smn}$	$\vec{H}(A, \theta, \phi)$	$Q^{(1)}_{smn}$ except possibly a single TM coefficient
Fig. 4.3	$Q^{(3)}_{smn}$	$\vec{E}(A, \theta, \phi)$	all $Q^{(3)}_{smn}$
Fig. 4.3	$Q^{(3)}_{smn}$	$\vec{H}(A, \theta, \phi)$	all $Q^{(3)}_{smn}$

4.2.3 Measurement of the radial field components

Measurement of the radial components of the field as the basis for determining the unknown test antenna coefficients T_{smn} has been described theoretically and experimentally by James and Longdon [5]. In the measurement procedure, the signal received in a radial electric and a radial magnetic dipole is recorded on the surface of a measurement sphere of radius A, $r_0 < A < \infty$ (see Fig. 4.4). It is assumed that there are no sources outside the test antenna minimum sphere.

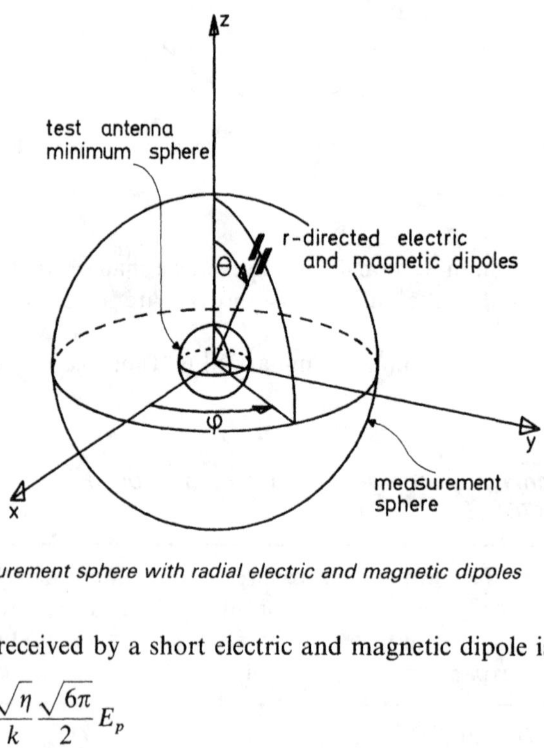

Fig. 4.4 *Measurement sphere with radial electric and magnetic dipoles*

The signal received by a short electric and magnetic dipole is given by

$$w^e = \frac{\sqrt{\eta}}{k}\frac{\sqrt{6\pi}}{2}E_p \tag{4.20}$$

and

$$w^m = \frac{1}{k\sqrt{\eta}}\frac{\sqrt{6\pi}}{2}H_p \tag{4.21}$$

(see Section 3.3.2).

In eqns (4.20) and (4.21), the subscript p indicates the field components parallel to the electric and magnetic dipole, respectively. Combining eqns (4.2) and (4.20), we have for the signal in the radial electric dipole

$$w_r^e(A, 0, \theta, \phi) = \frac{\sqrt{6\pi}}{2}v \sum_{s=1}^{2} \sum_{n=1}^{N} \sum_{m=-n}^{n} T_{smn}\vec{F}_{smn}^{(3)}(A, \theta, \phi) \cdot \hat{r}. \tag{4.22}$$

The angle χ (the probe orientation in the tangential plane) is not relevant for a radial dipole and has been set equal to zero in the argument list on the left-hand side. In order to solve eqn (4.22) for the unknown coefficients T_{smn}, we shall use the orthogonality integral (A1.68). Multiplying both sides of eqn (4.22) by $\vec{F}^{(3)}_{2,-\mu,\nu}(A, \theta, \phi) \cdot \hat{r}$ and integrating over θ and ϕ gives

$$\int_{\phi=0}^{2\pi} \int_{\theta=0}^{\pi} w_r^e(A, 0, \theta, \phi) \vec{F}^{(3)}_{2,-\mu,\nu}(A, \theta, \phi) \cdot \hat{r} \sin\theta \, d\theta \, d\phi$$

$$= \int_{\phi=0}^{2\pi} \int_{\theta=0}^{\pi} \frac{\sqrt{6\pi}}{2} v \sum_{s=1}^{2} \sum_{n=1}^{N} \sum_{m=-n}^{n} T_{smn} \{\vec{F}^{(3)}_{smn}(A, \theta, \phi) \cdot \hat{r}\}$$

$$\{\vec{F}^{(3)}_{2,-\mu,\nu}(A, \theta, \phi) \cdot \hat{r}\} \sin\theta \, d\theta \, d\phi$$

$$= \frac{\sqrt{6\pi}}{2} v T_{2\mu\nu} (-1)^{\mu} \nu(\nu+1) \left(\frac{h_n^{(1)}(kA)}{kA} \right)^2. \quad (4.23)$$

Substituting for convenience in eqn (4.23) the indices $(2, \mu, \nu)$ by $(2, m, n)$ and rearranging, we obtain

$$v T_{2mn} = \frac{2}{\sqrt{6\pi}} (-1)^m \frac{1}{n(n+1)} \left(\frac{kA}{h_n^{(1)}(kA)} \right)^2$$

$$\int_{\phi=0}^{2\pi} \int_{\theta=0}^{\pi} w_r^e(A, 0, \theta, \phi) \vec{F}^{(3)}_{2,-m,n}(A, \theta, \phi) \cdot \hat{r} \sin\theta \, d\theta \, d\phi \quad (4.24)$$

whereby the TM-coefficients T_{2mn} are found and expressed in terms of measured and known quantities only.

In determining the TE-coefficients, we must repeat the experiment using a radial magnetic dipole. Combining eqns (4.3) and (4.21) leads to the following equation for the signal in the radial magnetic dipole

$$w_r^m(A, 0, \theta, \phi) = -i \frac{\sqrt{6\pi}}{2} v \sum_{s=1}^{2} \sum_{n=1}^{N} \sum_{m=-n}^{n} T_{smn} \vec{F}^{(3)}_{3-s,m,n}(A, \theta, \phi) \cdot \hat{r} \quad (4.25)$$

from which a similar procedure as outlined above for the electric dipole leads to

$$v T_{1mn} = \frac{2i}{\sqrt{6\pi}} (-1)^m \frac{1}{n(n+1)} \left(\frac{kA}{h_n^{(1)}(kA)} \right)^2$$

$$\int_{\phi=0}^{2\pi} \int_{\theta=0}^{\pi} w_r^m(A, 0, \theta, \phi) \vec{F}^{(3)}_{2,-m,n}(A, \theta, \phi) \cdot \hat{r} \sin\theta \, d\theta \, d\phi. \quad (4.26)$$

As evident from the above discussion, measurement of one of the radial field components alone is not sufficient for determining the complete field. Both components are needed. On the other hand, information about partial fields (TM or TE) is obtainable from measurements where only one of the two elementary types of dipoles is employed.

A serious drawback of the method is that, for increasing r, the radial field components tend to zero at least as fast as r^{-2} while the tangential field components discussed in the next section decay as r^{-1}. The presence of relatively strong tangential components may put unrealistic requirements on the accuracy of radial probe alignment. Note, that measurements using radial dipoles alone are not sufficient in cases where both inward and outward propagating waves are present.

The results of this section are summarized in Table 4.2.

Table 4.2 *Summary of coefficients determined by the orthogonality integral, eqn (A1.68).*

Unknown coefficients present in expansion	Measurement for all (θ, ϕ) of	Coefficients obtainable from the orthogonality integral (A1.68)
T_{smn}	$E_r(A, \theta, \phi)$	T_{2mn} (the TM-coefficients)
T_{smn}	$H_r(A, \theta, \phi)$	T_{1mn} (the TE-coefficients)

4.2.4 Measurement of two tangential field components

This scheme, proposed by A. C. Ludwig [7], is based upon the orthogonality integral (A1.69). During the measurement, the signals received in theta- and phi-directed electric dipoles are recorded; see Fig. 4.5. Using eqns (4.2) and (4.20), we

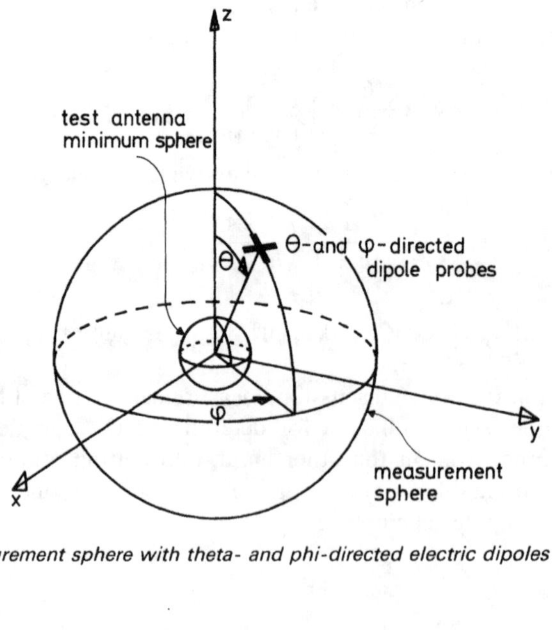

Fig. 4.5 *Measurement sphere with theta- and phi-directed electric dipoles*

have

$$w_\theta^e(A, 0, \theta, \phi) = \frac{\sqrt{6\pi}}{2} v \sum_{s=1}^{2} \sum_{n=1}^{N} \sum_{m=-n}^{n} T_{smn} \vec{F}_{smn}^{(3)}(A, \theta, \phi) \cdot \hat{\theta} \tag{4.27}$$

$$w_\phi^e\left(A, \frac{\pi}{2}, \theta, \phi\right) = \frac{\sqrt{6\pi}}{2} v \sum_{s=1}^{2} \sum_{n=1}^{N} \sum_{m=-n}^{n} T_{smn} \vec{F}_{smn}^{(3)}(A, \theta, \phi) \cdot \hat{\phi}. \tag{4.28}$$

Multiplying these equations by $\vec{F}_{\sigma,-\mu,\nu}^{(3)}(A, \theta, \phi) \cdot \hat{\theta}$ and $\vec{F}_{\sigma,-\mu,\nu}^{(3)}(A, \theta, \phi) \cdot \hat{\phi}$, respectively, integrating over θ and ϕ and adding the resulting equations, yields

$$\int_{\phi=0}^{2\pi} \int_{\theta=0}^{\pi} \{w_\theta^e \vec{F}_{\sigma,-\mu,\nu}^{(3)} \cdot \hat{\theta} + w_\phi^e \vec{F}_{\sigma,-\mu,\nu}^{(3)} \cdot \hat{\phi}\} \sin\theta \, d\theta \, d\phi = \frac{\sqrt{6\pi}}{2} v T_{\sigma\mu\nu}(-1)^\mu \{R_{\sigma\nu}^{(3)}(kA)\}^2. \tag{4.29}$$

Substituting for convenience in eqn (4.29) the indices (σ, μ, ν) by (s, m, n) and rearranging, we obtain

$$v T_{smn} = \frac{2}{\sqrt{6\pi}} (-1)^m \{R_{sn}^{(3)}(kA)\}^{-2} \int_{\phi=0}^{2\pi} \int_{\theta=0}^{\pi} \{w_\theta^e \hat{\theta} + w_\phi^e \hat{\phi}\} \cdot \vec{F}_{s,-m,n}^{(3)}(A, \theta, \phi) \sin\theta \, d\theta \, d\phi. \tag{4.30}$$

This equation expresses the test antenna transmitting coefficients T_{smn} in terms of measured and known quantities only.

If, instead of tangential electric dipoles we had used tangential magnetic dipoles, this would have lead to the alternative equation

$$v T_{smn} = \frac{2i}{\sqrt{6\pi}} (-1)^m \{R_{3-s,n}^{(3)}(kA)\}^{-2} \int_{\phi=0}^{2\pi} \int_{\theta=0}^{\pi} \{w_\theta^m \hat{\theta} + w_\phi^m \hat{\phi}\} \cdot \vec{F}_{3-s,-m,n}^{(3)}(A, \theta, \phi) \sin\theta \, d\theta \, d\phi. \tag{4.31}$$

As may be seen by comparing eqn (4.31) to eqn (4.30), there is, in principle, no difference in the information obtained about the test antenna whether one is

Table 4.3 *Summary of coefficients determined by the orthogonality integral, eqn (A1.69).*

Unknown coefficients present in expansion	Measurement for all (θ, ϕ) of	Coefficients obtainable from the orthogonality integral (A1.69)
T_{smn}	$E_\theta(A, \theta, \phi)$ and $E_\phi(A, \theta, \phi)$	T_{smn}
T_{smn}	$H_\theta(A, \theta, \phi)$ and $H_\phi(A, \theta, \phi)$	T_{smn}

using tangential electric or tangential magnetic dipoles. Both types of dipoles yields the complete field provided both orientations (θ and ϕ) are used. Note, that measurements using two tangential dipoles alone are not sufficient in cases where both inward and outward propagating waves are present.

The results of this section are summarized in Table 4.3.

4.2.5 *Wood's method*

Under certain circumstances in practical measurements, it may prove inconvenient or impossible to mount the antenna close to the centre of the measurement sphere. In such cases, there is a need for an ability to specify the origin of the spherical wave expansion of the test antenna independently of the measurement sphere. Wood [8, 9] has developed an interesting theory capable of handling the general situation, illustrated schematically in Fig. 4.6(b). The formulation is based on the reciprocity integral (A1.74) and requires, in principle, measurement of two electric and two magnetic tangential field components.

Let us first consider the concentric measurement set-up shown in Fig. 4.6(a). From eqns (4.2–4.3) and (4.20–4.21), we have for the four measured signals in this situation

$$w_\theta^e(A, 0, \theta, \phi) = \frac{\sqrt{6\pi}}{2} v \sum_{s=1}^{2} \sum_{n=1}^{N} \sum_{m=-n}^{n} T_{smn} \vec{F}_{smn}^{(3)}(A, \theta, \phi) \cdot \hat{\theta} \tag{4.32}$$

$$w_\phi^e\left(A, \frac{\pi}{2}, \theta, \phi\right) = \frac{\sqrt{6\pi}}{2} v \sum_{s=1}^{2} \sum_{n=1}^{N} \sum_{m=-n}^{n} T_{smn} \vec{F}_{smn}^{(3)}(A, \theta, \phi) \cdot \hat{\phi} \tag{4.33}$$

$$w_\theta^m(A, 0, \theta, \phi) = -i\frac{\sqrt{6\pi}}{2} v \sum_{s=1}^{2} \sum_{n=1}^{N} \sum_{m=-n}^{n} T_{smn} \vec{F}_{3-s,m,n}^{(3)}(A, \theta, \phi) \cdot \hat{\theta} \tag{4.34}$$

$$w_\phi^m\left(A, \frac{\pi}{2}, \theta, \phi\right) = -i\frac{\sqrt{6\pi}}{2} v \sum_{s=1}^{2} \sum_{n=1}^{N} \sum_{m=-n}^{n} T_{smn} \vec{F}_{3-s,m,n}^{(3)}(A, \theta, \phi) \cdot \hat{\phi} \tag{4.35}$$

Equations (4.32–4.35) are now multiplied by components of an inward travelling test field; we choose $\vec{F}_{3-\sigma,-\mu,\nu}^{(4)} \cdot \hat{\phi}$, $\vec{F}_{3-\sigma,-\mu,\nu}^{(4)} \cdot \hat{\theta}$, $\vec{F}_{\sigma,-\mu,\nu}^{(4)} \cdot \hat{\phi}$ and $\vec{F}_{\sigma,-\mu,\nu}^{(4)} \cdot \hat{\theta}$ respectively. By superposition with the proper signs of the resulting four equations, making use of the reciprocity integral (A1.74) and substituting the Greek letter indices (σ, μ, ν) by (s, m, n), we finally obtain

$$\begin{aligned} vT_{smn} = \frac{ik^2}{\sqrt{6\pi}}(-1)^m \int_{\phi=0}^{2\pi} \int_{\theta=0}^{\pi} \{[w_\theta^e \hat{\phi} - w_\phi^e \hat{\theta}] \cdot \vec{F}_{3-s,-m,n}^{(4)}(A, \theta, \phi) \\ + i[w_\theta^m \hat{\phi} - w_\phi^m \hat{\theta}] \cdot \vec{F}_{s,-m,n}^{(4)}(A, \theta, \phi)\} \, A^2 \sin\theta \, d\theta \, d\phi \end{aligned} \tag{4.36}$$

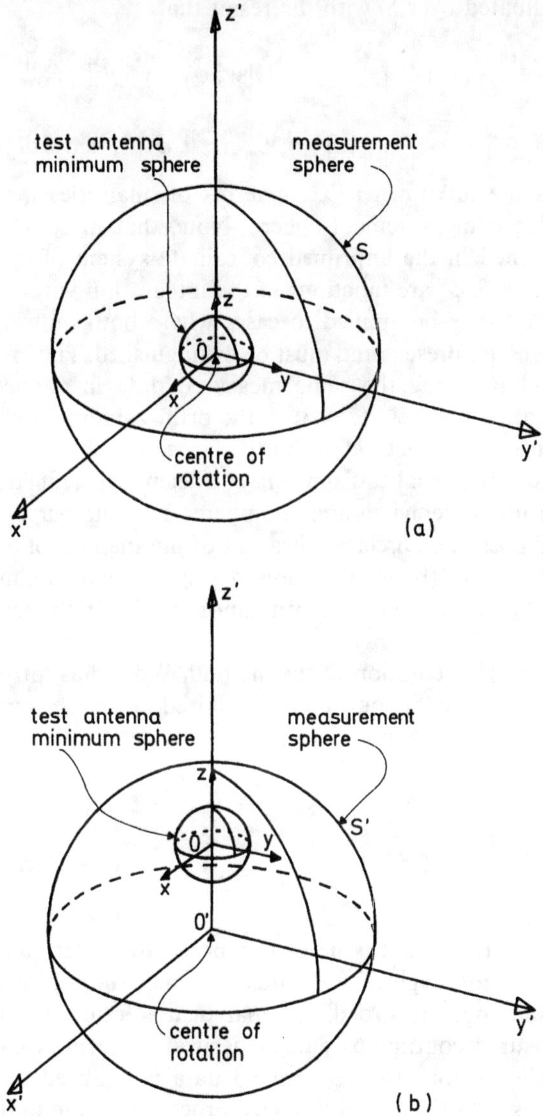

Fig. 4.6 *(a) Measurement sphere S and test antenna minimum sphere. (b) Displaced measurement sphere S′ and test antenna minimum sphere*

where the integration is carried out over the measurement surface S. In Fig. 4.6(b), the measurement situation is the same, apart from a displacement (e.g. along the z-axis, as shown) of the measurement surface S to a new relative position S' such that the test antenna minimum sphere is contained in both S and S'. According to Section A1.4.5, the orthogonality integral may then be equally well evaluated over S' with the result that

$$vT_{smn} = \frac{ik^2}{\sqrt{6\pi}}(-1)^m \int_{\phi'=0}^{2\pi} \int_{\theta'=0}^{\pi} \{[w^e_{\theta'}\hat{\phi}' - w^e_{\phi'}\hat{\theta}'] \cdot \vec{F}^{(4)}_{3-s,-m,n}(r,\theta,\phi)$$

$$+ i[w^m_{\theta'}\hat{\phi}' - w^m_{\phi'}\hat{\theta}'] \cdot \vec{F}^{(4)}_{s,-m,n}(r,\theta,\phi)\} A^2 \sin\theta' \, d\theta' \, d\phi' \tag{4.37}$$

which expresses the unknown vT_{smn} in terms of quantities measured on the eccentricially placed measurement sphere. Note, that in eqn (4.37) the wave functions are defined in the unprimed coordinate system of the test antenna. Their arguments (r, θ, ϕ) are functions of the integration variables (θ', ϕ').

Wood's method may be applied to cases where both outward and inward propagating waves are present and must be distinguished. The inward travelling modes are found by using the same measured data in connection with an outward propagating test field ($\gamma = 3$) in the orthogonality integral.

The importance of the method lies in the fact that even in eccentric set-ups, eqn (4.37) provides the usual transmitting coefficients as defined relative to the centre O of the test antenna minimum sphere. The number of coefficients is then as small as possible. Direct application of the method of Section 4.2.4 to the geometry in Fig. 4.6(b) would imply a test antenna minimum sphere of unnecessarily large radius, hence requiring more modes in the representation of the field.

In a practical implementation of the method, Wood has introduced certain approximations whereby the four dipoles required in the measurements can be substituted by two Huygens sources as probes. For details, the reader is referred to the original work [8].

4.3 Measurement with probe correction

4.3.1 Introduction

We shall now turn to a main issue of this book: the determination of transmitting coefficients for a given test antenna when the signal received by a directive probe has been recorded. An analytical solution in terms of certain integrals of measured continuous data is derived in Section 4.3.2. The corresponding formulation for discrete sampled data is deduced in Section 4.3.3. Section 4.3.4 deals with the straightforward processes of computing directivity, gain and radiated fields when the transmitting coefficients have been found.

4.3.2 Analytical solution of the transmission formula

The objective of this section is to solve the transmission formula for the unknowns, i.e. the test antenna transmitting coefficients T_{smn}. The transmission formula

$$w(A, \chi, \theta, \phi) = \frac{v}{2} \sum_{\substack{smn \\ \sigma\mu\nu}} T_{smn} \, e^{im\phi} \, d^n_{\mu m}(\theta) \, e^{i\mu\chi} \, C^{sn(3)}_{\sigma\mu\nu}(kA) R^p_{\sigma\mu\nu} \tag{4.38}$$

is a finite series representation of the signal received by an arbitrary probe with known receiving coefficients $R^p_{\sigma\mu\nu}$. Introducing the *probe response constants* (see Section 3.3.1)

$$P_{s\mu n}(kA) = \frac{1}{2}\sum_{\sigma\nu} C^{sn(3)}_{\sigma\mu\nu}(kA)R^p_{\sigma\mu\nu} \tag{4.39}$$

where $C^{sn(3)}_{\sigma\mu\nu}(kA)$ are translation coefficients (ref. Appendix A3) we obtain the shorter form

$$w(A, \chi, \theta, \phi) = v \sum_{\substack{smn\\ \mu}} T_{smn}\, e^{im\phi} d^n_{\mu m}(\theta)\, e^{i\mu\chi} P_{s\mu n}(kA). \tag{4.40}$$

In (4.40), w is the signal received in position (θ, ϕ) on the measurement sphere of radius A, v is the input signal to the test antenna and χ is the probe rotation angle. The products, $e^{im\phi} d^n_{\mu m}(\theta)\, e^{i\mu\chi}$, are rotation coefficients of the spherical wave functions (see Appendix A2). The probe response constants, $P_{s\mu n}(kA)$, express the probe's sensitivity at the location $(A, \chi, \theta, \phi) = (A, 0, 0, 0)$ to a test antenna mode (s, μ, n) of unity amplitude.

A given probe will in general need many probe coefficients $R^p_{\sigma\mu\nu}$, where $\sigma = 1, 2$, $\mu = -\nu_{max}, \ldots, 0, \ldots, \nu_{max}$ and $\nu = |\mu|, \ldots, \nu_{max}$, $(\nu \neq 0)$, for its adequate characterization. The maximum value of μ depends on the degree of azimuthal symmetry of the probe. However, μ can never exceed $\nu_{max} \cong kr_p + 10$ where r_p is the radius of the probe minimum sphere. Hence, a probe which is small in terms of the wavelength can only possess a few harmonics in its azimuthal pattern. The indices (s, m, n) of the test antenna transmitting coefficients T_{smn} are limited to $s = 1, 2$, $m = -N, \ldots, 0, \ldots, N$ and $n = |m|, \ldots, N$, $(n \neq 0)$, where $N \cong kr_t + 10$, r_t being the radius of the test antenna minimum sphere.

The solution is modelled after the methods proposed independently by Wacker [1] and Jensen [10]. It is carried out in three steps. We exploit the orthogonality of the exponential function

$$\int_0^{2\pi} e^{i(m-m')\psi}\, d\psi = 2\pi\delta_{mm'} \tag{4.41}$$

in the first two steps and the orthogonality (A2.10) of the rotation coefficient $d^n_{\mu m}(\theta)$ in the third step.

Let us first rewrite the transmission formula (4.40) slightly

$$w(A, \chi, \theta, \phi) = \sum_{\mu=-\nu_{max}}^{\nu_{max}} w_\mu(A, \theta, \phi)\, e^{i\mu\chi} \tag{4.42}$$

where

$$w_\mu(A, \theta, \phi) = v \sum_{s=1}^{2} \sum_{n=1}^{N} \sum_{m=-n}^{n} T_{smn}\, e^{im\phi} d^n_{\mu m}(\theta) P_{s\mu n}(kA). \tag{4.43}$$

As seen from (4.42), $w_\mu(A, \theta, \phi)$ is a coefficient in the expansion of w into a finite Fourier series in χ. We may solve eqn (4.42) for $w_\mu(A, \theta, \phi)$ by multiplying both sides by $e^{-i\mu'\chi}$ and integrating from 0 to 2π with respect to χ

$$\int_{\chi=0}^{2\pi} w(A, \chi, \theta, \phi)\, e^{-i\mu'\chi}\, d\chi = \int_{\chi=0}^{2\pi} \left\{ \sum_{\mu=-\nu_{max}}^{\nu_{max}} w_\mu(A, \theta, \phi)\, e^{i\mu\chi} \right\} e^{-i\mu'\chi}\, d\chi$$

$$= 2\pi w_{\mu'}(A, \theta, \phi). \tag{4.44}$$

Substituting for convenience μ' by μ and rearranging gives the transformed data

$$w_\mu(A, \theta, \phi) = \frac{1}{2\pi} \int_{\chi=0}^{2\pi} w(A, \chi, \theta, \phi)\, e^{-i\mu\chi}\, d\chi. \tag{4.45}$$

As expressed in this equation, $w_\mu(A, \theta, \phi)$ is the Fourier transform of the measured data $w(A, \chi, \theta, \phi)$. The transformed data $w_\mu(A, \theta, \phi)$ equals the transformed right-hand side of the transmission formula; see eqn (4.43).

In the next step we rewrite the transformed transmission formula (4.43) as follows

$$w_\mu(A, \theta, \phi) = \sum_{m=-N}^{N} w_{\mu m}(A, \theta)\, e^{im\phi} \tag{4.46}$$

where

$$w_{\mu m}(A, \theta) = v \sum_{s=1}^{2} \sum_{\substack{n=|m| \\ (n \neq 0)}}^{N} T_{smn}\, d^n_{\mu m}(\theta) P_{s\mu n}(kA). \tag{4.47}$$

As evident from (4.46), $w_{\mu m}(A, \theta)$ is a coefficient in the expansion of w_μ into a finite Fourier series in ϕ. We may solve eqn (4.46) (of the same form as eqn (4.42)) for $w_{\mu m}$ with the result

$$w_{\mu m}(A, \theta) = \frac{1}{2\pi} \int_{\phi=0}^{2\pi} w_\mu(A, \theta, \phi)\, e^{-im\phi}\, d\phi. \tag{4.48}$$

Equations (4.47) and (4.48) are two-dimensional Fourier transforms of the right-hand side of the transmission formula and the measured data, respectively.

The last step consists in rewriting the twice-transformed transmission formula (4.47) in this way

$$w_{\mu m}(A, \theta) = \sum_{\substack{n=|m| \\ (n \neq 0)}}^{N} w^n_{\mu m}(A)\, d^n_{\mu m}(\theta) \tag{4.49}$$

where

$$w^n_{\mu m}(A) = v \sum_{s=1}^{2} T_{smn} P_{s\mu n}(kA). \tag{4.50}$$

Equation (4.49) can be solved by applying the orthogonality (A2.10). The result is the following integral over the twice-transformed data

$$w^n_{\mu m}(A) = \frac{2n+1}{2} \int_{\theta=0}^{\pi} w_{\mu m}(A, \theta)\, d^n_{\mu m}(\theta) \sin\theta\, d\theta. \tag{4.51}$$

We have now completed the three steps needed to solve the original transmission formula (4.40). Writing eqn (4.50) explicitly we obtain

$$vT_{1mn}P_{1\mu n}(kA) + vT_{2mn}P_{2\mu n}(kA) = w^n_{\mu m}(A). \tag{4.52}$$

For each pair of indices (m, n) where $m = -N, ..., 0, ..., N$ and $n = |m|, ..., N$, $(n \neq 0)$, (4.52) constitutes as many equations in the two unknowns vT_{1mn} and vT_{2mn} as the number of different values of the probe azimuthal index μ one takes into account.

Following an idea of P. F. Wacker [2], a probe with indices $\mu = \pm 1$ only may be chosen. Such a probe can be constructed, e.g. as a conical horn. Also, at large measuring distances, the $\mu = \pm 1$ probe modes will dominate over any other probe mode that might be present. This is obvious from eqn (4.39) and the asymptotic form of the translation coefficients; see eqns (A3.22–3.24). Therefore, a probe which may not be completely rotational symmetric, e.g. a rectangular horn, will approximate a $\mu = \pm 1$ probe as long as it is applied not too closely to the test antenna. When a $\mu = \pm 1$ probe is selected, eqn (4.52) reduces to two equations in the two unknowns vT_{1mn} and vT_{2mn}.

$$vT_{1mn}P_{11n}(kA) + vT_{2mn}P_{21n}(kA) = w^n_{1m}(A) \tag{4.53}$$

$$vT_{1mn}P_{1,-1,n}(kA) + vT_{2mn}P_{2,-1,n}(kA) = w^n_{-1m}(A). \tag{4.54}$$

In rare cases where $\mu = \pm 1$ is insufficient in describing the probe, more than two equations in the two unknowns vT_{1mn} and vT_{2mn} may be derived from eqn (4.52). We shall not go into detail here with the question of dealing with this situation.

The analytical closed-form solution developed in this section forms the basis for the derivation in the next section of a discrete solution of the transmission formula. Assuming a $\mu = \pm 1$ probe, Section 4.3.3 is devoted to calculating the right-hand side of eqn (4.52) from the measured signal $w(A, \chi, \theta, \phi)$.

4.3.3 Discrete solution of the transmission formula

4.3.3.1 Introduction: In Section 4.3.2 we have derived equation (4.52) for the determination of the test antenna transmitting coefficients T_{smn}. As we have seen, the right-hand side is expressed as a three-fold integral transform of the measured data $w(A, \chi, \theta, \phi)$; see eqns (4.51), (4.48) and (4.45), respectively,

$$w^n_{\mu m}(A) = \frac{2n+1}{2}\int_{\theta=0}^{\pi} w_{\mu m}(A, \theta) d^n_{\mu m}(\theta) \sin\theta \, d\theta$$

$$n = 1, 2, ..., N; m = -n, ..., 0, ..., n; \mu = -v_{max}, ..., 0, ..., v_{max} \tag{4.55}$$

where

$$w_{\mu m}(A, \theta) = \frac{1}{2\pi}\int_{\phi=0}^{2\pi} w_\mu(A, \theta, \phi)\, e^{-im\phi}\, d\phi$$

$$m = -N, ..., 0, ..., N; \mu = -v_{max}, ..., 0, ..., v_{max} \tag{4.56}$$

with

$$w_\mu(A, \theta, \phi) = \frac{1}{2\pi} \int_{\chi=0}^{2\pi} w(A, \chi, \theta, \phi)\, e^{-i\mu\chi}\, d\chi$$

$$\mu = -\nu_{max}, \ldots, 0, \ldots, \nu_{max} \qquad (4.57)$$

where $w(A, \chi, \theta, \phi)$ is the received signal (measured data).

In this section we shall consider the question of calculating these integrals by numerical methods. In numerical integration schemes, samples of the function to be integrated are taken at certain points in the interval of integration and are multiplied by certain weights accompanying these points. The sum of the products of the sampled values of the function and the weights is then an approximation to the value of the integral. The choice of sampling intervals and weights is determined by consideration of the function to be integrated, the required accuracy, etc.

The three integrals we are dealing with here are, however, particularly simple. The chi and phi integrals are both Fourier integrals of periodic, band-limited functions. Such integrals may be evaluated exactly by the Discrete Fourier Transform (DFT) technique for which an efficient algorithm (the Fast Fourier Transform, FFT) exists. The theta integral may also be evaluated exactly by Fourier techniques [11, 2].

The chi integral (4.57) is calculated both by the DFT and in a more direct manner in Section 4.3.3.2. In Section 4.3.3.3, the DFT is used for the evaluation of the phi integral (4.56) and in Sections 4.3.3.4 and 4.3.3.5 the theta integral (4.55) is treated. In all cases, the simplicity of assuming a $\mu = \pm 1$ probe (see Section 4.3.2) is exploited. For a brief account of the Discrete Fourier Transform (DFT) and its inverse transform (IDFT), the reader is referred to Appendix A4.2. Considerations of reconstruction by sampling and of the evaluation of Fourier integrals are given in Appendix A4.3.

4.3.3.2 Measurement in chi: The chi integral, eqn (4.57), is repeated below for convenience

$$w_\mu(A, \theta, \phi) = \frac{1}{2\pi} \int_{\chi=0}^{2\pi} w(A, \chi, \theta, \phi)\, e^{-i\mu\chi}\, d\chi$$

$$\mu = -\nu_{max}, \ldots, 0, \ldots, \nu_{max} \qquad (4.58)$$

where $w_\mu(A, \theta, \phi) = 0$ for $|\mu| > \nu_{max}$ due to the band limitation (see eqn (4.42)). With a $\mu = \pm 1$ probe, only the $w_1(A, \theta, \phi)$ and $w_{-1}(A, \theta, \phi)$ modes of w are present.

Nevertheless, in calculating the integral (4.58), let us first make use of the results of Appendix A4.5.2, where a periodic function, e.g. w, is assumed to contain the $2N + 1$ modes of orders $-N, \ldots, 0, \ldots, N$ and where these are determined from integrals of the form (4.58). As shown in A4.5.2 these integrals are readily calculated in closed form by applying an IDFT of order J_χ to a sequence of J_χ samples of the function, where $J_\chi \geqslant 2N+1$. Assuming $\mu = \pm 1$, we have here $N = 1$ and $J_\chi \geqslant 3$. Three samples of the measured signal w (equally spaced in the interval $0 \leqslant \chi < 2\pi$) are therefore sufficient for determining $w_1(A, \theta, \phi)$, $w_{-1}(A, \theta, \phi)$ and $w_0(A, \theta, \phi)$.

During the measurement in chi (chi scanning), theta and phi as well as the measurement distance are kept constant. The samples are taken by rotating the probe to angular positions with separation $\Delta\chi$ (see Fig. 4.7) where $\Delta\chi = 2\pi/J_\chi$, $J_\chi = 3$. Denoting the samples by $w(A, j\Delta\chi, \theta, \phi)$, $j = 0, 1, 2$ we then have

$$\{w_\mu(A, \theta, \phi) \,|\, \mu = 0, 1, -1\} = \mathrm{IDFT}\{w(A, j\Delta\chi, \theta, \phi) \,|\, j = 0, 1, 2\}. \quad (4.59)$$

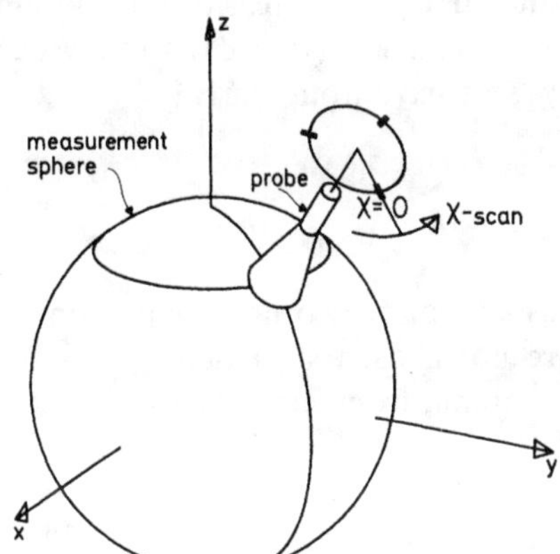

Fig. 4.7 *Measurement in chi*

Instead of $J_\chi = 3$, one might choose $J_\chi = 4$. This choice corresponds to the perhaps more natural sampling increment $\pi/2$ and represents only a slight oversampling. Denoting the samples by $w(A, j\Delta\chi, \theta, \phi)$, $j = 0, 1, 2, 3$ we have in this case

$$\{w_\mu(A, \theta, \phi) \,|\, \mu = 0, 1, \pm 2, -1\} = \mathrm{IDFT}\{w(A, j\Delta\chi, \theta, \phi) \,|\, j = 0, 1, 2, 3\} \quad (4.60)$$

where $w_2(A, \theta, \phi)$ and $w_{-2}(A, \theta, \phi)$ share the third position of the output sequence from the IDFT as indicated in the parameter list on the left-hand side of eqn (4.60).

With a $\mu = \pm 1$ probe the received signal $w(A, \chi, \theta, \phi)$ contains only two modes

$$w(A, \chi, \theta, \phi) = w_1(A, \theta, \phi)\, e^{i\chi} + w_{-1}(A, \theta, \phi)\, e^{-i\chi} \quad (4.61)$$

from which it follows that $w(A, \chi, \theta, \phi) = -w(A, \chi + \pi, \theta, \phi)$. Therefore, of four samples with $\Delta\chi = \pi/2$, only those with $\chi = 0$ and $\chi = \pi/2$ need to be actually measured. The two remaining may be generated in the computer. We have in this case the relationship

$$\{0, w_1(A, \theta, \phi), 0, w_{-1}(A, \theta, \phi)\}$$
$$= \mathrm{IDFT}\left\{w(A, 0, \theta, \phi), w\left(A, \frac{\pi}{2}, \theta, \phi\right), -w(A, 0, \theta, \phi), -w\left(A, \frac{\pi}{2}, \theta, \phi\right)\right\} \tag{4.62}$$

between the output sequence (left-hand side) and the input sequence (right-hand side) in the IDFT, ref. eqn (4.60).

As an alternative to the DFT procedure, a direct solution of eqn (4.61) for the two harmonics $w_1(A, \theta, \phi)$ and $w_{-1}(A, \theta, \phi)$ is easily derived. Choosing two sample points χ_1 and χ_2, we have from (4.61)

$$w(A, \chi_1, \theta, \phi) = w_1 \, e^{i\chi_1} + w_{-1} \, e^{-i\chi_1} \tag{4.63}$$

$$w(A, \chi_2, \theta, \phi) = w_1 \, e^{i\chi_2} + w_{-1} \, e^{-i\chi_2}. \tag{4.64}$$

The determinant of this system of two linear equations is different from zero provided χ_1 and χ_2 are not chosen such that $\chi_1 = \chi_2$ or $\chi_1 = \chi_2 + \pi$. With $\chi_1 = 0$ and $\chi_2 = \pi/2$ we obtain from eqns (4.63) and (4.64)

$$w_1(A, \theta, \phi) = \frac{1}{2}\left\{w(A, 0, \theta, \phi) - iw\left(A, \frac{\pi}{2}, \theta, \phi\right)\right\} \tag{4.65}$$

$$w_{-1}(A, \theta, \phi) = \frac{1}{2}\left\{w(A, 0, \theta, \phi) + iw\left(A, \frac{\pi}{2}, \theta, \phi\right)\right\}. \tag{4.66}$$

The chi integration is hereby accomplished.

4.3.3.3 Measurement in phi: The phi integral, eqn (4.56), is given by

$$w_{\mu m}(A, \theta) = \frac{1}{2\pi} \int_{\phi=0}^{2\pi} w_\mu(A, \theta, \phi) \, e^{-im\phi} \, d\phi \tag{4.67}$$

where $w_{\mu m}(A, \theta) = 0$ for $|m| > N$ because of band limitation; see eqn (4.46). We shall now determine $w_{\mu m}(A, \theta)$ by evaluating the integral (4.67) from samples in ϕ of $w_\mu(A, \theta, \phi)$. The latter function is known for $\mu = \pm 1$ from the previous section.

During the measurement in phi (phi scanning), chi, theta and the measurement distance A are kept constant. The samples are taken by moving the probe to angular positions with separation $\Delta\phi$ (see Fig. 4.8) where $\Delta\phi = 2\pi/J_\phi$, J_ϕ being the number of samples in $0 \leqslant \phi < 2\pi$. The samples are denoted by $w_\mu(A, \theta, j\Delta\phi)$, $j = 0, 1, \ldots, J_\phi - 1$. Assuming J_ϕ to equal the number of

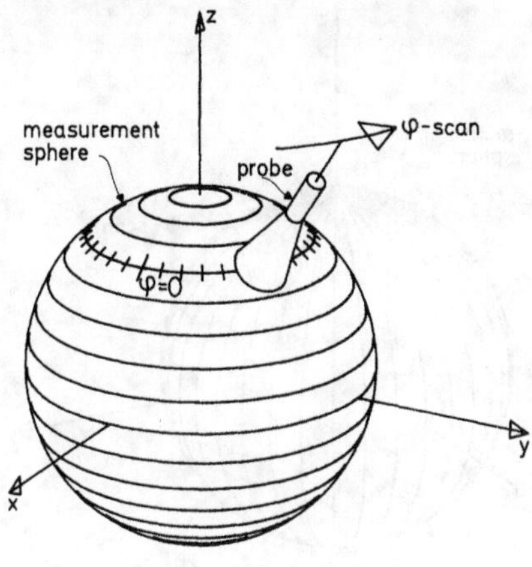

Fig. 4.8 *Measurement in phi*

harmonics $2N + 1$, we have from Appendix A4.5.2

$$\{w_{\mu m}(A, \theta) \,|\, m = 0, 1, \ldots, N, -N, \ldots, -1\} = \mathrm{IDFT}\{w_\mu(A, \theta, j\Delta\phi) \,|\, j = 0, 1, \ldots, J_\phi - 1\} \tag{4.68}$$

valid for $\mu = \pm 1$. The phi integration is hereby accomplished. The Fast Fourier Transformation algorithm is most efficient for an even number of input values which furthermore should contain small prime factors only. Some amount of oversampling is needed in order to fulfil this requirement.

4.3.3.4 Measurement in theta: The theta integral, eqn (4.55), is

$$w^n_{\mu m}(A) = \frac{2n+1}{2} \int_{\theta=0}^{\pi} w_{\mu m}(A, \theta) d^n_{\mu m}(\theta) \sin\theta \, d\theta \tag{4.69}$$

where $w^n_{\mu m}(A) = 0$ for $|n| > N$ because of the band limitation of $w_{\mu m}(A, \theta)$ expressed through eqn (4.49) and the finite Fourier series of the rotation coefficients

$$d^n_{\mu m}(\theta) = i^{\mu - m} \sum_{m'=-n}^{n} \Delta^n_{m'\mu} \Delta^n_{m'm} \mathrm{e}^{-im'\theta} \tag{4.70}$$

where $\Delta^n_{m'\mu}$ and $\Delta^n_{m'm}$ are constants, and where the range of the angular variable is unlimited; see Appendix A2. We shall now determine $w^n_{\mu m}(A)$ by evaluating the integral (4.69) from samples in θ of $w_{\mu m}(A, \theta)$ [2]. The latter function is known for $\mu = \pm 1$ and $|m| = 0, 1, \ldots, N$ from Section 4.3.3.3.

During the measurement in theta (theta scanning), chi, phi and the measurement distance A are kept constant. The samples are taken by moving the probe in theta to positions with angular separation $\Delta\theta$ (see Fig. 4.9) where $\Delta\theta = 2\pi/J_\theta$, J_θ being the number of samples on a complete great circle. A complication arises

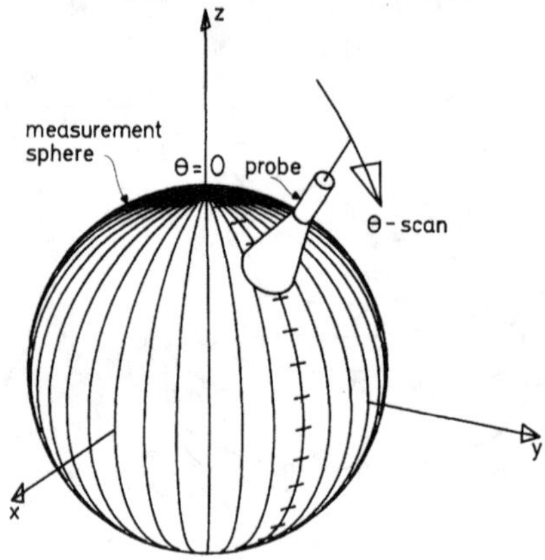

Fig. 4.9 *Measurement in theta*

here because the range of the spherical coordinate theta as an argument in $w_{\mu m}(A, \theta)$ is only $0 \leqslant \theta \leqslant \pi$. The integrand in (4.69) is therefore not a priori a periodic function and the integral cannot be computed similarly to the chi and phi integrals. However, the twice-Fourier-transformed transmission formula

$$w_{\mu m}(A, \theta) = \sum_{\substack{n=|m| \\ (n \neq 0)}}^{N} w_{\mu m}^{n}(A) d_{\mu m}^{n}(\theta) \tag{4.71}$$

cf. eqn (4.49), implies that $w_{\mu m}(A, \theta)$ can be continued in a natural way outside the range $0 \leqslant \theta \leqslant \pi$. The rotation coefficients $d_{\mu m}^{n}(\theta)$ are periodic in θ with the period 2π. They are, furthermore, even functions about $\theta = \pi$ if the number $(\mu - m)$ is even, and odd functions if $(\mu - m)$ is odd. These properties are independent of n. This means that the right-hand side of eqn (4.71) is also a periodic function (with the period 2π) which is either even or odd about $\theta = \pi$, dependent on the parity of $(\mu - m)$. Hence, the left-hand side, i.e. $w_{\mu m}(A, \theta)$ can be extended into $\pi < \theta < 2\pi$ as a function of θ with the same parity about $\theta = \pi$ as $(\mu - m)$. We define the extended data as

$$\tilde{w}_{\mu m}(A, \theta) = \begin{cases} w_{\mu m}(A, \theta), & 0 \leqslant \theta \leqslant \pi \\ w_{\mu m}(A, 2\pi - \theta), & \pi < \theta < 2\pi, \quad (\mu - m) \text{ even} \\ -w_{\mu m}(A, 2\pi - \theta), & \pi < \theta < 2\pi, \quad (\mu - m) \text{ odd.} \end{cases} \tag{4.72}$$

We note in passing that the function under the integration sign in (4.69) is the product of two functions of θ with the same parity and a third function, $\sin \theta$, which is odd. The value of the integral if computed with the upper limit extended to 2π and the extended data (4.72) used in place of $w_{\mu m}(A, \theta)$ is therefore zero.

The extended data function $\tilde{w}_{\mu m}$ is periodic in θ with the period 2π. It may therefore be expanded into the finite Fourier series

$$\tilde{w}_{\mu m}(A, \theta) = \sum_{l=-N}^{N} b_l^{\mu m} e^{il\theta} \qquad 0 \leqslant \theta < 2\pi \tag{4.73}$$

with the same number of terms as in (4.71). Insertion of (4.70) and (4.73) into (4.69) gives

$$w_{\mu m}^{n}(A) = \frac{2n+1}{2} \int_{\theta=0}^{\pi} \sum_{l=-N}^{N} b_l^{\mu m} e^{il\theta} i^{\mu-m} \sum_{m'=-n}^{n} \Delta_{m'\mu}^{n} \Delta_{m'm}^{n} e^{-im'\theta} \sin\theta \, d\theta \tag{4.74}$$

$$= \frac{2n+1}{2} i^{\mu-m} \sum_{l=-N}^{N} b_l^{\mu m} \sum_{m'=-n}^{n} \Delta_{m'\mu}^{n} \Delta_{m'm}^{n} \int_{\theta=0}^{\pi} e^{i(l-m')\theta} \sin\theta \, d\theta. \tag{4.75}$$

The integral in (4.75) is denoted by $G(l-m')$. We have

$$G(l-m') = \int_{\theta=0}^{\pi} e^{i(l-m')\theta} \sin\theta \, d\theta$$

$$= \begin{cases} \pm i\dfrac{\pi}{2}, & (l-m') = \pm 1 \\ 0, & |l-m'| = 3, 5, 7, \ldots \\ \dfrac{2}{1-(l-m')^2}, & |l-m'| = 0, 2, 4, \ldots . \end{cases} \tag{4.76}$$

In (4.75), the only unknowns on the right-hand side are now the Fourier coefficients $b_l^{\mu m}$. These may be found by inverting eqn (4.73) using the DFT relationship, ref. Appendix A4.3.3, between $b_l^{\mu m}$ and equispaced samples in $0 \leqslant \theta < 2\pi$ of the extended data $\tilde{w}_{\mu m}(A, \theta)$. With $J_\theta = 2N + 1$ samples we have

$$\{b_l^{\mu m} \mid l = 0, 1, \ldots, N, -N, \ldots, -1\}$$
$$= \mathrm{IDFT}\{\tilde{w}_{\mu m}(A, j\Delta\theta) \mid j = 0, 1, \ldots, J_\theta - 1\} \tag{4.77}$$

where $\Delta\theta = 2\pi/J_\theta$. Again, the Fast Fourier Transformation algorithm is most efficient for an even number of input values containing small prime factors only. Some amount of oversampling is needed to fulfil this requirement. Although the theta integral has now been reduced to discrete operations on the data available from the phi integral, further reductions are possible. These will be discussed in the next section.

4.3.3.5 Further reduction of the theta integral: In this section we shall show how the theta integral can be further reduced such that an efficient algorithm for its evaluation can be constructed. Equation (4.75) reads

$$w^n_{\mu m}(A) = \frac{2n+1}{2} i^{\mu-m} \sum_{l=-N}^{N} b^{\mu m}_l \sum_{m'=-n}^{n} \Delta^n_{m'\mu} \Delta^n_{m'm} G(l-m')$$

$$n = 1, \ldots, N;\ m = -n, \ldots, 0, \ldots, n;\ \mu = \pm 1 \qquad (4.78)$$

where the symbols have been defined earlier.

It is observed that

$$b^{\mu m}_l = (-1)^{\mu+m} b^{\mu m}_{-l} \qquad (4.79)$$

which is easily seen from eqns (4.72) and (4.73). Also, we have the relationship

$$\Delta^n_{m'\mu} \Delta^n_{m'm} = (-1)^{\mu+m} \Delta^n_{-m'\mu} \Delta^n_{-m'm} \qquad (4.80)$$

which is a consequence of eqn (A2.28).

From eqns (4.79) and (4.80) it then follows that

$$b^{\mu m}_l \Delta^n_{m'\mu} \Delta^n_{m'm} = b^{\mu m}_{-l} \Delta^n_{-m'\mu} \Delta^n_{-m'm}. \qquad (4.81)$$

Since further

$$G(l-m') = -G(m'-l), \qquad (l-m') = \pm 1 \qquad (4.82)$$

the terms in eqn (4.78) with $(l-m') = \pm 1$ cancel and we may write

$$w^n_{\mu m}(A) = \frac{2n+1}{2} i^{\mu-m} \sum_{m'=-n}^{n} \Delta^n_{m'\mu} \Delta^n_{m'm} \sum_{l=-N}^{N} \Pi(l-m') b^{\mu m}_l$$

$$n = 1, \ldots, N;\ m = -n, \ldots, 0, \ldots, n;\ \mu = \pm 1 \qquad (4.83)$$

where

$$\Pi(l-m') = \begin{cases} 0, & (l-m') \text{ odd} \\ \dfrac{2}{1-(l-m')^2}, & (l-m') \text{ even.} \end{cases} \qquad (4.84)$$

Note that $\Pi(l-m') = \Pi(m'-l)$.

The l-summations in eqn (4.83)

$$K(m') = \sum_{l=-N}^{N} \Pi(m'-l) b^{\mu m}_l, \qquad -N \leqslant m' \leqslant N \qquad (4.85)$$

are independent of n and must be carried out before the m'-summation in order to avoid unnecessary operations. Evaluation of $K(m')$ for all m' can be done in a straightforward manner by multiplication of a square matrix by a column matrix.

As shown by Larsen [3], considerable savings in computer time for evaluation of eqn (4.85) can be obtained by exploiting the special form of the Π-matrix. The right-hand side of eqn (4.85) resembles, for each value of m', a convolution between two sequences; see Appendix A4.2.2. Neither of the sequences are periodic and the DFT cannot immediately be used for calculation of the sum. However, the two sequences can be made periodic in the following way.

We define a new sequence $\tilde{\Pi}(j)$ obtained by truncation of the infinite Π sequence

$$\tilde{\Pi}(j) = \Pi(j), \qquad -2N < j \leqslant 2N \tag{4.86}$$

and define it to be periodic with a period of $4N$ so that $\tilde{\Pi}(j) = \tilde{\Pi}(j + c4N)$, $c = 0, \pm 1, \pm 2, \ldots$.

Likewise, let us introduce the sequence $\tilde{b}_l^{\mu m}$ obtained by extending the finite $b_l^{\mu m}$ sequence with zeros

$$\tilde{b}_l^{\mu m} = \begin{cases} b_l^{\mu m}, & -N \leqslant l \leqslant N \\ 0, & -2N < l < -N \quad \text{and} \quad N < l \leqslant 2N \end{cases} \tag{4.87}$$

and define it to be periodic with the period of $4N$ so that $\tilde{b}_l^{\mu m} = \tilde{b}_{l+c4N}^{\mu m}$, $c = 0, \pm 1, \pm 2, \ldots$.

It is now relatively easy to show that replacement of the original sequences by the periodic sequences does not change the l-sum for m' in the interval $-N \leqslant m' \leqslant N$

$$K(m') = \sum_{l=0}^{4N-1} \tilde{\Pi}(m' - l)\tilde{b}_l^{\mu m} \qquad -N \leqslant m' \leqslant N \tag{4.88}$$

and this expression has the form of a discrete convolution which can be calculated as the IDFT of the term by term product of two DFTs.

$$\begin{aligned} K(m') = \text{IDFT}\{&\text{DFT}\{\tilde{\Pi}(i) \mid i = 0, 1, \ldots, 4N\} \\ &\text{DFT}\{\tilde{b}_j^{\mu m} \mid j = 0, 1, \ldots, 4N\}\} \end{aligned} \tag{4.89}$$

cf Appendix A4.2.2. The DFT of the $\tilde{\Pi}$-sequence needs of course only to be calculated once. Furthermore, the $\tilde{\Pi}$-sequence is even, and

$$\tilde{b}_l^{\mu m} = (-1)^{\mu + m} \tilde{b}_{-l}^{\mu m}. \tag{4.90}$$

Since these properties are retained after taking discrete Fourier transforms, we conclude that

$$K(m') = (-1)^{\mu + m} K(-m') \tag{4.91}$$

a property which is used later.

As pointed out by Lewis [12], the fact that the sequences keep their parity through the discrete Fourier transform of eqn (4.77) can be used to cut the number of DFTs in half. A sequence with index m and a sequence with index $(m + 1)$ will always be of opposite parity. They can therefore be taken through the operations of eqn (4.77) simultaneously. The sequences are first added, and then, after the operations have been carried out on their sum, they are separated again; see Appendix A4, eqns (A4.9–4.10). Furthermore, as shown by Larsen [4], since the sequences keep their parity even through the convolution (4.89), it is possible here as well to cut the number of DFTs in half by the same technique.

An alternative way of obtaining the same reduction in the computations in the convolution (4.89) is to use the prior knowledge of the sequences being even (m odd) and odd (m even) [13]. For even and odd sequences, their lengths in the convolution can be reduced from $4N$ to $2N$.

Having performed the l-summations of eqn (4.83), we now consider the remaining sum

$$w^n_{\mu m}(A) = \frac{2n+1}{2} i^{\mu - m} \sum_{m'=-n}^{n} \Delta^n_{m'\mu} \Delta^n_{m'm} K(m')$$

$$n = 1, \ldots, N;\ m = -n, \ldots, 0, \ldots, n;\ \mu = \pm 1. \qquad (4.92)$$

It is observed that

$$\Delta^n_{-m'\mu} \Delta^n_{-m'm} K(-m') = \Delta^n_{m'\mu} \Delta^n_{m'm} K(m') \qquad (4.93)$$

which is a consequence of eqns (4.80) and (4.91). The summation range in eqn (4.92) can therefore be reduced to $0 \leqslant m' \leqslant n$ by multiplication by two for $m' \neq 0$.

The calculation of the deltas requires some attention. A useful recurrence relation is the following

$$\sqrt{(n+m'+1)(n-m')}\, \Delta^n_{m'+1,m} + \sqrt{(n+m')(n-m'+1)}\, \Delta^n_{m'-1,m} + 2m\Delta^n_{m'm} = 0 \qquad (4.94)$$

in which m' is the only varying parameter. The formula is numerically stable going backwards from $m' = n$ to $m' = 0$. The starting value can be given in closed form

$$\Delta^n_{nm} = 2^{-n} \sqrt{\frac{2n(2n-1)\ldots(n-m+1)}{(n-m)!}} \qquad (4.95)$$

ref. Appendix A2.4. The main advantage of (4.94) is that the deltas are calculated in the same order as they are needed. In this way intermediate storage of the three-dimensional data structure (the 'delta pyramid', Fig. A2.2) of the deltas is avoided. Notice that the recurrence for $\Delta^n_{m'\mu}$ and $\Delta^n_{m'm}$ can be done simultaneously, since the square roots are the same. The square roots of each parenthesis under the square root signs may conveniently be precomputed in a table giving $\sqrt{p}$ for $p = 1, \ldots, 2N+1$.

The above computations of $w^n_{\mu m}(A)$ must be carried out for all n, m and μ. This completes the discussion of the evaluation of the theta integral. The findings of Section 4.3.3 are summarized later (Section 4.4.2). They form the basis for the construction of the computer program outlined in Sections 4.4.4–4.4.6.

4.3.4 *Field transformations and radiated fields*

4.3.4.1 Input and output probes: In Sections 4.3.3.1–4.3.3.5, the test antenna transmitting coefficients T_{smn} have been found through a discrete, exact solution or inversion of the transmission formula

$$w(A, \chi, \theta, \phi) = \sum_{\substack{smn \\ \mu=\pm 1}} vT_{smn} e^{im\phi} d^n_{\mu m}(\theta) e^{i\mu\chi} P_{s\mu n}(kA). \qquad (4.96)$$

In this equation, v is the signal applied to the test antenna and w the signal (the *input signal*) received by the measurement probe (or *input probe*). The probe is located at a point (A, θ, ϕ) on the measurement sphere (the *input sphere*) and rotated by an angle χ as described in Chapter 3. The quantities $P_{s\mu n}(kA)$ are the probe response constants while the remaining factors on the right-hand side constitute a rotation coefficient for spherical wave functions; see Appendix A2.

The transmitting coefficients are sufficient to characterize the test antenna. However, the real interest is usually in gain and directivity or in components of the radiated field either at finite ranges or in the far field. We therefore need, as a final step in the complete measurement procedure, to derive a suitable formulation for these purposes.

A straightforward way of obtaining fields from known coefficients is to compute the relevant field expansion. The field at a finite distance can be found from the common spherical wave expansion. If the far field is wanted, a far-field expansion can be used.

However, an alternative method exists. As shown in Section 3.3.2, the determination of a field component by evaluation of a field expansion is equivalent to computing the signal received by a properly oriented short dipole located at the field point. Instead of evaluating field expansions, we may therefore equally well use the transmission formula and compute the signal received by a short dipole. We need only to insert on the right-hand side the known test antenna coefficients as well as the response constants for the short dipole.

The described way of computing a field component can be looked upon as a special case of a more general procedure. With reference to Fig. 4.10, let us assume that in addition to the input probe another probe (the *output probe*) is placed at (θ, ϕ) on an *output sphere* of radius A'. Let this probe be rotated by an angle χ and let $P'_{s\mu n}(kA')$ denote its response constants. The transfer of signals (i.e. coupling) between the given test antenna and the output probe is a quantity of general interest which can easily be found from the transmission formula. We need only to insert the various factors, which are all known, under the summation sign in eqn (4.96) and calculate the sum w' (the *output signal*).

$$w'(A', \chi, \theta, \phi) = \sum_{\substack{smn \\ \mu=\pm 1}} vT_{smn} e^{im\phi} d^n_{\mu m}(\theta) e^{i\mu\chi} P'_{s\mu n}(kA') \qquad (4.97)$$

assuming the output probe to be a $\mu = \pm 1$ probe.

The calculation of the signal received by an output probe requires that the

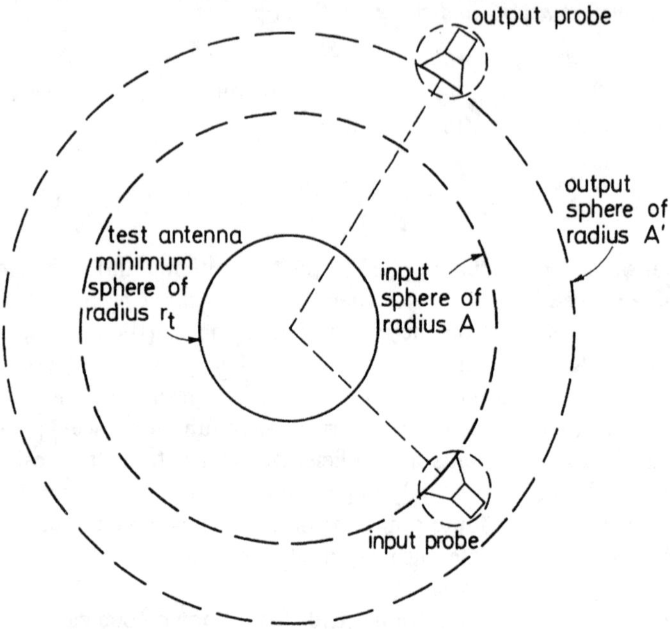

Fig. 4.10 *Illustrating a test antenna minimum sphere (radius r_t), an input sphere (radius A) with input probe and an output sphere (radius A') with output probe. The probe minimum spheres are also indicated*

minimum spheres of the test antenna and the output probe be separated by at least a few wavelengths. Otherwise numerical problems may arise in the calculation of translation coefficients. If the radius A' of the output sphere is smaller than the radius A of the input sphere, the number of modes retained when evaluating the right-hand side of the transmission formula must be carefully considered. On one hand all the significant contributions to the received output signal w' should be included. On the other hand too many modes can cause noise present in the input signal to be magnified and destroy the accuracy of the calculated result [14].

The calculations required for obtaining w' from eqn (4.97) are of course rather different from those needed for the evaluation of T_{smn} from eqn (4.96). However, due to the presence of the rotation coefficients in both equations, it is advantageous to combine the calculations in a single integrated *transformation algorithm.* Furthermore, as opposed to individual algorithms for eqns (4.96) and (4.97), a combined algorithm is structural symmetric and has the flexibility required for the solution of a variety of theoretical and experimental problems. A general transformation algorithm which transforms an input signal w defined on an input sphere to an output signal w' defined on an output sphere is outlined in Section 4.4.

In near-field to far-field transformations, the output sphere has infinitely large radius. In far-field to near-field transformations, the input sphere has infinitely large radius. Sometimes, e.g. for filtering purposes, a far-field to far-field transformation is required in which case both the input and the output sphere has infinitely large radius. The form taken by eqns (4.96–4.97) in case of infinitely large radii is discussed in Section 4.3.4.2.

4.3.4.2 Infinitely remote probes: In the special cases where the radius A of the input sphere or the radius A' of the output sphere, or both, are infinitely large, it is convenient to introduce a separate formulation. For an input probe which is linearly x'-polarized in the principal planes and where the azimuthal index μ is limited to ± 1, we have

$$P_{s,\pm 1,n}(kA) = \frac{1}{2}\sum_{\sigma\nu} C^{sn}_{\sigma,\pm 1,\nu}(kA) R^{p}_{\sigma,\pm 1,\nu} \tag{4.98}$$

where

$$P_{s,-1,n}(kA) = (-1)^{s+1} P_{s1n}(kA) \tag{4.99}$$

as described in Section 3.3.1. Introducing the large argument approximation (see eqns (A3.22–3.24)) of the translation coefficients in eqn (4.98) yields

$$P_{s,\pm 1,n}(kA) \to \frac{e^{ikA}}{kA} P^{\infty}_{s,\pm 1,n} \qquad kA \to \infty \tag{4.100}$$

where the *normalized far-field probe response constants* $P^{\infty}_{s,\pm 1,n}$ are given by

$$P^{\infty}_{s1n} = \tfrac{1}{4}\sqrt{2n+1}\; i^{-n-1} \sum_{\nu=1}^{\nu_{max}} \sqrt{2\nu+1}\; i^{\nu}\{R^{p}_{11\nu} + R^{p}_{21\nu}\} \tag{4.101}$$

and

$$P^{\infty}_{s,-1,n} = (-1)^{s+1} P^{\infty}_{s1n}. \tag{4.102}$$

In the important case where the input probe is a short x'-directed electric dipole, equations (4.101–4.102) reduce to

$$P^{e,\infty}_{s1n} = -\frac{\sqrt{6}}{8}\sqrt{2n+1}\; i^{-n} \tag{4.103}$$

$$P^{e,\infty}_{s,-1,n} = (-1)^{s}\frac{\sqrt{6}}{8}\sqrt{2n+1}\; i^{-n}. \tag{4.104}$$

For an x'-directed magnetic dipole, we state without proof

$$P^{m,\infty}_{s1n} = \frac{\sqrt{6}}{8}\sqrt{2n+1}\; i^{-n+1} \tag{4.105}$$

$$P^{m,\infty}_{s,-1,n} = (-1)^{s}\frac{\sqrt{6}}{8}\sqrt{2n+1}\; i^{-n+1}. \tag{4.106}$$

Note that the x'-directed magnetic dipole is y'-polarized.

The purpose of introducing the far-field probe response constants is of course to provide a means of dividing out the factor e^{ikA}/kA on both sides of the transmission formula. On the right-hand side of the transmission formula, $P^{\infty}_{s,\pm 1,n}$ will replace $P_{s\mu m}(kA)$ while on the left-hand side, the probe signal $w(A, \chi, \theta, \phi)$ is substituted by the *normalized far-field probe signal* $W(\chi, \theta, \phi)$ defined as

$$W(\chi, \theta, \phi) = \lim_{kA \to \infty} \left[w(A, \chi, \theta, \phi) \frac{kA}{e^{ikA}} \right]. \tag{4.107}$$

The far-field form of the transmission formula (4.96) becomes

$$W(\chi, \theta, \phi) = \sum_{\substack{smn \\ \mu = \pm 1}} v T_{smn} e^{im\phi} d^n_{\mu m}(\theta) e^{i\mu\chi} P^{\infty}_{s\mu n}. \tag{4.108}$$

The above equations pertaining to infinitely remote input probes apply equally well to infinitely remote output probes. The far-field form of the transmission formula (4.97) is

$$W'(\chi, \theta, \phi) = \sum_{\substack{smn \\ \mu = \pm 1}} v T_{smn} e^{im\phi} d^n_{\mu m}(\theta) e^{i\mu\chi} P'^{\infty}_{s\mu n}. \tag{4.109}$$

4.3.4.3 Relative measurements: In experimental antenna work, it is common practice to take relative pattern measurements only. Two different levels of relativity may be distinguished in this connection. Patterns can be measured relative to some unknown but fixed reference given, e.g. by an arbitrary setting of the measurement receiver amplification. They may also be made relative to the specific but unknown test antenna input power. Hypothetically, this may be assumed to be equally distributed in all directions hence defining the *isotropic level* or 0 dB level in the radiation pattern.

Absolute measurements are seldom required. If they are needed, e.g. for the determination of the *equivalent isotropic radiated power* (EIRP), the above relative measurements must be supplemented by an additional absolute measurement to determine the level in dBm of the applied signal source.

The formulations discussed until this point have all been in terms of absolute quantities. The purpose of the present and the following sections is to show how antenna parameters can actually be determined from relative measurements. The general tool employed in spherical near-field measurements is a transformation algorithm based upon formulas (4.96–4.97) and (4.108–4.109). A simple diagram for such an algorithm is shown in Fig. 4.11(a). More details of a particular algorithm are contained in Section 4.4.

The input to the algorithm consists of samples of the signal w measured by a calibrated input probe. If the input sphere is specified to have infinitely large radius, the input must instead be the normalized far-field probe signal W. The output signal w' is computed at points on the output sphere and is a prediction

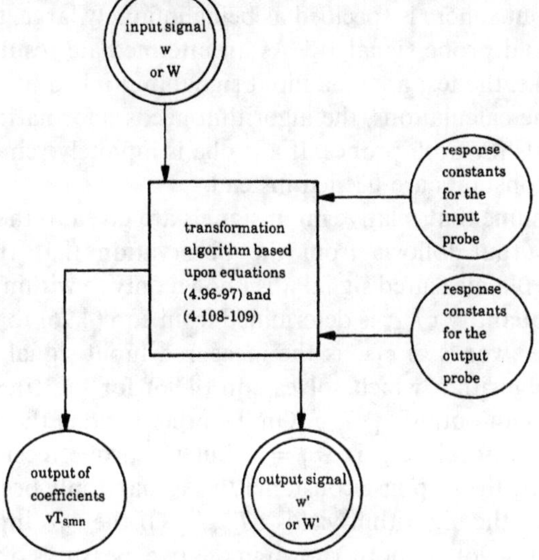

Fig. 4.11(a) *Simple diagram for a transformation algorithm. Inputs are the measured finite range data w or the far-field data W. Outputs are the finite range signal w' or the far-field signal W'. The algorithm needs information as to the response constants of the input and output probes and supplies the test antenna coefficients as an additional output*

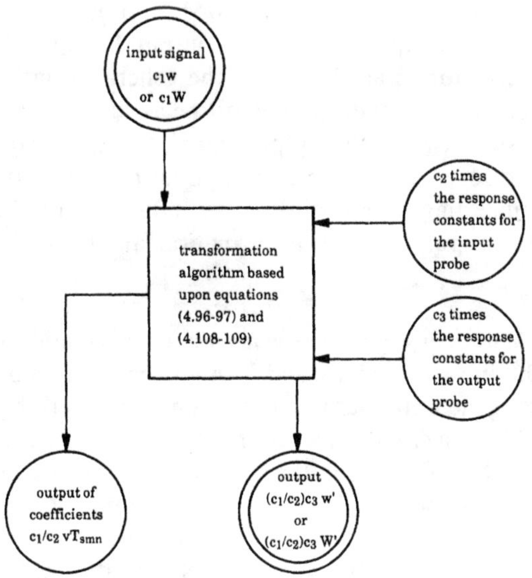

Fig. 4.11(b) *Illustrating the algorithm of Figure 4.11(a) when relative input quantities are applied*

of the signal that would be measured there using a calibrated (known) output probe. If the output sphere is specified as being infinitely large, the output is the normalized far-field probe signal W'. As an intermediate result, the algorithm computes vT_{smn}, i.e. the test antenna input signal multiplied by the transmitting coefficients. In the calculations, the algorithm needs information on the probe response constants for both probes. If a probe is infinitely remote, the far-field probe response constants are needed instead.

Let us now assume that relative input signals are given to the algorithm. The corresponding output follows from the observation that the transmission formula is linear. If a measured signal w is known only to within a multiplicative constant c_1, the quantity vT_{smn} is determined from eqn (4.96) to within the same constant. In other words, if $c_1 w$ is the measured input signal, the part of the transformation algorithm which solves eqn (4.96) for the transmitting coefficients produces as an output $c_1 vT_{smn}$. On the other hand, if the measured signal has been measured absolutely, i.e. $c_1 = 1$, but the probe receiving coefficients $R^p_{\sigma,\pm 1,\nu}$ and thereby the response constants $P_{s,\pm 1,n}$ have only been determined to within a factor c_2, the algorithm yields vT_{smn}/c_2. If the two input quantities w and $P_{s,\pm 1,n}$ of eqn (4.96) are both known in relative measures only, i.e. to within the factors c_1 and c_2, respectively, the computations lead to $vT_{smn}c_1/c_2$.

Similar considerations can be made in connection with eqn (4.97) for the output signal w' (or eqn (4.109) for W'). Here, the known quantities are vT_{smn} and $P'_{s,\pm 1,n}$ (or $P'^{\infty}_{s,\pm 1,n}$). Of these, vT_{smn} may have been determined apart from a factor c_1/c_2 as discussed above. The output probe response constant $P'_{s,\pm 1,n}$ is of course well known in such computations where a theoretical antenna like an electric short dipole is used as the output probe. In other cases, the same probe may have been chosen for input as well as for output. Here $P'_{s,\pm 1,n}$ will be known to within the same factor c_2 as $P_{s,\pm 1,n}$. In the general situation, however, the response constants for the output probe must be assumed unknown to within a multiplicative constant c_3. As the output signal, therefore, eqn (4.97) and the corresponding part of the transformation algorithm provides the quantity $w'(c_1/c_2)c_3$. A simple diagram illustrating the influence of unknown constants on the input and output of the transformation algorithm is shown in Fig. 4.11(b).

4.3.4.4 Determination of antenna parameters: In this section we shall address the question of obtaining directivity and gain for the test antenna, as well as its radiated fields, from the data measured in a spherical near-field set-up.

Let us first deal with a situation where relative measurements are made. The signal received by the probe during the spherical near-field scan is then $c_1 w$ as discussed in the previous section. The response constants of the input probe are obtained in a probe calibration scheme and are assumed to be $c_2 P_{s,\pm 1,n}$. For the output probe let us select an infinitely remote electric dipole. For such a probe, the far-field response constants can be found from eqns (4.103–4.104) without unknown constants, $c_3 = 1$. With the given inputs, the outputs of the transformation algorithm are $vT_{smn}c_1/c_2$ and $W'c_1/c_2$. The directivity in the direction (θ, ϕ) of that polarization component specified by the output probe orientation angle χ can now be found from Friis' transmission formula (3.71)

$$\frac{G_p D_t}{4} = \frac{\frac{1}{2}|W'|^2}{\frac{1}{2}\sum_{smn} |vT_{smn}|^2} \tag{4.110}$$

$$= \frac{\frac{1}{2}\left|W'\frac{c_1}{c_2}\right|^2}{\frac{1}{2}\sum_{smn}\left|vT_{smn}\frac{c_1}{c_2}\right|^2}. \tag{4.111}$$

In other words, expressed in measured and computed relative quantities, the test antenna directivity in the required polarization component is given by

$$D_t = \frac{8}{3}\frac{\left|W'\frac{c_1}{c_2}\right|^2}{\sum_{smn}\left|vT_{smn}\frac{c_1}{c_2}\right|^2} \tag{4.112}$$

since the probe gain G_p for a short electric dipole is $\frac{3}{2}$.

As a second example, we consider a measurement where the signal received during the near-field scanning is measured relative to the test antenna input power. First, a direct connection between the signal source and the receiver is established. Taking into account the various mismatch factors, an unknown constant c_1 times the test antenna input signal v is thereby found. Next, the near-field scanning is performed with the receiver settings untouched. This yields the signal received by the input probe to within the same constant $c_1 w$. Let us assume that the response constants of the input probe have been found absolutely by applying a suitable calibration procedure, i.e. $c_2 = 1$. For the output probe, we select an infinitely remote electric dipole. The response constants of a dipole are well known so that $c_3 = 1$. With the inputs mentioned, the outputs of the transformation algorithm are $c_1 W'$ and $c_1 vT_{smn}$. We now have from eqn (3.69)

$$\frac{G_p G_t}{4} = \frac{\frac{1}{2}|W'|^2}{\frac{1}{2}|v|^2} \tag{4.113}$$

$$= \frac{\frac{1}{2}|c_1 W'|^2}{\frac{1}{2}|c_1 v|^2}. \tag{4.114}$$

Expressed in measured and computed quantities, the test antenna gain in the specified polarization component is given by

$$G_t = \frac{8}{3}\frac{|c_1 W'|^2}{|c_1 v|^2} \tag{4.115}$$

since the gain of a short electric dipole probe is $\frac{3}{2}$. Note, that since $c_1 v$ is known from a separate measurement, the test antenna coefficients T_{smn} can also be determined from the algorithm output. If the equivalent isotropic radiated power (EIRP) is sought, an additional absolute measurement of $\frac{1}{2}|v|^2$ is required and

$$\text{EIRP}(\theta, \phi) = \tfrac{1}{2}|v|^2 G_t(\theta, \phi). \tag{4.116}$$

As the third example, we seek the absolute far-field pattern $\vec{K}(\theta, \phi)$ of the test antenna. From eqn (3.54), this is given in terms of the normalized far-field signal as follows

$$\vec{K}(\theta, \phi) = \frac{2\sqrt{6}}{3v}\left\{W'(0, \theta, \phi)\hat{\theta} + W'\left(\frac{\pi}{2}, \theta, \phi\right)\hat{\phi}\right\} \tag{4.117}$$

where $W'(0, \theta, \phi)$ and $W'(\pi/2, \theta, \phi)$ are the normalized far-field output probe signals in a theta- and a phi-oriented short electric dipole, respectively. To determine $\vec{K}(\theta, \phi)$, the knowledge of v and W' to within the same constant c_1 is obviously sufficient. Hence, the same measurements as in the previous example can be used. W' is required for both orientations $\chi = 0$ and $\chi = \pi/2$ of the output probe.

Instead of using an output probe in the shape of an electric dipole, a general directive output probe may be applied. It is clear from the Friis transmission formula (eqn (3.57)) that *any* remote linearly polarized probe of gain G_p will receive a signal proportional to the test antenna far field and

$$\vec{K}(\theta, \phi) = \frac{2\sqrt{6}}{3v}\left\{W'(0, \theta, \phi)\hat{\theta} + W'\left(\frac{\pi}{2}, \theta, \phi\right)\hat{\phi}\right\}\sqrt{\frac{G_e}{G_p}} \tag{4.118}$$

where $G_e = \frac{3}{2}$ is the gain of a short electric dipole.

The amplitude of the absolute far-field pattern $\vec{K}(\theta, \phi)$ equals unity at the isotropic level, i.e. where $G_t(\theta, \phi)$ is equal to one. If this feature is not required, a relative far-field pattern $\vec{K}_{\text{rel}}$ may be sufficient. We define

$$\vec{K}_{\text{rel}}(\theta, \phi) = c\left\{W'(0, \theta, \phi)\hat{\theta} + W'\left(\frac{\pi}{2}, \theta, \phi\right)\hat{\phi}\right\} \tag{4.119}$$

where c is an arbitrary constant.

Finally, let us consider the tangential components of the electric and magnetic fields at a point (A', θ, ϕ) of the test antenna field. We have

$$\vec{E}_{\text{tang}}(A', \theta, \phi) = \frac{2k}{\sqrt{6\pi\eta}}\left\{w'^e(A', 0, \theta, \phi)\hat{\theta} + w'^e\left(A', \frac{\pi}{2}, \theta, \phi\right)\hat{\phi}\right\} \tag{4.120}$$

$$\vec{H}_{\text{tang}}(A', \theta, \phi) = \frac{2k\sqrt{\eta}}{\sqrt{6\pi}}\left\{w'^m(A', 0, \theta, \phi)\hat{\theta} + w'^m\left(A', \frac{\pi}{2}, \theta, \phi\right)\hat{\phi}\right\} \tag{4.121}$$

cf eqns (3.45–3.48). Obviously, a determination of $\vec{E}_{\text{tang}}$ and $\vec{H}_{\text{tang}}$ requires that the output probe signals w' are known absolutely without multiplicative constants. We shall not go into detail here with the various schemes that can be devised for this purpose.

4.4 Outline of a spherical near-field transformation algorithm

4.4.1 Introduction

This section presents an outline of a general computer code for spherical near-field transformations with probe correction [4]. The code has already been touched upon briefly in Section 4.3.4, where the concepts of the input probe, the output probe, the input sphere and the output sphere were introduced.

For easy reference, a summary of the theory underlying the code is given in Section 4.4.2. Some remarks regarding the number of sample points follow in Section 4.4.3 and a detailed block diagram for the code is then described in Section 4.4.4. In Section 4.4.5 the numerical processes generated by the code are illustrated step by step for the simple but non-trivial case where an artificial input signal is sampled by an infinitely remote Huygens source probe. Section 4.4 concludes with some results from a test case (Section 4.4.6) that illustrate the accuracy of the computer code.

4.4.2 Summary of the theory underlying a transformation algorithm

The present section summarizes the equations behind an integrated spherical near-field algorithm. As discussed in Section 4.3.4.1, the functions of such an algorithm are twofold. In one part of the algorithm, the unknown test antenna coefficients T_{smn} are determined from the transmission formula (ref. e.g. Section 4.3.4.1)

$$w(A, \chi, \theta, \phi) = \sum_{\substack{smn \\ \mu=\pm 1}} v T_{smn} \, e^{im\phi} \, d^n_{\mu m}(\theta) \, e^{i\mu\chi} \, P_{s\mu n}(kA) \tag{4.122}$$

by a series of discrete linear operations on the signal $w(A, \chi, \theta, \phi)$ as received by the measurement probe (input probe) on the measurement sphere (input sphere). These operations have been studied in great detail in Sections 4.3.3.1–4.3.3.5. In a second, and much simpler part of the algorithm, the coefficients are reinserted into the transmission formula. This permits the evaluation of an output signal $w'(A',\chi,\theta,\phi)$ as it would be measured on the output sphere by the output probe. The output signal is given by

$$w'(A', \chi, \theta, \phi) = \sum_{\substack{smn \\ \mu=\pm 1}} v T_{smn} \, e^{im\phi} \, d^n_{\mu m}(\theta) \, e^{i\mu\chi} \, P'_{s\mu n}(kA') \tag{4.123}$$

ref. Section 4.3.4.1. Situations where either the radius A of the measurement sphere or the radius A' of the output sphere, or both, are infinite, must be treated as special cases. The governing equations here which replace (4.122) and (4.123) are

$$W(\chi, \theta, \phi) = \sum_{\substack{smn \\ \mu=\pm 1}} v T_{smn} e^{im\phi} d^n_{\mu m}(\theta) e^{i\mu\chi} P^{\infty}_{s\mu n} \tag{4.124}$$

and

$$W'(\chi, \theta, \phi) = \sum_{\substack{smn \\ \mu=\pm 1}} v T_{smn} e^{im\phi} d^n_{\mu m}(\theta) e^{i\mu\chi} P'^{\infty}_{s\mu n} \tag{4.125}$$

respectively, as discussed in Section 4.3.4.2. The two equations (4.124–4.125) are the limiting forms of (4.122–4.123) when the radii A and A', respectively, tend to infinity, and the factors e^{ikA}/kA and $e^{ikA'}/kA'$, respectively, are cancelled on both sides of the equation signs. The resulting normalized far-field probe response constants $P^{\infty}_{s\mu n}$ and $P'^{\infty}_{s\mu n}$ as well as the normalized far-field probe signals W and W' have been introduced and discussed in Section 4.3.4.2.

In brief, a transformation algorithm takes as its input either a signal w measured at a finite range A or a normalized signal W measured in the far field. The output is either a signal w' evaluated at a finite range A' or a normalized signal W' evaluated at infinity. Hence, four different modes of operation of the algorithm are possible. In all four, the transmitting coefficients T_{smn} for the test antenna are computed in the first part of the algorithm.

The calculations required in the four modes are very similar. For brevity, therefore, only the mode where both radii are finite will be considered in detail here.

First part of the algorithm (*evaluation of transmitting coefficients*): The radius A of the measurement sphere must satisfy the requirement that $kA > kr_t + kr_p + 10$ where r_t and r_p are the radii of the test antenna and the probe minimum spheres, respectively. The indices (s, m, n) of the test antenna coefficients T_{smn} are then limited to $s = 1, 2$, $m = -N, \ldots, 0, \ldots, N$ and $n = |m|, \ldots, N$, $(n \neq 0)$, where $N = kr_t + 10$. Even though the measurement radius A is a constant, we use the notation $w(A, \chi, \theta, \phi)$ for the received signal (the measured data) as a function of (χ, θ, ϕ) on the measurement sphere. The probe is assumed to be a symmetric x'-polarized, $\mu = \pm 1$ probe.

In the spherical scanning process, the first step is at an arbitrary measurement point (θ, ϕ) to rotate the probe in χ and take two samples of w; one for $\chi = 0$ and one for $\chi = \pi/2$. From this data, we are able to compute (Section 4.3.3.2) the transformed data $w_{\pm 1}(A, \theta, \phi)$ by

$$w_{\pm 1}(A, \theta, \phi) = \frac{1}{2}\left\{w(A, 0, \theta, \phi) \mp iw\left(A, \frac{\pi}{2}, \theta, \phi\right)\right\}. \tag{4.126}$$

The next step is to repeat the chi measurement at J_ϕ equispaced points $\phi = j\Delta\phi$, $j = 0, 1, \ldots, J_\phi - 1$ in the range $0 \leqslant \phi < 2\pi$, where $\Delta\phi = 2\pi/J_\phi$ is the sampling interval in ϕ. Based upon the results obtained (Section 4.3.3.3) we calculate the twice-transformed data $w_{\pm 1m}(A, \theta)$, $m = -N, ..., 0, ..., N$, by

$$\{w_{\pm 1m}(A, \theta) \,|\, m = 0, 1, ..., N, -N, ..., -1\}$$
$$= \text{IDFT}\{w_{\pm 1}(A, \theta, j\Delta\phi) \,|\, j = 0, 1, ..., J_\phi - 1\}. \qquad (4.127)$$

Finally, the above two-step procedure is repeated at equispaced points in theta. The data is collected in the natural interval $0 \leqslant \theta \leqslant \pi$. However, an extension of the data into an extended range $0 \leqslant \theta < 2\pi$ must be defined, as described in Section 4.3.3.4. This is done in the computer for each sample point, according to eqn (4.72). The extended data samples $\tilde{w}_{\pm 1m}(A, j\Delta\theta)$, $j = 0, 1, ..., J_\theta - 1$, where $\Delta\theta = 2\pi/J_\theta$, are used to compute the Fourier coefficients $b_l^{\pm 1m}$

$$\{b_l^{\pm 1m} \,|\, l = 0, 1, ..., N, -N, ..., -1\}$$
$$= \text{IDFT}\{\tilde{w}_{\pm 1m}(A, j\Delta\theta) \,|\, j = 0, 1, ..., J_\theta - 1\} \qquad (4.128)$$

for all $m = -N, ..., 0, ..., N$. With the Fourier coefficients at hand we are able to compute from equation (4.83)

$$w^n_{\pm 1m}(A) = \pm \frac{2n+1}{2} i^{1-m} \sum_{m'=-n}^{n} \Delta^n_{m',\pm 1} \Delta^n_{m'm} \sum_{l=-N}^{N} \Pi(l-m') b_l^{\pm 1m}$$
$$m = -N, ..., 0, ..., N;\; n = |m|, ..., N, (n \neq 0). \qquad (4.129)$$

To this end, the l-summation is recast in the shape of a convolution summation

$$K(m') = \sum_{l=0}^{4N-1} \tilde{\Pi}(l-m') \tilde{b}_l^{\pm 1m} \qquad (4.130)$$

where $\tilde{\Pi}(l-m')$ and $\tilde{b}_l^{\pm 1m}$ are sequences obtained from $\Pi(l-m')$ and $b_l^{\pm 1m}$, respectively, by truncations and period extensions as detailed in Section 4.3.3.5. The convolution summation can be efficiently calculated by the Fast Fourier Transform

$$K(m') = \text{IDFT}\{\text{DFT}\{\tilde{\Pi}(i) \,|\, i = 0, 1, ..., 4N\}$$
$$\text{DFT}\{\tilde{b}_j^{\pm 1m} \,|\, j = 0, 1, ..., 4N\}\}. \qquad (4.131)$$

The remaining m'-summation

$$w^n_{\mu m}(A) = \frac{2n+1}{2} i^{\mu-m} \sum_{m'=-n}^{n} \Delta^n_{m'\mu} \Delta^n_{m'm} K(m'), \qquad \mu = \pm 1 \qquad (4.132)$$

is then computed using a recurrence in m' for the deltas.

The two linear equations (4.53–4.54)

$$vT_{1mn} P_{11n}(kA) + vT_{2mn} P_{21n}(kA) = w^n_{1m}(A) \qquad (4.133)$$

$$vT_{1mn} P_{1,-1,n}(kA) + vT_{2mn} P_{2,-1,n}(kA) = w^n_{-1m}(A). \qquad (4.134)$$

in the two unknowns vT_{1mn} and vT_{2mn} may now be solved for all values of (m, n) in the ranges $m = -N, ..., 0, ..., N$ and $n = |m|, ..., N$, $(n \neq 0)$. In (4.133–4.134),

v denotes the test antenna input signal amplitude and $P_{s\mu n}$ are the probe response constants; see eqn (4.39). In order to find T_{1mn} and T_{2mn}, the signal v applied to the test antenna must be measured separately and divided out. However, a knowledge of the product vT_{smn} is sufficient in many cases, as discussed in Section 4.3.4.3.

Second part of the algorithm (*evaluation of output signal*): Insertion of vT_{smn} and evaluation of the right-hand side of eqn (4.123) is straightforward. Let us rewrite eqn (4.123) by introducing the Fourier series representation of the rotation coefficients (see eqn (A2.12))

$$w'(A', \chi, \theta, \phi) = \sum_{\mu=\pm 1} e^{i\mu\chi} \sum_{m=-N}^{N} e^{im\phi} \sum_{m'=-N}^{N} e^{im'\theta}$$

$$\sum_{n=\max(|m'|,|m|,1)}^{N} \Delta^{n}_{m'\mu} \Delta^{n}_{m'm} i^{m-\mu} \sum_{s=1}^{2} vT_{smn} P'_{s\mu n}(kA'). \qquad (4.135)$$

The computations can conveniently be carried out from right to left. It is noted that the delta products needed for each (m, n) are the same as those used in the computation of the coefficients vT_{smn} in the first part of the algorithm. In a computer code, it is therefore efficient to make use of the delta product by computing the relevant term of the n-summation in eqn (4.135) before proceeding in the first step of the code to the computation of T_{smn} for the next value of n. The code outlined in Sections 4.4.4–4.4.5 utilizes this principle of integrating the computations in the two steps. The m'-, m- and μ-summations may all be carried out efficiently for the required number of directions (θ, ϕ) and probe orientations χ, respectively, by using DFT techniques; see Appendix A4.

4.4.3 On the number of sample points

Input sampling: The process summarized in Section 4.4.2 completely defines the *input sampling* scheme which must be followed during the scanning of the test antenna near field. The sampling is equidistant in theta and phi, and is done with two probe orientations ($\chi = 0$ and $\chi = \pi/2$) at each sampling point. Assuming a full mode set (s, m, n) for the test antenna, where $s = 1, 2$, $n = 1, 2, \ldots, N$ and $m = -n, \ldots, 0, \ldots, n$ the total number of modes is given by (cf Appendix A1.1.2)

$$N_0 = 2N(N + 2). \qquad (4.136)$$

The number of samples required for the determination of all mode coefficients may be estimated as follows.

The minimum number of theta-samples J_θ which must be available is $J_\theta = 2N + 1$ (see Section 4.3.3.4). Of these, however, only $N + 1$ samples are real measurements taken in the natural interval $0 \leqslant \theta \leqslant \pi$. The remaining N samples are generated in the computer. The minimum number of phi-samples is $J_\phi = 2N + 1$ taken in $0 \leqslant \phi < 2\pi$. The minimum number of chi-samples is 2.

Hence, the total number of samples on the measurement sphere is

$$N_1 = 2(2N + 1)(N + 1). \tag{4.137}$$

For N large this is approximately equal to twice the total number of modes N_0 in the full mode set. This may be surprising, but the reason for the low sampling efficiency (~ 50 per cent) is that we have assumed $J_\phi = 2N + 1$ for all theta. This leads to an unnecessarily high density of sample points near the poles $\theta = 0$ and $\theta = \pi$.

An alternative sampling strategy is to apply some degree of thinning of the closely spaced samples near the poles. If instead of $2N + 1$ samples we use $(2N + 1)\sin\theta$ samples on each circle of latitude, the interspacing (measured in wavelengths) between sampling points in phi remains constant for all values of theta. Since the average value of $\sin\theta$ in the interval $0 \leqslant \theta \leqslant \pi$ is $2/\pi$, the average number of samples on the circles of latititude becomes $(2N + 1)2/\pi$ in this way. With thinned sampling the total number of samples on the sphere is therefore given by

$$N_2 = \frac{2}{\pi}N_1 = \frac{4}{\pi}(2N + 1)(N + 1). \tag{4.138}$$

Of course, with $(2N + 1)\sin\theta$ samples on a circle of latitude, only harmonics with $|m| < N\sin\theta$ can be determined from the phi integral. On the other hand, due to the cut-off property of the associated Legendre functions (see Section 2.2.3) the remaining harmonics have low amplitudes and contribute negligibly to the test antenna's near and far fields.

The mode limit N corresponds to a minimum sphere of radius $r_t \cong N/k$ for the test antenna. The above sampling criteria yields a distance of $\lambda/2$ between sampling points if these are projected on the minimum sphere.

Further reduction in the number of samples may be obtained in cases where a priori information of the test antenna is available. Let us first assume an antenna geometry which is such that it can be enclosed by a *minimum cylinder* with its axis coinciding with the z-axis and with a radius r_c which is smaller than the radius of the minimum sphere. The cylindrical mode expansion involves exponential functions $\exp(\pm im\phi)$ where ϕ is the cylindrical azimuth coordinate. This expansion may be truncated at $|m| = M$, where $M = kr_c + 10$. However, the same limit applies for the spherical modes which also have the exponential dependence $\exp(\pm im\phi)$. Hence, with $M = kr_c + 10$ we have for the total number of samples on the measurement sphere:

$$N_3 = 2(2M + 1)(N + 1). \tag{4.139}$$

Another case in which the mode content for the test antenna is limited at $|m| = M$, where $M < N$, is the rotational symmetric antenna where the axis of rotation coincides with the z-axis and where the excitation contains the harmonics $\exp(\pm im\phi)$, $m = 0, 1, \ldots, M$ only. The rotational symmetric antenna structure will not create components of the spherical wave field other than those already present in the excitation. The total number of input samples on the measurement sphere is again given by eqn (4.139).

Output sampling: The output signal may of course be evaluated in any number of points and no sampling criteria exist in this connection. A natural procedure, however, is to compute the output for sufficiently many points on the output sphere to retain all information (i.e. all modes) present in the input signal. In this way, the output may replace the input signal in the computer memory. If, at a later stage, a special representation in the shape of, for example, a contour

Table 4.4 *Input parameters for the computer code SNIFT performing spherical wave transformations.*

J_θ	Number of samples in $0 \leqslant \theta < 2\pi$ of the input signal w (or W).
J_ϕ	Number of samples in $0 \leqslant \phi < 2\pi$ of the input signal w (or W).
A/λ	Radius of input sphere (measurement distance) in wavelengths. If the input signal is W (far field measurement) this parameter is not used by the code.
ν_{max}	Maximum value of ν for the receiving coefficients of the input probe.
N	Maximum value of n in the coefficients T_{smn} of the spherical wave expansion to be fitted to the input data. N must be specified in the interval $1 \leqslant N \leqslant (J_\theta - 1)/2$.
M	Maximum value of m in the coefficients T_{smn} of the spherical wave expansion to be fitted to the input data. M must be specified in the interval $0 \leqslant M \leqslant \min((J_\phi - 1)/2, N)$.
J'_θ	Number of samples in $0 \leqslant \theta < 2\pi$ of the output signal w' (or W').
J'_ϕ	Number of samples in $0 \leqslant \phi < 2\pi$ of the output signal w' (or W').
A'/λ	Radius of output sphere (evaluation distance) in wavelengths. If the output signal is W' (far-field evaluation) this parameter is not used by the code.
ν'_{max}	Maximum value of ν' for the receiving coefficients of the output probe.

Control parameter for selecting one of the four modes of operation

(1) $w \to w'$, (near field to near field)	(2) $w \to W'$, (near field to far field)	(3) $W \to w'$, (far field to near field)	(4) $W \to W'$ (far field to far field)

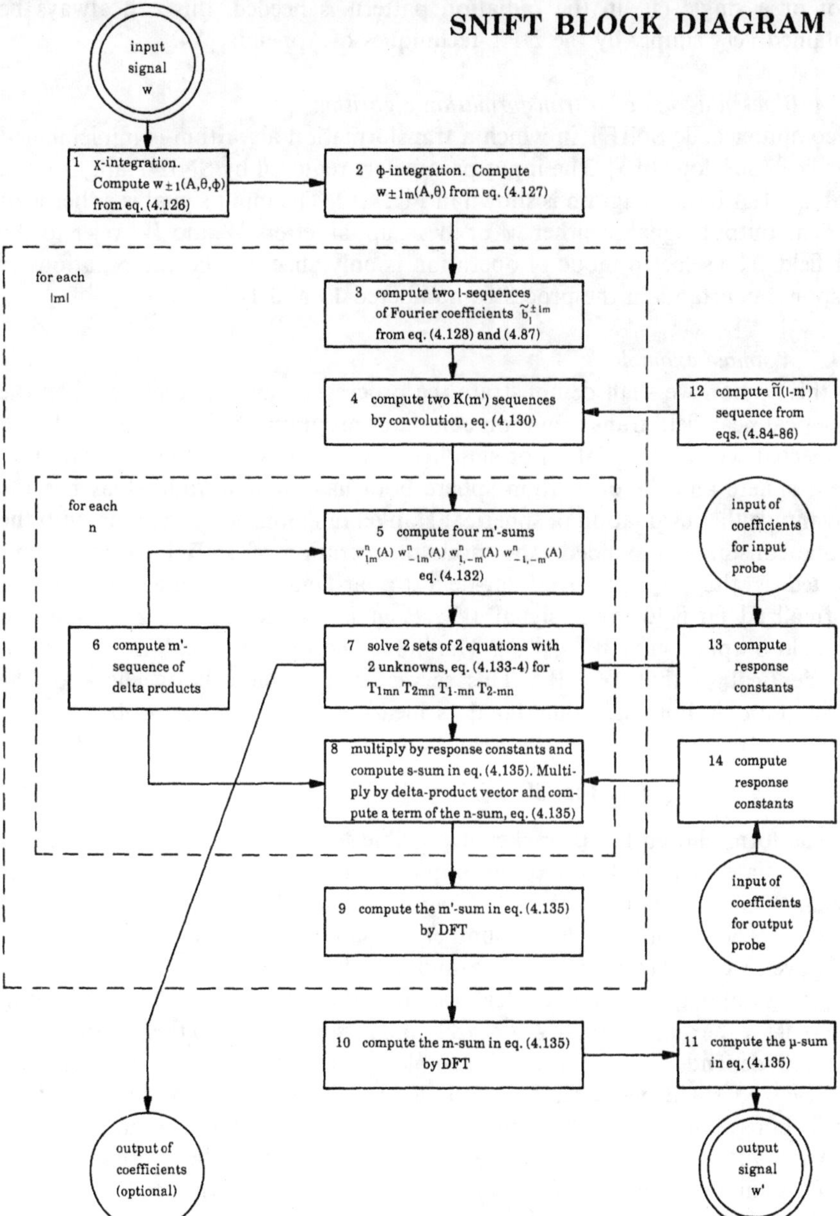

Fig. 4.12 *Block diagram for spherical near-field transformation computer code SNIFT Even though the near-field mode of operation ($w \rightarrow w'$) is shown, the diagram is representative for all four cases*

plot or a single cut in the radiation pattern is needed, this can always be obtained very simply by the DFT-techniques of Appendix A4.

4.4.4 Block diagram of a transformation algorithm

A computer code SNIFT in which a transformation algorithm is implemented has been developed [3]. The input parameters required by SNIFT are listed in Table 4.4. A block diagram is shown in Fig. 4.12. The input signal is either w or W. The output signal is either w' or W'. Capital letters W and W' refer to the far field. The selected mode of operation is only affecting the computations of response constants in the processes numbered 13 and 14.

4.4.5 A simple example

In this section we shall demonstrate the processes which take place when the spherical near-field transformation computer program SNIFT is executed with a selected set of input data. For simplicity, we shall assume that the measurement sphere and the evaluation sphere both have infinite radii. This has the advantage that evaluation of spherical Hankel functions in connection with the probe constants is avoided. The mode of operation of SNIFT to be demonstrated is thus a non-trivial far-field to far-field transformation where a normalized far-field input signal $W(\chi, \theta, \phi)$ is transformed to a normalized far-field output signal $W'(\chi, \theta, \phi)$. The input and output probes are chosen to be identical so that $W = W'$. This serves as a useful validity check of the computations. For the input signal as measured by the input probe we shall choose a trigonometric function continuous on the measurement sphere

$$W(\chi, \theta, \phi) = (20\cos 2\theta + 32\cos\theta + 12)\cos(\chi + \phi). \tag{4.140}$$

The form chosen for W makes it possible to express all data samples and intermediate results solely in terms of integers and square roots of integers. The signal W is band limited with $N = 2$ and $M = 1$ so relatively few harmonics are present. The theoretical least number of samples in theta is $2N + 1 = 5$. However, we shall choose here six samples in theta ($0 \leqslant \theta < 360°$) corresponding to a sampling interval of 60 degrees. Four of these samples fall in the natural interval for the angle theta $0 \leqslant \theta \leqslant 180°$ and form the input to the program. For the phi dependence, $2M + 1 = 3$ samples in the interval $0 \leqslant \phi < 360°$ are sufficient. A slight oversampling shall also be applied in phi, however, and we shall choose four samples here. Finally, in chi, we choose the two samples $\chi = 0$ and $\chi = 90°$ which are sufficient when a $\mu = \pm 1$ input probe is used.

The sampling intervals, 60° in theta and 90° in phi, yield an *even length* for all DFTs and IDFTs. In the previous sections, the number of samples in theta and phi has always been the theoretical least, i.e. odd. In the example here it is felt, however, that even length sampling sequences would be appropriate. The reason is that practical near-field measurements almost exlusively employ even length sequences. This is due to the fact that the FFT algorithm is most efficient for an even number of samples. The total number of input samples becomes 32. Let us arrange the input in a table as shown in Table 4.5(a). With this set of input data, we shall now go through the program step by step following the block diagram in Fig. 4.12. First, the chi integration or conversion to circular polarization components is performed in step 1. The results are given in Table 4.5(b). The phi integration yields the results given in Table 4.5(c).

The $|m|$-loop of Fig. 4.12 is now entered with the initial value $|m| = 1$. Each of the $m = \pm 1$ columns in Table 4.5(c) should be treated separately. However, we need only consider $m = 1$, i.e. $W_{11}(\theta)$, since the column $W_{-1,-1}(\theta)$ gives exactly

Table 4.5(a) $W(\chi, \theta, \phi)$.

$\phi =$	0°		90°		180°		270°	
$\chi =$	0°	90°	0°	90°	0°	90°	0°	90°
$\theta = 0°$	64	0	0	−64	−64	0	0	64
60°	18	0	0	−18	−18	0	0	18
120°	−14	0	0	14	14	0	0	−14
180°	0	0	0	0	0	0	0	0

Table 4.5(b) $W_\mu(\theta, \phi)$.

$\phi =$	0°		90°		180°		270°	
$\mu =$	1	−1	1	−1	1	−1	1	−1
$\theta = 0°$	32	32	$32i$	$-32i$	−32	−32	$-32i$	$32i$
60°	9	9	$9i$	$-9i$	−9	−9	$-9i$	$9i$
120°	−7	−7	$-7i$	$7i$	7	7	$7i$	$-7i$
180°	0	0	0	0	0	0	0	0

Table 4.5(c) $W_{\mu m}(\theta)$.

$m =$	0		1		±2		−1	
$\mu =$	1	−1	1	−1	1	−1	1	−1
$\theta = 0°$	0	0	32	0	0	0	0	32
60°	0	0	9	0	0	0	0	9
120°	0	0	−7	0	0	0	0	−7
180°	0	0	0	0	0	0	0	0

the same result. In step 3, the following operations on $W_{11}(\theta)$ are performed. We have

$$\{W_{11}(j60°) \mid j = 0, 1, 2, 3\} = \{32, 9, -7, 0\}. \tag{4.141}$$

Extension into the range $0 \leqslant \theta < 360°$ gives

$$\tilde{W}_{11} = \{32, 9, -7, 0, -7, 9\}. \tag{4.142}$$

The IDFT hereof is†

$$b_l^{11} = \{6, 8, 5, 0, 5, 8\} \tag{4.143}$$

which is subsequently extended with zeros to obtain the l-sequence

$$\tilde{b}_l^{11} = \{6, 8, 5, 0, 0, 0, 0, 0, 0, 0, 5, 8\}. \tag{4.144}$$

In step 12, the following sequence of length 12 is generated

$$\tilde{\Pi} = \{2, 0, -\tfrac{2}{3}, 0, -\tfrac{2}{15}, 0, -\tfrac{2}{35}, 0, -\tfrac{2}{15}, 0, -\tfrac{2}{3}, 0\}. \tag{4.145}$$

In step 4, the convolution of (4.144) and (4.145) is performed as follows. The DFTs of (4.144) and (4.145) are, respectively,

$$\{32, 11 + 8\sqrt{3}, 9, -4, -7, 11 - 8\sqrt{3}, 0, 11 - 8\sqrt{3}, -7, -4, 9, 11 + 8\sqrt{3}\} \tag{4.146}$$

and

$$\{\tfrac{36}{105}, \tfrac{160}{105}, \tfrac{288}{105}, \tfrac{328}{105}, \tfrac{288}{105}, \tfrac{160}{105}, \tfrac{36}{105}, \tfrac{160}{105}, \tfrac{288}{105}, \tfrac{328}{105}, \tfrac{288}{105}, \tfrac{160}{105}\}. \tag{4.147}$$

Note, that (4.146) is nothing but an interpolation in theta of the previous data W_{11}, ref. eqn (4.141). Multiplying the two DFTs together and taking the IDFT of the product gives

$$\{\tfrac{16}{3}, \tfrac{32}{3}, \tfrac{16}{3}, -\tfrac{32}{5}, -\tfrac{464}{105}, -\tfrac{32}{21}, -\tfrac{176}{105}, -\tfrac{32}{21}, -\tfrac{464}{105}, -\tfrac{32}{5}, \tfrac{16}{3}, \tfrac{32}{3}\}. \tag{4.148}$$

The first three and the last two numbers form the $K(m')$ sequence with $m = 1$

$$\{K(m') \mid m' = 0, 1, 2, -2, -1\} = \{\tfrac{16}{3}, \tfrac{32}{3}, \tfrac{16}{3}, \tfrac{16}{3}, \tfrac{32}{3}\}. \tag{4.149}$$

Due to the symmetry, only the values $0 \leqslant m' \leqslant 2$ are needed in the m'-summation. Multiplication by two for $m' > 0$ gives the *reduced* $K(m')$ sequence obtained with $m = 1$

$$\{\tfrac{16}{3}, \tfrac{64}{3}, \tfrac{32}{3}\}. \tag{4.150}$$

Treatment of $W_{-1,-1}(\theta)$ gives an identical sequence for $m = -1$.

† Note that the sequences in eqn (4.143) could have been generated from only three samples in theta (i.e. $\theta = 0, 90°, 180°$) by application of the 'splitting' techniques, eqn (A4.32). This is due to the fact that (4.143) is an even sequence. Had eqn (4.140) contained $m = 0$ or $|m| = 2$ contributions, the sequences corresponding to (4.143) would have been odd, and the splitting techniques would not have worked. Since the splitting technique is not fully general, it has not been utilized here.

Before entering the n-loop of Fig. 4.12, let us define the input and output probes and evaluate their normalized far-field response constants. For both probes, we shall apply an x'-polarized maximum directivity antenna with $N = 2$; see Section 2.4.4. The main direction of both probes is towards the test antenna, i.e. in the direction of the negative z'-axis of the probe coordinate system. Without proof, we shall state

$$\left.\begin{aligned} P^{\infty}_{111} &= P^{\infty}_{211} = P^{\infty}_{1,-1,1} = -\frac{\sqrt{6}}{2}i \\ P^{\infty}_{2,-1,1} &= \frac{\sqrt{6}}{2}i \\ P^{\infty}_{112} &= P^{\infty}_{212} = P^{\infty}_{1,-1,2} = -\frac{\sqrt{10}}{2} \\ P^{\infty}_{2,-1,2} &= \frac{\sqrt{10}}{2} \end{aligned}\right\} \qquad (4.151)$$

for the input probe. The same constants apply for the output probe, so $P'^{\infty}_{s\mu n} = P^{\infty}_{s\mu n}$.

Let us also state the delta products required for evaluating the reduced m'-sums in step 5. The reduced summation range in m' is $0 \leqslant m' \leqslant n$ so we only need the following products (cf Appendix A2.6)

$$\left.\begin{aligned} &\Delta^1_{01}\Delta^1_{01} = \tfrac{1}{2}, \quad \Delta^2_{01}\Delta^2_{01} = 0 \\ &\Delta^1_{11}\Delta^1_{11} = \tfrac{1}{4}, \quad \Delta^2_{11}\Delta^2_{11} = \tfrac{1}{4} \\ &\qquad\qquad\quad \Delta^2_{21}\Delta^2_{21} = \tfrac{1}{4} \end{aligned}\right\} \qquad (4.152)$$

The n-loop of Fig. 4.12 is now entered. For $n = 1$, we easily obtain in step 5: $W^1_{11} = 12$, $W^1_{-1,1} = W^1_{1,-1} = 0$, $W^1_{-1,-1} = 12$. The two sets of two equations with two unknowns become, for this value of n,

$$\begin{Bmatrix} -\frac{\sqrt{6}}{2}i & -\frac{\sqrt{6}}{2}i \\ -\frac{\sqrt{6}}{2}i & \frac{\sqrt{6}}{2}i \end{Bmatrix} \begin{Bmatrix} vT_{111} \\ vT_{211} \end{Bmatrix} = \begin{Bmatrix} 12 \\ 0 \end{Bmatrix} \qquad (4.153)$$

and

$$\begin{Bmatrix} -\frac{\sqrt{6}}{2}i & -\frac{\sqrt{6}}{2}i \\ -\frac{\sqrt{6}}{2}i & \frac{\sqrt{6}}{2}i \end{Bmatrix} \begin{Bmatrix} vT_{1,-1,1} \\ vT_{2,-1,1} \end{Bmatrix} = \begin{Bmatrix} 0 \\ 12 \end{Bmatrix}. \qquad (4.154)$$

In step 7 these equations are solved. Four test antenna coefficients are thereby found as $vT_{111} = vT_{211} = vT_{1,-1,1} = -vT_{2,-1,1} = (\sqrt{6}/2)i$. With these coefficients, the first part of the algorithm is left for a while and the computation of the output signal is begun in step 8. We still have $n = 1$ and the first term of the n-summation in eqn (4.135) is now computed. The output probe constants as well as the delta products are already available for this purpose. The results are given in Table 4.6(a). Note that the table is made to the same size as Table 4.5. The

Table 4.6(a) *First term of n-summation.*

$m =$	0		1		± 2		-1	
$\mu =$	1	-1	1	-1	1	-1	1	-1
$m' = 0$			6	0			0	6
1			3	0			0	3
2			0	0			0	0
3								

Table 4.6(b) *Second term of n-summation.*

$m =$	0		1		± 2		-1	
$\mu =$	1	-1	1	-1	1	-1	1	-1
$m' = 0$			0	0			0	0
1			5	0			0	5
2			5	0			0	5
3								

Table 4.6(c) *Resulting n-sum.*

$m =$	0		1		± 2		-1	
$\mu =$	1	-1	1	-1	1	-1	1	-1
$m' = 0$			6	0			0	6
1			8	0			0	8
2			5	0			0	5
3								

tables generated in the first part of the code may thus be overwritten as soon as a column has been processed.

The *n*-loop is now re-entered with $n = 2$. Here we find in step 5: $W^2_{\bar{1}1} = W^2_{-1,-1} = 20$ and $W^2_{-1,1} = W^2_{\bar{1},-1} = 0$. Two sets of two linear equations with two unknowns are solved in step 7 with the results $vT_{112} = vT_{212} = vT_{1,-1,2} = -vT_{2,-1,2} = -2\sqrt{10}$, and the second term in the *n*-summation is computed. The results are given in Table 4.6(b) and since there is no higher *n*, the completed *n*-summation can be stored, ref. Table 4.6(c).

With the *n*-loop completed for the current value of $|m|$, i.e. $|m| = 1$, the m'-summation in step 9 can now be made. The two non-zero columns in Table

Table 4.7(a) *The m′-sum, $W'_{\mu m}(\theta)$, of eqn (4.135).*

$m =$	0		1		± 2		1	
$\mu =$	1	−1	1	−1	1	−1	1	−1
$\theta = 0°$	0	0	32	0	0	0	0	32
60°	0	0	9	0	0	0	0	9
120°	0	0	−7	0	0	0	0	−7
180°	0	0	0	0	0	0	0	0

Table 4.7(b) *The m-sum, $W'_{\mu}(\theta, \phi)$, of eqn (4.135).*

$\phi =$	0°		90°		180°		270°	
$\mu =$	1	−1	1	−1	1	−1	1	−1
$\theta = 0°$	32	32	32*i*	−32*i*	−32	−32	−32*i*	32*i*
60°	9	9	9*i*	−9*i*	−9	−9	−9*i*	9*i*
120°	−7	−7	−7*i*	7*i*	7	7	7*i*	−7*i*
180°	0	0	0	0	0	0	0	0

Table 4.7(c) *The μ-sum, $W'(\chi, \theta, \phi)$, of eqn (4.135).*

$\phi =$	0°		90°		180°		270°	
$\chi =$	0°	90°	0°	90°	0°	90°	0°	90°
$\theta = 0°$	64	0	0	−64	−64	0	0	64
60°	18	0	0	−18	−18	0	0	18
120°	−14	0	0	14	14	0	0	−14
180°	0	0	0	0	0	0	0	0

4.6(c) give identical results. Extension to negative m' and inclusion of an extra zero to obtain a sequence of six elements give

$$\{6, 8, 5, 0, 5, 8\}. \tag{4.155}$$

Taking the DFT now provides the m'-summation which is a set of six equispaced data values in theta

$$\{32, 9, -7, 0, -7, 9\}. \tag{4.156}$$

Only the first four elements, corresponding to $\theta = 0, 60°, 120°, 180°$, respectively, need to be retained. The $|m|$-loop has now been completed for $|m| = 1$ with the resulting four numbers valid for both $m = 1$ and $m = -1$. Further repetitions of the $|m|$-loop are unnecessary as all data for $m = 0$ and $m = \pm 2$ are zero, ref. Table 4.5(c). Upon exit from the $|m|$-loop we have, therefore, the results given in Table 4.7(a).

In step 10, the m-summation of eqn (4.135) is performed by applying DFTs to the entries of Table 4.7a. The outcome is shown in Table 4.7(b). Finally, upon performing the μ-summation, we obtain the output signal W' as listed in Table 4.7(c). As can be seen, W' checks with the input signal W as expected. Of course, the output signal could have been evaluated in any given number of directions. As a matter of convenience, we have chosen here the same directions for evaluating W' as for sampling of the input signal W, ref. Section 4.4.3.

The test antenna considered in the present example is a maximum directivity antenna with $N = 2$. Assuming that the input and output signals are measured in watt$^{\frac{1}{2}}$, the test antenna radiates a total power of 83 watts. A relative far-field pattern is depicted in Fig. 2.16.

4.4.6 A test case [3]

The test runs described below for the computer code SNIFT were made in order to demonstrate the accuracy of the code for the case of a fairly large test antenna, i.e. with a diameter of fifty wavelengths. For the test antenna as well as for the measurement probe, exact theoretical models in the form of arrays of short dipoles were chosen; see Fig. 4.13. All dipoles of the test antenna are excited with the same amplitude and phase. The near field as well as the far field for the test antenna array was evaluated by superposition of the contributions from each element. The signal received by the two-element probe was calculated as the sum of the signals in the elements.

For comparison with the exact far field, the near-field signal was transformed by SNIFT. In the transformation, probe correction for the pattern of the two-element probe was included. Sampling schemes with equal increments Δ in theta and phi were used. The number J of samples in $0 \leqslant \phi < 2\pi$ was varied in the range $260 \leqslant J \leqslant 440$, where $J = 2\pi/\Delta$. For each value of J, as many modes as possible were computed using SNIFT, and the corresponding far field was found using the same increments for the output sampling as for the input sampling.

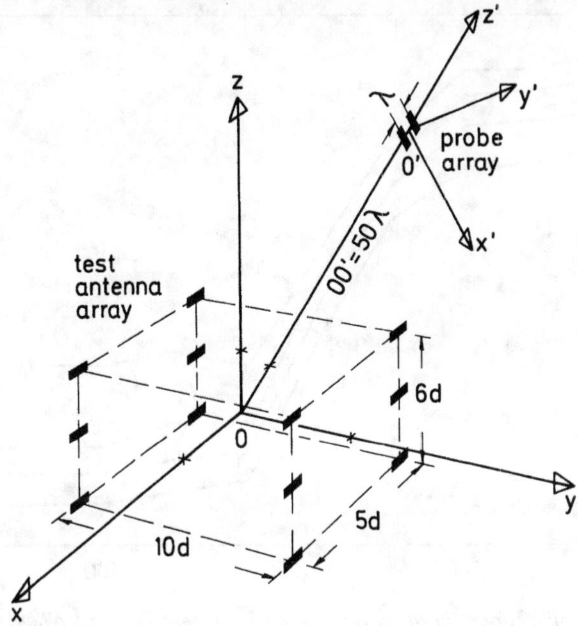

Fig. 4.13 *Geometry of the probe array and the test antenna array used in test runs for the spherical near-field to far-field transformation code SNIFT*
The dimension *d* has been chosen such that the test antenna minimum sphere has a diameter of 50λ

The radiation from the array of twelve widely spaced elements is characterized by the presence of several main lobes (grating lobes). At each far-field point, the relative error was found according to the formula

$$\varepsilon = 20\log\left\{\frac{|\vec{K}(\theta,\phi) - \vec{K}_{\text{exact}}(\theta,\phi)|}{\max|\vec{K}|}\right\}\text{dB.} \tag{4.157}$$

The measurement distance was $A = 50\lambda$.

The results are shown in Fig. 4.14 where the relationship between sampling density and accuracy can be studied. Two kinds of errors are present; these are sampling errors and rounding errors. As represented through the finite number of samples, the near field becomes a strictly band-limited function of the angular coordinates. The sampling errors indicate the approximation herein and depend upon the sampling density and the size of the antenna. The formulas for transforming the band-limited input signal to the far field are exact, so rounding errors in the computer are the only remaining errors.

With J sampling points on a great circle, modes of order n up to $(J-1)/2$ are correctly treated by the program. It is known from spherical wave theory (see

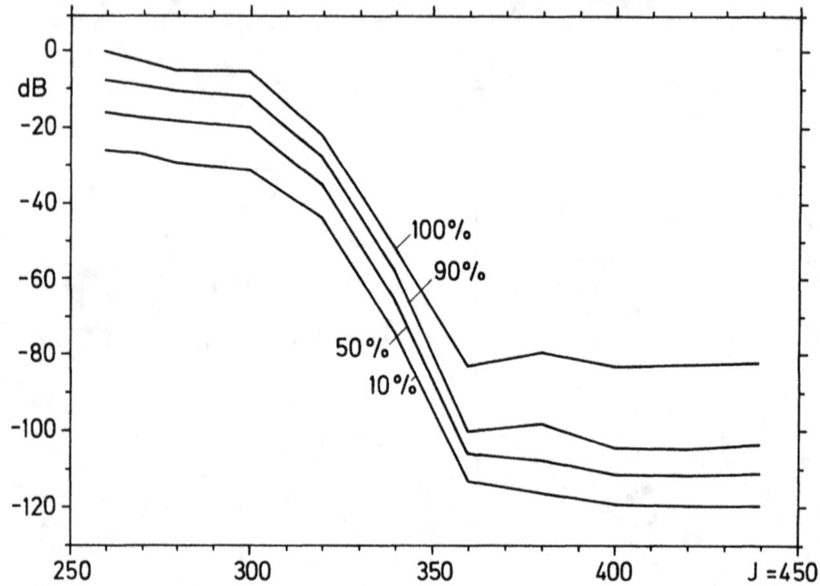

Fig. 4.14 *Percentile curves for the errors of a spherical near-field to far-field transformation using the computer program SNIFT on the geometry of Fig. 4.13*
The four curves show the error (eqn 4.157) as a function of the number of samples J on a great circle. For each curve is indicated the percentage of all far-field points for which the error is equal to or smaller than the values represented by the curve

Chapter 2) that the maximum mode order N for an antenna with radius r_t is roughly $kr_t + 10$. In this case, $r_t = 25\lambda$, or $kr_t + 10 = 167$, and a value of 335 is needed for J. Thus, the sampling error is dominant for $J < 335$. On the other hand, for $J > 360$ there is no effect of increasing the number of samples as can be seen from Fig. 4.14. The errors here are rounding errors (single precision arithmetic).

From the test runs described it is concluded that the errors originated in the transformation are of the order of -50 dB or lower, if the test antenna field is sampled in accordance with the rule $N = kr_t + 10$.

4.5 Review of spherical near-field transformations

In this section we shall present a brief review of the mathematical methods that have been applied in spherical near-field testing. As in any near-field scheme, large amounts of data are produced, and efficient data reduction becomes of paramount importance. Extraction of spherical wave coefficients from the measured data is the main objective of the software and is normally done by numerical integration. Much effort has been devoted to the development and improvement of numerical methods appropriate for the particular forms of integrals occurring in spherical near-field testing. The phi and theta integrals discussed in the preceding sections may serve as prototypes for such integrals.

In the methods used by James and Longdon [5] in connection with the measurement of radial field components (see Section 4.2.3), the integrals were solved numerically by Simpson's rule in both theta and phi. The practical cases investigated were restricted to spherical mode orders up to $N = 3$.

For larger values of N, it is imperative to take advantage of the fact that the theta and phi integrals can be carried out separately. The phi integrals are independent of the index n and should therefore be carried out first. For the phi integrals which have periodic, band-limited integrands, an exact algorithm is avalable—see Appendix A4.5.2. The integration rule in question, the trapezoidal rule, is here equivalent to a discrete Fourier transform. This is simpler than Simpson's rule and can furthermore be applied to the phi integral for all values of m simultaneously. The phi integrals are of the form

$$I_\phi = \frac{1}{2\pi}\int_0^{2\pi} \mathrm{f}(\phi)\,\mathrm{e}^{-im\phi}\,\mathrm{d}\phi, \qquad m = -N, \ldots, 0, \ldots, N. \tag{4.158}$$

The required number of samples in the trapezoidal rule equals the number of harmonics, $2N + 1$.

For the probe-corrected case, the theta integral has the form (ref. Section 4.3.3.4),

$$I_\theta = \int_0^{\pi} g_{\mu m}(\theta)\mathrm{d}^n_{\mu m}(\theta)\sin\theta\,\mathrm{d}\theta$$

$$\mu = \pm 1;\ m = -n, \ldots, 0, \ldots, n;\ n = 1, 2, \ldots, N. \tag{4.159}$$

One could hope for the existence of an exact algorithm for this integral as well. Neither Simpson's rule nor the trapezoidal rule can be expected to provide exact results for theta integrals, since the integrand is not periodic. Accurate results may of course be achieved by these rules if a sufficient number of sample points are used. Ludwig [15] has used the trapezoidal rule for the integral in Section 4.2.4 and Wood [16] applied Simpson's rule for the integral in Section 4.2.5. Ludwig has furthermore developed an iterative improvement scheme [17] to correct for the imperfections of the quadrature rule. The iterations also correct for rounding errors in the first steps of the calculations, so the final results can be more accurate than those obtained by the so-called exact algorithms. An example of an exact algorithm for the integral in eqn (4.159) is the Gaussian integration rule. With p sampling points, this type of integration is exact for a polynomium of degree $2p - 1$. Apart from the $\sin\theta$ factor, the integrand in eqn (4.159) will always be a polynomial in $\cos\theta$. The degree of this polynomial is less than or equal to $2N$ assuming a band limit of N for the data function $g_{\mu m}(\theta)$. This can be deduced from eqns (A2.5) and (A2.8). Therefore, a Gaussian quadrature rule for the integral (4.159) with $\cos\theta$ as the integration variable and

$N+1$ sampling points over the interval $-1 \leqslant \cos\theta \leqslant 1$ gives exact values of the integral and thereby exact values of the spherical wave coefficients. A disadvantage of the method in experimental work is that the samples have to be taken at specific non-equidistant values of theta.

In search for an exact method based on equidistant samples, Ricardi and Burrows [11] and, independently, Wacker [2] discovered the possibility of applying a Fourier transformation in theta. It is not trivial to see this since the data $g_{\mu m}(\theta)$ is only taken in the interval $0 \leqslant \theta \leqslant \pi$ and is not a periodic function. However, as explained in Section 4.3.3.4, the data can be artificially extended with the correct parity into the interval $\pi < \theta < 2\pi$ such that the extended data function, denoted $\tilde{g}_{\mu m}(\theta)$, is a band-limited function, periodic with the period 2π. A finite Fourier series therefore exists

$$\tilde{g}_{\mu m}(\theta) = \sum_{l=-N}^{N} b_l e^{il\theta} \qquad 0 \leqslant \theta < 2\pi. \tag{4.160}$$

Ricardi and Burrows insert this expansion into the theta integral with the result

$$I_\theta = \sum_{l=-N}^{N} \int_0^\pi e^{il\theta} d^n_{\mu m}(\theta) \sin\theta \, d\theta \; b_l \tag{4.161}$$

where the integral can be evaluated analytically with the aid of certain recurrence relations. In the method of Wacker, the extended data as well as the product $d^n_{\mu m}(\theta)\sin\theta$ is expanded into a finite Fourier series and inserted into the theta integral. Also in this case the resulting integral can be solved directly in closed form. The *Wacker algorithm* is a key to exact and efficient handling of the theta integral.

Further improvements have been introduced by Larsen [3, 4]. Instead of applying a Fourier series to the product $d^n_{\mu m}(\theta)\sin\theta$, Larsen used the Fourier expansion (ref. Appendix A2) of $d^n_{\mu m}(\theta)$ and kept the factor $\sin\theta$ in the remaining integral so that

$$I_\theta = i^{\mu-m} \sum_{l=-N}^{N} \sum_{m'=-n}^{n} \Delta^n_{m'\mu} \Delta^n_{m'm} \left\{ \int_0^\pi e^{i(l-m')\theta} \sin\theta \, d\theta \right\} b_l. \tag{4.162}$$

This facilitates the calculation of the far field where the Fourier coefficients of $d^n_{\mu m}$ can be reused. Also, as shown by Larsen, considerable savings can be made by carrying out the l-summation before the m'-summation. This is due to the fact that the l-summation is independent of n and, furthermore, can be computed as a convolution summation. In addition, a recurrence formula for the deltas in eqn (4.162) was found with the summation index m' as the only varying parameter. The features mentioned are all important for an efficient implementation of Wacker's idea. The Wacker–Larsen approach was described in detail in Section 4.3.

The three Fourier transform methods for the theta integral are exact for band-limited functions. They are also numerically well conditioned in cases where the data function, e.g. due to noise, is not strictly band limited. The methods of Section 4.3 have been used extensively at TUD since 1978 for far-field computations from measured or numerically generated near-field data. This has proven reliable for a number of applications [18, 19]. Examples are provided in Chapter 5.

Recent developments have shown that the number of additions and multiplications in spherical near-field algorithms can be further reduced at the cost of some complexity in the formulation.

In order to explain how this reduction is obtained, we shall compare the scheme with probe correction to the non-corrected cases where either the radial field components or two tangential components are measured, respectively.

In the probe-corrected scheme, there are two scalar integrals (4.159) to calculate for each (m, n), i.e. one with $\mu = +1$ and one with $\mu = -1$. In the expansion (4.161), the terms are generally non-zero. The same is true for the delta products in eqn (4.162).

In data reduction for the case where the measured signal is a radial field component, ref. Section 4.2.3, the data functions w_r^e and w_r^m can be expanded into scalar spherical harmonics. The resulting two theta integrals have the same form as (4.161) with $\mu = 0$. As it turns out, however, every other term for varying l is zero [11]. If the theta integrals are further reduced to a form similar to (4.162) with $\mu = 0$, every other delta product for varying m' is zero as well; see eqn (A2.38).

In data reduction for the case of measurements of two tangential field components, as described in Section 4.2.4, the data functions are tangential vectors and must therefore be expanded into tangential wave functions. The functions $\vec{K}_{smn}(\theta, \phi)$ (see Appendix A1.3.2) can be used for this purpose. According to eqns (A2.17–2.23) this is equivalent to employing the functions

$$\frac{d^n_{0m}(\theta)}{\sin\theta} \quad \text{and} \quad \frac{d}{d\theta} d^n_{0m}(\theta) \tag{4.163}$$

multiplied by the tangential unit vectors. For each (m, n) there are two vector dot products to integrate. This yields four scalar theta integrals of a form similar to (4.161) with $d^n_{\mu m}(\theta)$ replaced by the functions (4.163). As it turns out, half of the terms in the equations corresponding to (4.161) vanish in this case as well. If the theta integrals are further reduced to forms similar to (4.162), every other delta product vanishes.

As indicated above, the probe-corrected case and the non-corrected case with two tangential components yield the same number of terms for the theta integrals. In the method where the radial field components are measured, half as many terms are needed for the theta integrals.

As pointed out by Yaghjian and Wittmann [20, 21, 22], the probe-corrected case can be formulated in terms of the tangential components of the wave functions. This follows from the relationship

$$d^n_{\pm 1m}(\theta) = \frac{-1}{\sqrt{n(n+1)}}\left(\frac{m d^n_{0m}(\theta)}{\sin\theta} \pm \frac{\mathrm{d}}{\mathrm{d}\theta} d^n_{0m}(\theta)\right) \tag{4.164}$$

in connection with eqns (A2.17–2.23). An explicit formula is eqn (3.38). Yaghjian [20] obtained this result by expressing the output of a directive probe in terms of spatial derivatives of the test antenna fields at the origin of the probe coordinate system. He was thereby able to avoid the translation coefficients of the transmission formula.

Wittmann [21] showed that the four scalar theta integrals arising in the formulation can be reduced to two integrals still involving $d^n_{0m}(\theta)$ such that every other term in the resulting two summations corresponding to eqn (4.162) is zero. Also, the recursion relation (A2.37) for the product $\Delta^n_{m'0}\Delta^n_{m'm}$ can be used in this connection.

It is interesting to note that the original paper by Ricardi and Burrows [11] also devises a way to cut the computations in half. They calculate the integrals containing $(\mathrm{d}/\mathrm{d}\theta)\, d^n_{0m}(\theta)$ from the integrals containing $d^n_{0m}(\theta)/\sin\theta$ by recursion. As an alternative way by which the halving of terms for the theta integral can be obtained, Lewis and Wittmann [23] have shown that the integrals

$$\int_0^\pi g_{\mu m}(\theta) d^n_{\mu m}(\theta) \sin\theta \, \mathrm{d}\theta, \qquad \mu = \pm 1 \tag{4.165}$$

can be evaluated recursively from the integrals

$$\int_0^\pi g_{\mu m}(\theta) d^n_{0m}(\theta) \sin\theta \, \mathrm{d}\theta, \qquad \mu = \pm 1 \tag{4.166}$$

for which every other term vanishes when expanded similarly to eqns (4.161–4.162).

In conclusion, the two integrals with $\mu = \pm 1$ (eqn (4.162)) in our formulation may be replaced by two integrals where every other term in the summation is zero. Thus the number of terms may be suppressed to the same number as is required for the theta integral in connection with the measurement of radial field components. On the other hand, the resulting algorithm is complicated by some extra calculations either before or after the m'-summation.

Finally, it should be mentioned that the computations can be further reduced if the input data are zero on a part of the measurement sphere. Lewis [12, 23] has described a scheme for the case where the data are zero on a hemisphere around one of the poles, e.g. for $\pi/2 \leqslant \theta \leqslant \pi$. When the test antenna's main beam is pointing in the direction of the equator of the measurement sphere, i.e. $\theta = \pi/2$, computational savings can be obtained by assuming the field to repeat itself with a certain period $2\pi/p$, where p is an integer [24, 25]. The latter approach is demonstrated in Chapter 6.

References

1. WACKER, P. F., 1974, 'Near-field antenna measurements using a spherical scan: Efficient data reduction with probe correction'. Conference on Precision Electromagnetic Measurements. *IEE Conference Publication No.* **113**, pp. 286–288, London, England.
2. WACKER, P. F., 1975, 'Non-planar near-field measurements: Spherical scanning', Report NBSIR **75-809**, Electromagnetics Division, Inst. for Basic Standards, Nat. Bur. Stds., Boulder, Colorado, June.
3. LARSEN, FL. HOLM, 1979, 'Improved algorithm for probe-corrected spherical near-field/far-field transformation'. *Electronics Letters*, Vol. **15**, No. 19, pp. 588–590, Sep.
4. LARSEN, FL. HOLM, 1980, 'Probe-corrected spherical near-field antenna measurements', Ph.D. Thesis, Electromagnetics Institute, Technical University of Denmark, Report **LD 36**, Dec.
5. JAMES, J. R. and LONGDON, L. W., 1969, 'Prediction of arbitrary electromagnetic fields from measured data', *Alta Frequenza*, Vol. **38**, Num. Spec., pp. 286–290, May.
6. STRATTON, J. A., 1941, *Electromagnetic Theory*, Chapter VII, McGraw-Hill, Inc., New York.
7. LUDWIG, A. C., 1971, 'Near-field far-field transformations using spherical-wave expansions', *IEEE Trans.*, Vol. **AP-19**, No. 2, pp. 214–220, Mar.
8. WOOD, P. J., 1977, 'The prediction of antenna characteristics from spherical near field measurements, Part I, theory', *Marconi Review*, Vol. **40**, No. 204, pp. 42–68.
9. WOOD, P. J., 1977, 'The prediction of antenna characteristics from spherical near field measurements, Part II, experimental validation', *Marconi Review*, Vol. **40**, No. 205, pp. 117–155.
10. JENSEN, F., 1975, 'On the probe compensation for near-field measurements on a sphere', *Archiv für Elektronik und Übertragungstechnik*, Vol. **29**, No. 7/8, pp. 305–308, July–Aug.
11. RICARDI, L. J. and BURROWS, M. L., 1972, 'A recurrence technique for expanding a function in spherical harmonics', *IEEE Trans.*, Vol. **C-21**, No. 6, pp. 583–585, June.
12. LEWIS, R. L., 1981, 'Spherical scanning data processing; An algorithm for halving the data processing effort when the radiation into the back hemisphere is negligible', *IEEE AP-S Symposium Digest*, Vol. **1**, pp. 246–249, Los Angeles, USA, June.
13. LEWIS, R. L., 1982, 'The aliasing error in near-field spherical scanning due to the FFT—equal to zero?', URSI National Radio Science Meeting, University of Colorado, p. 145, Boulder, Colorado, USA, Jan.
14. CABAYAN, H. S., MURPHY, R. C. and PAVLASEK, T. J. F., 1973, 'Numerical stability and near-field reconstruction', *IEEE Trans.*, Vol. **AP-21**, No. 3, pp. 346–351, May.
15. LUDWIG, A. C., 1970, 'Calculation of scattered patterns from asymmetrical reflectors', Ph.D. dissertation, Jet Propulsion Laboratory, Technical Report **32-1430**, Pasadena, California, Feb.
16. WOOD, P. J., 1980, *Reflector antenna analysis and design*, Peter Peregrinus, London.
17. LUDWIG, A. C., 1972, 'Calculation of orthogonal-function expansions from imperfect data', Laboratory of Electromagnetic Theory, Technical University of Denmark, Report **R102**.
18. LEMANCZYK, J. H., HANSEN, J. E. and LARSEN, FL. HOLM, 1984, 'Evaluation of the spherical near field range at the Technical University of Denmark', *IEEE AP-S Symposium Digest*, Vol. **2**, pp. 682–685, Boston, June.
19. HANSEN, J. E., 1984, 'The TUD-ESA spherical near-field antenna test facility, Lyngby, Denmark', Publication **BR-19**, European Space Agency.
20. YAGHJIAN, A. D., 1984, 'Simplified approach to probe-corrected spherical near-field scanning', *Electronics Letters*, Vol. **20**, No. 5, pp. 195–196, Mar.
21. WITTMANN, R. C., 1984, 'Probe correction in spherical near-field scanning, viewed as an ideal probe measuring an effective field', *IEEE AP-S Symposium Digest*, Vol. **2**, pp. 674–677, Boston, June.
22. YAGHJIAN, A. D. and WITTMANN, R. C., 1985, 'The receiving antenna as a linear differential operator: Application to spherical near-field scanning', *IEEE Trans.*, Vol. **AP-33**, No. 11, pp. 1175–1185, Nov.

23. LEWIS, R. L. and WITTMANN, R. C., 'Improved spherical and hemispherical scanning algorithms', *IEEE Trans. Antennas Propagat.* To be published.
24. SHAPIRA, J., 1979, 'Sampling and processing in antenna pattern measurement', Record of the 11th Convention of IEEE in Israel, Paper No. **B2-5**, Tel Aviv, Israel, Oct.
25. HESS, D. W. and TAVORMINA, J. J., 1982, 'An algorithm with enhanced efficiency for computing the far field from near-field data on a partial spherical surface', URSI National Radio Science Meeting, University of Colorado, p. 185, Boulder, Colorado, USA, Jan.

Chapter 5

Measurements

5.1 Introduction

This chapter deals with the practical aspects of spherical near-field antenna testing. Section 5.2 on measurement probes provides a transition from the theoretical chapters to the experimental techniques beginning in Section 5.3. Section 5.3 on probe corrected measurements contains a section (5.3.2) where the spherical near-field test range is examined from a system point of view. In another section (5.3.3), the test range is described from the point of view of the various measurement procedures that can be carried out on the range and that constitute a complete measurement of a given test antenna. Some measurement results obtained at the TUD test range are also presented (Section 5.3.4). As was shown in Chapter 4, the basic spherical near-field measurement not only yields the radiation patterns of the test antenna but also its directivity. If the test antenna gain is wanted, some additional measurements must be carried out. These are dealt with in Section 5.4.

5.2 Probes for near-field scanning

5.2.1 Introduction

A probe for measurements of electromagnetic fields is usually thought of as a small device designed for the detection, at a single point, of electric field components, magnetic field components or a combination of these [1, 2]. However, in near-field scanning of antennas, advantage is taken of the fact that the sources for the field of interest are located to one side of the probe only. Thus, a directive probe like a horn antenna can be used to suppress signals from sources outside the test zone; it will also receive a stronger signal from the test antenna. This corresponds to what is done to avoid the ground reflection in far-field measurements. The directivity implies that the receiving pattern in the solid angle of interest will be rather independent of the structure on which the probe is mounted. The pattern can therefore be measured in a separate set-up, and the results can be used for corrections in the data processing when the probe is employed in the near-field scanner. Because of the directive probes and the existence of good absorbers, near-field measurements on microwave antennas can be orders of magnitude more accurate than field strength measurements at lower frequencies. Although the directive probe senses a combination of the E-field and the H-field, each of the individual field components can be found by data processing after the near-field scanning.

The above considerations could be interpreted as if an optimum probe would have a constant pattern in the directions of the test antenna and zero outside, as indicated in Fig. 5.1. However, to shape such a pattern even approximately would require a large probe which would cause a high level of multiple reflections between the probe and the test antenna. Therefore it is better to accept a certain amplitude taper over the test zone as indicated in Fig. 5.2, and to use the probe correction to correct for it.

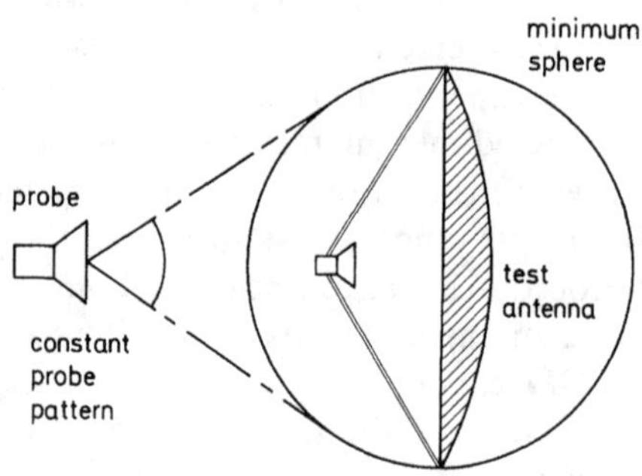

Fig. 5.1 *Probe having a constant pattern over the minimum sphere*

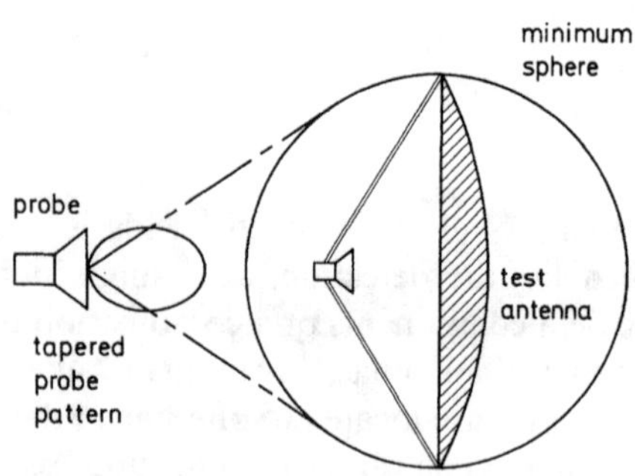

Fig. 5.2 *Probe having a tapered pattern over the minimum sphere*

In Chapter 4 we found that in order to obtain a practical data-processing scheme for spherical near-field measurements, it was necessary to restrict the probe pattern to contain spherical modes with $\mu = \pm 1$ only. This requirement does not put a limit on the directivity of the probe since the latter is related to the maximum value of the polar index ν (ref. Section 2.4.4). The $\mu = \pm 1$ requirement in spherical scanning is satisfied by rotationally symmetric horns fed by the fundamental TE_{11} mode in a circular waveguide. It has been found useful to confine the choice of probes to rotationally symmetric structures although the patterns of these are restricted relative to the general $\mu = \pm 1$ patterns.

However, this restriction allows a model to be formulated for a dual-polarized probe where the on-axis polarizations of the two ports are allowed to be different. An important feature of this probe system [3], which is described in the present chapter, is that critical parts of the calibration are not carried out until the probe has been mounted in its final position and connected to the receiver. In this way the calibration does not depend on reproducibility of cable connections and reflection coefficients. Because the field of the test antenna in spherical near-field measurements is rotated relative to the polarization vector of the probe, determination of cross polarization is very sensitive to the phase and amplitude difference between the measured field components. Accordingly, the calibration method mentioned above is important for spherical near-field measurements. In fact the dual polarized probe system can be used for all scan geometries with the following advantages:

1. The two polarization components are measured simultaneously at the same scan coordinates.
2. Only one scan rather than two is needed.
3. No mechanical rotation of the probe is needed.
4. Critical parts of the receiver (like the mixer) can be time-shared between the two channels such that amplitude and phase drift affect the two polarization components in the same way [4].
5. The pattern calibration only requires measurement of the copolar patterns in two planes which simplifies the calibration relative to a full pattern measurement [5, 6].
6. The polarization calibration takes care of mismatch between the probe and the receiver cables as well as coupling between the two ports of the probe and differences between the two channels of the receiver.

Sections 5.2.2 to 5.2.5 outline the theory behind the calibrations and the corrections pertaining to the dual-polarized probe.

5.2.2 Rotationally symmetric probes

5.2.2.1 Influence of rotational symmetry on patterns: The general form of the far-field pattern from a spherical wave expansion that only contains modes with $\mu = \pm 1$ can be written in terms of $\cos\phi$ and $\sin\phi$ as

$$\begin{aligned}\vec{K}(\theta,\phi) &= \vec{K}(\theta,0^\circ)\cos\phi + \vec{K}(\theta,90^\circ)\sin\phi \\ &= \{K_\theta(\theta,0^\circ)\hat{\theta} + K_\phi(\theta,0^\circ)\hat{\phi}\}\cos\phi \\ &\quad + \{K_\theta(\theta,90^\circ)\hat{\theta} + K_\phi(\theta,90^\circ)\hat{\phi}\}\sin\phi. \end{aligned} \tag{5.1}$$

In order to obtain such a pattern, the probe is usually designed as a rotationally symmetric antenna excited by the TE_{11} mode in a circular waveguide. However, the rotational symmetry of the structure implies a field expression more simple than eqn (5.1). In order to see this we can decompose the general (elliptically polarized) TE_{11} mode field into an x-polarized TE_{11} mode and a y-polarized TE_{11} mode with reference to the coordinate system shown in Fig. 5.3. The x-polarized mode must have a radiation pattern which is x-polarized in the principal planes. Thus $K_\phi(\theta, 0°) = 0$ and $K_\theta(\theta, 90°) = 0$. In addition, the y-polarized mode must have the same radiation pattern only rotated 90° in ϕ.

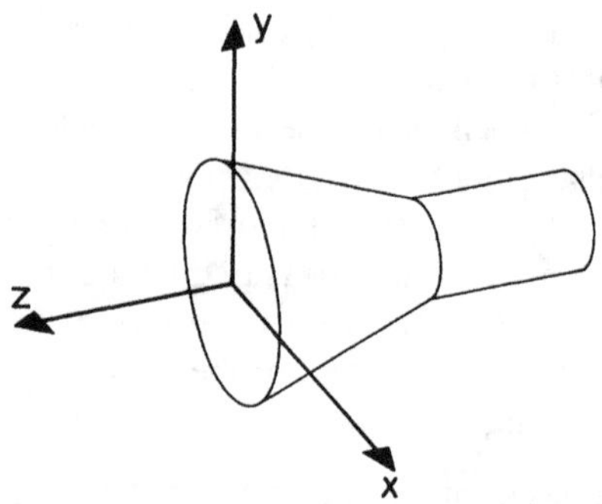

Fig. 5.3 *Rotationally symmetric probe*

With p being the *complex excitation ratio*

$$p = \frac{E_y}{E_x} \tag{5.2}$$

between the field components E_x and E_y of the two waveguide modes on the z-axis, the rotational symmetry gives the far-field pattern

$$\begin{aligned}\vec{K}(\theta, \phi) &= K_\theta(\theta, 0°)\cos\phi\hat{\theta} + K_\phi(\theta, 90°)\sin\phi\hat{\phi} \\ &\quad + p\{K_\theta(\theta, 0°)\sin\phi\hat{\theta} - K_\phi(\theta, 90°)\cos\phi\hat{\phi}\} \\ &= \{f_1(\theta)\hat{\theta} + pf_2(\theta)\hat{\phi}\}\cos\phi + \{pf_1(\theta)\hat{\theta} - f_2(\theta)\hat{\phi}\}\sin\phi.\end{aligned} \tag{5.3}$$

Thus, the full pattern is determined by the *copolar* E-*plane pattern* $f_1(\theta) = K_\theta(\theta, 0°)$, the *copolar* H-*plane pattern* $f_2(\theta) = -K_\phi(\theta, 90°)$ and the excitation ratio p which is seen to be the same as the linear polarization ratio of the field on axis, i.e. at $\theta = 0°$ in Fig. 5.3. In analysis of conical horn antennas it is customary to assume that only the x-polarized mode exists in the circular waveguide, thus $p = 0$. However, for accurate measurements, it must be taken into account that the on-axis field might not be purely x-polarized. Uniqueness of the field expression requires

$$f_1(\theta = 0°) = f_2(\theta = 0°) \tag{5.4}$$

and

$$f_1(\theta = 180°) = -f_2(\theta = 180°). \tag{5.5}$$

Note that the pattern of the rotationally symmetric probe contains two pattern functions while the general $\mu = \pm 1$ field equation (5.1) allows for four independent pattern functions.

Relations can be derived for the copolar and the cross-polar field components for the pattern in eqn (5.3) with the x-polarized mode excitation

$$\vec{K}^x(\theta, \phi) = \mathrm{f}_1(\theta)\cos\phi\hat{\theta} - \mathrm{f}_2(\theta)\sin\phi\hat{\phi}. \tag{5.6}$$

Multiplying by the complex conjugate of the polarization vectors after Ludwig's third definition, ref. eqns (2.190–2.191), we obtain

$$K^x_{\mathrm{co}}(\theta, \phi) = \mathrm{f}_1(\theta)\cos^2\phi + \mathrm{f}_2(\theta)\sin^2\phi \tag{5.7}$$

$$K^x_{\mathrm{cross}}(\theta, \phi) = \{\mathrm{f}_1(\theta) - \mathrm{f}_2(\theta)\}\cos\phi\sin\phi. \tag{5.8}$$

Thus, it is seen that the cross-polar field has maxima in the $\phi = 45°$ and $\phi = 135°$ planes that are related to a difference in amplitude and phase between the E-plane and the H-plane patterns as pointed out by Ludwig [7, 8].

5.2.2.2 Influence of rotational symmetry on coefficients: In a similar fashion, the rotational symmetry imposes certain restrictions upon the transmitting or receiving coefficients of the probe. For purely x-polarized excitation, the current distribution is symmetrical about the $\phi = 0°$ plane, and eqn (A1.89) yields

$$T^x_{11\nu} = T^x_{1,-1,\nu} \tag{5.9}$$

$$T^x_{21\nu} = -T^x_{2,-1,\nu} \tag{5.10}$$

Using the reciprocity relation (2.104) we get for the receiving coefficients

$$R^x_{11\nu} = R^x_{1,-1,\nu} \tag{5.11}$$

$$R^x_{21\nu} = -R^x_{2,-1,\nu} \tag{5.12}$$

These relations were imposed on the probe treated in Section 3.3.1.

For purely y-polarized excitations, equation (A1.89) gives

$$T^y_{11\nu} = -T^y_{1,-1,\nu} \tag{5.13}$$

$$T^y_{21\nu} = T^y_{2,-1,\nu} \tag{5.14}$$

The coefficients for the total pattern (5.3) are proportional to the x-coefficients plus p times the y-coefficients. For these coefficients, eqns (5.9–5.14) combine to

$$T_{11\nu} = QT_{1,-1,\nu} \tag{5.15}$$

$$T_{21\nu} = -QT_{2,-1,\nu} \tag{5.16}$$

where Q is the circular polarization ratio, ref. eqn (2.199), for $\theta = 0°$.

$$Q = \frac{K_R}{K_L} = \frac{K_x - iK_y}{K_x + iK_y} = \frac{1 - ip}{1 + ip} \tag{5.17}$$

5.2.2.3 Equal E- *and* H-*plane patterns*: For a rotationally symmetric x-polarized antenna where the E-plane pattern and the H-plane pattern are the same in amplitude and phase

$$f_1(\theta) = f_2(\theta) \tag{5.18}$$

eqns (5.7) and (5.8) show that the copolar field pattern is rotational symmetric and the cross-polar field is zero. A comparison with eqn (5.5) shows that such a pattern must have a null on the negative z-axis.

The implication of eqn (5.18) on the coefficients is

$$T^x_{11\nu} = T^x_{21\nu} = T^x_{1,-1,\nu} = -T^x_{2,-1,\nu}. \tag{5.19}$$

Examples of such patterns are the Harrington maximum directivity antennas discussed in Section 2.4.4. For a given ν, equal weighting and proper phasing of the TE and the TM modes as in eqn (5.19) give a field with equal E- and H-plane patterns. Such a field can be called a hybrid mode $HE_{1\nu}$ in the spherical waveguide, corresponding to the nomenclature used in corrugated horns [9].

It is also of interest to consider the field, (5.7) and (5.8), assuming opposite E- and H-plane patterns

$$f_1(\theta) = -f_2(\theta). \tag{5.20}$$

Such a field has a null on the positive z-axis and the relationship

$$T^x_{11\nu} = -T^x_{21\nu} = T^x_{1,-1,\nu} = T^x_{2,-1,\nu} \tag{5.21}$$

among the coefficients. In eqn (5.21), the TM coefficients have changed sign relative to eqn (5.19). For a given ν, the four modes correspond to an $EH_{1\nu}$ mode in the spherical waveguide.

However, the transformation from eqn (5.19) to (5.21) can also be carried out by reversing the z-axis by rotating it 180° about the y-axis. This can be verified by inspection of the rotation coefficients for spherical waves (A2.16). Therefore, a spherical $EH_{1\nu}$ mode referred to the positive z-axis is the same as an $HE_{1\nu}$ mode referred to the negative z-axis. The sign combination for the four mode coefficients with a given ν is distinctive for the boresight direction and the x- or y-polarization of a hybrid mode field, as summarized in Table 5.1.

5.2.2.4 Circularly polarized excitation: We shall now consider circularly polarized excitation in the waveguide. For right-hand circular excitation, indicated below by a superscript R, we have

$$p = i \tag{5.22}$$

in eqn (5.3). Multiplying by the complex conjugate of the $\hat{\imath}_{co,RC}$ reference vector

Table 5.1 *Sign combination for hybrid mode field.*

Boresight	Polarization	Relations between coefficients
$\theta = 0°$	x	$T^x_{11\nu} = T^x_{21\nu} = T^x_{1,-1,\nu} = -T^x_{2,-1,\nu}$
$\theta = 0°$	y	$T^y_{11\nu} = T^y_{21\nu} = -T^y_{1,-1,\nu} = T^y_{2,-1,\nu}$
$\theta = 180°$	x	$T^x_{11\nu} = -T^x_{21\nu} = T^x_{1,-1,\nu} = T^x_{2,-1,\nu}$
$\theta = 180°$	y	$-T^y_{11\nu} = T^y_{21\nu} = T^y_{1,-1,\nu} = T^y_{2,-1,\nu}$

in eqn (2.192), we obtain

$$K^R_R(\theta, \phi) = \frac{1}{\sqrt{2}}\{\mathrm{f}_1(\theta) + \mathrm{f}_2(\theta)\} \tag{5.23}$$

and similarly, with $\hat{\imath}_{\mathrm{co},LC}$ from eqn (2.194),

$$K^R_L(\theta, \phi) = \frac{1}{\sqrt{2}}\{\mathrm{f}_1(\theta) - \mathrm{f}_2(\theta)\}\,\mathrm{e}^{i2\phi}. \tag{5.24}$$

As for the linearly polarized case, we find that the cross polarization is related to a difference between the E- and H-plane patterns. This time the amplitude is independent of ϕ, but the phase referred to the reference vector in Ludwig's third definition is not. The left-hand component is zero when $\mathrm{f}_1(\theta) = \mathrm{f}_2(\theta)$. It is also zero at $\theta = 0°$.

For left-hand circular excitation we similarly obtain

$$p = -i \tag{5.25}$$

$$K^L_L(\theta, \phi) = \frac{1}{\sqrt{2}}\{\mathrm{f}_1(\theta) + \mathrm{f}_2(\theta)\} \tag{5.26}$$

and

$$K^L_R(\theta, \phi) = \frac{1}{\sqrt{2}}\{\mathrm{f}_1(\theta) - \mathrm{f}_2(\theta)\}\,\mathrm{e}^{-i2\phi} \tag{5.27}$$

5.2.3 Polarization correction

5.2.3.1 Definitions: Having explored the properties of rotationally symmetric probes, we shall now derive the equations that correct for non-perfect polarizations of the two ports in a dual polarized probe. We assume that the probe is rotationally symmetric and that the circular waveguide is connected to an orthomode transducer (OMT) with two nominally linear perpendicular polarizations. The fields excited from the two ports of the OMT will in practice be slightly, and differently, elliptically polarized; the amplitudes and the phases in particular will be different. When the probe is receiving, the signal in one port will therefore not precisely correspond to the signal in the other port when the probe is rotated 90°, as it is assumed in the usual near-field probe-correction algorithms.

One basic assumption for the following equations is that the two ports are separated from the flared part of the probe by a circular waveguide of sufficient length such that only two perpendicular TE_{11} modes can pass through. As indicated in Fig. 5.4, the port with nominal x-polarization will be denoted 'port I',

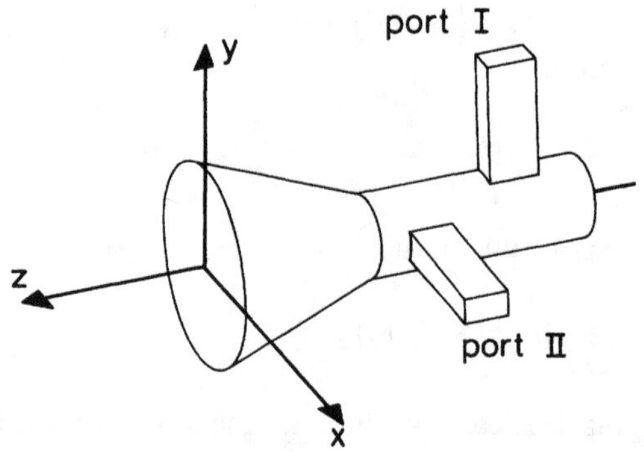

Fig. 5.4 *Dual-polarized probe. Excitation of port I yields a field on the z-axis which is essentially x-polarized. Excitation of port II yields a field on the z-axis which is essentially y-polarized*

while the y-polarized port is 'port II'. In this way, the excited port index is not later confused with the x- and y-component indices. The polarization properties of the two ports are described either by the linear polarization ratios p^{I} and p^{II} or by the circular polarization ratios Q^{I} and Q^{II}. In addition an *amplitude phase factor* $A_{\theta\phi}$ between the two ports is introduced. We have

$$p^{\mathrm{I}} = \frac{E_y^{\mathrm{I}}}{E_x^{\mathrm{I}}} = i\frac{Q^{\mathrm{I}} - 1}{Q^{\mathrm{I}} + 1} \tag{5.28}$$

$$p^{\mathrm{II}} = -\frac{E_x^{\mathrm{II}}}{E_y^{\mathrm{II}}} = i\frac{Q^{\mathrm{II}} + 1}{Q^{\mathrm{II}} - 1} \tag{5.29}$$

$$Q^{\mathrm{I}} = \frac{E_x^{\mathrm{I}} - iE_y^{\mathrm{I}}}{E_x^{\mathrm{I}} + iE_y^{\mathrm{I}}} = \frac{1 - ip^{\mathrm{I}}}{1 + ip^{\mathrm{I}}} \tag{5.30}$$

$$Q^{\mathrm{II}} = \frac{E_x^{\mathrm{II}} - iE_y^{\mathrm{II}}}{E_x^{\mathrm{II}} + iE_y^{\mathrm{II}}} = -\frac{1 - ip^{\mathrm{II}}}{1 + ip^{\mathrm{II}}} \tag{5.31}$$

$$A_{\theta\phi} = \frac{E_\theta^{\mathrm{I}}}{E_\phi^{\mathrm{II}}} = \frac{E_x^{\mathrm{I}}}{-E_y^{\mathrm{II}}}. \tag{5.32}$$

The quantities E_x^{I}, E_y^{I}, E_x^{II} and E_y^{II} are field components on the probe axis when the probe is transmitting from the port given by its upper index. Lower indices x and y refer to the coordinate system in Fig. 5.5(a). During measurements on the probe, it is convenient to have the z-axis pointing out of the probe as in Fig. 5.5(a). However, in the transformation algorithm described in Chapter 4, the z-axis is directed into the probe as in Fig. 5.5(b). In addition, the polarization components corresponding to $\chi = 0°$ and $\chi = 90°$ are directed as the $\hat{\theta}$ and $\hat{\phi}$ unit vectors in Fig. 5.5(b); hence $E_\phi^{\mathrm{II}} = -E_y^{\mathrm{II}}$, ref. eqn (5.32) which defines $A_{\theta\phi}$. The strategy chosen here is to use the unprimed coordinate system for the correction for the non-perfect polarization of the probe and then, reverse the coordinate system such that the receiving coefficients used for probe pattern correction are referred to the primed coordinate system in Fig. 5.5(b).

The parameters in eqns (5.28–5.32) have been chosen such that a perfect polarized probe has $p^{\mathrm{I}} = 0$, $p^{\mathrm{II}} = 0$, $Q^{\mathrm{I}} = 1$, $Q^{\mathrm{II}} = -1$ and $A_{\theta\phi} = 1$.

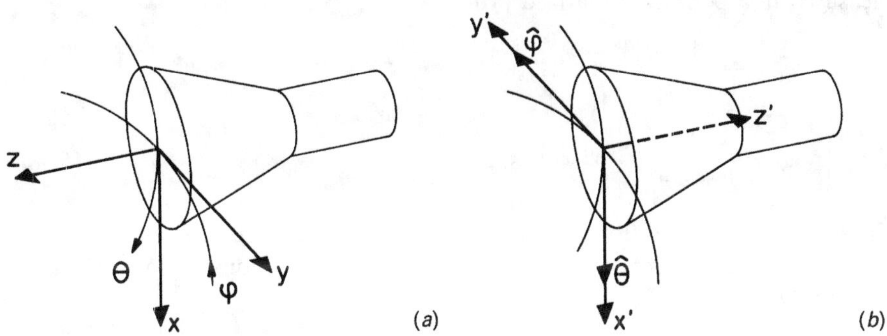

Fig. 5.5 *Probe polarization correction*
(*a*) Probe coordinate system (*x, y, z*) used during measurement
(*b*) Probe coordinate system (*x', y', z'*) used in transformation
The Greek letter coordinates and unit vectors belong to the test antenna coordinate system

5.2.3.2 Derivation of the polarization correction equations: The transverse components of the transmitted E-field on the z-axis can now be written in terms of the above parameters. If we transmit through port I, we get

$$E_x^{\mathrm{I}}\hat{x} + E_y^{\mathrm{I}}\hat{y} = E_x^{\mathrm{I}}(\hat{\theta} - p^{\mathrm{I}}\hat{\phi}). \tag{5.33}$$

If we excite port II with the same amplitude and phase, we get

$$E_x^{\mathrm{II}}\hat{x} + E_y^{\mathrm{II}}\hat{y} = \frac{E_x^{\mathrm{I}}}{A_{\theta\phi}}(p^{\mathrm{II}}\hat{\theta} + \hat{\phi}). \tag{5.34}$$

When a field is incident on the probe, it will create a theta-directed TE_{11} mode field and a phi-directed TE_{11} mode field in the circular waveguide. Each of the two modes will couple to each of the two ports. We define w_θ to be the signal received in port I from the theta-directed mode. Further let w_ϕ be defined as the signal that would have been received from the phi-directed mode if port I were rotated 90° clockwise as seen from the test antenna. Hereby the ratio between w_θ and w_ϕ equals the amplitude ratio and phase difference of the two mode fields in the circular waveguide. Because of the rotational symmetry of the probe, the relative couplings from the two modes to the two ports will be the same as the signals that would be received from two small electric dipoles placed on the z-axis. This allows us to apply reciprocity to eqns (5.33) and (5.34). The signals received in the two ports become

$$w^{\mathrm{I}} = w_\theta - p^{\mathrm{I}} w_\phi \tag{5.35}$$

$$w^{\mathrm{II}} = \frac{1}{A_{\theta\phi}}(p^{\mathrm{II}} w_\theta + w_\phi). \tag{5.36}$$

These equations are solved for w_θ and w_ϕ which are the signals that would be measured in a perfectly polarized probe.

$$w_\theta = \frac{1}{1 + p^{\mathrm{I}} p^{\mathrm{II}}}(w^{\mathrm{I}} + p^{\mathrm{I}} A_{\theta\phi} w^{\mathrm{II}}) \tag{5.37}$$

$$w_\phi = \frac{1}{1 + p^{\mathrm{I}} p^{\mathrm{II}}}(-p^{\mathrm{II}} w^{\mathrm{I}} + A_{\theta\phi} w^{\mathrm{II}}). \tag{5.38}$$

These are the polarization correction equations. They can also be written in terms of circular components by substitution of equations (4.65–4.66) and (5.28–5.29)

$$w_{\mu=+1} = \frac{1}{2}\frac{1}{Q^{\mathrm{I}} - Q^{\mathrm{II}}}\{(Q^{\mathrm{I}} + 1)w^{\mathrm{I}} - A_{\theta\phi} i(1 - Q^{\mathrm{II}})w^{\mathrm{II}}\} \tag{5.39}$$

$$w_{\mu=-1} = \frac{1}{2}\frac{1}{Q^{\mathrm{I}} - Q^{\mathrm{II}}}\{-(Q^{\mathrm{I}} + 1)Q^{\mathrm{II}} w^{\mathrm{I}} + A_{\theta\phi} i Q^{\mathrm{I}}(1 - Q^{\mathrm{II}})w^{\mathrm{II}}\} \tag{5.40}$$

5.2.3.3 Discussion: Although reciprocity entered the derivation of the above equations, they can be used for non-reciprocal probes as well, provided the polarization parameters are determined in the proper transmit/receive mode. For a receiving probe, $E^{\mathrm{I,II}}_{x,y}$ in the definitions (5.28) to (5.32) can be taken as the signal received in the proper port from a dipole with the indicated polarization.

A significant observation [3] about the polarization correction equations is that they only require a linear relationship between the two pure waveguide modes and the signals w^{I} and w^{II}. Therefore w^{I} and w^{II} can be defined as readings on the receiver with the entire RF-system assembled as described in Section 5.3.2.2. In that case, the polarization parameters are measured for the integrated system and with that, eqns (5.37) to (5.40) will also correct for mismatch in connectors and cables, bad isolation in the OMT or the switch, and cross talk, gain and phase differences between the two channels of the receiver. In most cases, including the measurement examples in the present chapter, only relative probe correction is needed since the total system sensitivity is determined in a separate measurement. However, the output of eqns (5.37) and (5.38) is defined in an absolute sense as the signal that would be measured with a perfect x-polarized probe having the same gain as the x-component of port I (cf. eqn 5.35).

Equations (5.37) to (5.40) can also be used for a single polarized probe since a 90° rotation between the two measurements results in $p^{\mathrm{II}} = p^{\mathrm{I}}$ and $Q^{\mathrm{II}} = -Q^{\mathrm{I}}$. A perfect rotary joint implies $A_{\theta\phi} = 1$.

5.2.4 *Polarization calibration*

5.2.4.1 Polarization scans: A polarization calibration of the probe comprises determination of the parameters used in the equations described in the preceding section. Usually, these measurements are carried out after the probe pattern measurement described in Section 5.2.5, but they are outlined here in order to shorten the treatment of on-axis probe polarization. The measurements are carried out with the probe integrated into the scanning system, and a second antenna with known on-axis polarization rotating in front of it. The second antenna can be a small symmetric pyramidal horn which typically will have an axial ratio better than 60 dB and therefore can be assumed to be linearly polarized, or its polarization can be determined in a *three-antenna polarization measurement* in which one of the antennas may be the probe.

One pair of antennas is mounted in front of each other at a time. By rotating one of the antennas about the common axis, one can determine the product of the two circular polarization ratios. This is done for all three pairs of antennas and the individual polarization ratios of the antennas can then be found by solving three equations with three unknowns.

Such polarization scans are used for polarization calibration as well as for the three-antenna measurement. The phi-scan for theta equal to zero in a spherical near-field measurement is an example of a polarization scan. We shall therefore consider this in some detail. We require a coordinate system being defined for each antenna with the z-axis pointing out of the aperture and the x-axis perpendicular to it. For this purpose, it is convenient to mount the horns on precision frames where a spirit level can be used to define horizontality of the x-axis.

Suppose two antennas a and b are mounted as shown in Fig. 5.6 with the observation point P being in the far field of both antennas. The field radiated by each antenna can then be decomposed into circular components at the point P.

The letter R denotes right- and L denotes left-hand circular polarization with respect to the directions of propagation z^a and z^b. If one antenna is transmitting to the other, and either antenna is rotated an angle u as defined in the figure, then the received signal can be written

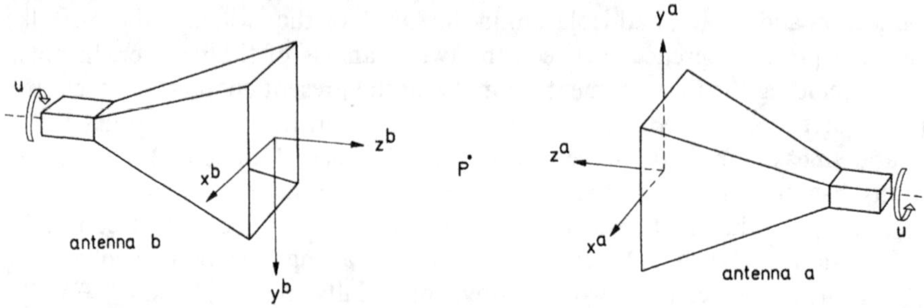

Fig. 5.6 *Polarization measurement*

$$\begin{aligned} w(u) &= C'(K_R^a K_R^b \mathrm{e}^{iu} + K_L^a K_L^b \mathrm{e}^{-iu}) \\ &= C\left(\frac{K_R^a}{K_L^a}\frac{K_R^b}{K_L^b}\mathrm{e}^{iu} + \mathrm{e}^{-iu}\right) \\ &= C(Q^a Q^b \mathrm{e}^{iu} + \mathrm{e}^{-iu}) \\ &= C(Q^{ab} \mathrm{e}^{iu} + \mathrm{e}^{-iu}) \end{aligned} \tag{5.41}$$

where

$$Q^a \equiv \frac{K_R^a}{K_L^a}, \qquad Q^b \equiv \frac{K_R^b}{K_L^b} \tag{5.42}$$

$$Q^{ab} \equiv Q^a Q^b \tag{5.43}$$

and where C and C' are constants.

Note that u is defined such that the signal is independent of which antenna is rotated. In addition, the symbols R and L are defined such that (5.41) is independent of which antenna is transmitting and which is receiving. A reciprocal antenna which is right-hand polarized when transmitting will also be right-hand polarized when receiving. The angle u is zero when x^a is parallel to x^b.

The phase variation with u can be explained in the following way: assume that the field from antenna a is purely left-hand polarized, $K_R^a = 0$. In this case, increasing u is equivalent to increasing the time t (see Fig. 5.6).

As our time factor is $\mathrm{e}^{-i\omega t}$, the received signal must depend on u as e^{-iu}. It is seen that the variation of the signal is determined by the product of the circular polarization ratios for the two antennas. This product can either be determined by a manual procedure [10] or from an automatic measurement [11].

5.2.4.2 Manual polarization measurements: The manual measurement involves

1. The maximum-to-minimum ratio of the signal in dB.
2. The position u_{min} of one of the minima.
3. The sense of the phase variation with increasing u.

The signal has a minimum when the e^{iu} term and the e^{-iu} term are in opposite phase. There is a maximum when they are in phase. This happens 90° from the minima. There are two maxima and two minima. Let $u_{\min}$ be one of the minima, then

$$\begin{aligned} \arg(Q^{ab}) + u_{\min} &= -u_{\min} + 180° + n360° \\ \arg(Q^{ab}) &= -2u_{\min} + 180° + n360°. \end{aligned} \tag{5.44}$$

Further we define the *maximum-to-minimum ratio*

$$S^{ab} \equiv \left|\frac{w_{\max}}{w_{\min}}\right| = \left|\frac{1 + |Q^{ab}|}{1 - |Q^{ab}|}\right| \tag{5.45}$$

which can be solved to give

$$|Q^{ab}| = \left\{\frac{S^{ab} + 1}{S^{ab} - 1}\right\}^{\pm 1}. \tag{5.46}$$

To determine whether the plus or the minus sign shall be applied, the phase variation of $w(u)$ near $u_{\min}$ is examined. The plus sign implies $|Q^{ab}| > 1$ whereby eqn (5.41) indicates that $\arg(w(u))$ will increase with increasing u near $u_{\min}$. The minus sign shall be used when $\arg(w(u))$ is decreasing. However, the above formulas assume a time factor of $e^{-i\omega t}$. In order to check the time factor of the measurement equipment, one can note the direction of phase change when the distance between the two antennas is slightly increased. A decreasing phase means an e^{-jkr} dependence of the measured signal and a time factor of $e^{j\omega t}$ such that the sign of the phase must be shifted before the above formulas can be applied.

5.2.4.3 *Automatic polarization measurement*: In the automatic polarization measurement procedure a full scan is taken for $\theta = 0°$, $0° \leqslant \phi < 360°$ and the phase and amplitude of the signal are sampled equidistantly. The angle ϕ is equivalent to u in Fig. 5.6. These data are then Fourier transformed. With J_ϕ being the number of samples, the Fourier components are calculated from

$$w_m = \frac{1}{J_\phi}\sum_{j=0}^{J_\phi - 1} w(\phi_j)\, e^{-im\phi_j}, \qquad \phi_j = j\frac{2\pi}{J_\phi} \tag{5.47}$$

corresponding to the Fourier series

$$w(\phi) = \sum_{m=-(J_\phi - 1)/2}^{(J_\phi - 1)/2} w_m\, e^{im\phi}. \tag{5.48}$$

Ideally, the signal should have the form

$$w(\phi) = w_{+1}\, e^{i\phi} + w_{-1}\, e^{-i\phi} \tag{5.49}$$

but in practice, other components will be present due to nonlinearity of the receiver, small phase changes during the rotation, misaligment and near-field effects. The polarization ratio product Q^{ab} is found directly as

$$Q^{ab} = \frac{W_{+1}}{W_{-1}}. \tag{5.50}$$

The procedure has the advantage that, apart from being automatic, Q^{ab} is found by averaging over many samples thereby increasing the accuracy of the measurement.

An extension of this method has been published, where both antennas are rotated such that antenna a rotates a full revolution for each position of antenna b [12].

5.2.4.4 Completing the three-antenna measurement: With three antennas, a, b and c, three sets of measurements are made and Q^{ab}, Q^{ac} and Q^{bc} are determined. From these quantities Q^a, Q^b and Q^c can be calculated.

$$Q^a = \pm\sqrt{\frac{Q^a Q^b Q^a Q^c}{Q^b Q^c}} = \pm\sqrt{\frac{Q^{ab} Q^{ac}}{Q^{bc}}} \tag{5.51}$$

$$Q^b = Q^{ab}/Q^a \tag{5.52}$$

$$Q^c = Q^{ac}/Q^a. \tag{5.53}$$

The sign ambiguity in eqn (5.51) means that the polarization can be calculated apart only for a constant of 90° in the tilt angle. The sign must be chosen from knowledge of one of the antennas. Near x-polarization (within 45°) implies that the real part of Q is positive.

From Q^a (the circular polarization ratio), other polarization parameters like axial ratio and tilt angle can be found. In measurements it is customary to state the axial ratio in dB and the sense explicitly. This is slightly different from the definition in eqn (2.201).

$$\text{Axial ratio} = 20\log\left|\frac{1+|Q^a|}{1-|Q^a|}\right| \text{dB} \tag{5.54}$$

$$\text{Tilt angle} = -\tfrac{1}{2}\arg(Q^a) \tag{5.55}$$

$$\text{Sense:} \quad \left.\begin{array}{ll} |Q^a| > 1 & \text{implies right-hand polarization} \\ |Q^a| = 1 & \text{implies linear polarization} \\ |Q^a| < 1 & \text{implies left-hand polarization} \end{array}\right\} \tag{5.56}$$

If a dual-polarized probe enters, two sets of three-antenna measurements have actually been made: one for each of the two ports I and II. Therefore two values of the circular polarization ratio can be calculated for each of the two other antennas. The inconsistency between the results indicates the accuracy of the measurement. Further redundancy can be added by including a fourth antenna [10].

In case the polarization ratio of the rotating antenna Q^a in front of the probe is known, polarization ratios Q^{I} and Q^{II} for the probe ports can be found from eqn (5.53). However, it is possible to calculate the ratio $Q^{\mathrm{I}}/Q^{\mathrm{II}}$ from a polarization scan without knowing Q^a, since

$$\frac{Q^{\mathrm{I}}}{Q^{\mathrm{II}}} = \frac{Q^{\mathrm{I}}Q^a}{Q^{\mathrm{II}}Q^a} = \frac{w^{\mathrm{I}}_{+1}w^{\mathrm{II}}_{-1}}{w^{\mathrm{I}}_{-1}w^{\mathrm{II}}_{+1}}. \tag{5.57}$$

Thus each time an antenna is rotated in front of the probe, one gets a value which can be checked for consistency with the polarization calibration.

5.2.4.5 Amplitude phase factor for dual-polarized probes: The polarization calibration also comprises determination of the amplitude phase factor between the two ports of the probe. As we shall see later, accuracy in this determination is highly important in spherical near-field measurements. We shall here derive a number of estimates that should all theoretically give the same result, but since no measurement is perfect, they will not. One will have to consider the experimental circumstances to choose among them since they are sensitive to different kinds of errors.

The amplitude phase factor $A_{\theta\phi}$ is defined in eqn (5.32). Comparison with Figs 5.5(a) and 5.6 shows that if the second antenna is linearly x-polarized, the ratio can be found from measurements at two angles u of port I and port II, respectively.

$$(A_{\theta\phi})_1 = \frac{w^{\mathrm{I}}(0°)}{w^{\mathrm{II}}(270°)}. \tag{5.58}$$

Similarly

$$(A_{\theta\phi})_2 = \frac{w^{\mathrm{I}}(180°)}{w^{\mathrm{II}}(90°)}. \tag{5.59}$$

Since a rotation of 180° ideally gives a phase change of 180°, various other combinations or mean values of the above expressions can be used. If a polarization scan has been taken with an x-polarized antenna and has been Fourier transformed, we find

$$(A_{\theta\phi})_3 = \frac{w^{\mathrm{I}}_{+1}\mathrm{e}^{i\cdot 0} + w^{\mathrm{I}}_{-1}\mathrm{e}^{-i\cdot 0}}{w^{\mathrm{II}}_{+1}\mathrm{e}^{-i(\pi/2)} + w^{\mathrm{II}}_{-1}\mathrm{e}^{i(\pi/2)}} = i\frac{w^{\mathrm{I}}_{+1} + w^{\mathrm{I}}_{-1}}{w^{\mathrm{II}}_{+1} - w^{\mathrm{II}}_{-1}} \tag{5.60}$$

If the antenna is not x-polarized, i.e. $Q^a \neq 1$, the ratio between w^{I}_{+1} and w^{I}_{-1} is changed by Q^a, since ideally

$$\frac{w^{\mathrm{I}}_{+1}}{w^{\mathrm{I}}_{-1}} = Q^{\mathrm{I}}Q^a \tag{5.61}$$

with a similar relation for port II. Therefore, if we multiply the measured w_{-1} by Q^a, we get the same relationship between w_{+1} and w_{-1} as we would have obtained for the perfect polarized antenna such that we can use eqn (5.60)

$$(A_{\theta\phi})_4 = i\frac{w^{\mathrm{I}}_{+1} + w^{\mathrm{I}}_{-1}Q^a}{w^{\mathrm{II}}_{+1} - w^{\mathrm{II}}_{-1}Q^a}. \tag{5.62}$$

A particular formula for a y-polarized antenna is obtained with $Q^a = -1$. Equation (5.62) is good for near-linear polarization, but it becomes inaccurate if antenna a is close to circular polarization. Right-hand polarization for example implies that Q^a is large such that the small numbers w^{I}_{-1} and w^{II}_{-1}, corresponding to left-hand circular polarization, are amplified in the expression. We can eliminate $w^{\mathrm{I}}_{-1}Q^a$ and $w^{\mathrm{II}}_{-1}Q^a$ by using the ideal formula (5.61) for port I and its equivalent for port II, respectively. In this way we obtain an expression which only involves the right-hand circular components

$$(A_{\theta\phi})_5 = i\frac{(Q^{\mathrm{I}} + 1)Q^{\mathrm{II}}w^{\mathrm{I}}_{+1}}{(Q^{\mathrm{II}} - 1)Q^{\mathrm{I}}w^{\mathrm{II}}_{+1}} = i\frac{(1 - ip^{\mathrm{II}})}{(1 - ip^{\mathrm{I}})}\frac{w^{\mathrm{I}}_{+1}}{w^{\mathrm{II}}_{+1}}. \tag{5.63}$$

We can also get a formula which only involves the left-hand circular components

$$(A_{\theta\phi})_6 = i\frac{(Q^{\mathrm{I}} + 1)}{(Q^{\mathrm{II}} - 1)}\frac{w^{\mathrm{I}}_{-1}}{w^{\mathrm{II}}_{-1}} = -i\frac{(1 + ip^{\mathrm{II}})}{(1 + ip^{\mathrm{I}})}\frac{w^{\mathrm{I}}_{-1}}{w^{\mathrm{II}}_{-1}}. \tag{5.64}$$

Finally, it is possible to find the amplitude phase factor from measurement of w^{I} and w^{II} with a linearly polarized antenna rotated 45°. If Q^a is the polarization ratio of the antenna before the 45° rotation, i.e. Q^a is close to 1, insertion of w^{I} and w^{II}, after the rotation, into the polarization correction equations (5.39) and (5.40) ideally yields a ratio $w_{\mu=+1}/w_{\mu=-1}$ equal to $-iQ^a$. Solving the resulting equation gives

$$(A_{\theta\phi})_7 = \frac{(1 - iQ^{\mathrm{II}}Q^a)(Q^{\mathrm{I}} + 1)w^{\mathrm{I}}(u = 45°)}{(i + Q^{\mathrm{I}}Q^a)(1 - Q^{\mathrm{II}})w^{\mathrm{II}}(u = 45°)}. \tag{5.65}$$

It should be made clear that the various estimates are not totally independent. Nevertheless, in practical measurements the final cross-polar results can depend on which one is used. Estimates 3, 4, 5 and 6 are averages over many samples. They require the phase measurement to be stable and the antenna not to deform during rotation. Estimates 1, 2, 3 and 4 are not very sensitive to small (unknown) ellipticities of the polarizations of either the ports or the rotating antenna. These estimates are very suitable in a calibration without a three-antenna measurement. They require, however, reasonably linear polarization of the small horn involved. Estimates 5 and 6 should be used for circularly polarized antennas. If the ratio $Q^{\mathrm{I}}/Q^{\mathrm{II}}$ determined from the measured data by eqn (5.57) is the same as the ratio between Q^{I} and Q^{II} used in 5 and 6, these estimates will give identical results. If furthermore, Q^a used in 4 satisfies (5.61), 4 will also give this result. Estimate 7 is not affected by changes during rotation such as deformation of a large antenna. This can be of interest when checking whether an amplitude phase factor has changed between calibration and near-field scanning of the test antenna.

Nonlinearity of the receiver will affect the polarization correction and the amplitude phase factor. In a polarization scan, nonlinearity can be observed as low-order harmonics different from ± 1. If they are too high compared to the ± 1 harmonics, signal power can be reduced or a linearity correction can be worked out.

5.2.5 Pattern calibration

After polarization correction, the near-field data for the test antenna can be interpreted as having been measured by a symmetric probe with linear polarization on the axis, i.e. a probe with a pattern as in eqn (5.6). Correction for this pattern is done using spherical receiving coefficients for the probe as described in Chapter 4. Only the copolar E-plane pattern $f_1(\theta)$ and the copolar H-plane pattern $f_2(\theta)$ are needed for determination of the coefficients. Although the probe pattern correction in the data processing takes place after the polarization correction, measurement of the probe pattern is actually done before the polarization calibration described in the previous section.

The measurements are usually carried out with the probe mounted as the test antenna on a two-axis positioner; see Fig. 5.11. A small pyramidal horn can be used as measuring antenna. The probe axis shall be coincident with the horizontal rotation axis (phi axis). The theta angle is varied by rotation about the vertical axis. Preferably this axis shall be in the aperture plane of the probe. This is not strictly necessary but the intersection point for the two axes defines the centre of the probe coordinate system. It is convenient to have this located at the probe aperture when the near-field measurement distance is to be determined later. There is no reason to search for a phase centre in this measurement since the phase pattern is measured relative to the probe coordinate system and is included in the probe pattern correction. The E-plane pattern is measured with the pyramidal horn horizontally polarized while the horn polarization must be vertical during H-plane measurements. The probe polarization shall also be horizontal and vertical, respectively, in the two measurements. This can be accomplished by rotating the probe 90° between the two measurements but for a dual-polarized probe it is preferable to switch the excitation from port I to port II. In this way the two patterns are measured in the same physical plane such that their phase difference is not affected by a small error in the intersection of the probe axis and the vertical axis. The pyramidal horn has still to be rotated. The phase difference is important for the determination of cross-polar lobes as can be seen from eqn (5.8). Lack of intersection will show up as an asymmetrical phase curve as explained in [13, p. 74]. The probe model, ref. eqn (5.3), assumes the patterns to be symmetric about $\theta = 0°$. Measured data can be slightly asymmetric. Therefore, if they have been measured all the way from $\theta = 0°$ to

$\theta = 360°$, one can calculate a symmetric pattern as

$$w_{\text{sym}}(\theta, \phi) = \tfrac{1}{2}\{w(\theta, \phi) + w(360° - \theta, \phi)\}. \tag{5.66}$$

Also, one can rotate the probe 180° in ϕ, measure the pattern, and subtract it from the first one to get an average pattern. The pattern shall be subtracted because the phase has changed 180°. Finally, the symmetric E- and H-plane patterns are normalized to have the same amplitude and phase at $\theta = 0°$.

An alternative to the above procedure is to measure the copolar and the cross-polar pattern in the $\phi = 45°$ plane [9, p. 179]. The functions $f_1(\theta)$ and $f_2(\theta)$ can then be found by solving eqns (5.7) and (5.8) with $\phi = 45°$.

An advantage of the direct measurement of E- and H-plane patterns is that the probe and the measuring antenna need not be perfectly linearly polarized on boresight. In addition the two patterns can be mutually normalized by comparison at the intersection point such that one does not have to keep the phase constant during polarization shift of the measuring antenna.

The full pattern of the probe with linear excitation can now be found by eqn (5.6). The pattern is inserted into the algorithm described in Chapter 4 which calculates spherical wave coefficients T_{smn}. Theoretically, one should apply the receiving coefficients for the measuring antenna in this calculation but in practice the probe and the measuring antenna are small compared to the measurement distance. Therefore, dipole coefficients can be used for the measuring antenna, ref. Section 3.3.2, and the iteration described in Section 3.2.5 is not necessary.

The final step in the calculation of probe coefficients is a reorientation of the coordinate system in Fig. 5.5(a). During the measurement of the probe pattern and the polarization measurements, it is convenient to have the z-axis pointing out of the probe as in Fig. 5.5(a). This is the normal convention in measurement set-ups. However, in the transmission formula for spherical near-field measurements with probe correction, the z'-axis points into the probe and away from the test antenna, as described in Chapter 4 and shown in Fig. 5.5(b). The reorientation is accomplished by rotating the coordinate system 180° about the x-axis. It is carried out by the following operation on the coefficients

$$T^{\text{rot}}_{smn} = (-1)^n T_{s,-m,n} \tag{5.67}$$

which can be derived from the rotation formulas for spherical waves, (A2.2) and (A2.16). Finally, receiving coefficients are found by the following formula valid for reciprocal antennas:

$$R_{smn} = (-1)^m T^{\text{rot}}_{s,-m,n}. \tag{5.68}$$

The R_{smn} terms for $m = \pm 1$ are the wanted probe coefficients. When the probe is moved to the probe tower after pattern calibration, the name of the coefficients is changed to $R^p_{\sigma\mu\nu}$, thereby leaving the indices smn for the test antenna modes. Because of the symmetry and the linear polarization of the probe pattern inserted into the algorithm, the calculated probe coefficients shall satisfy the relations (5.11–5.12).

5.3 Probe-corrected measurements

5.3.1 Introduction

We shall now deal with the experimental side of spherical near-field measurements. A spherical near-field range, like other types of antenna test ranges, is a fairly complex arrangement consisting of several more or less independent systems. Important among these are the mechanical system, the radio-frequency system and the systems for data collection and control and for data processing. Section 5.3.2 gives a general description of the spherical near-field test range from a system point of view. The emphasis is on the mechanical system which characterizes the spherical near-field test range relative to other near-field test ranges.

Section 5.3.3 covers the basic measurement procedures. The spherical near-field range is described here in terms of the sequence of processes which form a complete measurement. The typical measurement for radiation pattern and directivity comprises alignment of the set-up (Sections 5.3.3.1–5.3.3.3 and 5.3.3.5), probe calibration (Section 5.3.3.4), sampling and scanning the test antenna (Sections 5.3.3.6–5.3.3.7) and data processing. The latter is mentioned in connection with the examples in Section 5.3.4. If measurement of gain is wanted, the required additional procedures are given in Section 5.4.

5.3.2 The spherical near-field test range

5.3.2.1 Mechanical system: The geometry of spherical near-field testing requires a mechanical set-up by which the probe can be placed at any point on the surface of a measurement sphere containing the test antenna. The probe must be pointing towards the centre of the measurement sphere at all times and is furthermore required to receive (or transmit) in two orthogonal polarizations.

Basically, the positioning mechanism must be able to provide three angular motions. These are the chi rotation, the theta rotation and the phi rotation corresponding to the three angular variables (χ, θ, ϕ) which together with the fixed measurement distance A describe the orientation and location of the probe relative to the test antenna.

Which of the two antennas that actually moves during the positioning is of course less important although in some cases, in particular in space applications, rotation of a test antenna is restricted due to gravity deformations. Each specific type of spherical test range is characterized by the mechanical interconnection of the three rotation axes and their relation to the permanent *laboratory coordinate system* (X, Y, Z).

When properly aligned (see Section 5.3.3), a spherical near-field test range provides a theta axis and a phi axis intersecting at right angles. Both axes are oriented and their intersection point O defines the centre of the measurement sphere. In addition, the test range provides a third axis, the chi axis or *line of sight* which passes through O and is orthogonal to the theta axis. The chi axis contains the measurement point O′ and is oriented in the direction from O towards O′. The measurement distance A is equal to the distance between O and O′. The basic geometry is illustrated in Fig. 5.7 [14]. The orientations of the

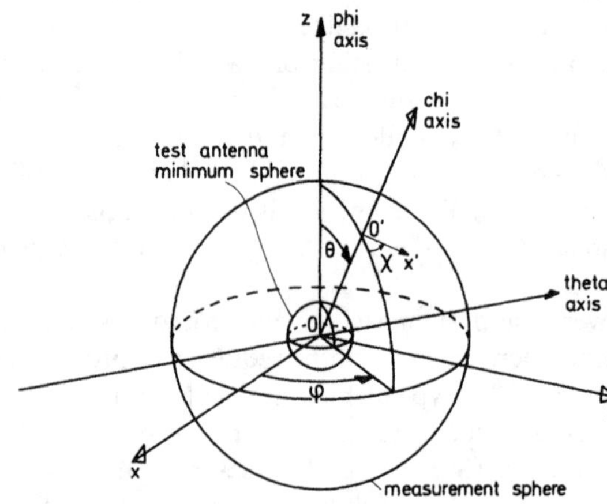

Fig. 5.7 *Illustrating the general spherical near-field set-up. The test antenna is located within the minimum sphere and is fixed with respect to the rectangular range coordinate system (x, y, z). The measurement point (or probe location) O′ and the probe orientation are defined by the range coordinates* (A, χ, θ, ϕ)

three rotation axes are such that for $\theta = \pi/2$ the chi, theta and phi axes, in that order, form a right-handed coordinate system. In principle, an angular display, or read-out, is connected to each of the rotation axes in order that the angles of rotation can be measured. The three angles, chi, theta and phi, respectively, are defined as increasing when the probe rotation, as seen from the test antenna, is in agreement with the direction of positive orientation of the corresponding axis of rotation (right-hand rule).

In the alignment process, ref. Section 5.3.3, the correct zero setting of the angular displays is made. The *range coordinates* (A, χ, θ, ϕ) of any given measurement point and probe orientation are thereby defined for the measurement set-up. The three rotation axes, the range coordinates and the associated rectangular *range coordinate system* (x, y, z) are illustrated schematically in Fig. 5.7.

At a test range, provisions for easy and reproducible mounting of the probe and the test antenna should be present. Various types of spherical ranges are shown in Figs. 5.9–5.13, which will be discussed later.

The probe, normally of the symmetric $\mu = \pm 1$ type (cf. Section 5.2), has its own coordinate system (x', y', z') in relation to which it is calibrated. During the alignment of the probe in the range, the probe axis (the z'-axis) is made coincident with the chi axis. The origin of the probe coordinate system is made coincident with the measurement point O′.

The test antenna usually has its rectangular system as well. Let us denote *the test antenna coordinate system* by (x'', y'', z'') for a moment. It may be defined with the aid of fiducial marks on the antenna or by an optical cube glued to it. If the test antenna is a rectangular horn, the coordinate system can be defined via the waveguide mounting flange or via the horn aperture. When mounting the test antenna in the set-up, one of two different orientations is normally chosen. In the first of these, the test antenna is 'polar pointing' (see Fig. 5.8(a)), with the main beam in the direction of the phi-axis of the range. In the second orientation, the test antenna is 'equator pointing' (see Fig. 5.8(b)), with the main beam orthogonal to the phi axis of the range.

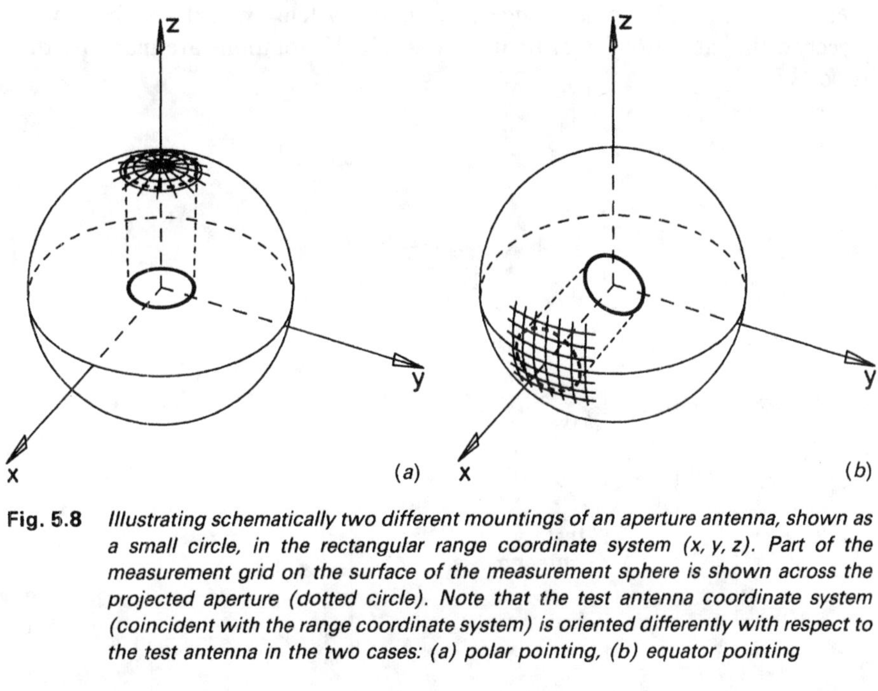

Fig. 5.8 *Illustrating schematically two different mountings of an aperture antenna, shown as a small circle, in the rectangular range coordinate system (x, y, z). Part of the measurement grid on the surface of the measurement sphere is shown across the projected aperture (dotted circle). Note that the test antenna coordinate system (coincident with the range coordinate system) is oriented differently with respect to the test antenna in the two cases: (a) polar pointing, (b) equator pointing*

A test antenna radiation pattern is naturally required in test antenna spherical coordinates. However, the test antenna will always be measured in range coordinates. The purpose of its alignment in the range is then to make its (x'', y'', z'') coordinate system coincident with the range system (x, y, z). If this can be accomplished with sufficient precision there is no need to distinguish between the range coordinates and the test antenna coordinates.

In practice, a precise alignment of the two coordinate systems is not required. Instead, their relative location and orientation is measured before the test antenna is demounted from the range. Using translation and rotation of coordinate systems, the measured pattern can then always be expressed in test antenna coordinates even though it has been measured in range coordinates. It is then clear that without any loss of generality we may assume in later discussions that the two systems are coincident.

It should be noted that in many circumstances, radiation patterns need not be expressed in any precisely defined test antenna coordinate system. This is typically the case when a particular test antenna can be electrically aligned at the site of its final application. In such situations, test range coordinates that have an unknown or partially known orientation with respect to the test antenna structure are often sufficient.

Examples of spherical near-field set-ups are shown in Figs 5.9–5.13. The set-up that most closely resembles the general, academic configuration of Fig. 5.7 is probably the *double gantry arm* mechanism shown in Fig. 5.9. Here, the test antenna and the rectangular range coordinate system (x, y, z) are fixed with respect to the laboratory coordinates (X, Y, Z). All rotations are made by the probe.

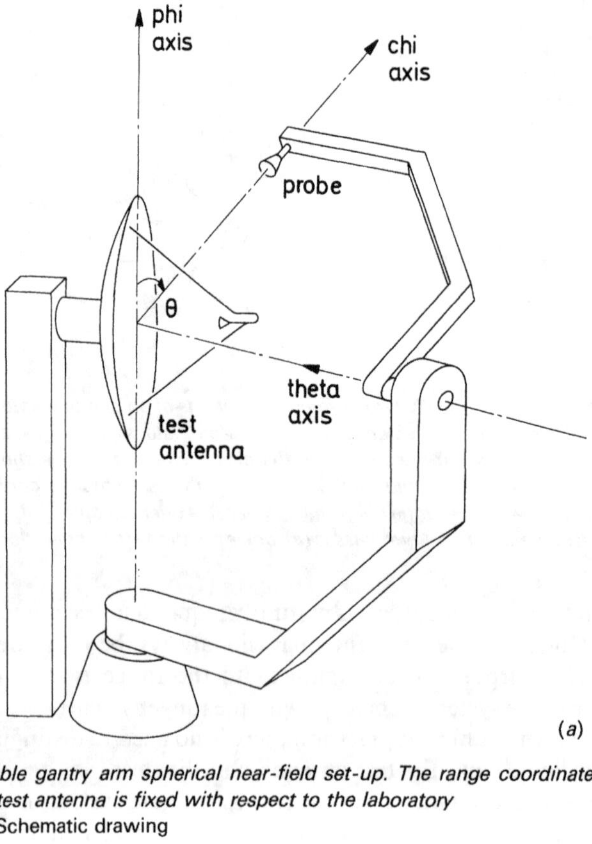

(*a*)

Fig. 5.9 *Double gantry arm spherical near-field set-up. The range coordinates system with the test antenna is fixed with respect to the laboratory*
(*a*) Schematic drawing

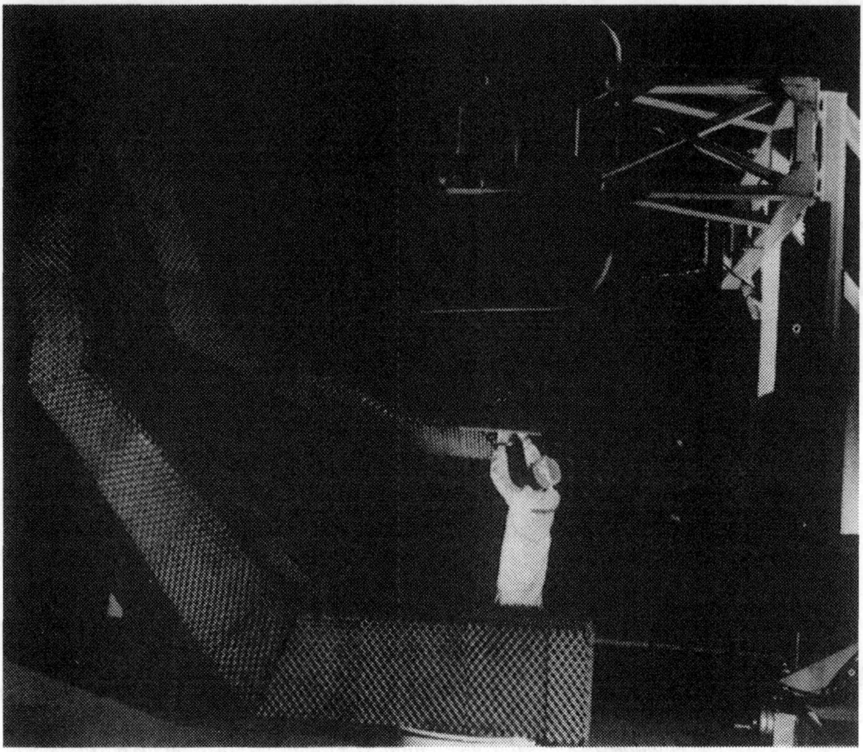

Fig. 5.9 (*b*) *Practical implementation.* (*Photo courtesy Aerospatiale, Cannes, France*)

For the *single gantry arm* set-up shown in Fig. 5.10 the phi rotation is made by the test antenna. The direction of rotation for increasing phi is then opposite to that corresponding to the orientation of the axis and is shown by the broad arrow. The theta and chi rotations are made by the probe and are therefore positive in agreement with the positive orientation of the theta and chi axes, respectively.

Both gantry arm set-ups may of course be applied in the polar-pointing as well as the equator-pointing mode. An advantage of these ranges is the fact that the test antenna is either not moving at all (Fig. 5.9) or is only rotating about a vertical axis (Fig. 5.10). A disadvantage is that a change of measurement distance is inconvenient.

The best known of the spherical near-field set-ups is presumably the mechanical system shown in Fig. 5.11 which is sometimes called an *elevation-over-azimuth* positioning system [15]. The radius of the measurement sphere can be easily changed in this set-up by moving the probe tower closer to or further away from the test antenna. Although a polar-pointing test antenna is shown, equator pointing can also be accommodated. Here, the only movement of the probe is the chi rotation. The test antenna, together with the rectangular range

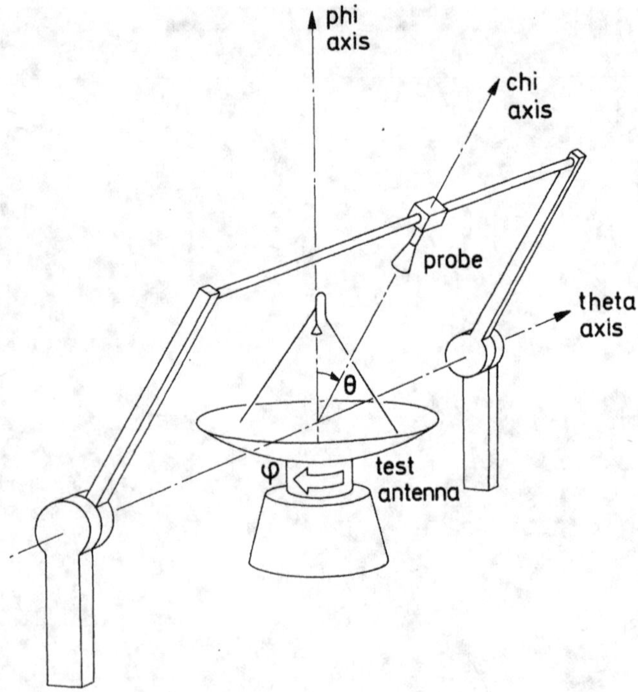

Fig. 5.10 *Single gantry arm spherical near-field set-up. The range coordinate system with the test antenna performs the phi rotation. The theta and chi rotations are done with the probe.*

coordinate system (x, y, z) is performing the rotations about the phi and theta axes. The directions of rotations for increasing phi and theta are then opposite to those corresponding to the positive orientations of the two axes and are shown by the broad arrows. Note, that both broad arrows are pointing clockwise when seen from the test antenna. In Section 5.3.3 we shall deal with the alignment of this type of test set-up.

As modern alternatives to the above spherical near-field set-ups, two examples of *robotic positioners* are shown in Figs 5.12–5.13. An advantage of using this type of positioner is the high degree of flexibility. Spherical, cylindrical and planar near-field scanning can thus be performed with a single robotic positioner.

Finally (see Fig. 5.14), a spherical near-field set-up with only one mechanical axis of rotation has been described [16]. The test antenna (shown equator-pointing) and the rectangular range coordinate system (x, y, z) are fixed with respect to the laboratory coordinates (X, Y, Z). A circular gantry arm with dual-polarized probes in the form of scattering dipoles provides the chi rotation as well as the theta rotation electrically while the rotation about the phi axis is mechanical. The advantage here is the high measurement speed.

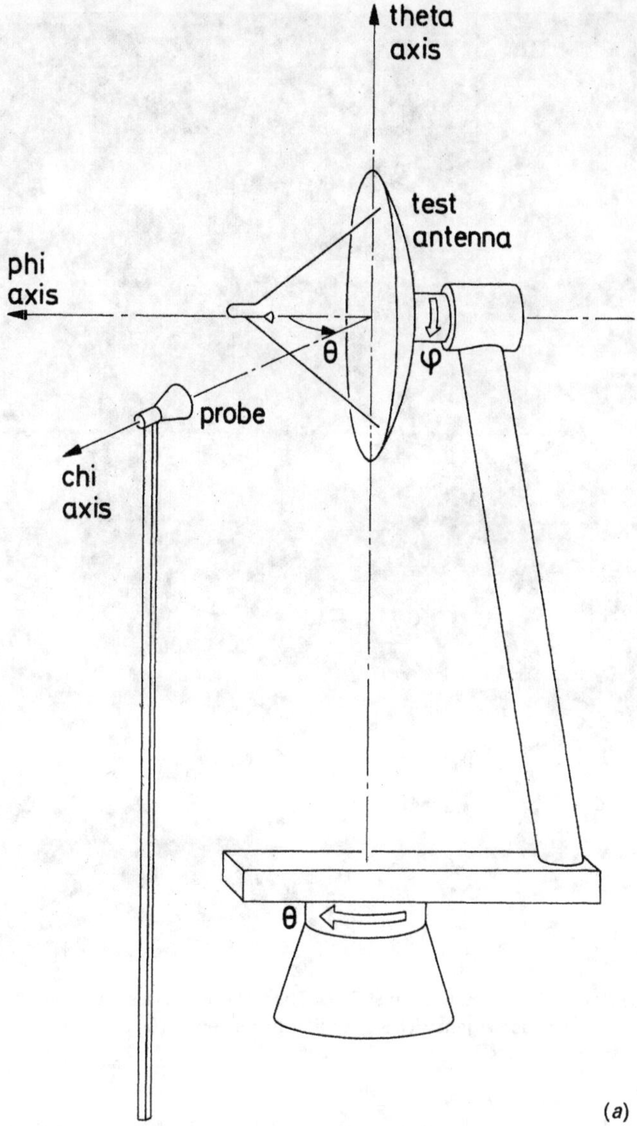

Fig. 5.11 *Elevation over azimuth spherical near-field set-up. The range coordinate system with the test antenna performs both the phi and the theta rotation. The chi rotation is done with the probe*
(*a*) Schematic drawing

Fig. 5.11 (*b*) Practical implementation at TUD. The test object is a spot-beam antenna for the European Space Agency (ESA) orbital test satellite (OTS)

5.3.2.2 Radio-frequency system: Spherical near-field testing relies, as other near-field scanning methods, upon an RF signal being transmitted between an antenna under test and a probe. At the receiving side, the signal is measured in amplitude and phase. A complete scan (ref. Section 5.3.3.7) of the measurement sphere by the probe could take from a few minutes to several days depending upon the number of sample points. This puts stringent requirements on the RF circuit which must be highly stable in attenuation and electrical path lengths.

The basic components of an RF system for near-field measurements comprise a signal source, an amplitude phase receiver, a probe (preferably dual polar-

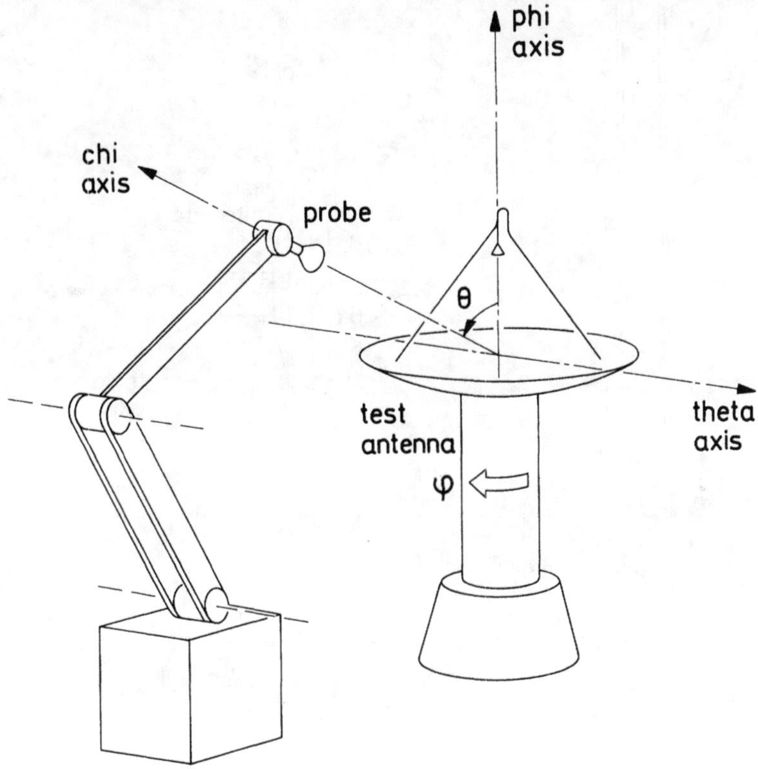

Fig. 5.12 *Robotic positioner*

ized) and the interconnecting circuitry. An RF system also includes a suitable anechoic room with a reasonable control of temperature.

Block diagrams for typical RF systems are shown in Fig. 5.15(a, b) [4, 17]. The set-ups have opposite directions of transmission and both employ a receiving system with harmonic mixing. This gives a wide frequency coverage using a single local oscillator, phase locked to the signal source. The governing equation is

$$f_S = pf_{LO} \pm f_{IF} \tag{5.69}$$

where f_S is the RF signal frequency, f_{LO} the local oscillator frequency and f_{IF} the intermediate frequency. The harmonic employed for the coverage of a particular frequency band is designated by the positive integer p. Both set-ups contain a signal (measurement) path consisting of an RF section of length l_1 and an IF section of length l_3. Besides the signal path, the set-ups contain a reference path with an RF section of length l_2 and an IF section of length l_4. The measurement path will normally consist of two time-shared polarization channels (not shown in Fig. 5.15), one for each of the two ports of a dual-polarized probe.

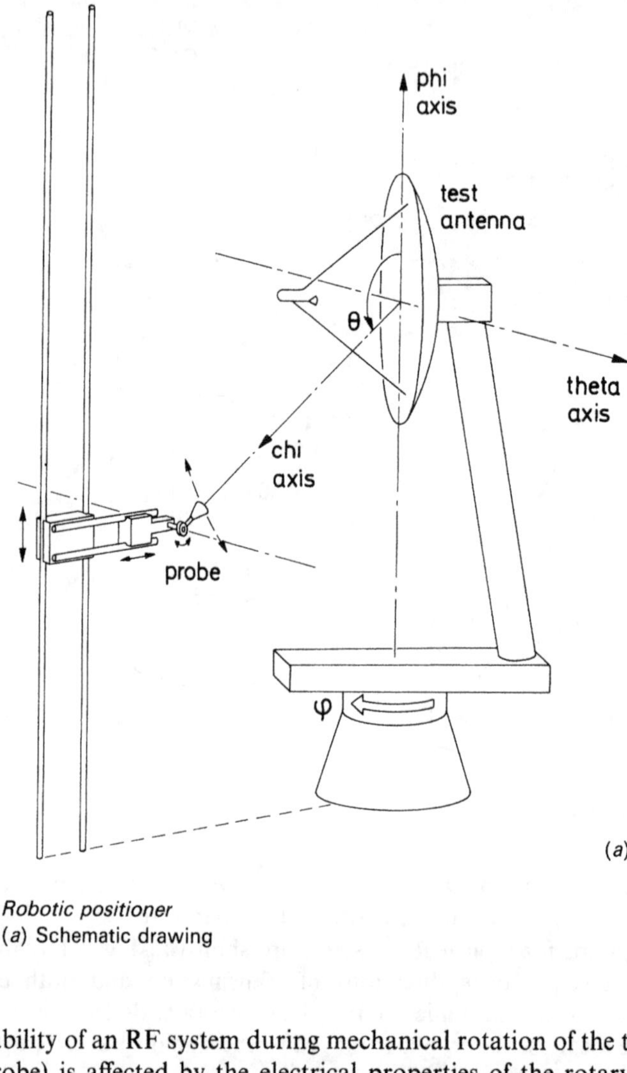

Fig. 5.13 *Robotic positioner*
(*a*) Schematic drawing

The stability of an RF system during mechanical rotation of the test antenna (or the probe) is affected by the electrical properties of the rotary joints employed. The phase shift through a coaxial rotary joint as a function of rotation is often rather independent of frequency and could typically amount to ± 0.1 degrees at f_{LO} as well as at f_S. In Fig. 5.15(a) the phase change introduced by the rotary joints is therefore typically a factor p (the LO harmonic number) smaller than the corresponding phase change in Fig. 5.15(b). Hence, from this point of view, the set-up in Fig. 5.15(a) is preferable. On the other hand, the length l_1 of the RF section in the signal path is considerably shorter in Fig. 5.15(b) than in Fig. 5.15(a) for the same measurement distance; this is a clear advantage at the highest frequency where cable attenuation may become a limiting factor. Cable attenuation should be kept as low as practical in order

Fig. 5.13 (*b*) *Details of probe carriage.* (Photo courtesy Ericsson Radio Systems, Gothenburg, Sweden)

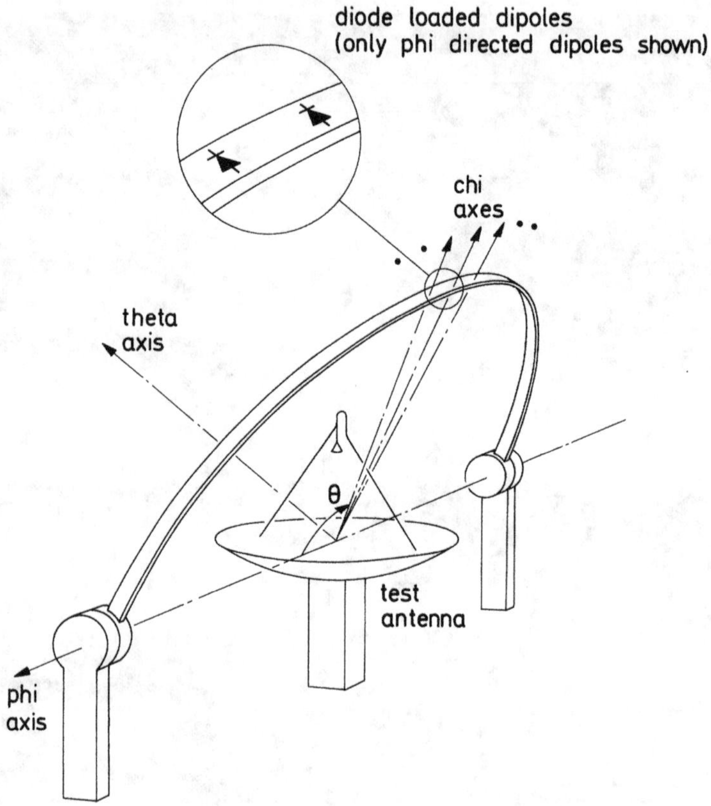

Fig. 5.14 *Spherical near-field set-up with circular gantry arm and scattering dipoles*

that small leakages at connectors carrying a high signal level should not interfere with the RF transmission between the test antenna and the probe.

The long-term stability of an RF system is dependent upon, for example, temperature drift and frequency drift of the various components. For a discussion of the influence of these factors, reference is made to Chapter 6. In order to make the phase measurement in Fig. 5.15 least sensitive to small changes in temperature (cable lengths) and frequency, the signal frequency phase shifts β_1 and β_2 (radians) in the sections l_1 and l_2, respectively, and the local oscillator frequency phase shifts β_3 and β_4 (radians) in the sections l_3 and l_4, respectively, should be related as follows

$$\beta_1 - p\beta_3 = \beta_2 - p\beta_4. \tag{5.70}$$

This requirement determines the optimum position of the local oscillator unit in the RF circuit under the condition of frequency drift.

Proper operation of the RF system is very critical for the accuracy of the measurements. With the test antenna oriented for maximum signal, the

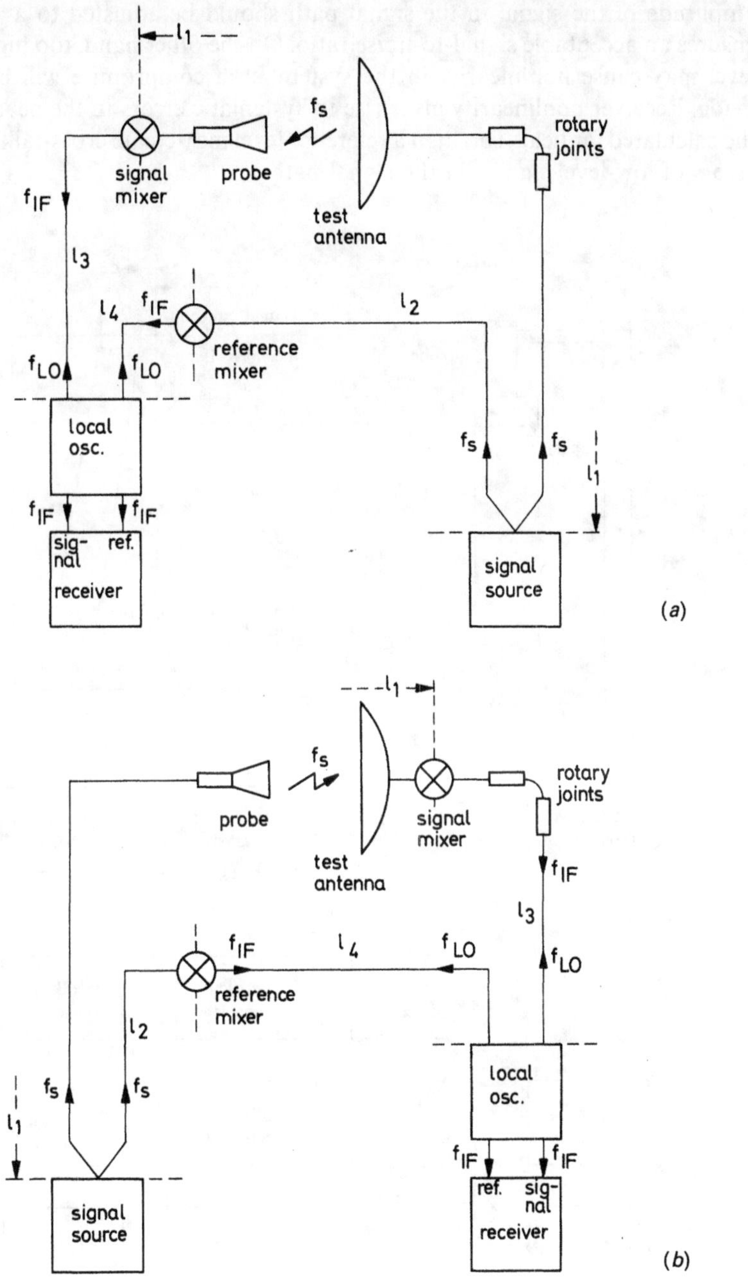

Fig. 5.15 *Examples of RF systems for near-field amplitude and phase measurements. Lengths of sections l_1, l_2, l_3, l_4 are measured between dashed lines*
(*a*) Test antenna transmitting, probe receiving
(*b*) Probe transmitting, test antenna receiving

amplitude of the signal in the signal path should be adjusted to a value that ensures an acceptable signal-to-noise ratio. On the other hand, too high a signal level may cause nonlinearity in the system, so a compromise will have to be made. Receiver nonlinearity gives rise to systematic errors in the near field and the calculated far field. Too high a reference level may create cross-talk, i.e. phase errors of low-level signals in the signal path.

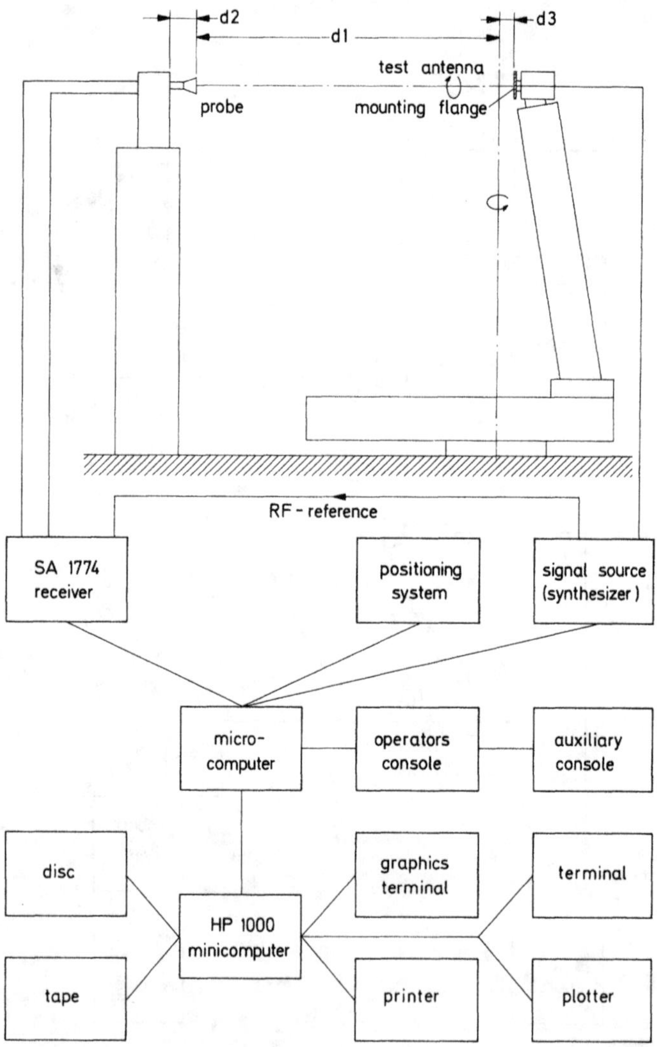

Fig. 5.16 *Schematic of the TUD spherical near-field test range. The measurement distance is indicated by d1*
The dash-dot lines indicate axes of rotation

5.3.2.3. Data collection and control: The large amount of data required in spherical near-field antenna testing can only be handled by automated, computer-controlled measurement systems. The near-field data must be digitally sampled and facilities for data storage must be present. The many instruments that comprise a near-field test range, and the large number of parameters that must be defined prior to any measurement series, require computer control. Complex and lengthy scanning sequences must be performed automatically. Data processing is essential in all near-field ranges where the test antenna far field is obtained through computation. For a measurement system to be efficient it must be able to process previously acquired data while new measurements are in progress. This puts certain requirements to the overall system. It is not the intention here to go into detail with antenna range

Table 5.2 *Software packages included in the data collection and control system of the TUD spherical near-field test range.*

Monitor program
— supervisory program from which the other programs may be invoked
— restarts the system after power failure or major error conditions
— restarts the system after operator intervention
— chains a sequence of test files and invokes the test file executor automatically

Keyboard module
— provides manual control (via terminal) of positioning and measuring equipment

Test file generator
— generates and edits test files
— offers interactive assistance to users
— default values are provided if desired
— the resulting test file is saved on disc
— multiple frequencies in a single test file, including automatic phase-lock control of receiver

Test file executor
— performs the measurements on the basis of labelled test files
— supervises the measurement equipment and collects data

Scan conversion program
— controls data quality on a full-scan basis; faulty scans are repeated
— performs data reduction on oversampled scans; tests for higher-order modes
— reference channel data filtering
— optional data normalization

instrumentation. Instead, as an example, the data system for the spherical near-field range at TUD [18] is briefly outlined below. This system is probably typical of many automated test range designs. A schematic of the system is shown in Fig. 5.16. As can be seen, the signal source, the receiver and the positioning system are tied to the microcomputer and the minicomputer by a data bus.

The control of the antenna tower movements, the setting of frequencies and power levels of the signal source and the receiver settings are done by the microcomputer. During the near-field scanning, the two angular coordinates (rotation about the horizontal and vertical axes) of each measurement point are measured by high-precision, backlash-free angular decoders and transmitted to the microcomputer which also reads the measured signal, i.e. amplitude and phase of each of the two polarization channels, as well as the reference signal amplitude. The microcomputer may hence be described to act as a preprocessor, relieving the minicomputer of the time-critical parts of the measurements. The minicomputer has the overall control and guides the measurements according to a predefined *test file*. It also makes all data reduction, scan-by-scan quality analysis, plotting of curves and contour maps, as well as the spherical near-field to far-field transformations, as concurrent tasks. The measurement system is operated from either of two parallel-coupled terminal consoles. A small hand-held control box is very convenient during the alignment process. Some of the software packages of the system are listed in Table 5.2.

Table 5.3 *Software packages included in the TUD spherical near-field test range.*

Transformation program (SNIFT)
— near-field to far-field transformation with probe correction
— efficient use of vector instructions
— efficient data segmentation
— underflow protection
— minimum increments $\Delta\theta = \Delta\phi = 0.18°$

Spherical near-field test facility simulation program (FACSIM)
— computes the influence of errors in spherical near-field facilities
— contains several test antenna models
— handles several types of error sources
— several output formats

Utility programs
— interpolation, contour plots, polar plots, colour graphics, etc.
— FORTRAN
— statistical-analysis programs

5.3.2.4 Data-processing system: The basic element of the data-processing system at the TUD spherical near-field range is the transformation algorithm SNIFT described in detail in Chapter 4. Besides SNIFT, a near-field facility simulation program FACSIM is very convenient in assessing the influence of various error sources in spherical near-field testing. FACSIM has been used extensively in the preparation of the material in Chapter 6. Finally, the TUD data-processing system contains various utility programs. The software packages are listed in Table 5.3.

5.3.3 Basic measurement procedures

5.3.3.1 Geometrical adjustments and alignment tools: Prior to any measurement, the mechanical set-up must be precisely aligned to conform with the geometrical constraints of spherical near-field testing. Adjustment possibilities for this purpose must be integrated into the system, and suitable alignment tools such as levellers and mirrors should be available. The various types of set-ups (see Section 5.3.2.1) may require their own specially designed alignment procedures [11, 14, 19, 20]. In the present section, however, only the mechanical system shown in Fig. 5.11 will be discussed. The TUD implementation (cf. Fig. 1.1) of this system is outlined in Fig. 5.17 and may serve as a prototype [11, 19]. The discussion in the remainder of this section pertains to the TUD spherical near-field scanner.

Its basic elements are the test antenna tower, or *antenna tower*, which provides the theta and phi axes and the movable *probe tower* which provides the chi axis. The test antenna can be mounted on a *mounting flange* on the horizontal phi axis for which the bearings are contained in the *model head* on top of the antenna tower. The vertical theta axis is based on preloaded high-precision ball bearings located in the *tower base.*

Intersection of the theta and phi axes can be obtained using a *translation shuttle* which moves the *swing arm* and the antenna tower relative to the tower base. The swing arm is fitted with a hand wheel by which the distance d3, the *flange offset* from the theta axis to the mounting flange can be varied. The dual-polarized probe (length d2) is mounted on an interchangeable *front plate* on a *probe frame* (see also Fig. 5.20) which ensures its precise and reproducible fixation on the probe tower. A system of adjustment screws and shuttles allows the probe frame to be placed as required during the alignment process. Besides the probe frame, mounted on a vertical translation shuttle, there is a one-second theodolite which is used extensively when the system is aligned.

In Table 5.4 are listed the adjustments A1–A13, not all of which are shown in Fig. 5.17. Also, a list of alignment tools (T1–T6) and a list of short designations (X1–X6) for various axes of the system, are given. The symbols are used in the next two sections where the alignment of the set-up is described.

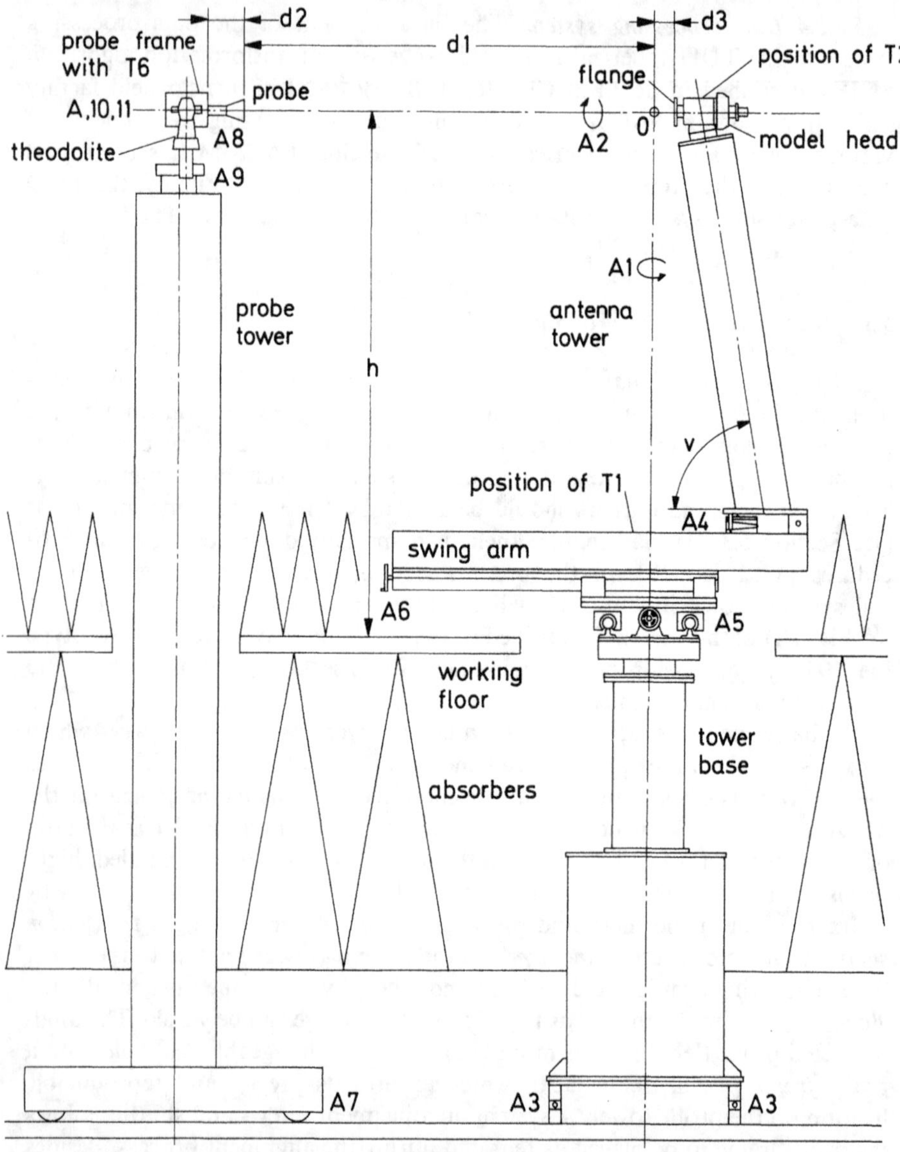

Fig. 5.17 *Outline of the TUD spherical near-field scanner*
The three distances, d1, d2 and d3 are the measurement distance, the probe length and the flange offset, respectively

5.3.3.2 Alignment of the antenna tower: The purpose of aligning the antenna tower is to arrive at a situation where two conditions are fulfilled:

1. The theta axis (X1) is vertical and intersects at right angles with the phi axis (X2). The intersection point is denoted by O.

Table 5.4 *List of adjustments, axes designations and alignment tools pertaining to the TUD spherical near-field scanner.*

Adjustments	
A1	Rotation about the theta axis (X1)
A2	Rotation about the phi axis (X2)
A3	Adjustment of tower base
A4	Tilting of the antenna tower
A5	Translation of antenna tower and swing arm (transverse)
A6	Translation of antenna tower and swing arm (longitudinal)
A7	Movement of probe tower
A8	Vertical translation of the theodolite
A9	Tilt of probe turntable and probe
A10	Probe turntable
A11	Three-dimensional translation of the probe
A12	Offset of theta angle indicator (not shown)
A13	Offset of phi angle indicator (not shown)
Axes designations	
X1	Theta axis
X2	Phi axis
X3	Optical axis of the theodolite
X4	Chi axis or probe axis
X5	Image of X3 in T4
X6	Polarization axis (x'-axis) of the probe
Alignment tools (see Fig. 5.20)	
T1	Leveller for alignment of the theta axis
T2	Leveller for alignment of the phi axis
T3	Theodolite
T4	Adjustable optical mirror, with centre cross, for mounting flange
T5	Optical mirror with centre cross
T6	Leveller for alignment of the probe frame

2. An optical mirror (T4) with centre cross is fixed to the mounting flange with its reflecting face perpendicular to X2 and its centre coinciding with O.

While the first condition ensures the alignment of the antenna tower per se, the second condition allows the antenna tower to be used in the process of aligning the probe tower; see Section 5.3.3.3. The alignment of the antenna tower comprises the following three steps:

Alignment of the theta axis (X1): The leveller (T1) is placed on the swing arm near the axis and the air bubble is observed at four fixed angular positions (90-degree intervals) of the swing arm. The four wedge blocks (A3) are adjusted until the bubble stands at the same point of the scale in the four positions. It is possible to align the theta axis to be vertical within 0.0005 degrees. The alignment has proven to be stable over a long period of time. Control of the alignment and possible readjustment of the theta axis is done as a routine each year.

Alignment of the phi axis ($X2$): 1. The adjustable mirror (T4) is mounted on the mounting flange and aligned to be perpendicular to the phi axis. This is done with the aid of the theodolite (T3) as follows (see Fig. 5.18). The internal reference plane of T3 is first carefully adjusted to be horizontal. Using the vertical and horizontal adjustments on T3 until the image (in T4) of its cross-hairs is viewed as coincident with the cross-hairs themselves makes the optical

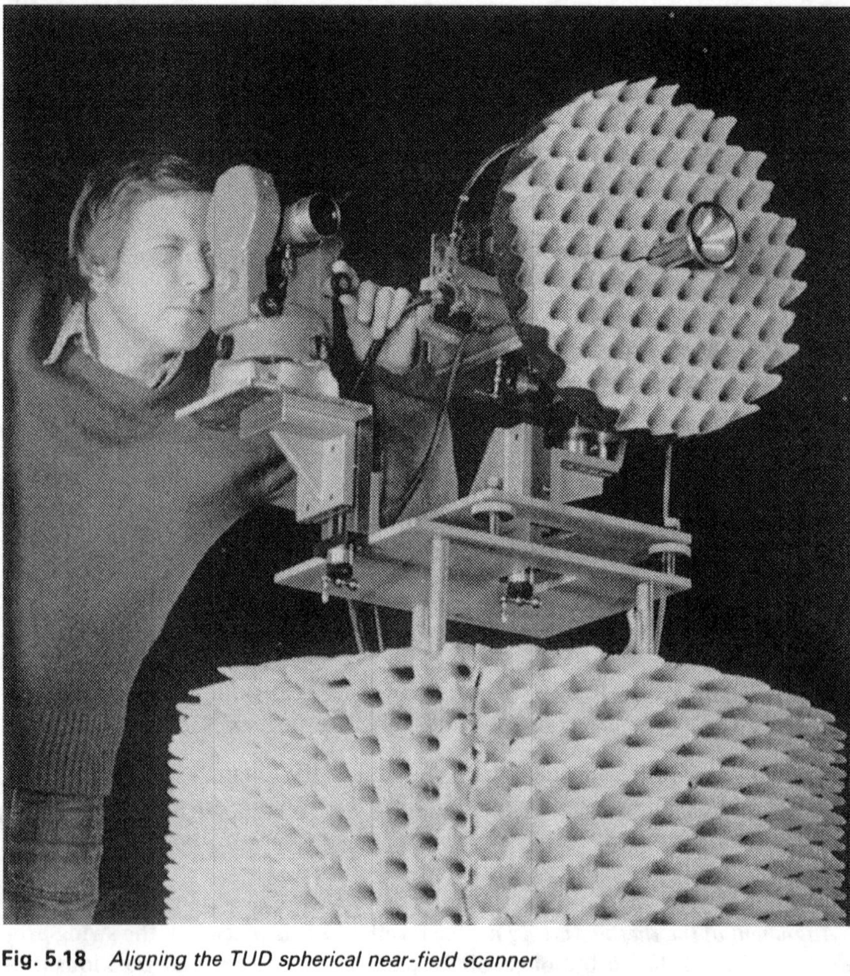

Fig. 5.18 *Aligning the TUD spherical near-field scanner*

axis (X3) of the theodolite perpendicular to T4. (This method of making X3 perpendicular to a plane mirror is known as *autocollimation.*) The horizontal and vertical scales of T3 are now read. After a rotation of T4 by 180.00° about the phi axis, a new autocollimation is made resulting in new readings of the horizontal and vertical scales of T3. The optical axis (X3) of the instrument is then set to the mean of the readings from the two autocollimations. A third autocollimation can now be made during which the adjustment screws on T4 are used while T3 is untouched. T4 is now perpendicular to the phi axis.

2. In order to locate the point where the phi axis passes the plane of T4, the theodolite is focused on and aimed at the centre cross of the mirror. The footprint of the phi axis is then found as the mid-point between two positions of the centre cross as observed before and after a rotation A2 of the flange with T4 by 180.00° about the phi axis. The optical axis X3 of the theodolite is directed towards the mid-point and the centre of the mirror T4 is moved to this position using the translation shuttles on the mirror. The phi axis now passes through the centre of the mirror. The process may be repeated in order to improve the accuracy.

3. With T4 aligned, the theodolite optical axis X3 is set horizontal and autocollimation in T4 is made by adjusting the antenna tower inclination angle v using A4 while T3 is untouched. The phi axis is now horizontal and the leveller T2 on the model head can be adjusted until the air bubble is in the centre position. Finally, the theodolite is moved by means of the vertical translation shuttle (A8) until the optical axis (X3) is at the same height as the centre of T4. The phi axis is hereby aligned.

Intersection of the theta axis (X1) and the phi axis (X2): 1. First, the theta axis is identified by the theodolite. This is accomplished by selecting a fixed point on the antenna tower and taking the average of two readings during which the tower is rotated about X1 to +90.00° and −90.00°, respectively.

2. The theodolite axis (X3) is now directed towards the theta axis and set horizontal. Autocollimation in the mirror (T4) is made by rotation about the theta axis, leaving the theodolite untouched. With T3 focused on the plane of T4, the centre of the latter is moved to the optical axis (X3) by use of the translation shuttle A5 beneath the swing arm.

3. The theta and phi axes now intersect. However, in order to check this and for the sake of the probe alignment in the next step, the antenna tower is moved in the direction of the phi axis (using the translation shuttle A6) until the centre of the mirror coincides with the intersection point. This may be seen by rotating the tower about 60° about the vertical axis such that the centre of the mirror can still be seen in the theodolite. If no movement of the centre of the mirror is observed upon subsequent rotations, the mirror centre is coincident with the intersection point. The intersection of the theta and phi axes can now be checked

by observing that the centre of the mirror does not move when rotated about either axis.

5.3.3.3 Completing the alignment: The purpose of the procedures in this section is to achieve the following conditions

1. The probe axis X4 (the chi axis) is horizontal and passes through the intersection point O. The probe polarization reference axis X6 (x'-axis, Fig. 5.7), defined to be parallel to the probe frame top face, is horizontal.
2. The theta angle zero setting and the measurement distance d1 are defined. The set-up is ready for setting of the wanted flange offset d3 and for mounting of a test antenna.

The alignment of the set-up is completed in the following three steps:

Alignment of the probe: The theodolite is aimed at the intersection point O and is, apart from focusing, not touched during the rest of the alignment procedure. At this point, the image (X5) of the optical axis (X3) in the mirror (T4) is for all settings of A1 horizontal and goes through the intersection point.

The probe axis (X4) is now aligned by making it coincident with X5. The probe mirror (T5) is mounted on a front plate on the probe frame and the antenna tower is rotated about the theta axis until the centre of the probe mirror can be seen in the theodolite. The probe mirror is now made perpendicular to X5 by indirect (two-mirror) autocollimation (see Fig. 5.19), using the probe frame turntable (A10) and the three adjustment screws (A9). At the same time, the polarization reference axis (X6), defined to be parallel to the probe frame top face, is made horizontal by using a leveller T6.

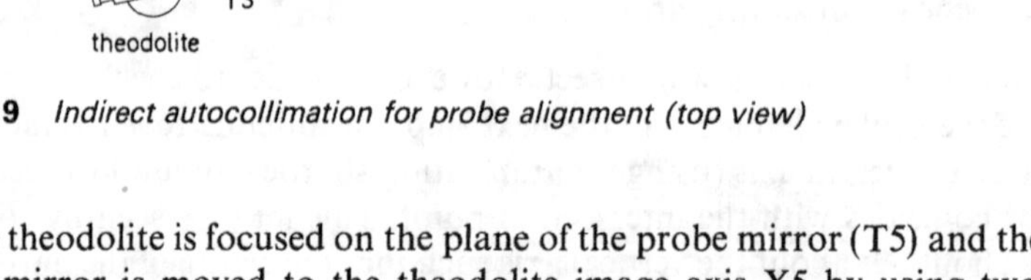

Fig. 5.19 *Indirect autocollimation for probe alignment (top view)*

The theodolite is focused on the plane of the probe mirror (T5) and the centre of the mirror is moved to the theodolite image axis X5 by using two of the translation shuttles below the probe frame. After some readjustment of A9 and A10, the axes X4 and X5 coincide and X6 is horizontal.

The probe frame is hereby aligned.

Zero setting of theta indicator: The angle display for the theta axis is read, and the antenna tower is rotated (A1) about this axis until autocollimation is obtained directly in the adjustable mirror (T4). The angle display is read again and the difference between the two readings gives the angle u in Fig. 5.19. After a rotation about the theta axis by $2u$ in the opposite direction the phi axis points towards the probe centre and the theta angle display can be set to zero using A12. The mirror with front plate can now be removed and a probe on a front plate put in its place.

Measurement distance and flange offset: The measurement distance is defined as the distance between the probe reference point O′, usually the centre of the probe aperture, and the intersection point of the theta and phi axes. It can be obtained by measuring the distance with a ruler between a reference point on the edge of the adjustable mirror mount (T4) and the probe frame front plate, adding the thickness of the mirror mount and subtracting the length d2 of the probe. It is important that the phi axis is pointing towards the probe centre during the measurement. As an alternative method of measuring the distance, a reference distance bar (Fig. 5.20) of length 0.50000 m can be mounted on the

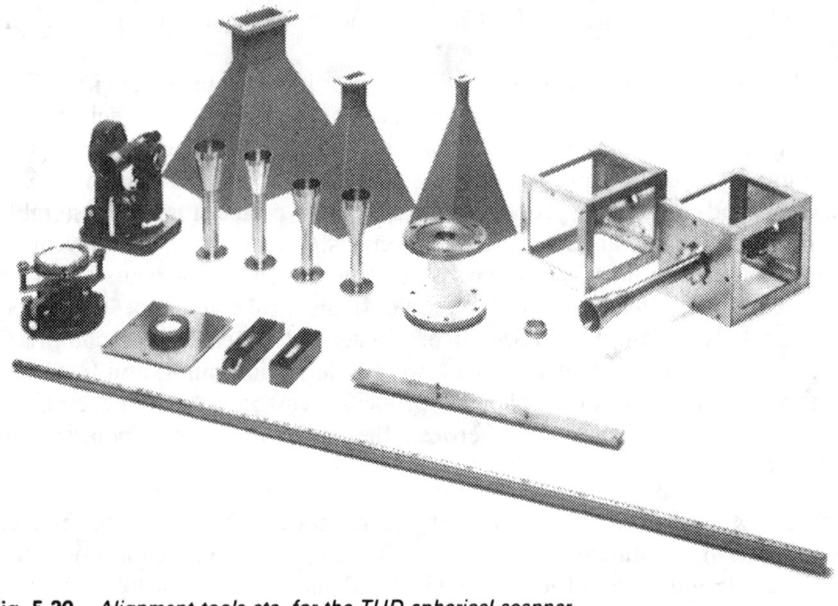

Fig. 5.20 *Alignment tools etc. for the TUD spherical scanner*
Front: 5 m extendible ruler for determination of the measurement distance
Second row: adjustable mirror (T4), mirror (T5) on probe frame front plate, levellers T1 and T2, 0.5 m reference bar
Third row: Kern DKM-2A theodolite, four probes, extension flange for antenna tower, ring for mounting probe frame on antenna tower
Fourth row: three standard gain horns, probe frame, probe frame with dual polarized probe on front plate

flange of the model head. By reading the subtended angle as seen from the theodolite, the measurement distance can be determined.

The measurement distance can be varied about 5 cm by using the axial translation shuttle A11. Large variations of the distance must be made by moving the probe tower on the rails (A7).

If the distance d3 from the flange on the model head to the intersection point (i.e. the phase reference point) is required to be different from the distance determined by the adjustable mirror, the handwheel on the swing arm (A6) can be used. It is important that the displacement is done with the adjustable mirror (T4) mounted in the antenna tower. In this way, possible deformations can be observed in the theodolite and corrected for by A4, A5, A11 and A12.

The measurement set-up is now fully aligned. The adjustable mirror (T4) can be removed and the antenna under test put in its place. Adjustment A4, together with T2, can be used to correct for possible bending of the antenna tower due to the weight of the test antenna.

Some of the alignment tools employed are shown in Fig. 5.20.

5.3.3.4 Probe calibration: With the spherical near-field set-up aligned, the next step in the sequence of measurements is the calibration of the probe. The purpose of the probe calibration is to generate the probe receiving coefficients required by the near-field to far-field transformation algorithm (ref. Section 4.4.4).

If the measurement distance is large enough or if the probe directivity is low, the coefficients of the physical probe may be substituted by the coefficients of a simple theoretical probe model such as a short dipole or a Huygens source. (The error involved in this approximation is discussed in Section 6.5.) At closer distances and at higher probe directivities, the coefficients of a suitable maximum-directivity antenna, (ref. Section 2.4.4) may serve as convenient theoretical substitutions for the probe coefficients. This is of course also an approximation. If full accuracy is required, the probe must be calibrated experimentally as described in Section 5.2. In general, the calibration comprises pattern calibration (probe in antenna tower), polarization calibration (probe in probe tower) and, for dual-polarized probe systems, a measurement of the complex amplitude phase ratio between the signal in the two polarization channels.

5.3.3.5 Alignment test: With the calibrated probe on the probe tower and with the given test antenna on the mounting flange of the antenna tower the so-called *flip test* is useful for checking the alignment. As seen from the test antenna, the two directions (θ, ϕ) and $(-\theta, \phi + \pi)$ are identical. Therefore we have for the received signal

$$w(\theta, \phi) = -w(-\theta, \phi + \pi) \tag{5.71}$$

so that

$$|w(\theta, \phi)| = |w(-\theta, \phi + \pi)|. \tag{5.72}$$

Hence, if two amplitude cuts are taken for example at $\phi = 0$ and $\phi = \pi$, they will appear identical if the latter cut is flipped, i.e. plotted against $-\theta$ instead of against θ. Figure 5.21(a) shows the result of a correct theta angle display setting. As can be seen, the curves with $\phi = 0$ and $\phi = \pi$ are identical. In Fig. 5.21(b) the effect of an error of 0.1 degrees in the setting of the theta angle display is clearly visible. The displacement between the curves is twice the error in the phi axis pointing. Once detected, the pointing error may easily be corrected for.

The flip test may also be used to detect possible deformations under rotation of the test antenna or its mount. If such deformations are present, the two curves in a flip test cannot be made coinciding by a displacement in theta.

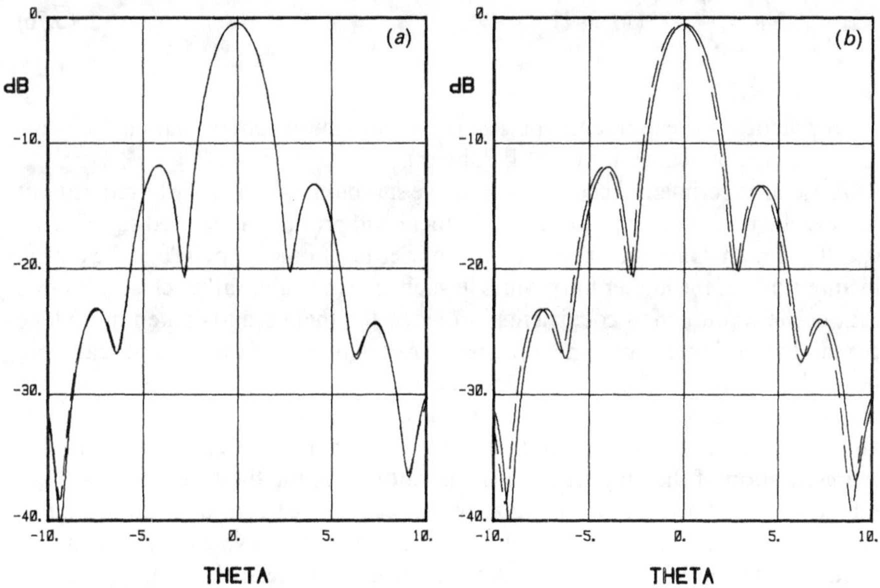

Fig. 5.21 *Flip test for a test antenna of diameter 27.4 wavelengths*
(*a*) Correct theta angle display setting
(*b*) Theta axis error of 0.1°

5.3.3.6 Sampling and sampling test: Prior to the near field scanning of the test antenna, the distribution of data points on the measurement sphere, i.e. the sampling scheme, must be decided upon according to the general sampling criteria.

The sampling of the test antenna near field follows the general principles outlined in Section 4.4.3. Usually, the test antenna is mounted so that the radius r_t of its minimum sphere is as small as possible. The highest significant wave mode present in the test antenna field is given by

$$N = kr_t + 10. \tag{5.73}$$

The maximum permissible sampling increment in theta is then

$$\Delta\theta = \frac{2\pi}{2N+1} \tag{5.74}$$

$$\cong \frac{\pi}{N} \qquad (N \gg 1). \tag{5.75}$$

In phi, the maximum sampling increment is the same. If the radius r_c of the smallest cylinder parallel to the z-axis and enclosing the test antenna is smaller than r_t we may choose

$$\Delta\phi \cong \frac{\pi}{M} \qquad (M \gg 1) \tag{5.76}$$

where $M \cong kr_c + 10$.

In practical measurements the size of the minimum sphere that encloses the currents of significance may not be obvious.

A useful experimental check of the above sampling criteria may be carried out as described below. A single full scan in theta will produce a received signal $w(\theta)$, $0 \leqslant \theta < 2\pi$, amenable to expansion in Fourier harmonics $\exp(im'\theta)$. The cut-off in amplitude of the higher harmonics in such an expansion rather closely follows that of the test antenna coefficients, provided the theta scan is taken in a plane containing the largest antenna dimension. Actually, one of the flip test scans can be used. In other words, a graph of the harmonics of $w(\theta)$ may be used to obtain an independent estimate of the number of significant spherical modes in the test antenna field. As an example of this test (the *N-test*), Fig. 5.22 shows a graphic representation of the amplitude of the harmonics of the theta cut (Fig. 5.21) in the near field of a reflector antenna. In this case the current distribution has a well-defined minimum sphere with a diameter of 27.4 wavelengths. As can be seen, a choice of $N = kr_t = 86$ is almost, but not quite, satisfactory. If, instead, $N = kr_t + 10 = 96$ is chosen, all significant modes are seen to be included. The graph was obtained by taking the IDFT of 360 sampled values of $w(\theta)$. The theoretical justification for the N-test can be examined by inserting the finite Fourier series (A2.12) for the rotation coefficient into the transmission formula (3.10).

5.3.3.7 Schemes for scanning: When the question of interspacing between measurement points has been settled a scanning scheme must be determined. This defines the precise order in which the individual points are to be taken. The measurement of the near-field signal can of course be made in a process where the probe is moved stepwise from point to point. For several reasons, however, continuous mechanical movements in the set-up are much more expedient, at least when a full measurement sphere shall be scanned. Assuming a sampling as given by equations (5.75–5.76) an obvious requirement is that the angular

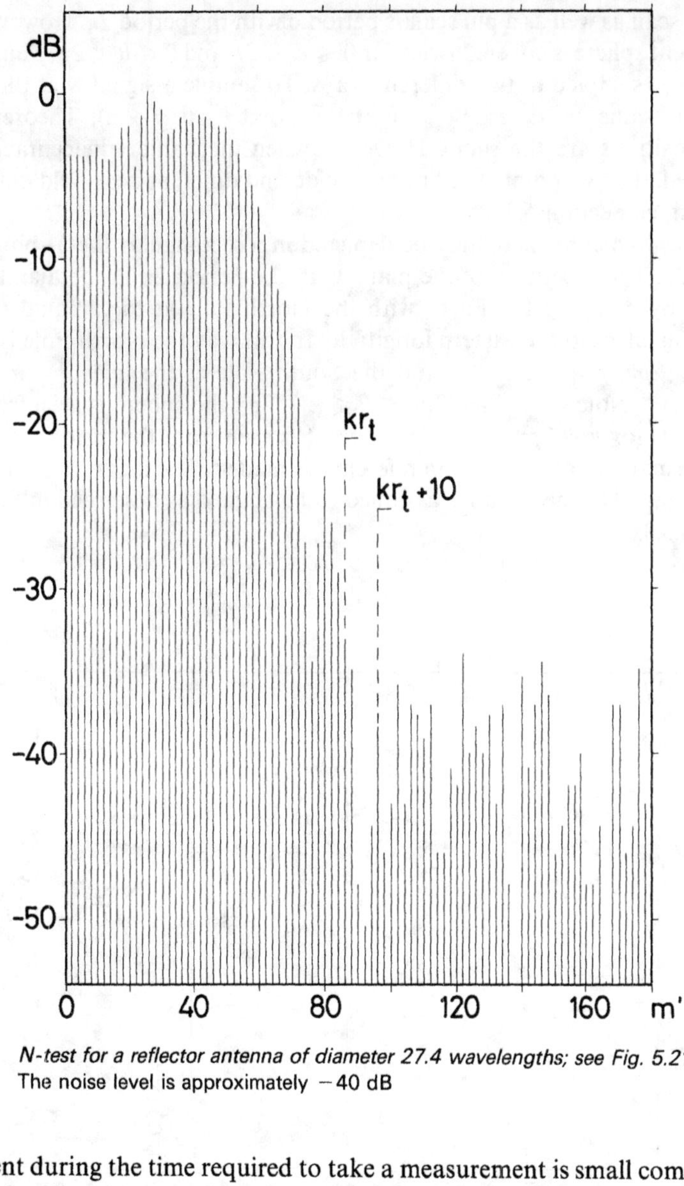

Fig. 5.22 *N-test for a reflector antenna of diameter 27.4 wavelengths; see Fig. 5.21*
The noise level is approximately −40 dB

movement during the time required to take a measurement is small compared to the angular distance between consecutive measurement points.

Of the various ways the measurement sphere may be mechanically scanned the most common are (1) scans in phi with steps in theta and (2) scans in theta with steps in phi. Less common is (3) scanning in both phi and theta, e.g. continuous *ball-of-yarn* scanning. At first sight, it may seem unimportant which scheme is chosen. There are differences between them, however, which will be described in the present section.

A theta scan as well as a phi scan is periodic with the period 2π. However, the measurement sphere is covered twice for $0 \leqslant \phi < 2\pi$ and $0 \leqslant \theta < 2\pi$. Thus, each point may be sampled in two different ways. To sample a signal w at the point (θ, ϕ) is the same as to sample $-w$ at the point $(-\theta, \phi + \pi)$. Theoretically, the two positions are the same. However, when mechanical inaccuracies are encountered, the two points need not coincide and different near-field data may be sampled, cf. Section 5.3.3.5.

The measurement sphere may be depicted in a rectangular theta–phi plane; see Fig. 5.23. The repetition of the plane with the period 2π in theta and phi is illustrated by mapping the Earth with the North Pole at theta equal to zero and phi equivalent to the eastern longitude. In this way the North Pole is at the lines $\theta = \pm 2p\pi$, $p = 0, 1, 2 \ldots$, and the South Pole at the lines $\theta = \pi \pm 2p\pi$, $p = 0, 1, 2 \ldots$. Note that when $2p\pi \leqslant \theta \leqslant \pi + 2p\pi$ the sphere is seen from the outside while for $\pi + 2p\pi \leqslant \theta \leqslant 2\pi + 2p\pi$ it is seen from the inside.

In the remainder of this section reference is made to the measurement set-up shown in Fig. 5.11. The set-up is assumed to be aligned as described in Sections 5.3.3.2–5.3.3.3.

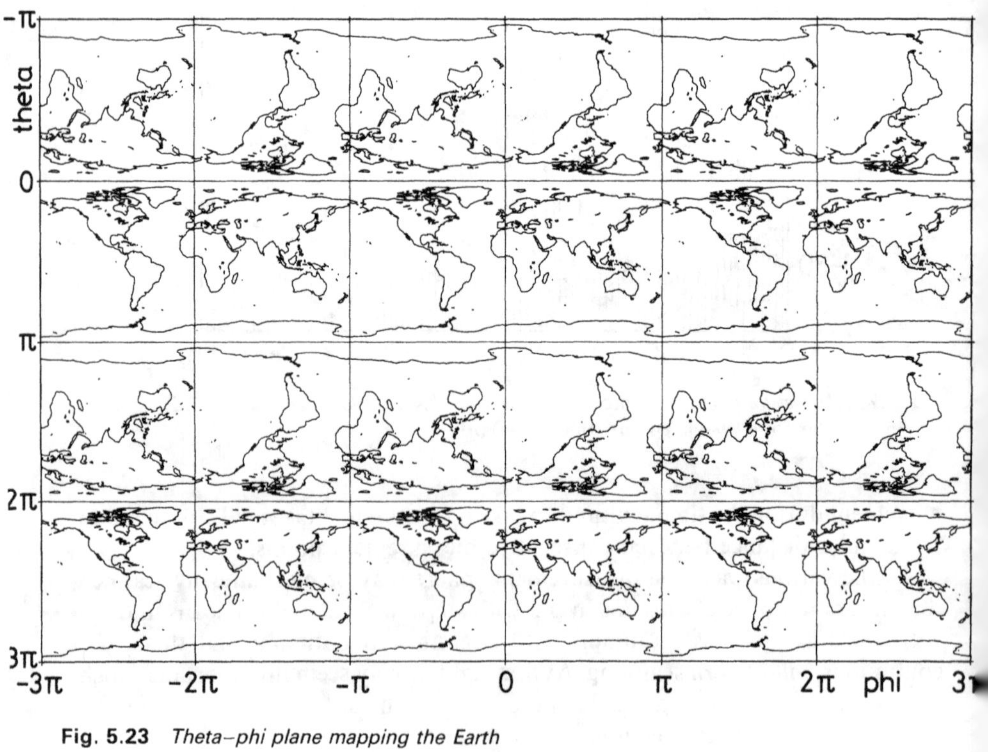

Fig. 5.23 *Theta–phi plane mapping the Earth*

Scan in phi, step in theta: In this case, the test antenna is first rotated about the theta axis to the first theta value, i.e. $\theta = 0$. Then, a continuous rotation about the phi axis will perform a scan in phi during which the near-field signal may be measured at equidistant points with increments $\Delta\phi$. A step $\Delta\theta$ is now made and a new phi scan can be taken; see Fig. 5.24. All scans are assumed to be in the interval $0 \leqslant \phi < 2\pi$. In order to maintain a continuous rotation in phi, it may be useful to scan in intervals $\phi_o \leqslant \phi < \phi_o + 2\pi$, where ϕ_o is changed from scan to scan. The data can then be translated to the interval $0 \leqslant \phi < 2\pi$ afterwards.

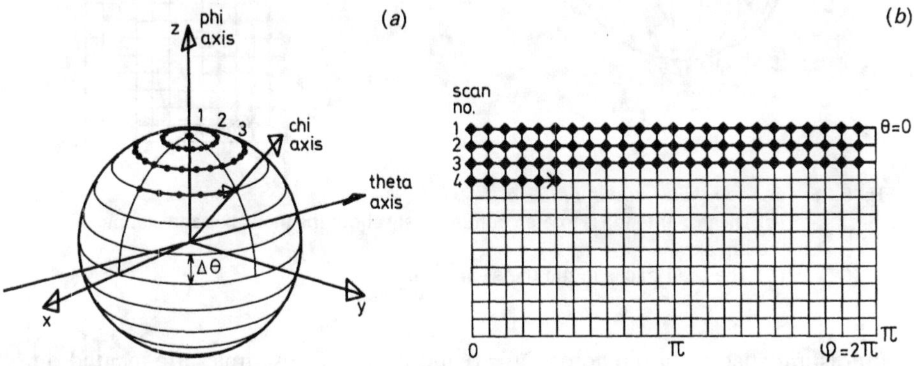

Fig. 5.24 *Scan in phi, step in theta*
(*a*) The test antenna is rotated continuously about the phi axis and stepped about the theta axis
(*b*) The scan traces in the theta–phi plane with the first few scans shown

Scan in theta, step in phi: When the test antenna is scanned in theta (see Fig. 5.25) the desired starting angle, i.e. $\phi = 0$ is first obtained by rotation about the phi axis. A continuous rotation about the theta axis will then give a scan in theta with $-\pi < \theta \leqslant \pi$ for a full revolution. The near-field signal is sampled at equispaced increments $\Delta\theta$ with the two poles included. Having completed the theta scan, a step $\Delta\phi$ is now made about the phi axis, and so on. An advantage for this scheme is that the points $\theta = 0$ and $\theta = \pi$ provide a possibility for comparison of the near-field signal since these two positions are common to all theta scans.

Ball-of-yarn scanning: This scanning strategy is illustrated in Fig. 5.26 where the version with a fast scan in theta simultaneous with a slow scan in phi is shown. The sampling points are then lying on the straight line $\phi = (\Delta\phi/2\pi)\,\theta$ in the theta–phi plane. In the figure, all scanning traces are displaced to the interval $0 \leqslant \theta < 2\pi$. The sampling is for $0 \leqslant \theta \leqslant \pi$. The sampling interval in phi is $0 \leqslant \phi < 2\pi$. The full rotation of 2π in theta is unavoidable for continuous mechanical scanning and the full rotation of 2π in phi is required in order to get

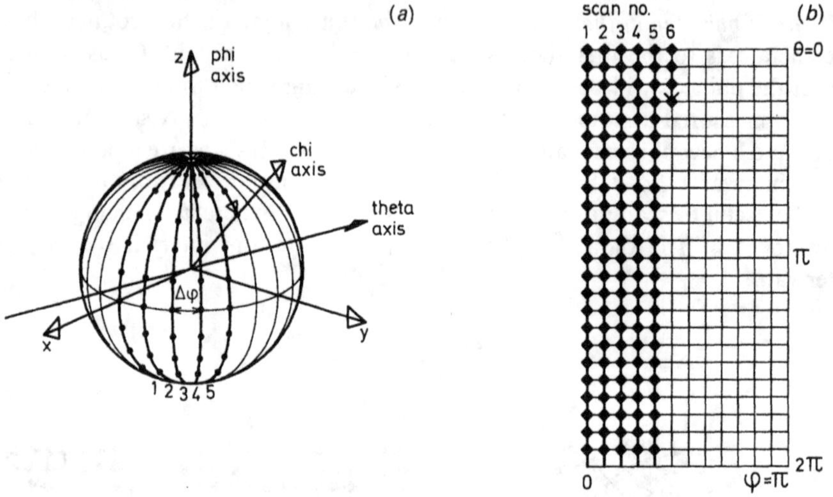

Fig. 5.25 *Scan in theta, step in phi*
(*a*) The test antenna is rotated continuously about the theta axis and stepped about the phi axis.
(*b*) The scan traces in the theta–phi plane.

equidistant spacing in phi across $\phi = 0$ and $\phi = \pi$. The samples are located on the circles of latitude $\theta = (n_\theta - 1)\Delta\theta$, $n_\theta = 1, 2, \ldots, J_\theta/2 + 1$ where $J_\theta = 2\pi/\Delta\theta$ is the number of samples for a full revolution $0 \leqslant \theta < 2\pi$. This means that the sample points are displaced in phi by $(n_\theta - 1)\Delta\phi/J_\theta$ in the scanned ϕ-interval. Since this interval is $0 \leqslant \phi < 2\pi$ then the samples may easily be shifted to ordinary scan positions $\phi = n_\phi \Delta\phi$, $n_\phi = 0, 1, 2 \ldots$ in the beginning of the near-field to far-field transformation, see Section A4.4.2.

The ball-of-yarn type of scanning is of relevance in cases where steps that may cause mechanical vibrations of the test antenna must be avoided. A variation of this technique has been considered by Mostafavi and Bolomey [16] who proposed an electronically scanned gantry arm; see Fig. 5.14.

Discussion: From Figs 5.24 and 5.25 it is seen that the scan areas in the theta–phi plane are different. Thus, the field is scanned over different halves of the double sphere, cf. Fig. 5.23. With mechanical inaccuracies in the scan set-up, the sample points (θ, ϕ) and $(-\theta, \phi + \pi)$ will not coincide. It may then be of importance where the cut separating the sampled sphere from the remaining part of the double sphere is made since a discontinuity in the sampled field may occur along this cut.

By scanning in theta, $-\pi < \theta \leqslant \pi$, phi is stepped for $\phi_o \leqslant \phi < \phi_o + \pi$ where any value of ϕ_o may be chosen. The cut in the double sphere is in this case the great circle $\phi = \phi_o, \phi_o + \pi$. Discontinuities along this cut will give rise to non-physical harmonics (especially harmonics in m) in the field.

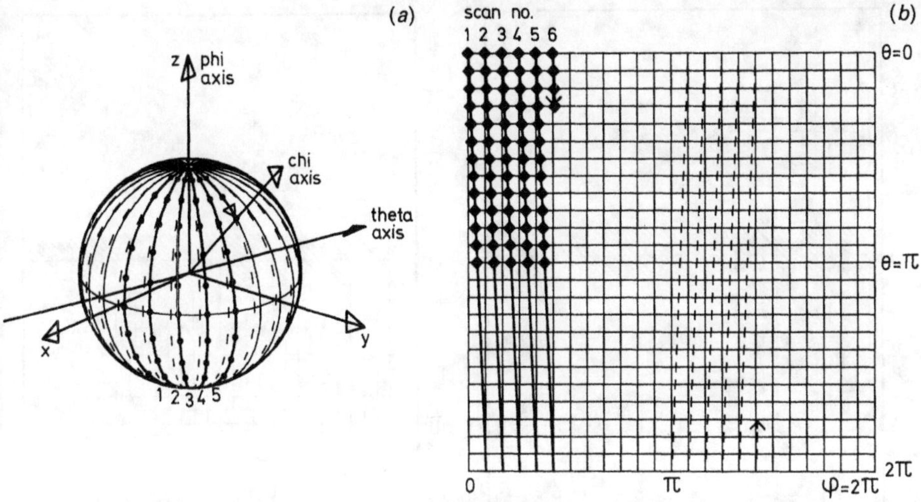

Fig. 5.26 *Ball-of-yarn scanning*
(*a*) The test antenna is making a fast scan about the theta axis and, simultaneously, a slow scan about the phi axis
(*b*) The scan traces in the theta–phi plane with the first few scans shown
The dotted lines are the equivalent positions in the interval $\pi < \phi < 2\pi$, cf. Fig. 5.23. They cannot be used because they do not supplement the solid lines to give equidistant points in the full interval $0 \leqslant \phi < 2\pi$

With scanning in phi, $0 \leqslant \phi < 2\pi$, and stepping in theta, $0 \leqslant \theta \leqslant \pi$, the sample surface is continuous over the sphere since the double sphere now is cut at the poles $\theta = 0, \pi$. The sampled field will then be non-physical at the poles only.

As mentioned in Section 5.3.2.1 the test antenna may be polar-pointing (main beam in direction $\theta = 0$) or equator-pointing (main beam in direction $\theta = \pi/2$, $\phi = 0$) independent of the mechanical set-up. For directive antennas, the measurements may be restricted to the region in front of the antenna (as illustrated in Fig. 5.8) and the set-up may be best suited for either polar-pointing or equator-pointing test antennas. Truncation of the measurement surface is further discussed in Chapter 6.

The difference may be of importance for non-ideal measurements. The bandwidth of the receiving system will normally set a limit for the scan speed. This is illustrated in Figs 5.27 and 5.28. Figure 5.27 shows the theta–phi plane for a polar-pointing test antenna. The locations of the beam on the measurement sphere are shaded. For scanning in theta (Fig. 5.27(a)), each scan passes through the aperture region. This results in a rapidly varying signal which may be deteriorated by a narrow receiver bandwidth. Further, each scan contributes to the aperture field, so long-term drift in the measurement system may influence the results.

Figure 5.27(b) shows the same situation for scanning in phi. The scans are

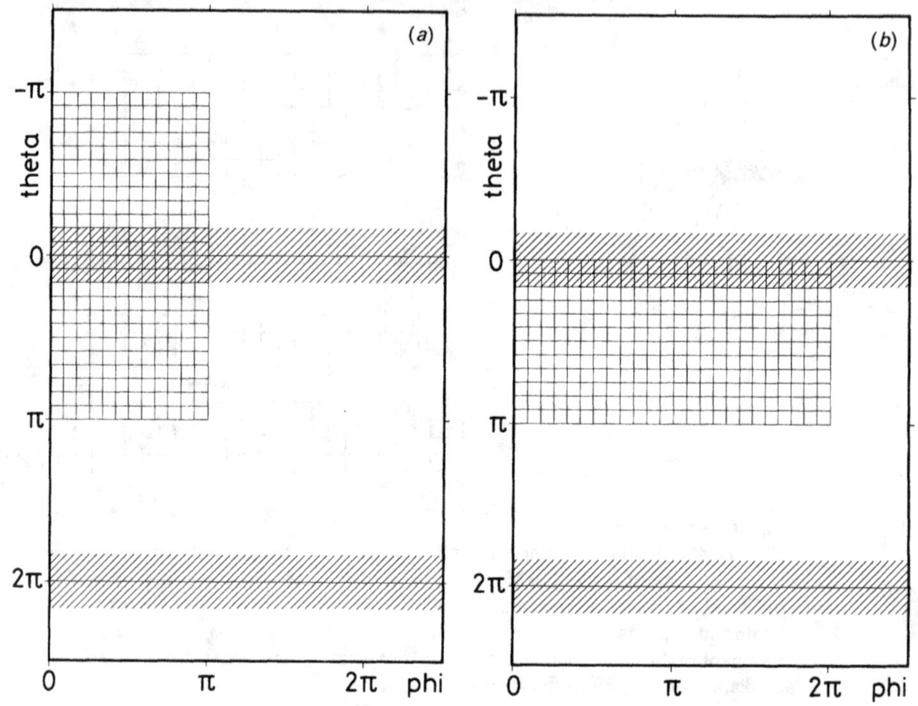

Fig. 5.27 *Theta–phi plane for a polar-pointing test antenna. The beam area is shown shaded.*
(*a*) Scanning in theta
(*b*) Scanning in phi

then either within the aperture or outside. This gives slowly varying signals, and a narrow bandwidth may be applied within the first few scans. Drift in the measurement system throughout the measurement period has only little influence.

For the equator-pointing test antenna, (Fig. 5.28) there is no fundamental difference between scanning in theta and scanning in phi. Some of the scans will pass through the aperture region. This results in a rapidly varying signal in the receiver demanding a wide bandwidth. Further, twice as many scans are required to cover the aperture as compared with phi scanning for the polar-pointing antenna (Fig. 5.27(b)). Thus, drift in the measurement system will influence correspondingly more.

Conclusion: For a given test antenna the choice of scanning strategy can be based upon the above discussion. Before measurements can be started and data taken, however, there are usually other considerations to be made. These are concerned with questions as scan speed, maximum permissible acceleration, criteria for rejection of single scans (inaccurate sampling positions), on-line

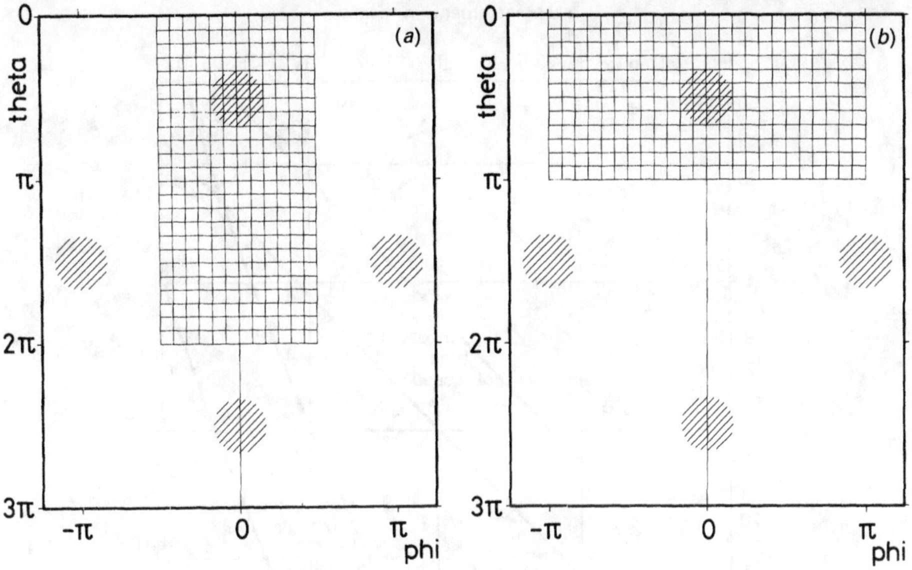

Fig. 5.28 *Theta–phi plane for a equator-pointing test antenna. The beam area is shown shaded.*
(*a*) Scanning in theta
(*b*) Scanning in phi

processing of phi scans (oversampling and Fourier transformation), mains power failure actions, etc. However, as these items depend very much on the test antenna, the required accuracy and on the individual range installations, details will be omitted here.

Upon completion of a spherical near-field scan the acquired data is processed by the transformation program. Figure 5.29 illustrates the measuring and computer time for the TUD facility as a function of the test antenna diameter or as a function of the step increment. As an example, a test antenna of diameter 110 wavelengths requires, in a normal scan-step procedure, a step increment of about 0.5 degree. At a fixed measurement speed, the measuring time is proportional to the number of scans. At a speed of 4 degrees per second, the total measuring time is about 10 hours. The transformation time is about half of the measurement time for this size of test antenna.

The time estimates of Fig. 5.29 may be lowered by applying the techniques of Section 6.6. See also Section 4.4.3.

5.3.4 Measurement examples

As an example, the three near-field patterns shown in Fig. 5.30(a) have been measured at 11.7 GHz for the 30-wavelengths front-fed reflector antenna in Fig. 5.11 at distances of 1.5 m, 3 m and 5 m, respectively. Figure 5.30(b) shows the three copolar and three cross-polar patterns after transformation to the far

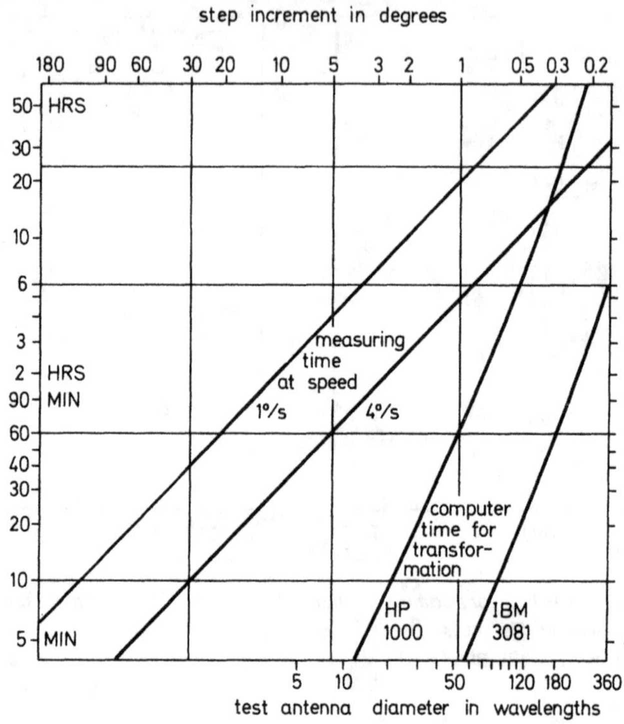

Fig. 5.29 *Measuring and computer time for the TUD spherical near-field facility*

field. The agreement between the curves serves as a consistency check on the measurement system and indicates the measurement accuracy.

The influence of the probe pattern correction on the copolar field is demonstrated in Fig. 5.30(c) where the correction has been omitted in obtaining the dashed curve. Probe polarization correction is included. At a distance of 1.5 m, the probe pattern has a taper of 2 dB over the reflector aperture and if this taper is not corrected for, one gets a broadening of the main lobe and a lower peak directivity. The influence on peak directivity and gain can be deduced from Table 5.6. While the first spherical near-field measurements with probe pattern correction were carried out on a low-gain antenna [21], the effect on a high-gain antenna was first demonstrated indirectly by comparison of results from a spherical measurement without probe correction to a far-field pattern obtained from planar near-field measurements [22]. That the probe pattern correction is actually able to adjust the main lobe correctly was first demonstrated for the curved coordinate in cylindrical near-field measurements [23] and later for spherical near-field measurements [11].

The cross-polar field depends on the value chosen for the polarization reference angle (ϕ_o in eqns 2.190–2.191). In order to display the actual axial

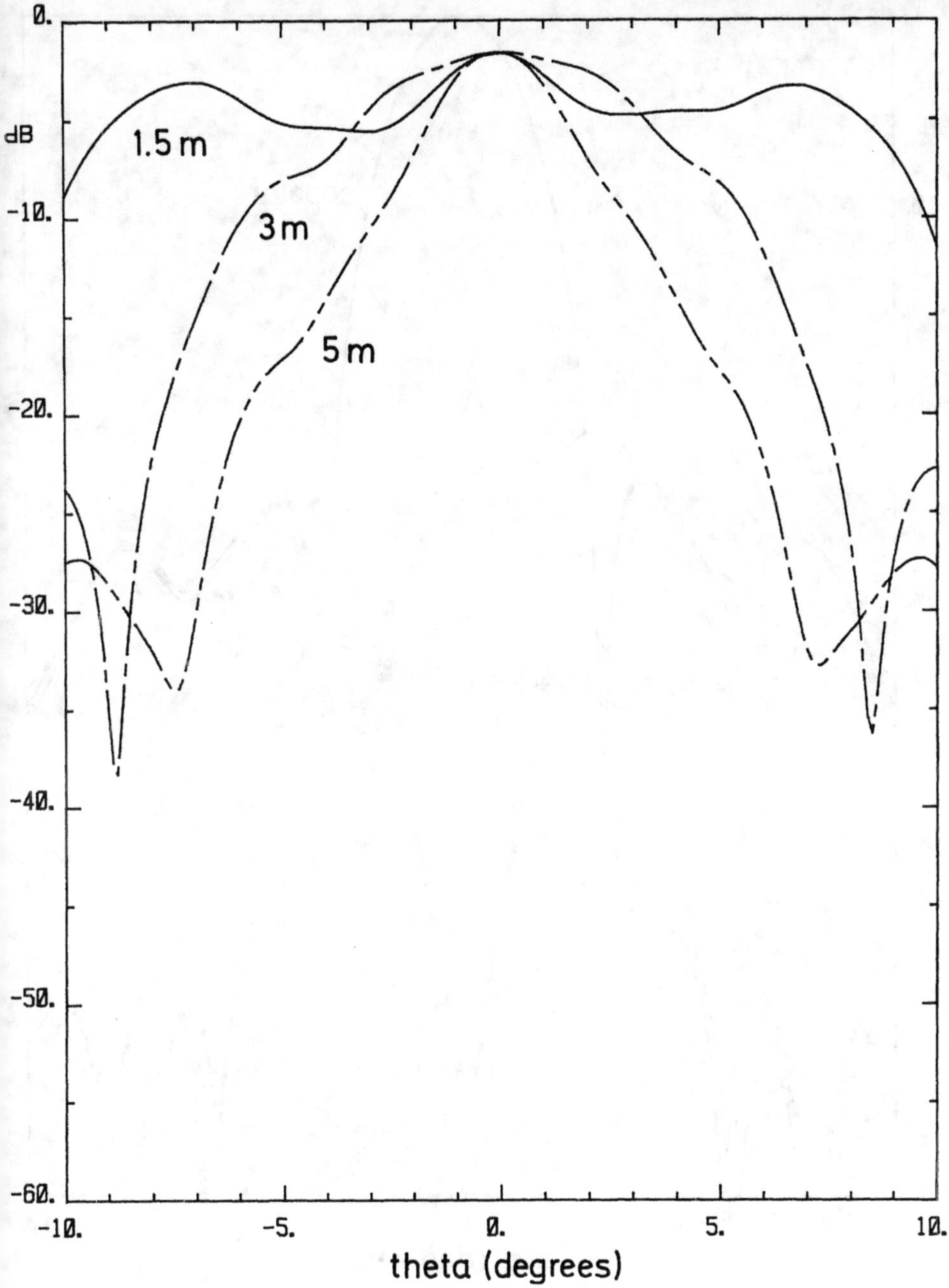

Fig. 5.30(*a*) *Front-fed reflector antenna of diameter 30λ at 11.7 GHz. Copolar near-field signal measured at the distances 1.5 m, 3 m and 5 m. The measurement plane is given by $\phi = 45°$*

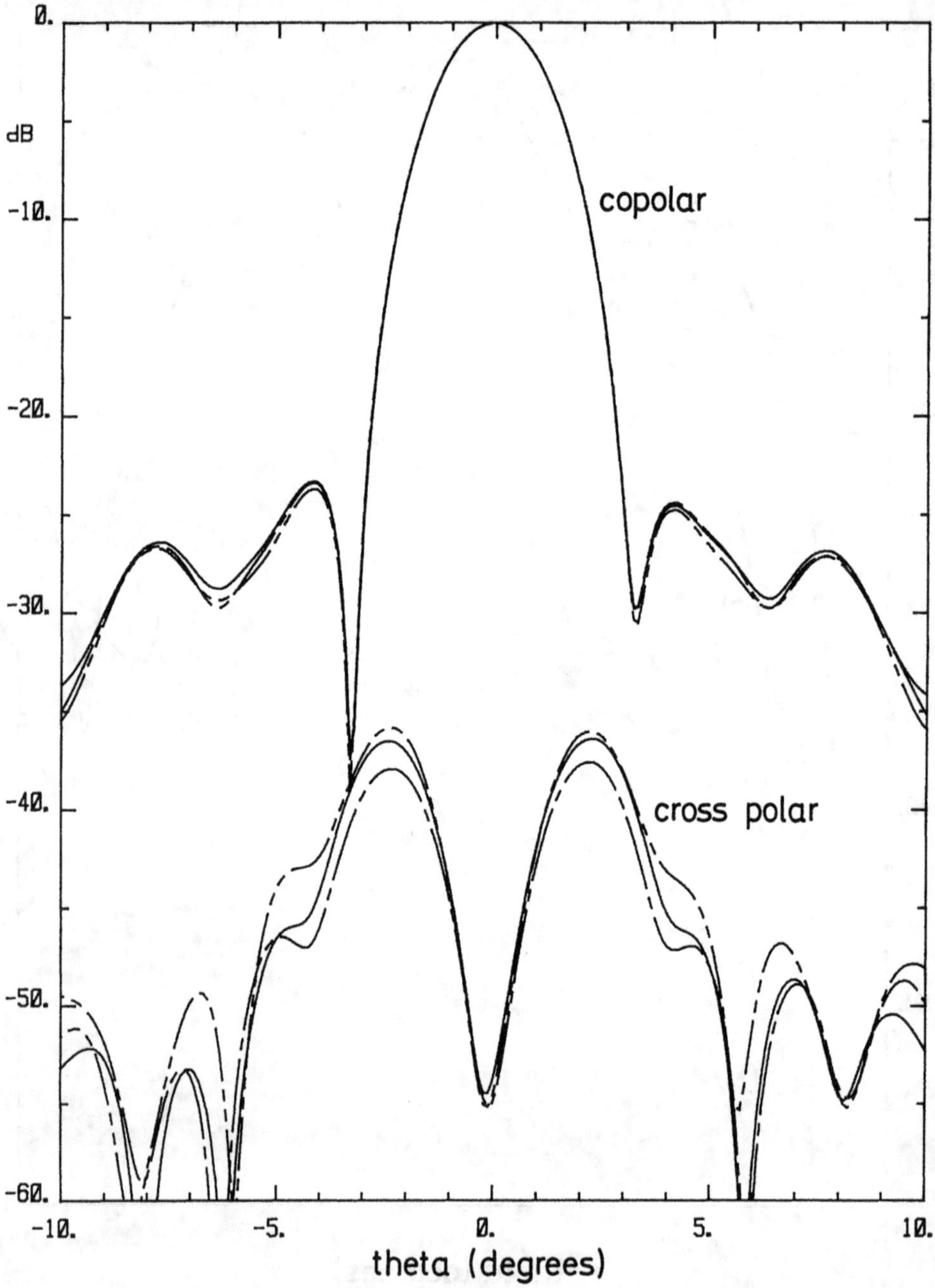

Fig. 5.30(*b*) *Front-fed reflector antenna of diameter 30λ at 11.7 GHz. Copolar and cross-polar far fields transformed from near fields measured at three distances; see Fig. 5.30(a). The measurement plane is given by $\phi = 45°$*

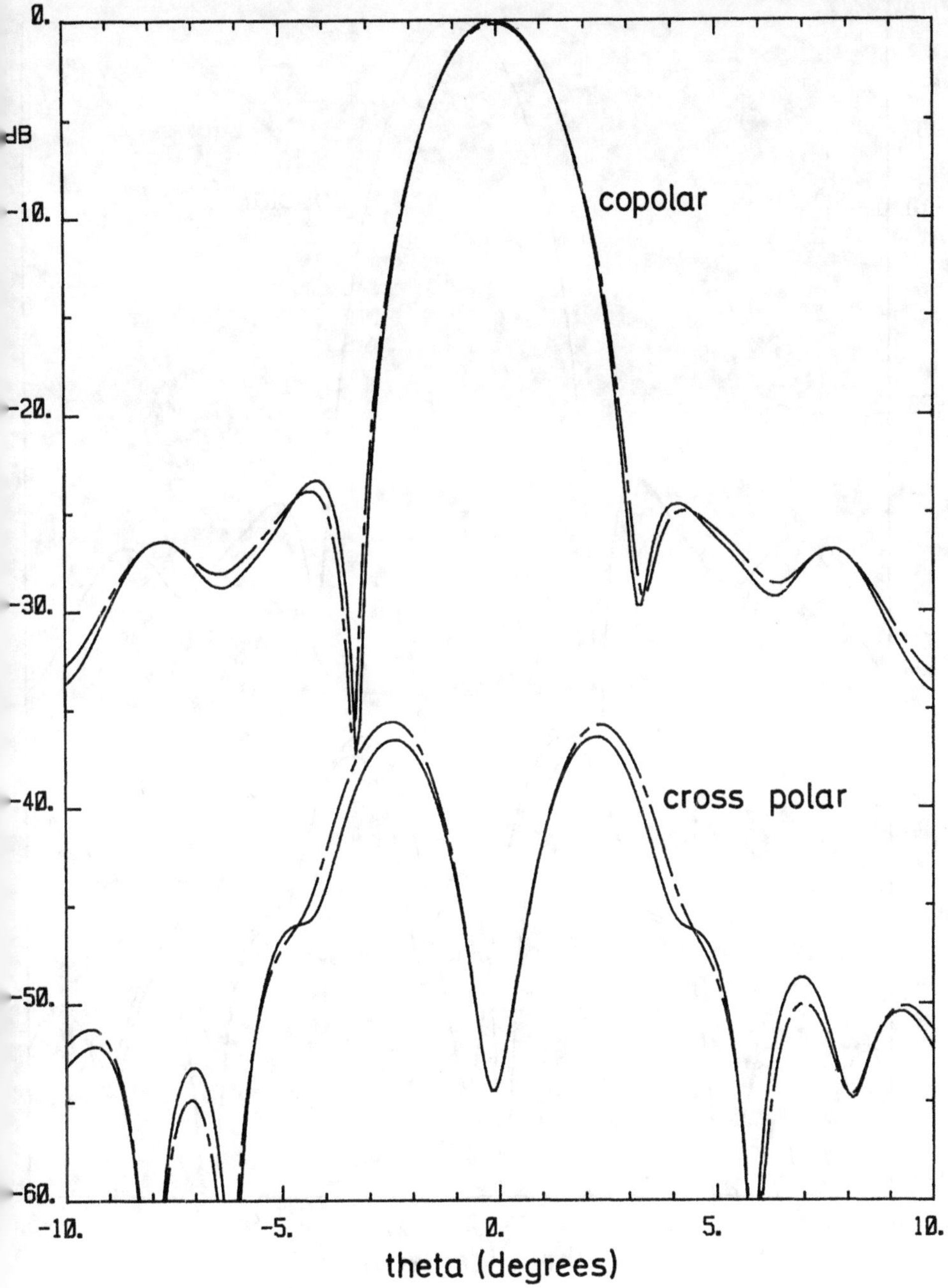

Fig. 5.30(*c*) *Front-fed reflector antenna of diameter 30λ at 11.7 GHz. Copolar and cross-polar far fields transformed from near field measured at 1.5 m*
Dotted line: probe pattern correction omitted
Full line: probe pattern correction included
The measurement plane is given by $\phi = 45°$

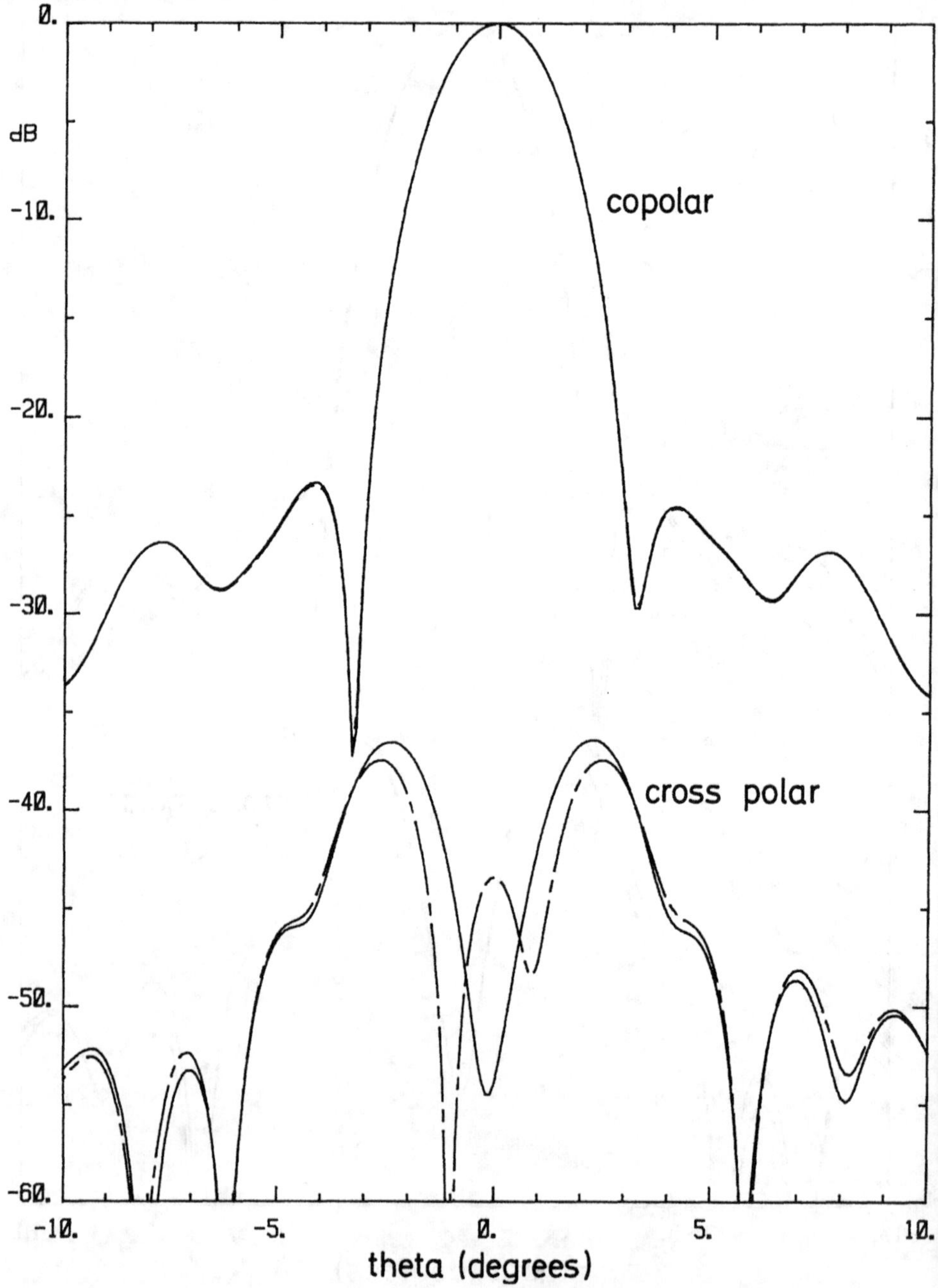

Fig. 5.30(*d*) *Front-fed reflector antenna of diameter 30λ at 11.7 GHz. Copolar and cross-polar far fields transformed from near field measured at 1.5 m*
Dotted line: without probe polarization correction
Full line: with probe polarization correction
The measurement plane is given by $\phi = 45°$

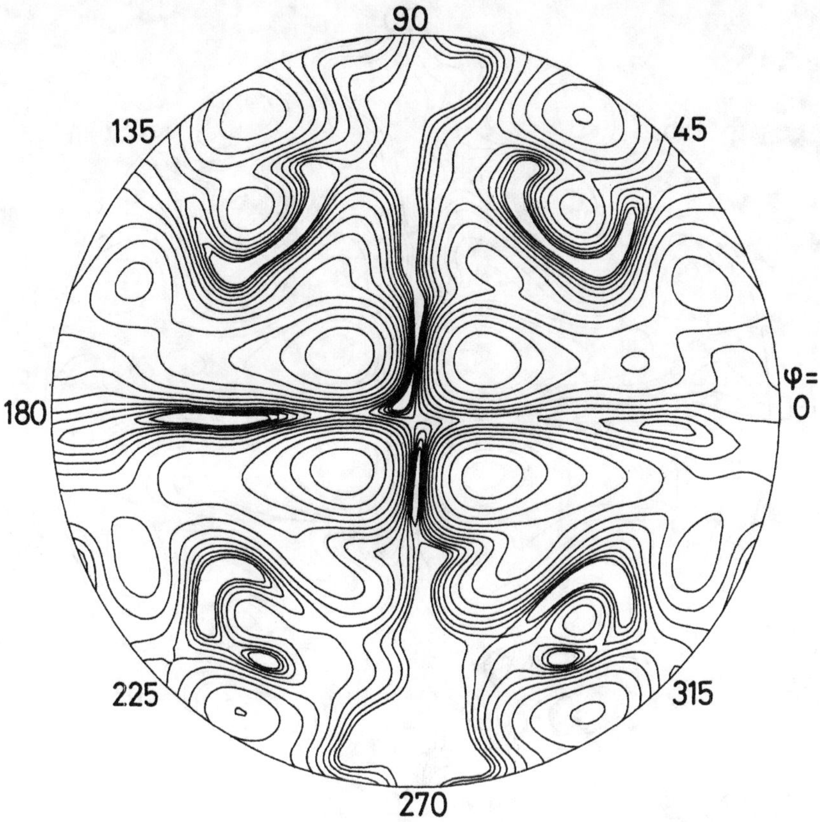

Fig. 5.30(e) *Front-fed reflector antenna of diameter 30λ at 11.7 GHz. Contour plot centred at $(\theta, \phi) = (0, 0)$ and of diameter 20° of the cross-polar far field transformed from the near field measured at 1.5 m. Polarization correction included; see Fig. 5.30(b) and (d). Levels in dB: −38, −40, ..., −60*

ratio of the far field on boresight, ϕ_o has been adjusted such that the cross-polar reference vector is parallel to the minor axis of the polarization ellipse at $\theta = 0°$. This corresponds to rotating the probe polarization in a far-field measurement. However, the same value of ϕ_o has been used for the three curves in Fig. 5.30(b). The fact that ϕ_o can be adjusted in the computer after the measurements have been carried out is an advantage of near-field scanning and transformation. A contour plot corresponding to the solid cross-polar pattern in Fig. 5.30(b) is shown in Fig. 5.30(e). By considering the measured phase, it can be deduced that the field is left-hand elliptical in the two lobes in the $\phi = 45°$ cut, but right-hand elliptical in the $\phi = 135°$ lobes. Consequently, there are borders separating the lobes where the field is perfectly linearly polarized.

However, as the tilt angle is slightly changing, one will not have exact zero cross polarization along these borders as we keep ϕ_o constant for the plot. At

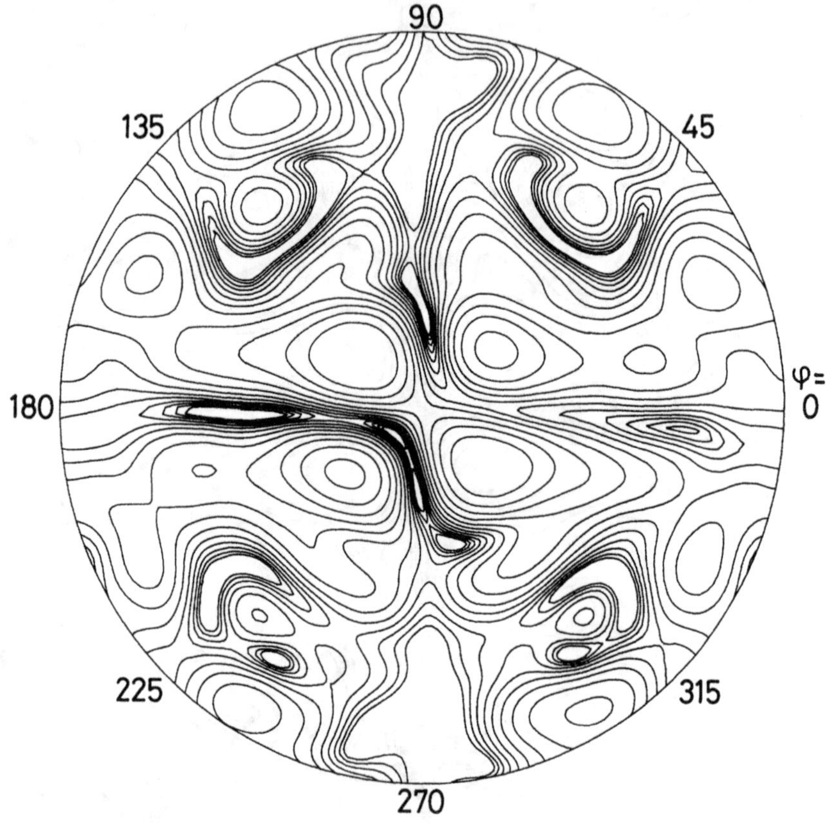

Fig. 5.30(*f*) *Front-fed reflector antenna of diameter 30λ at 11.7 GHz. Contour plot centred at (θ, φ) = (0, 0) and of diameter 20° of the cross-polar far field transformed from near field measured at 1.5 m. No polarization correction, see Fig. 5.30(d). Levels in dB: −38, −40, ..., −60*

$\theta = 0°$, the field is left-hand elliptical with an axial ratio of 54 dB in accordance with the two left-hand lobes somewhat melting together in Fig. 5.30(e).

The probe polarization correction mainly influences the cross-polar fields at boresight. This can be seen in Figs 5.30(d) and 5.30(f) where the dashed curve and the contour plot shows the cross-polar field calculated without polarization correction. Probe pattern correction is included. The two probe ports had an average polarization of 43 dB right-hand elliptical and this changes the determined on-axis polarization of the test antenna to be right-hand elliptical if it is not corrected for. The pattern correction can also affect the cross-polar lobes but this is not pronounced in the present example because the cross-polar field of the horn probe is small within the test zone between the 2 dB points. The discrepancy of the cross-polar lobes in Fig. 5.30(b) is mainly related to the amplitude phase factor between the two probe channels since the lobes are very sensitive to even small changes in this factor.

As another example, Figs 5.31 show directivity patterns for a 34 by 19 wavelengths offset reflector. The antenna has a contoured beam formed by four feed horns [18]. Figure 5.31(a) shows a comparison between far fields transformed from near-field measurements made at 1.4 m, 2.5 m and 4.9 m. Figure 5.31(b) shows a contour plot of the copolar pattern as transformed from the near-field measurement taken at 2.5 m.

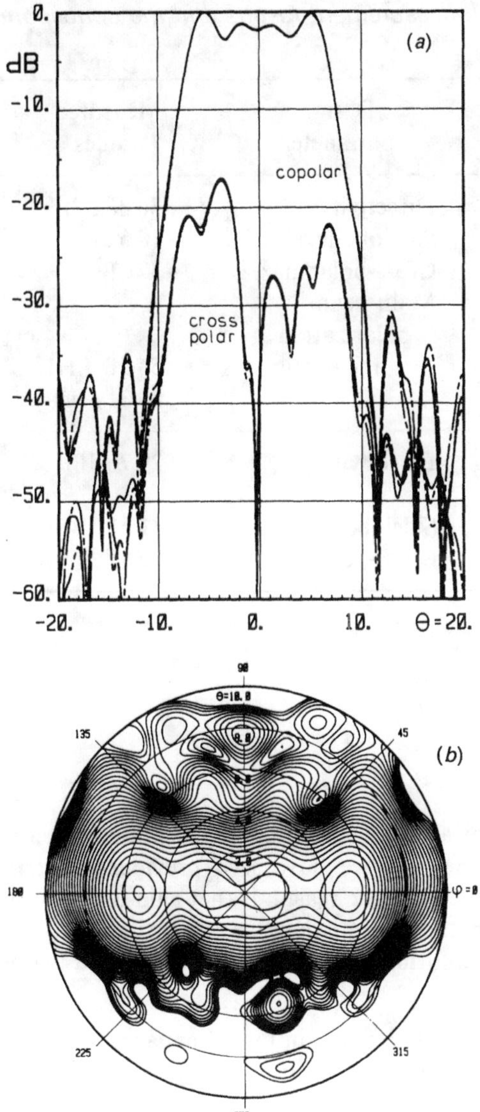

Fig. 5.31 *Radiation patterns for a 34 × 19 wavelengths offset reflector*
(a) Copolar and cross-polar pattern for $\phi = 0°$
(b) Copolar contour pattern for $\theta \leqslant 10°$

The accuracies of the above two spherical near-field measurement examples are summarized in Table 5.5. The accuracies quoted in Table 5.5 are estimates based upon repeated measurements of the same patterns, but with measurement parameters such as distance, speed, scan type etc. varied in a systematic manner [24].

Table 5.5 *Typical measurement results and the corresponding 1σ accuracies.*

Antenna type	Design parameter	Measured values	Accuracy (1σ)
Symmetrical front-fed 30λ reflector	Directivity	36.86 dBi	±0.03 dB
	Sidelobe level	−18.5 dB	±0.2 dB
	Cross-polar max.	−37 dB	±2.0 dB
	Main beam polarization		
	axial ratio	54 dB LHE	±2.0 dB
	tilt angle	89.31 deg	±0.03 deg
Elliptical offset reflector (34λ × 19λ)	Directivity	29.70 dBi	±0.04 dB
	Sidelobe level	−31.51 dB	±0.43 dB
	Cross-polar max.	−20.85 dB	±0.16 dB

5.4 Determination of gain

5.4.1 Introduction

While directivity for a test antenna can be calculated directly from a spherical near-field measurement as we have seen previously, determination of gain requires some extra effort. The reason is that gain involves the power accepted by the antenna from the transmission line rather than the power radiated by the antenna. The extra information needed can be obtained in one of the following ways:

1. Calculation of the relative loss in the antenna, ref. eqn (2.212).
2. Connection of the signal generator to the receiver through a device with known attenuation.
3. Substitution of the test antenna by an antenna with known loss or known gain.

The first method cannot be used in general. The second is not convenient when a calibrated dual-polarized probe is used since it would imply a disconnecting of one of the ports of the probe that could disturb the phase and amplitude factor. It would also require the gain of the probe to be known as discussed in Section 4.3.4.4. The third method therefore, is adopted here and will be described below.

Usually, a standard gain horn is used in the substitution measurement. Three steps are involved.

1. Near-field scanning of the test antenna.
2. Comparison of the near-field signal from the test antenna with the near-field signal from the horn.
3. Near-field scanning of the standard gain horn.

In addition, the loss in the standard gain horn must be estimated. Mismatch correction can be added if necessary.

The measurement of the standard gain horn together with the loss calculation implies that the gain of the horn does not have to be known in advance. In fact, the spherical near-field measurement of the standard gain horn constitutes a gain calibration in itself. With proper care, the accuracy one can obtain in this calibration is comparable to the best of other methods available.

Therefore, we shall treat the horn gain calibration separately in Section 5.4.2 and the gain determination by substitution in Section 5.4.3.

5.4.2 Gain calibration of horns

The gain calibration of horns by a spherical near-field measurement is based on the assumption that the loss in the horn is small. For an X-band standard gain horn, the loss can be calculated to values between 0.01 dB and 0.02 dB, mainly arising from the waveguide section of the horn. Thus, inaccuracies in the loss calculation will not contribute significantly to the overall error budget. Therefore, when the directivity D has been determined, the gain can be found by the formulas

$$G(\text{dB}) = D(\text{dB}) - \text{loss}(\text{dB}). \tag{5.77}$$

The calculation of directivity from a spherical measurement is described in previous chapters. For standard gain horns it is of interest to optimize the measurement parameters for accurate determination of the copolar far field at the boresight and the total radiated power. The measurement time should also be as short as possible since calibration is often wanted at many frequencies. For example, the determination of probe polarization ratios can in general be omitted.

The spherical near-field transformation with probe correction takes care of the near-field correction corresponding to the distance correction calculated by Chu and Semplak [25]. An advantage of the transformation is that the correction is based on measured patterns rather than on an idealized model. Correction factors based on measured patterns have earlier been published by Ludwig and Norman

[26]. However, many other error sources must be controlled to give an accurate directivity value. In order of importance, they are

1. Multiple reflections between horn and probe.
2. Non-linearity of the receiver.
3. Noise and drift.
4. Scattering and absorption by the antenna tower.
5. Reflections from the surroundings.
6. Leakage from connectors.
7. Sampling errors.
8. Malfunction of a component or an instrument.

Scattering from the antenna tower is made small by mounting the horn well in front of the tower head. One cannot place a large absorber behind the horn since this would reduce the total radiated power.

Reflections from the floor and the walls can be reduced by shortening the measurement distance for a fixed height of the towers and by increasing the directivity of the probe [27]. On the other hand, this increases *multiple reflections* between the horn and the probe. In principle, therefore, a tradeoff must be made. However, for measurements in an anechoic chamber with towers of reasonable height, multiple reflections will dominate. The level of multiple reflections can be measured by varying the measurement distance over a short interval. Figure 5.32 shows an example of the signal between two X-band standard gain horns at 11.7 GHz with variation of the measurement distance around 2.7 m. Multiple reflection causes a variation of the signal with a period of $\lambda/2$. The peak-to-peak variation is about 0.11 dB in the example but can be larger at other frequencies. At longer ranges, ground reflections can also be identified by varying the separation between the antennas but this requires measurements over a longer interval than above [8]. One way of reducing the error from multiple reflections is to take the average of two measurements differing by $\lambda/4$ in distance.

Standard gain horns have been used as probes for gain calibration at TUD because one then has two similar antennas such that the iteration procedure for calculation of probe coefficients in Section 3.2.5 could be carried out [28]. However, it has been found that, by using a conical horn probe with a gain of 15 dBi rather than a standard gain horn with 22 dBi gain and increasing the measurement distance to 5 m, the influence of multiple reflections in the example above is reduced to 0.01 dB. Additional advantages of the conical probe is that it can be dual polarized and that it is in general more symmetrically and accurately made than a conventional standard gain horn. A disadvantage is that two probes are needed to cover a waveguide band, e.g. the X-band.

For standard gain horns, it is advantageous to scan the radiated field in theta. First, the horn is radiating in the z-direction implying that a denser *sampling* in theta than in phi is needed. Also the fact that only accurate field values close to $\theta = 0$ are wanted adds to this conclusion. It has been found that $\Delta\phi = 10°$ is

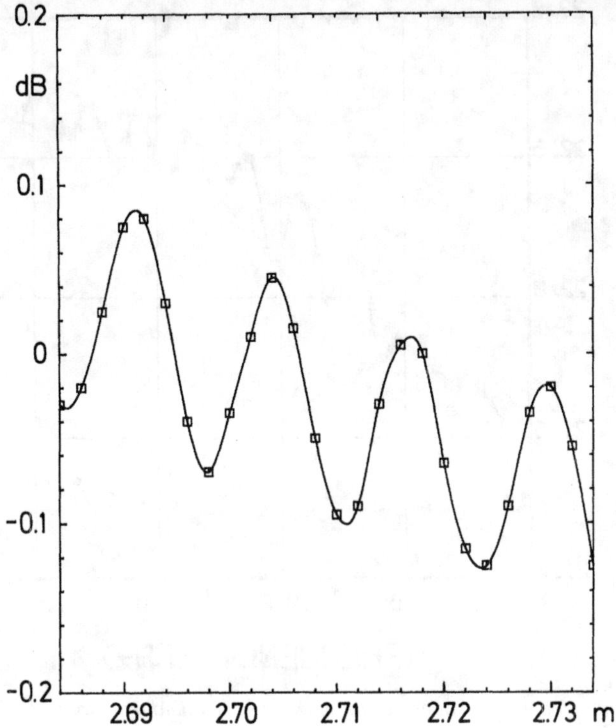

Fig. 5.32 *Multiple reflections recorded by changing the distance between two X-band standard gain horns around 2.7 m*

sufficient. Secondly, scanning in theta means that long-term drift in amplitude will affect the peak far-field value and the total radiated power in the same way. The reason is that the peak far-field value in the transformation is found by summing over mainly one point ($\theta = 0°$) from each scan and this value will follow the power contribution from the scan. With phi scanning, the data for the peak far-field value would be determined in the beginning of the measurement and drift would only affect the power. Thus the influence of long-term amplitude drift on the gain value is small. Thirdly, scanning in theta means that the number of modes required in the transformation can be determined by the N-test described in Section 5.3.3.6. Thus, noise reduction can be obtained in the transformation by oversampling in theta. Optimum location of the rotation centre will be somewhere inside the horn behind the aperture.

The influence of *noise and drift* can be investigated by repeated measurements with other measurement parameters unchanged. Another consistency check is to measure two spheres as described in Section 5.3.3.7 and compare the results.

Using the above precautions, the major error sources will be the remaining

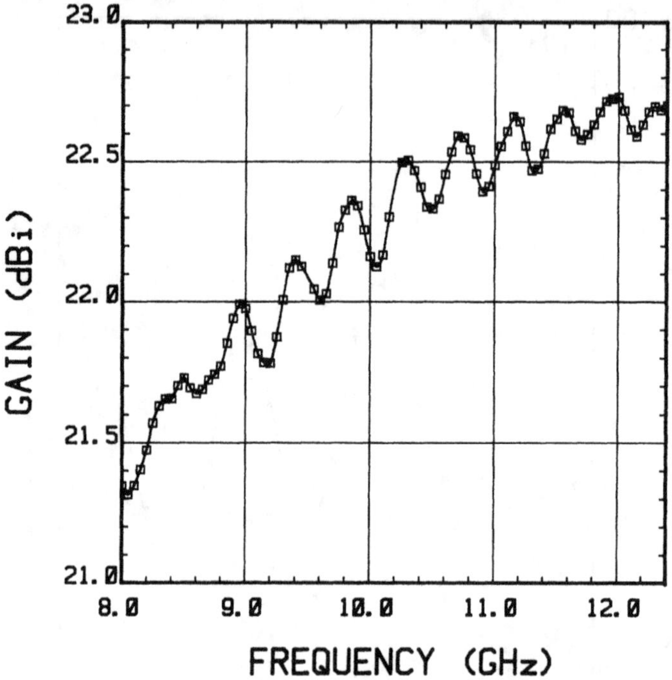

Fig. 5.33 *Calibration at 85 frequencies for an X-band standard gain horn [29], measured at TUD*

influence of multiple reflections and the *receiver non-linearity*. The latter can be reduced by receiver calibration, but this is a difficult and tedious process.

As an example of a calibration, Fig. 5.33 shows a calibration curve for an x-band standard gain horn of NRL design [29]. The curve was obtained by spherical near-field measurements at 85 frequencies equidistantly spaced over the band, each measurement taking 45 minutes. The curve exhibits a number of ripples with frequency, arising from field contributions being diffracted from the aperture edges and reflected from the interior of the horn [30]. Because of small differences in the dimensions of horns of the same type, the location of the ripples can change from one individual horn to the next. Therefore, at frequencies where the slope is steep, the curve can only be used for the particular horn measured. Based upon an error budget, the accuracy of the measurement in Fig. 5.33 is $\pm 0.033\,\mathrm{dB}(1\sigma)$.

5.4.3 Gain determination by substitution

When the losses in the test antenna are unknown, gain must be determined by substitution with an antenna of known losses as illustrated in Fig. 5.34 [31]. This antenna does not have to be a standard gain horn but in practice such a horn is a good choice. From the discussion in connection with eqn (4.112) we can

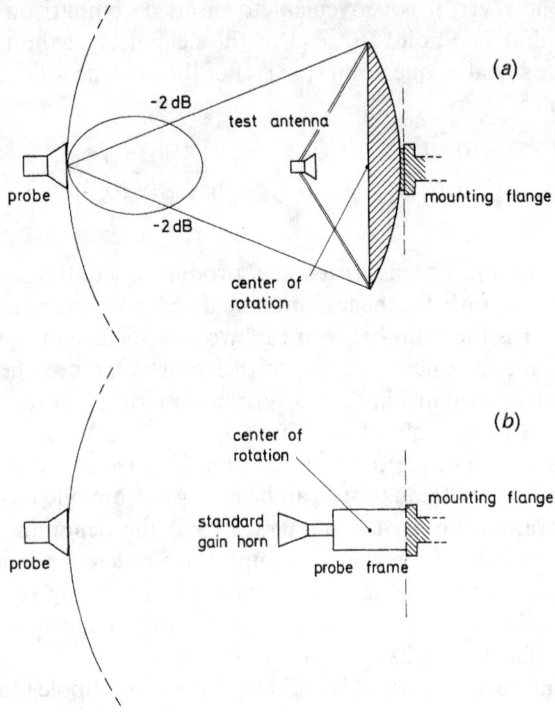

Fig. 5.34 *Gain measurement by near-field substitution. (a) The signal received from the test antenna and (b) The signal received from a standard gain horn are compared. The probe pattern has a 2 dB taper over the test antenna aperture*
Note the position of the centre of rotation relative to the standard gain horn

calculate directivity as

$$D_t(\theta, \phi) = \frac{8}{3} \frac{\frac{1}{2}|W_t(\theta, \phi)|^2}{P_{rad}} \tag{5.78}$$

where W_t is the far-field output signal from the transformation with a short dipole as output probe. P_{rad} is calculated as one half times the sum of the squares of the coefficients in the same transformation. By definition

$$G_t(\theta, \phi) = \frac{8}{3} \frac{\frac{1}{2}|W_t(\theta, \phi)|^2}{P_{rad} + P_{loss}} = \frac{8}{3} \frac{\frac{1}{2}|W_t(\theta, \phi)|^2}{P_{acc}} \tag{5.79}$$

where P_{acc} is the power accepted by the antenna.

Disregarding the influence of mismatch, one can find P_{acc} by exchanging the test antenna by a horn. The horn is then scanned and the data transformed without any changes in gain settings or data-processing parameters. Assuming no loss in the horn, P_{acc} is found in proper scale from the spherical wave coefficients.

In practice, however, it is convenient to insert a comparison measurement where the signal at one point (θ_1, ϕ_1) in the near field of the test antenna is compared to the signal at one point (θ_2, ϕ_2) of the horn near field. Formally we can extend eqn (5.79) as

$$G_t(\theta, \phi) = \frac{\frac{1}{2}|W_t(\theta, \phi)|^2}{\frac{1}{2}|w_t(\theta_1, \phi_1)|^2} \frac{\frac{1}{2}|w_t(\theta_1, \phi_1)|^2}{\frac{1}{2}|w_h(\theta_2, \phi_2)|^2} \frac{\frac{1}{2}|w_h(\theta_2, \phi_2)|^2}{\frac{1}{2}|W_h(0, 0)|^2} \frac{8}{3} \frac{\frac{1}{2}|W_h(0, 0)|^2}{P_{\text{acc}}}. \tag{5.80}$$

In this way, the gain has been written as a product where the first factor is the far-field to near-field ratio for the test antenna determined in the transformation. The second factor is the ratio between the two near-field signals determined in the comparison measurement and the third factor is the near-field to far-field ratio for the horn determined in the transformation for the horn. The last factor can be recognized as the gain of the horn.

The advantage in splitting the measurement into factors in this way is that receiver gain and cable attenuations can be changed from one factor to another so long as the ratio between the numerator and the denominator of the individual factors is not affected. For example, two different semirigid antenna cables can be used for the test antenna and the horn during scanning of each antenna and a stable flexible cable can be used in the comparison measurement where the antennas do not have to be rotated.

It is a prerequisite for the use of eqn (5.80) that a short dipole is used as output probe in the transformations for both the test antenna and the horn. Also, the input probe coefficients must be the same in the two transformations. Concerning the near-field values, it must be noted that the polarization correction eqns (5.37–5.38) shall be applied to either both or none of two corresponding values. This means that either the polarization correction must be included in the near-field far-field ratio or it must be applied to the comparison measurement. For linearly polarized antennas where the component selected to enter the comparison measurement is parallel to one port of the probe, polarization correction will not be very significant as long as the same port is used for test antenna and horn. If different ports are used, it is important to compensate for the amplitude difference.

The last factor in eqn (5.80) can in principle be replaced with a true (precalibrated) gain value for the standard gain horn [32, 33]. However, it is still necessary to scan the horn, and therefore a gain value can always be calculated. One can argue that it is advantageous to use the latter gain value even if it has been measured under less ideal circumstances than described in the previous section and is therefore less accurate. Using the latter gain value, the two values of $W_h(0, 0)$ in eqn (5.80) are identical such that one effectively divides by the accepted power. Using the true gain value, one divides by the peak far field. Multiple reflections are a significant error source. They will vary during scanning such that they affect the total power less than the peak far-field value.

If deemed necessary, mismatch can be corrected for by measuring the an-

Table 5.6 *Calculation of gain values for a test antenna, ref. Fig. 5.11(b), by near-field substitution with standard gain horn (SGH). The values both with and without probe pattern correction are given for each of the two distances. The corresponding directivity values are given in the bottom line.*

	Near-field distance: 1.5 m		Near-field distance: 3 m	
	Without correction	With correction	Without correction	With correction
(+) Far field, test antenna	13.27 dB	13.87 dB	8.24 dB	8.40 dB
(−) Near field, test antenna	0	0	0	0
(+) Signal with test antenna	−0.23	−0.23	0.21	0.21
(−) Signal with SGH	0.13	0.13	−4.80	−4.80
(+) Near field SGH	0	0	0	0
(−) Far field SGH	−0.68	−0.64	−0.60	−0.58
(+) Gain SGH	22.60	22.54	22.68	22.67
(=) Gain of test antenna	36.19 dB	36.69 dB	36.53 dB	36.66 dB
Directivity of test antenna	36.66 dBi	36.82 dBi	36.83 dBi	36.87 dBi

tenna reflection coefficients and the reflection coefficient of the cable used in the comparison measurement as described for far-field substitution measurements in [13].

Table 5.6 shows the application of equation (5.80) on measurements on the reflector antenna described in Section 5.3.4. Scanning and substitution has been carried out at two distances 1.5 m and 3.0 m. Transformations have been carried out both with and without probe correction, i.e. with the actual probe and a short electric dipole, respectively, as input probe. It is seen that all near-field values have been normalized to 0 dB before the transformations. The reflector antenna and the horn have very different gains in the far field. The difference between the near-field signals at 1.5 m is quite small because the beam width for the horn corresponds to the width of the reflector aperture. Note that the gain values used for the standard gain horn are different. Probe pattern correction gives a change of 0.5 dB in test antenna gain but only 0.16 dB in directivity. These values apply to the 1.5 m measurement where the probe pattern has a taper of 2 dB over the test antenna aperture, as indicated in Fig. 5.34. Comparison between gain and directivity shows that internal loss in the antenna is between 0.1 and 0.2 dB.

References

1. BASSEN, H. I. and SMITH, G. S., 1983, 'Electric field probes—A review', *IEEE Trans.*, Vol. **AP-31**, pp. 710–718, Sep.
2. KANDA, M., LARSEN, E. B., BORSERO, M., GALLIANO, P. G., YOKOSHIMA, I. and NAHMAN, N. S., 1986, 'Standards for electromagnetic field measurements', *Proc. IEEE*, pp. 120–128, Jan.
3. LARSEN, FL. HOLM and HANSEN, J. E., 1979, 'A dual-polarized probe system for near-field measurements', *IEEE AP-S Symposium Digest*, pp. 557–560, Seattle, USA, June.
4. Scientific-Atlanta, 1977, 'Model 1774 three-channel programmable microwave receiver, instruction manual', Aug.
5. NEWELL, A. C., 1985, 'Planar near-field measurements', Lecture Notes, National Bureau of Standards, Boulder, Colorado, June.
6. NEWELL, A. C., FRANCIS, M. H. and KREMER, D. P., 1984, 'The determination of near-field correction parameters for circularly polarized probes', Antenna Measurement Techniques Association Meeting, Paper No. **3A3**, pp. 1–29, San Diego, Oct.
7. LUDWIG, A. C., 1967, 'Gain computations from pattern integration', *IEEE Trans.*, Vol. **AP-15**, pp. 309–311, Mar.
8. LUDWIG, A. C., HARDY, J. and NORMAN, R. 1972, 'Gain calibration of a horn antenna using pattern integration', Technical Report **32-1572**, Jet Propulsion Laboratory, California, Oct.
9. CLARRICOATS, P. J. B. and OLVER, A., 1984, *Corrugated horns for microwave antennas*, Peter Peregrinus Ltd., London.
10. NEWELL, A. C., 1975, 'Improved polarization measurements using a modified three-antenna technique', *IEEE AP-S Symposium Digest*, pp. 337–340, Illinois, June.
11. CHRISTENSEN, E. L., HANSEN, J. E., HANSEN, C. K., HANSEN, S. B., LARSEN, Fl. HOLM and LEMANCZYK, G. H., 1981, 'Facility description and measurement results', Experimental spherical near-field antenna test facility, phase 2. Final report Vol. 1, ESA Contract No. 3916/79/NL/DG. Electromagnetics Institute, Technical University of Denmark, Report **R248**, Dec.

12. JONES, J. R. and HESS, D. W., 1985, 'Automated three-antenna polarization measurements using digital signal processing', Antenna Measurement Techniques Association Meeting, Paper No. **6**, pp. 1–11, Florida, Oct.
13. IEEE, 1979, 'IEEE Standard test procedures for antennas', The Institute of Electrical and Electronics Engineers, Inc., *IEEE Std.* **149**.
14. HESS, D. W., 1982, 'Series 2020 spherical near-field range configuration', Scientific-Atlanta, Inc. Report **41A004Z**, Apr.
15. HOLLIS, J. S., LYON, T. J. and CLAYTON, Jr., L., (editors), 1970, *Microwave antenna measurements*, Scientific-Atlanta, Inc., Atlanta, Georgia, USA, July.
16. MOSTAFAVI, M. and BOLOMEY, J. C., 1985, 'Correction scheme for continuous rotation of the test antenna using fast near-field measurement techniques', *IEEE Trans.*, Vol. **AP-33**, No. 8, pp. 922–924, Aug.
17. BOYLES, J. W., 1986, 'Long range antenna measurements with the HP 8510 using harmonic mixers', Antenna Measurement Techniques Association Meeting, pp. 19–24, Ottawa, Sep.
18. HANSEN, J. E., 1984, 'The TUD–ESA spherical near-field antenna test facility, Lyngby, Denmark', Publication **BR-19**, European Space Agency.
19. BACH, H., CHRISTENSEN, E. LINTZ, HANSEN, J. E., JENSEN, F. and LARSEN, FL. HOLM, 1979, 'Facility design and test runs', Experimental spherical near-field antenna test facility. Final Report, Vol. I (summary of work). ESA Contract No. 3337/NL/AK. Electromagnetics Institute, Technical University of Denmark, Report **R208**, Apr.
20. ZIEG, S. W., 1982, 'A precision optical range alignment technique', Antenna Measurement Techniques Association Meeting, Paper No. **14**, pp. 1–20, New Mexico, Oct.
21. LARSEN, FL. HOLM, 1977, 'Probe correction of spherical near-field measurements', *Electronics Letters*, Vol. **13**, No. 14, pp. 393–395, July.
22. NEWELL, A. C. and REPJAR, A. G., 1977, 'Results of spherical near-field measurements on narrow-beam antennas', *IEEE AP-S Symposium Digest*, pp. 382–385, Stanford, USA, June.
23. STUBENRAUCH, C. F., 1979, 'Some recent near-field measurements at NBS', Antenna Measurement Techniques Association Meeting, Atlanta, Georgia, USA, Oct.
24. LEMANCZYK, J. H., HANSEN, J. E. and LARSEN, FL. HOLM, 1984, 'Evaluation of the spherical near-field range at the Technical University of Denmark', *IEEE AP-S Symposium Digest*, Vol. **2**, pp. 682–685, Boston, June.
25. CHU, T. S. and SEMPLAK, R. A., 1965, 'Gain of electromagnetic horns', *Bell Syst. Tech. J.*, Vol. **44**, pp. 527–537, Mar.
26. LUDWIG, A. C. and NORMAN, R. A., 1973, 'A new method for calculating correction factors for near-field gain measurements', *IEEE Trans.*, Vol. **AP-21**, pp. 623–628, Sep.
27. HANSEN, P. C. and LARSEN, FL. HOLM, 1984, 'Suppression of reflections by directive probes in spherical near-field measurements', *IEEE Trans.*, Vol. **AP-32**, pp. 119–125, Feb.
28. LEMANCZYK, J. H., LARSEN, FL. HOLM, HANSEN, J. E. and AASTED, J., 1981, 'Calibration of standard gain antennas using a spherical near-field technique', Conference on Antennas and Propagation, *IEE Conference Publication No.* **195**, Part 1, pp. 241–245, York, England, Apr.
29. SLAYTON, W. T., 1954, 'Design and calibration of microwave antenna gain standards, U.S. Naval Res. Lab., Report **4433**, Nov.
30. JULL, E. V., 1973, 'Errors in the predicted gain of pyramidal horns', *IEEE Trans.*, Vol. **AP-21**, pp. 25–31, Jan.
31. LARSEN, FL. HOLM, 1983, 'SNIFTD—a new computer program for spherical near-field transformation', Antenna Measurement Techniques Association Meeting, Paper No. **20**, pp. 1–20, Annapolis, Sep.
32. HESS, D. W. and JONES, J. R., 1983, 'Gain comparison measurements in spherical near-field scanning', Antenna Measurement Techniques Association Meeting, Paper No. **22**, pp. 1–28, Annapolis, Sep.
33. HESS, D. W., JONES, J. R., GREEN, C., MELSON, B. and PROCTOR, J., 1985, 'Spherical near-field antenna measurements improve through expanded software features', *Microwave Systems News*, Vol. **15**, No. 3, pp. 64–83, Mar.

Chapter 6

Error analysis of spherical near-field measurements

6.1 Introduction

Near-field testing involves a numerical transformation of the measured data to the far field of the antenna. Thus, the influence of measurement inaccuracies may be found difficult to evaluate directly and it may be an advantage to apply computer simulations of the measurements [1]. The principle and results of simulations for the spherical technique are presented in the following sections. The influence of probe correction is analyzed and the effects of truncation of the measurement sphere will be considered. Finally, simulated inaccuracies are compared to inaccuracies of real measurements.

The simulations are all carried out for directive spot beam antennas with emphasis on prediction of their far fields within the region of the main beam and the first side lobe. For other types of test antennas, such as contoured beam antennas or omnidirectional antennas mounted on a scattering body, the nature of the near field is quite different from the quasi-plane wave in front of the aperture of the spot-beam antenna. Hence, the results of simulations presented here should be taken only as guidelines. For evaluation of a particular antenna measurement, simulations with the actual parameters are desirable.

6.2 Near-field measurements

The principle of near-field simulations is to numerically sample a modelled antenna near field with a well-known, simulated measurement inaccuracy. The test antenna model must have a well-defined near field that can be easily generated at all realistic probe positions. The most simple probe model is a short electric dipole which, as its output, directly gives a signal proportional to the electric field component in the direction of the dipole, ref. eqns (3.45–3.46).

In the spherical measurement set-up, the field is probed at points given by spherical coordinates (A, θ, ϕ) where A is the radius of the measurement sphere, and θ and ϕ vary in equidistant steps. At these points, the theta and phi components of the field are measured, and the sampled signals are proportional to

$$E_\theta = \vec{E}(A, \theta, \phi) \cdot \hat{\theta} \tag{6.1a}$$

$$E_\phi = \vec{E}(A, \theta, \phi) \cdot \hat{\phi} \tag{6.1b}$$

where $\vec{E}(A, \theta, \phi)$ is the electric field of the test antenna.

When simulating mechanical inaccuracies, the positions of the sampling probe are slightly modified to (A', θ', ϕ') according to the inaccuracies simulated. The probe orientation may be inaccurate as well. Calling the probe orientation $\hat{\theta}'$ and $\hat{\phi}'$ when the probe is intended to sample the theta and phi components of the field, respectively, we get for the sampled signals

$$E'_\theta = \vec{E}(A', \theta', \phi') \cdot \hat{\theta}' \tag{6.2a}$$

$$E'_\phi = \vec{E}(A', \theta', \phi') \cdot \hat{\phi}'. \tag{6.2b}$$

The position of the sample point (A', θ', ϕ') may be different for the two polarization samples if these are not sampled simultaneously.

When electrical inaccuracies are simulated, the sampled signals are disturbed to

$$E''_\theta = \mathrm{f}_1(E'_\theta, E'_\phi) \tag{6.3a}$$

$$E''_\phi = \mathrm{f}_2(E'_\theta, E'_\phi) \tag{6.3b}$$

where f_1 and f_2 are functions describing the receiver non-linearities, time variation, etc. The two functions will in general be different when a two-channel receiver system is applied. For the ideal electrical system (mechanical inaccuracies only), we have

$$E''_\theta = E'_\theta \tag{6.4a}$$

$$E''_\phi = E'_\phi \tag{6.4b}$$

and for the ideal mechanical system (receiver inaccuracies only), we have

$$E''_\theta = \mathrm{f}_1(E_\theta, E_\phi) \tag{6.4c}$$

$$E''_\phi = \mathrm{f}_2(E_\theta, E_\phi). \tag{6.4d}$$

When the sampled near field has been generated, the far field can be determined by the near-field far-field transformation and the inaccuracies of the far-field pattern may be assessed by comparing to an accurate far-field reference. This reference field may be obtained by a transformation of the accurate near field.

The model of the test antenna shall be correct in the sense that the set of field data satisfies Maxwell's equations. This can be obtained by defining the antenna through a set of coefficients to a spherical wave expansion. The near field may then be determined from this expansion. A correct model may also be obtained if the antenna is an array of short dipoles. The near field is then determined by superposition of the near fields from each of the individual dipoles.

However, these models have a drawback. The inaccurate near field must be determined at points that in general are not placed in a regular grid at a fixed measurement distance. The near-field generation in the computer may therefore require considerably more time than the transformation to the far field. The time for the near-field generation increases drastically with the antenna diameter; for practical purposes, this may give an upper limit for the size of the test antenna to be simulated.

An approximate antenna model with a fast field determination should then be sought. Such an approximate model may be applied as long as its precision is better than that of the set-up to be simulated. This can be tested by displacing the measurement grid with respect to the antenna, e.g. by moving the point of rotation. The displacement must be comparable in size to the inaccuracies to be simulated. If the antenna model is correct (satisfying Maxwell's equations), the far-field parameters remain unchanged. When the antenna model is approximate, small changes may occur. These changes in the far-field parameters due to the approximate antenna model must be small compared to the field changes determined by the simulations.

A high-gain test antenna model which is fast and still has a high accuracy may be based on the geometrical theory of diffraction (GTD). The test antenna can then be modelled as a paraboloidal reflector antenna [2, 3]. The GTD model only needs to include the reflected ray and two diffracted rays; see Fig. 6.1. The on-axis caustic from the reflector rim may be ignored if the simulated mechanical inaccuracies are sufficiently small. Numerical experiments have shown that the radial displacement of grid points near $\theta = 0$ due to inaccuracies must be small compared to the wavelength, and the transverse displacement must be small compared to the grid size in theta.

Actual simulations have been carried out; they will be described in the following sections. The simulations are carried out by the program package FACSIM [4, 5]. As test antenna, a front-fed reflector antenna with a diameter $D = 30\lambda$ and focal length $f = 15\lambda$ has been chosen. As the GTD model is not sufficiently accurate for the probe evaluations in Section 6.5, the antenna model is defined through a set of coefficients to the spherical wave expansion. These coefficients are obtained from a reference near field based on the GTD model of Fig. 6.1. The antenna parameters are chosen close to those of the reflector antenna for which measurements were presented in Section 5.3.4. The feed is linearly polarized and the reflector edge illumination is -15 dB. The near field is sampled with a theta spacing of 1.25° and a phi spacing of 22.5°. The antenna far-field characteristics are given at the top of Table 6.1 in the next section.

6.3 Mechanical inaccuracies

Simulations of mechanical inaccuracies may be carried out for any one of the measurement set-ups discussed in Section 5.3.2.1, taking into account the different mechanical inaccuracies present in the actual set-up. The influence of deformations due to gravity has been discussed by Frandsen [6].

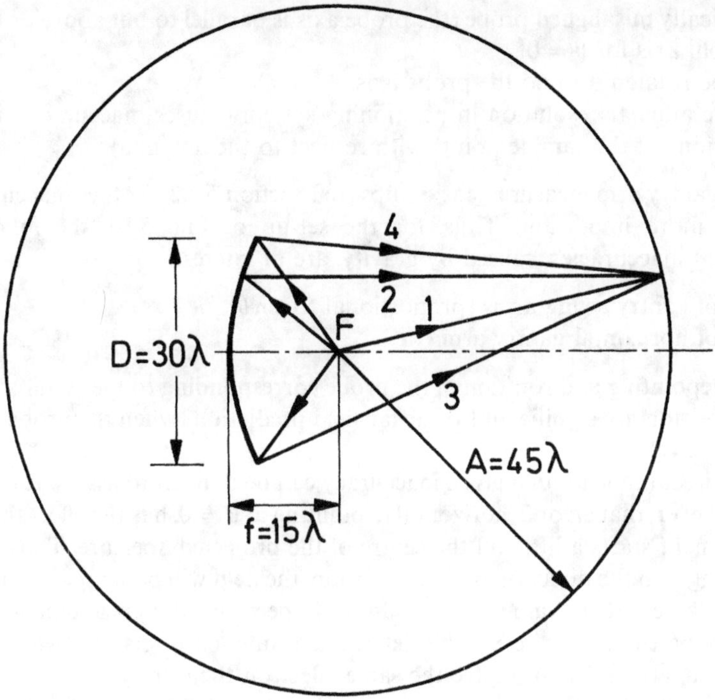

Fig. 6.1 *Test-antenna model and measurement sphere. The model is a front-fed reflector antenna with focus at F, focal length f, diameter D*
GTD rays included are (1) direct ray from feed, (2) reflected ray, (3) and (4) edge diffracted rays

When simulating mechanical inaccuracies, the measurement set-up and the alignment procedure must be carefully evaluated. For the elevation-over-azimuth set-up (Fig. 5.11) the dominating inaccuracies are listed below. The test antenna is assumed to be polar-pointing and the theta axis is assumed vertical. In the description of each of the inaccuracies, they are assumed to occur one at a time:

1. Non-intersecting axes of rotation (phi axis laterally displaced). That other inaccuracies are not occurring implies here that the probe for $\theta = 0$ is not at the phi axis.
2. Horizontal depointing at $\theta = 0$ (theta offset, phi axis is for $\theta = 0$ not pointing at the probe).
3. Vertical depointing at $\theta = 0$ (phi axis is not horizontal).
4. Inaccurate measurement distance inserted in the transformation formula.
5. Horizontally misaligned probe (the probe is translated in the direction perpendicular to the direction to the test antenna, thus the probe axis is parallel to but alongside the phi axis for $\theta = 0$).

6. Vertically misaligned probe (the probe axis is parallel to but above or below the phi axis for $\theta = 0$).
7. Probe rotated around the probe axis.
8. Finite angular resolution in positioning system (causes inaccuracies in the positions of the sample points with respect to the antenna).

For the gantry arm measurement set-ups (ref. Section 5.3.2.1) other inaccuracies may be more important. Thus, for the set-up of Fig. 5.10, the following additional inaccuracies, caused by gravity, are of interest.

9. Sag of gantry swing arms (proportional to $\sin \theta$).
10. Sag of horizontal gantry arm.

The depointing and rotation of the probe corresponding to these shifts of the probe position are significant for the far-field prediction (when the probe is not very directive).

The direct influence of a given inaccuracy can be difficult to assess. It may be seen, however, that error 2, horizontal depointing at $\theta = 0$, has the effect that the measurement starts a little off the centre of the projected aperture. Thus, if the depointing is in the direction of theta rotation, the field will be sampled at points with too large a theta value and the sampled aperture will appear smaller than the true aperture. As a result the determined antenna directivity will be too small. Also, errors 1 and 5 have the same effect on the directivity.

Error 7, in which the probe is rotated around the probe axis, causes a part of the unwanted field component to be sampled unintentionally. Then, for a linearly polarized test antenna, too high an on-axis cross polarization will be predicted.

It shall be noted that for several of the mechanical inaccuracies, a correction is possible if the amount of the inaccuracy can be determined. Thus, for error 7, the test antenna coordinate system will often be defined by the on-axis linear polarization direction. Hence the probe rotation will, to some extent, be compensated for by a corresponding rotation of the test antenna reference system and the cross polarization caused by the probe rotation will vanish in the on-axis direction.

For actual simulations, results are given in Table 6.1. Inaccuracies in lengths are chosen as 0.1λ and in angles, as 0.1° apart from the angular resolution errors which are $\pm 0.03^\circ$ (rectangularly distributed). The results are obtained by simulating the inaccuracies individually.

The simulations have shown that most of the far-field inaccuracies of Table 6.1 have a linear dependence on the inaccuracies of the set-up but in some cases, the dependence is quadratic, cf. Section 6.7. It is thus realistic to scale the table values to smaller mechanical inaccuracies than stated, while an extrapolation to higher inaccuracies cannot be recommended. Furthermore, the results of simulations have shown that the contribution to the far-field inaccuracies arising from several individual mechanical inaccuracies in general may be found as the root sum square of the individual inaccuracies.

Table 6.1 *Influence on far-field pattern performances of typical mechanical inaccuracies (reflector antenna, D = 30λ). Dashes (—) indicate values below the level of numerical accuracy. The sign of the inaccuracies will in some cases change when the sign of the simulated error is changed.*

Mechanical inaccuracies	Main beam			First null	First side lobe	Cross-polar lobe
	Directivity	Cross-polar level	Beam width	Position	Level	Level
Reference far-field values	38.4 dBi	$-\infty$ dB	2.25°	2.95°	−26.8 dB	−48.2 dB
	Change in dB	Increased to (dB)	Change in degrees	Change in degrees	Change in dB	Change in dB
Elevation-over-azimuth set-up						
1. Non-intersecting axes (0.1λ)	0.07	—	0.02	0.02	0.07	0.05
2. Horizontal depointing (0.1°) at $\theta = 0$	0.03	—	—	—	1.2	0.03
3. Vertical depointing (0.1°) at $\theta = 0$	—	−75	—	—	—	—
4. Wrong measurement distance (0.1λ)	—	—	—	—	0.14	—
5. Horizontal probe misalignment (0.1λ)	0.04	—	—	0.01	1.6	0.03
6. Vertical probe misalignment (0.1λ)	—	—	—	—	—	—
7. Probe rotated (0.1°) around probe axis	—	−55	—	—	—	0.15
8. Angular resolution (±0.03°)	—	—	—	—	0.11	0.09
Especially for single gantry arm set-up:						
9. Sag of gantry swing arms (0.1° maximum)	0.01	—	0.01	—	0.24	—
10. Sag of horizontal gantry arm (0.1λ)	—	—	—	—	—	0.01

The reference field has a null for the cross polarization in the main beam direction. The table values show, relatively to the peak directivity, the level to which the cross polarization increases when the inaccuracies are introduced.

Note that horizontal and vertical has different meaning with respect to test antenna orientation for the elevation-over-azimuth set-up than for the gantry arm set-up, cf. Figs 5.10–5.11.

The results are found reliable with an error level below −70 dB for the actual test antenna. Minor changes in the test antenna f/D ratio or feed taper can change the influence on the pattern parameters and for practical evaluations of an actual measurement set-up, the stated pattern inaccuracies shall be taken as guidelines only.

From the values in Table 6.1, it can thus be concluded that good directivity measurement requires small tolerances on the intersection of the axes (error 1) and on the horizontal alignment (2 and 5). As a result of the test antenna being polar-pointing, the horizontal alignment of the probe (5) is much more critical than the vertical alignment (6); the former contributes most significantly to the inaccuracy of the side lobe level. The cross-polar contribution due to slight rotation of the probe (7) increases the on-axis cross-polar level with an equal amount, $20\log(\sin 0.1°) = -55$ dB. The cross-polar lobe itself is, in spite of the low level at −48 dB, only slightly affected by the mechanical inaccuracies.

For the single gantry arm set-up, the sag of the swing arms (9) results in a varying phase error over the projected aperture (polar-pointing test antenna, Fig. 5.10). This effect is also observed at case 4, wrong measurement distance, and causes an inaccurate prediction of the side-lobe level. The sag of the horizontal gantry arm (10) is, however, assumed to be vertical and independent of the swing arm position given by theta. It will, over the projected aperture, cause a constant phase error with little effect on the far-field pattern.

As a result of the applied geometry, depointing of the main beam does not occur. This parameter is therefore not shown in the table. Prediction of pattern inaccuracies for test antennas or mechanical set-ups different from those investigated here requires separate investigations. It has, however, been found from simulations of other antennas that the best measurement accuracies are obtained for antennas with high directivity. The larger the antenna aperture within which the field has nearly constant amplitude, the less sensitive to mechanical inaccuracies is the sampling.

6.4 Receiver inaccuracies

When the signal from the test antenna has been sampled by the measuring probe, it is interpreted by the receiver system and stored for later numerical processing. This signal interpretation depends partly on the type of mechanical scanning. The limited band width of the receiver and the drift in the measurement system may result in a preferred scan technique, cf. Section 5.3.3.7. In the following simulations, a polar-pointing test antenna is assumed scanned in phi with step in theta, cf. Fig. 5.27(b).

Electrical inaccuracies are considered as arising in the receiver only. Other electrical inaccuracies, e.g. instabilities in the signal source, can be simulated as corresponding receiver instabilities and thus need not be separately simulated. Further examples are bending of cables (position dependent) and change of cable length with temperature (time dependent) which may both be simulated as phase errors. Frequency variations have not been considered since the frequency stability usually is very high (thus a frequency change of 10^{-7} over a range length of 280λ, 8 metres at X-band, results in a phase change of 0.01°, a factor 10 below the phase resolution).

The receiver inaccuracies are modelled by disturbing the signals in the two channels according to evaluated inaccuracies in an actual set-up or according to the specifications for the receiver. The latter are discussed below for Scientific-Atlanta Inc. Series 1770 receivers.

Amplitude inaccuracies

The receiver may be applied in two modes, a logarithmic and a linear amplitude mode.

In the logarithmic mode, the error typically consists of a one-sided level-dependent term, 0.05 dB/10 dB, a one-sided level-independent term, 0.1 dB (assumed to be linearly decreasing to zero when the signal increases from −10 dB to 0 dB; since the maximum per definition is 0 dB a systematic error signal is not possible here). The resolution is ±0.1 dB (round-off error). These error specifications are depicted, against the signal level, in Fig. 6.2.

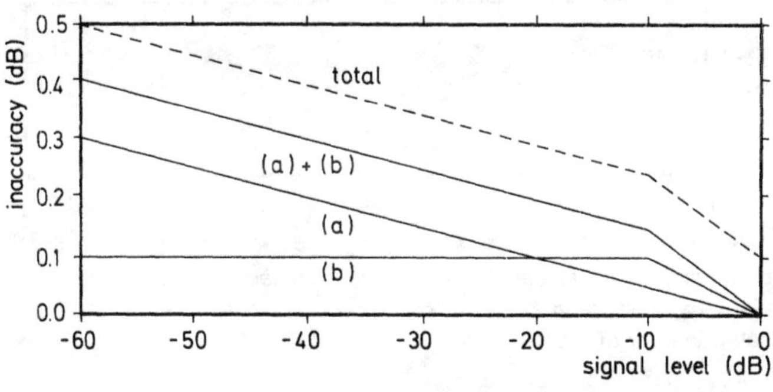

Fig. 6.2 *Specification limits for receiver amplitude inaccuracies in dB (logarithmic mode)* Level-dependent term (*a*), level-independent term (*b*), and total inaccuracy specification; the latter includes a resolution error of ±0.1 dB

In the linear mode, the error consists of the one-sided level-dependent term 0.05 dB/10 dB, a one-sided level-independent term, 0.5 per cent, (assumed linearly decreasing to zero when the signal decreases from 15 per cent to 0 per cent and also when the signal increases from 95 per cent to 110 per cent (maximum)), and a resolution of 0.1 per cent, cf. Fig. 6.3.

The one-sided terms are simulated to act in the same direction in order to simulate worst-case conditions.

For both modes, the dynamic range is simulated as 60 dB in the way that signals below −60 dB are set to this level.

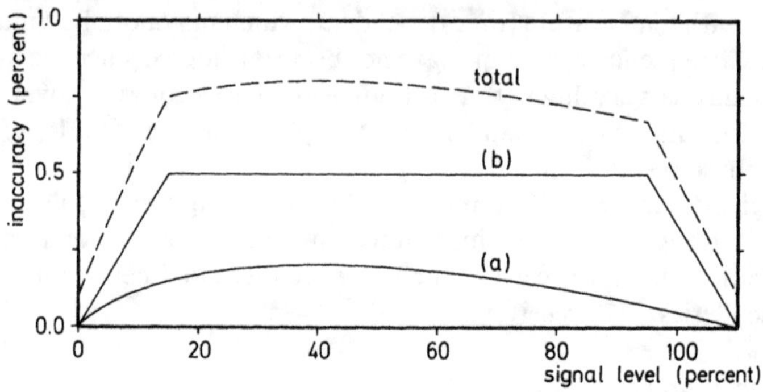

Fig. 6.3 *Specification limits for receiver amplitude inaccuracies (linear mode)*
Level-dependent term (*a*), level-independent term (*b*), and total inaccuracies specification; the latter includes a resolution error of $\pm 0.1\%$

Phase nonlinearities
The phase error may consist of a one-sided level-dependent term, $0.04°/\text{dB}$, and a resolution of $0.1°$ as illustrated in Fig. 6.4.

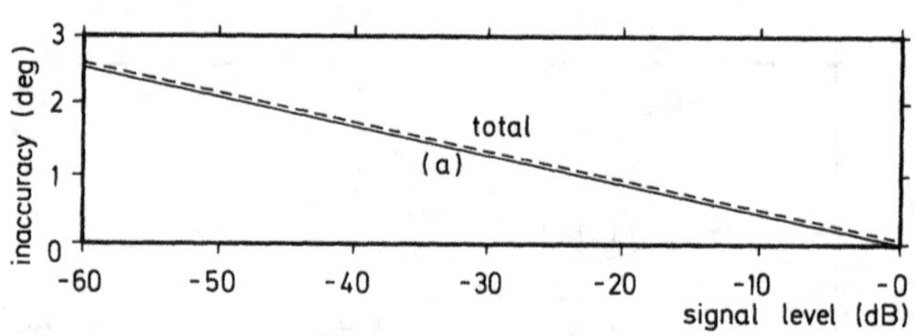

Fig. 6.4 *Specification limits for receiver phase inaccuracies*
Level-dependent term (*a*), and total inaccuracy specification; the latter includes a resolution error of $\pm 0.1°$

Thermal noise
The thermal noise is Rayleigh-distributed at a noise level 80 dB below maximum signal.

The phase of the thermal noise is uniformly distributed between $-180°$ and $+180°$.

Cross-talk
The cross-talk coupling between any of the three channels is -80 dB, the phase being arbitrary.

Drift
Drift can be caused by several sources. Aging is modelled as a linear drift, 0.1 dB/24 h in amplitude and 0.8°/24 h in phase. Temperature variations are

assumed sinusoidal, causing amplitude inaccuracies, 0.05 dB/°C, and phase inaccuracies, 0.2°/°C. The correlation between the channels are assumed to be good, i.e. the relative drifts as well as the caused inaccuracies for the three channels are the same within ±20 per cent.

Measurement delay

The field is usually probed during a continuous scan starting at the theoretical measurement point and lasting 10–15 ms, i.e. the measurement is carried out over a certain angular region. This is simulated by delaying the amplitude measurement by 5 ms and the phase measurement by 7.25 ms which corresponds to the average time delays.

Measurement response time

The limited receiver bandwidth causes a finite response time of the receiver. If the true signal is $w(\phi)$, the received response $w'(\phi)$ is

$$w'(\phi) = \int_{-\infty}^{\phi} \frac{1}{v\tau} e^{-(\phi-\phi_i)/v\tau} w(\phi_i)\, d\phi_i \tag{6.5}$$

when ϕ is the scan angle. The sampling rate is given by the angular scan velocity v and τ is the response time, 9 ms for amplitude and 5.8 ms for phase measurements.

Simulations of the receiver inaccuracies according to the above specifications yield the results stated in Table 6.2. For the drift, the measurements are simulated to last 24 h with a sinusoidal temperature change of ±1°C over a period of 10 h. For the measurement delay and the measurement response time the test antenna is assumed scanned in phi with an angular velocity of 1 rpm.

At the bottom of Table 6.2, the inaccuracy of the total receiver system is given as the result of the simulations and as the algebraic sum of the separate inaccuracies in the table. Good agreement is found here which shows that the inaccuracies are additive.

Care must be taken, however, in adding the individual contributions. For example, consider the main-beam directivity for which the amplitude error in linear mode is 0.005 dB and the phase error is −0.005 dB. When simulated together, the error vanishes, i.e. the two inaccuracies seem to cancel each other. This is not, however, necessarily the case in actual measurements since the inaccuracies here depend on the side to which the one-sided terms are actually occurring. In other cases, as for measurement delay and measurement response for which each of the errors tends to increase the predicted level of cross polarization, the two errors partly cancel the effect of each other when acting at the same time. The reason for this cancelling is that while the former error causes a delay of the measurement, the latter, due to the finite bandwidth in the receiver, causes a delay of the signal through the receiver. Taking these effects into consideration, the worst-case sums are given at the bottom of Table 6.2.

Table 6.2 *Pattern inaccuracies caused by receiver (Scientific-Atlanta, Inc., Series 1770). Reflector antenna, D = 30λ. Dashes (—) indicate values below the level of numerical accuracy. The sign of the inaccuracies will in some cases change when the sign of the simulated error is changed. (See also the footnotes to Table 6.1.)*

Receiver inaccuracies	Main beam			First null	First side lobe	Cross-polar lobe
	Directivity	Cross-polar level	Beam width	Position	Level	Level
Reference far-field values	38.4 dBi	−∞ dB	2.25°	2.95°	−26.8 dB	−48.2 dB
Elevation over azimuth set-up	Change in dB	Increased to (dB)	Change in degrees	Change in degrees	Change in dB	Change in dB
Amplitude, logarithmic mode	0.01	−70	−0.01	−0.02	0.3	0.1±0.2
linear mode	0.005	−80	−0.01	−0.02	0.3	0.2±0.4
Phase	−0.005	—	—	—	−0.04	0.1
Thermal noise	—	—	—	—	—	−0.01±0.03
Cross-talk	—	—	—	—	—	—
Linear drift	−0.005	—	—	—	—	—
Temperature drift	—	—	—	—	0.02	0.5
Scan velocity 6° s^{-1}: measurement delay	—	−65	—	—	—	±0.03
measurement response	—	−60	—	—	—	0.02±0.06
Total logarithmic mode						
simulated	−0.005	−76	−0.01	−0.02	0.3	0.5±0.2
algebraic sum	—	−78	−0.01	−0.02	0.3	0.7±0.2
worst-case sum	0.02	−62	−0.01	−0.02	0.3	0.7±0.3
Total linear mode						
simulated	−0.005	−70	−0.01	−0.02	0.3	0.7±0.2
algebraic sum	−0.005	−70	−0.01	−0.02	0.3	0.8±0.4
worst-case sum	0.015	−65	−0.01	−0.02	0.4	0.8±0.5

In a similar way as for the mechanical inaccuracies, the receiver inaccuracies have been found to be proportional to the receiver errors when these are small. For the measurement delay and measurement response, the inaccuracies are proportional to the sampling rate (scan velocity).

Furthermore, as the mechanical errors and the receiver errors are independent, it is expected that the consequences of mechanical and electrical errors acting in the same measurement will give inaccuracies of the far-field pattern which can be predicted as the sum of the inaccuracies caused individually by the mechanical and the electrical inaccuracies. This has been verified by simulations of typical mechanical and electrical inaccuracies.

Main-beam depointing has not been observed for any of the simulated cases, and this parameter has not therefore been discussed.

In the same way as for the mechanical errors, the receiver simulations themselves are quite accurate but the results can only be taken as guidelines since the determined accuracies depend on the actual test antenna and on the test set-up.

From the tabulated inaccuracies, it is concluded that the two different amplitude modes are of the same accuracy. While the mechanical simulations showed the highest inaccuracy on the prediction of the side-lobe level, the receiver inaccuracies also influences the cross-polarization levels.

It is difficult to relate the various receiver inaccuracies to the inaccuracies on the far-field pattern. The only exception is the change in cross-polarization level caused by measurement delay and response time. Since the simulated measurements are scanned in phi, these inaccuracies will have the effect that the sample points are slightly shifted in phi. This gives an apparent rotation of the test antenna in the opposite sense and since the field is linearly polarized, this results only in a polarization shift. This might be avoided by scanning in theta, in which case the polarization only changes slowly with the scan angle. But in that case, the amplitude variation with the scan angle is large, and inaccuracies will then occur on the copolar pattern parameters.

The choice of the scanning scheme with respect to the double sphere, cf. Section 5.3.3.7, also has influence on the electrical inaccuracies. Thus, a slow time-varying drift in the system will not be serious when scanning in phi. The aperture field, which is most important, is then determined during one unbroken part of the measurements. If the measurements were performed with scans in theta, the aperture would be measured as part of each scan during the whole measurement period and long-term drift is influencing. Furthermore, a drift will cause a field discontinuity when the first scan at $\phi = \phi_o$ shall be matched to the last scan at $\phi = \pi + \phi_o - \Delta\phi$.

Finally, it shall again be stressed that electrical inaccuracies other than those caused by the receiver will exist and need to be evaluated separately.

6.5 Probe correction

With respect to inaccuracies, the probe correction may conceptually be separated as polarization correction and pattern correction.

While probe pattern correction is important only for near-field measurements the probe polarization correction is required for all antenna measurements when the probe polarization is not ideal. Thus, it is well known from far-field measurements that the polarization cannot be measured with better accuracy than that of the polarization of the probe. This is also the case for near-field measurements and inaccurate polarization correction will not be further discussed.

It should, however, be mentioned that if the test antenna is circularly polarized, the probe can with advantage also be circularly polarized. If the probe is linearly polarized, the cross polarization of the test antenna shall be determined as the difference between two large, linearly polarized signals which nearly cancel each other. This requires high accuracy in the measurements to obtain a good cross-polarization determination.

In the following, the previously defined linearly polarized test antenna will be sampled by ideally linearly polarized probes and the effect of probe pattern correction will be analyzed. In the transformation the input probe is varied, while the output probe always is a short electric dipole, cf. Section 4.3.4.1.

Probe pattern correction concerns the pattern directivity of the probe. If the field components are to be sampled directly, a small probe, (e.g. an open-ended waveguide) shall be applied. A small probe with a wide beam has, however, a low gain and will collect stray fields from the measurement site; see Section 5.2.1. It is thus advantageous to apply a directive probe, and probe pattern correction may be required in the near-field far-field transformation.

With respect to the influence of measurement errors, simulations have shown that the far-field inaccuracies are independent of the probe directivity. Thus, the previous results for the influence of mechanical errors and of receiver errors are unchanged.

In the simulations of the influence of the probe correction, it is necessary to have a probe model which is correct (i.e. satisfying Maxwell's equations). An end-fire array of six short electric dipoles is chosen as probe. This array has $\mu = \pm 1$ as required by the algorithm. The directivity is 12.1 dBi and the 3 dB beamwidth is 40° and 43° in the *E*- and *H*-planes, respectively.

Each dipole element will sample the electrical field component along the dipole at the dipole position. The response of the six-element dipole probe can then be determined as the sum of six field components. In the summation, the field components must be weighted by the excitation coefficients of the probe array.

Of the actual simulations, we first consider the inaccuracies introduced when the input probe is directive while the probe correction is carried out as if the input probe is a short electric dipole. This is referred to as the case of no probe correction. The inaccuracies will depend on the degree to which the directive probe pattern deviates from the electric dipole pattern over that angular region in which the probe receives from the test antenna. As this angular region narrows with increasing measurement distance, the deviation of the probe pattern from the pattern of the electric dipole will decrease and therefore, the inaccuracy introduced by omitting probe correction will decay for increasing measurement distance. This is in agreement with the fact that probe correction is not required for far-field measurements.

For the previously defined test antenna, the inaccuracy by omitting probe correction is shown in Fig. 6.5 as a function of the measurement distance A. The pattern inaccuracy (deviation of patterns when these are expressed in dB) is shown for main-beam directivity, side-lobe level and cross-polarization level. A distinct $1/A^2$ dependence of the pattern inaccuracies is observed.

High inaccuracies may be observed, especially for the side-lobe level and the on-axis directivity. The inaccuracy of the cross-polar level is relatively small

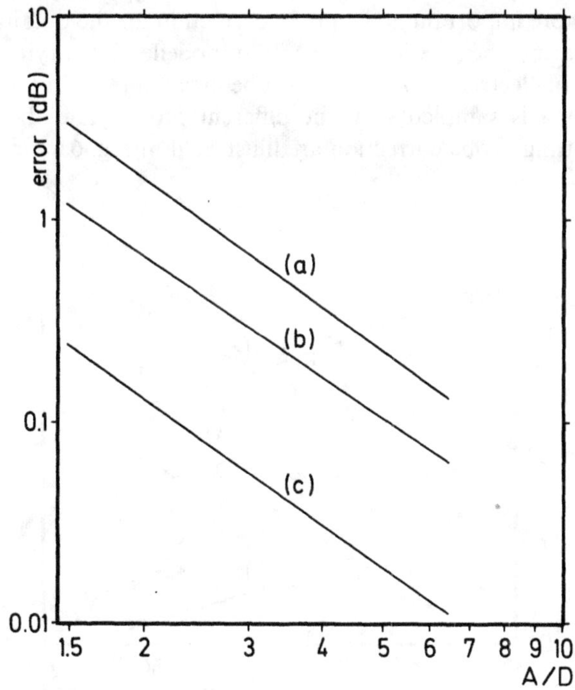

Fig. 6.5 *Error when probe correction is omitted versus relative measurement distances A/D. Probe directivity is 12.1 dBi*
(*a*) Inaccuracy of first side-lobe level (at −26.8 dB)
(*b*) Inaccuracy of cross-polar lobe (at −48.2 dB)
(*c*) Inaccuracy of main-beam directivity (at 38.4 dBi)
Both scales are logarithmic. Error in dB is seen to be proportional to $1/A^2$

when taking into account the low value of this level. That the cross polarization is almost not affected is due to the fact that the simulated directive probe has a cross-polarization behaviour which is not far from that of the single electric dipole. The inaccuracies in on-axis directivity and side-lobe level are, however, caused directly by the probe directivity. When the probe is directly in front of the test antenna, a large signal is received. When the probe moves towards the edge of the test antenna projected aperture, the signal will decrease and will decrease all the more as the probe directivity increases. This is due to the fact that the quasi-plane aperture field from the test antenna will then illuminate the probe from an off-axis direction. This will represent too high an aperture taper of the test antenna and thus will result in too low values for its directivity and side-lobe level when not probe corrected.

The inaccuracy resulting from omission of probe correction is seen to be significant for short measurement distances, i.e. in cases for which the probe pattern distinctly differs from the pattern of the electric dipole over the region illuminating the test antenna. With this background, it is also expected that the higher the directivity of the applied probe is (or, more precisely, the narrower its beam is), the more important is the probe correction for the obtained accuracy.

Probes of various beam widths have been modelled by varying the number (from 4 to 10) of electric dipoles in the probe array applied in the simulations. The test antenna is sampled with the different probes, and the inaccuracies caused by omitting probe correction are illustrated in Fig. 6.6 for the measure-

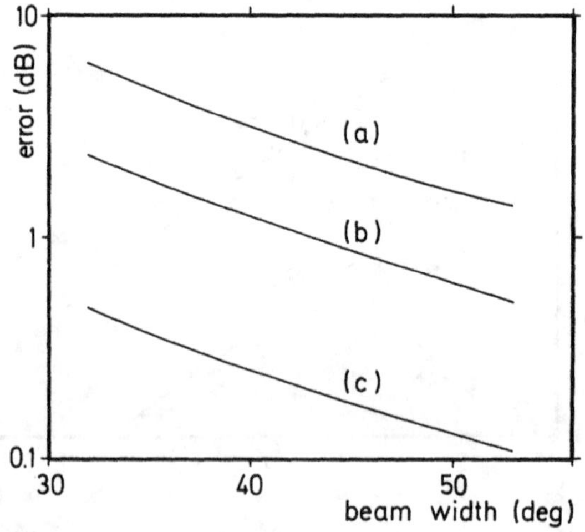

Fig. 6.6 *Error when probe correction is omitted versus E-plane beam width of sampling probe. Measurement distance is $A = 45\lambda$*

(*a*) Inaccuracy of first side-lobe level (at −26.8 dB)

(*b*) Inaccuracy of cross-polar lobe (at −48.2 dB)

(*c*) Inaccuracy of main-beam directivity (at 38.4 dBi)

ment distance $A = 45\lambda$. It is again seen that the effect is most serious for the prediction of the side-lobe level and on-axis directivity. Even for a probe with a relatively wide beam, i.e. a small probe, the side-lobe level is inaccurate by more than 1 dB, and the main-beam directivity is inaccurate by more than 0.1 dB. In this case the measurement distance is such that the test antenna rim subtends an angle of 30° as seen from the probe.

In Fig. 6.6, the E-plane half-power beam width θ_E has been applied as an independent variable. If the H-plane beam width θ_H of the probe is known, this may be related to the E-plane beam width θ_E. For the simulated probes, the E-plane pattern equals the H-plane pattern multiplied by $\cos\theta$, and the central part of the probe pattern may be approximated by a paraboloid. This yields

$$\theta_E^2 \simeq \frac{\sqrt{2}-1}{\sqrt{2}-\cos\theta_H}\theta_H^2. \tag{6.6}$$

Thus, probe correction is required for short measurement distances and for directive probes if good far-field prediction is to be obtained. However, the correction need not be carried out by using data for the actual probe. If, for example, an antenna is tested at different frequencies and high accuracy is required, the probe needs to be calibrated for each of the applied frequencies. For many applications, the accuracy may be sufficiently high if the probe is calibrated at a single frequency, and the probe coefficients hereby obtained are applied for the probe correction at all frequencies.

This is illustrated in Fig. 6.7 for inaccuracies in main-beam directivity and in Fig. 6.8 for inaccuracies in side-lobe level and in level of cross-polar lobe. The test antenna near field is sampled at a nominal frequency f_0. By scaling the probe size ± 20 per cent, the probe receiving pattern changes corresponding to frequencies within $f_0 \pm 20$ per cent. The near field is then transformed to far fields by applying the various probe patterns in the probe correction. The inaccuracies in the far fields are depicted in the figures. At f_0 the probe correction is exact. At the low frequencies the input probe in the transformation is modelled as smaller, i.e. having a wider beam (right-hand side of the figures), than the probe that was used when sampling the near field. As previously discussed, sampling with a probe with a more narrow beam than the input probe in the transformation will cause the test antenna to have an apparently higher aperture taper. Thus the main-beam directivity and the side-lobe level will be predicted to have too low values. This is also seen in Figs 6.7–6.8.

The conclusions are that probe pattern correction is important but may be omitted when the measurement distance is large or the probe directivity is low. The accuracy of the predicted far field may, however, be improved by an order of magnitude if an approximate probe pattern is applied as a basis for the probe correction. These qualitative results are found very useful for the understanding of the spherical scanning technique. The quantitative results are, however, derived for a specific simulated test antenna (diameter 30 wavelengths) and care

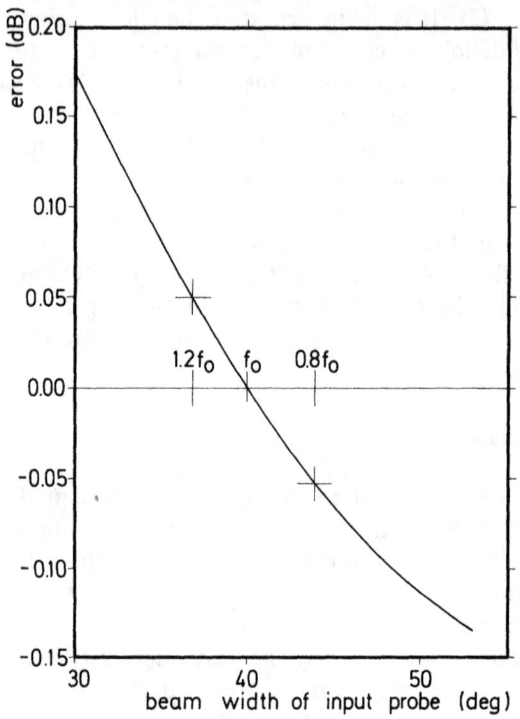

Fig. 6.7 *Error in main-beam directivity (at 38.4 dBi) when transformation probe is different from sampling probe, sample distance $A = 45\lambda$*
Within the two horizontal lines the simulated probe changes are obtained by a frequency shift of $\pm 20\%$ from the nominal frequency f_0 of a probe of 6 hertzian dipoles. Outside the lines, the probe is changed by the number of elements in the probe

must be taken in an attempt to draw quantitative conclusions for test antennas quite different from the one applied here.

6.6 Truncation

6.6.1 Introduction

The measurement time may be reduced considerably if the near field is sampled over only a part of the measurement surface. The remaining field values are then, in the transformation to far field, assumed to be zero.

The truncation principle is that the near field must be measured over at least that part of the surface which includes all rays which may radiate from any part of the test antenna to any of the desired far-field directions [7, 8]. This is schematically shown in Fig. 6.9. The near field shall be sampled up to $\theta = \theta_t$ if

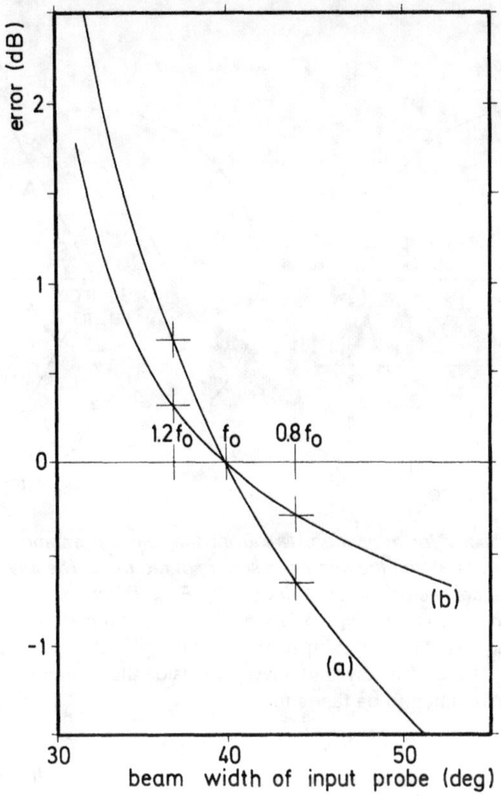

Fig. 6.8 *Error when transformation probe is different from sampling probe, sample distance $A = 45\lambda$. Probe as in Fig. 6.7*
(*a*) Inaccuracy of first side-lobe level at -26.8 dB)
(*b*) Inaccuracy of cross-polar lobe (at $-48{\cdot}2$ dB)

the far field is sought for $\theta < \theta_{tf}$. For the general case, Fig. 6.9(b), it is found that

$$A \sin(\theta_t - \theta_{tf}) = r_0$$

when A is the radius of the measurement sphere and r_0 is the radius of the minimum sphere surrounding the test antenna and concentric with the measurement surface. Thus

$$\theta_{tf} = \theta_t - \arcsin(r_0/A); \tag{6.7}$$

θ_{tf} should be positive for the truncation to yield valid far-field data. This general formula may be too pessimistic if the test antenna is not centrally positioned in the minimum sphere. In that case, a geometrical consideration may give a larger value for θ_{tf} for a given truncation angle θ_t than stated by eqn (6.7); cf. Fig. 6.9(a).

It should be noted that the rule for truncation does not involve the magnitude

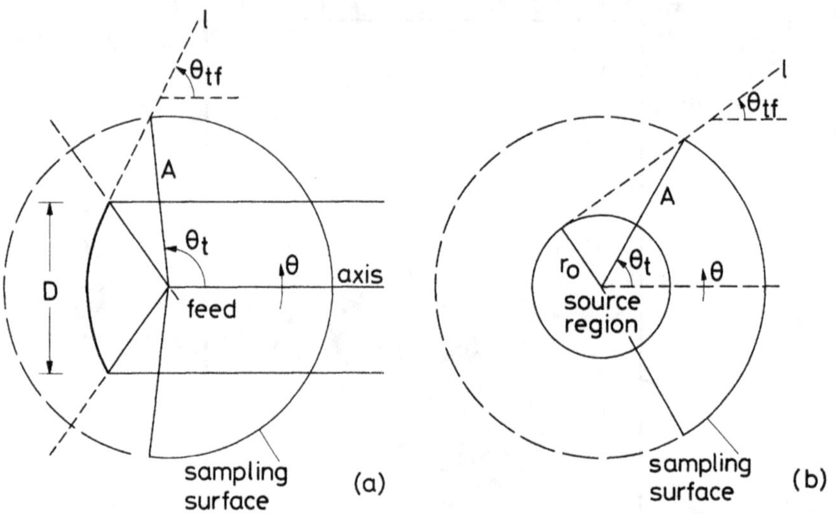

Fig. 6.9 *Regions of validity for truncation. (a) Front-fed reflector antenna and (b) general case with sources within the minimum sphere of radius r_0. The near field is sampled over the spherical sector, radius A, for $\theta < \theta_t$. Edge diffracted rays are encountered for $\theta < \theta_{tf}$ which limits the region in which the far field can be predicted*
In the specific case (*a*), the path of the limiting diffracted ray *l* is known. In the general case (*b*) the path may start anywhere inside the minimum sphere to which *l* then must be assumed to be tangential

of the field. This is not required to be negligible over the neglected part of the measurement surface.

Depending on the direction of the lobes of interest with respect to the spherical measurement grid, the truncation may be carried out in theta (measurements over a polar segment) or in phi (measurements over an equatorial sector) or both. This will be discusssed further below.

6.6.2 Polar truncation

The simulations for polar truncations are carried out with a test antenna which, in order not to have rotational symmetry, is an array of $4 \times 4 \times 4$ y-directed electrical short dipoles (the array is centred in the origin, element spacings are $5\lambda/6$, $15\lambda/16$ and $3\lambda/8$ along x, y and z, respectively, fed with a phase progression of $-11\pi/12$ along z). The radius of the minimum sphere is $r_0 = 1.96\lambda$. The spherical wave expansion may be then truncated at $N = kr_0 + 10 = 22.3$. We will apply $N = 24$ corresponding to sample spacings $\Delta\theta = \Delta\phi = 7.5°$. The near field is sampled at $A = 8\lambda$; see Fig. 6.10.

The near field has been truncated at $\theta_t = 30°$ and $45°$ for which $\theta_{tf} = 16°$ and $31°$, respectively. The obtained far fields are shown together with the reference far field in Fig. 6.11. The agreement between the patterns is found to be good for $\theta < \theta_{tf}$. The fields are given as directivity patterns. Thus, the reference far-field

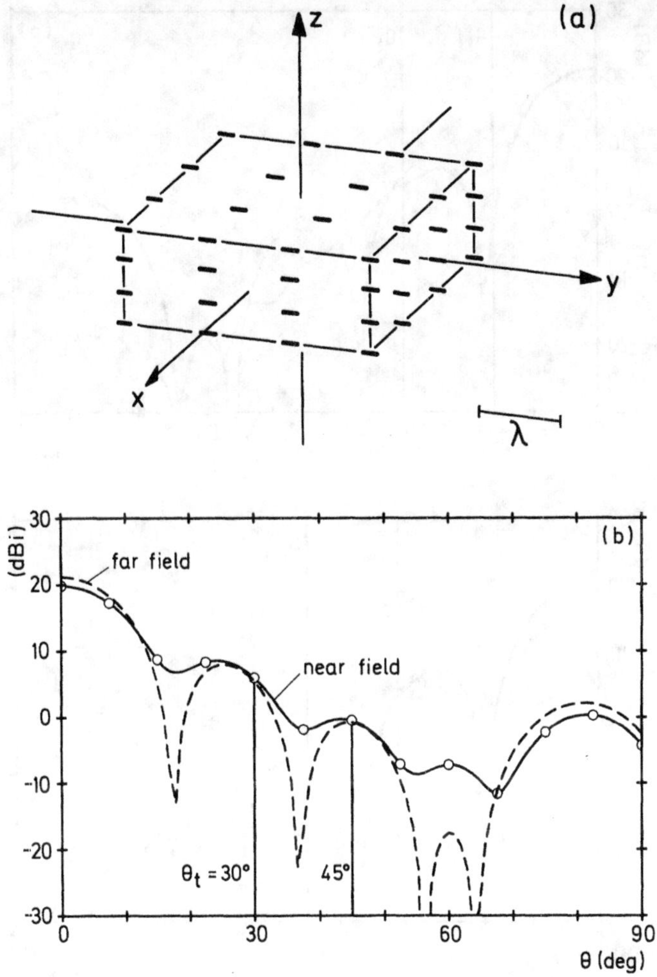

Fig. 6.10 *Test antenna applied in simulations of polar truncations. (a) The test antenna is an array of 64 dipoles, measurement radius is $A = 8\lambda$. (b) Near-field and far-field H-plane patterns ($\phi = 0$) of the test antenna. The dots indicate the points at which the near field is sampled. Truncation angles are at θ_t*

pattern is normalized to radiated power relative to isotropic radiation. The truncated far-field patterns are normalized with the same normalization factor as the reference pattern since the radiated (but not the sampled) power is independent of truncation.

Due to the cut-off properties, cf. Section 2.2.3, spherical harmonics with $m > n_{max} \sin \theta_t$ will not contribute to the field for $\theta < \theta_t$. It is thus expected that these harmonics are not detected when the field is truncated at θ_t. To illustrate

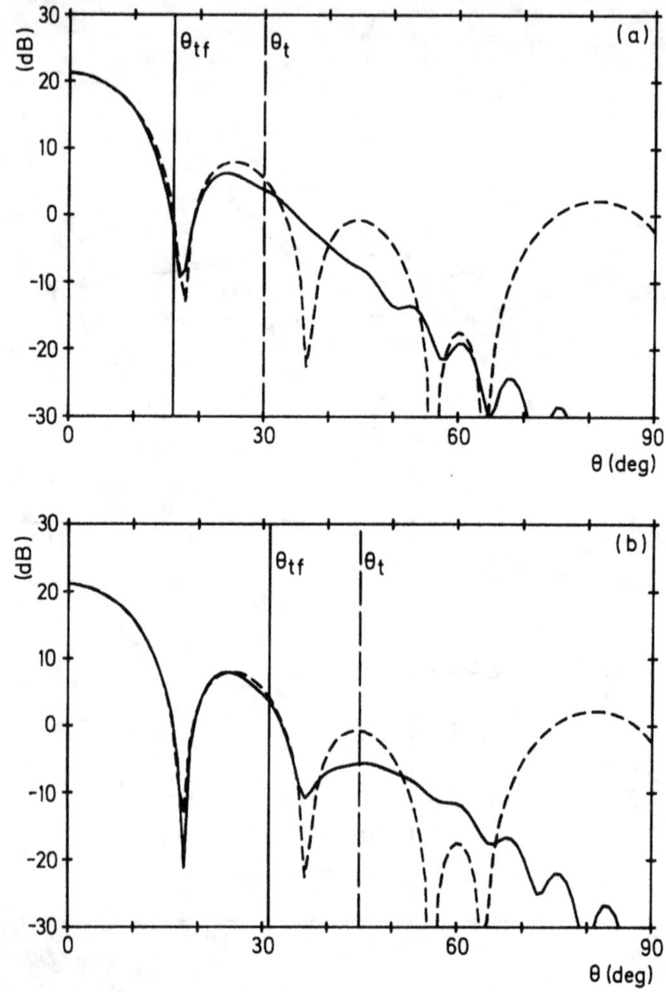

Fig. 6.11 *Far-field patterns, $\phi = 0$ when the near field is truncated at (a) $\theta_t = 30°$ and (b) $\theta_t = 45°$*

——— Far field obtained from the near field truncated at θ_t

- - - - - - - - Reference far field, normalized to dBi

this we will consider the power radiated by all modes with the same n

$$P^{(n)}_{\text{rad}} = \frac{1}{2} \sum_{s=1,2} \sum_{m=-n}^{n} |Q_{smn}|^2 \tag{6.8}$$

and the power radiated by all modes with the same $|m|$

$$P^{(m)}_{\text{rad}} = \frac{1}{2} \sum_{s=1,2} \sum_{m'=-m,m} \sum_{n=m}^{n_{\text{max}}} |Q_{sm'n}|^2 \qquad m \geqslant 0 \tag{6.9}$$

in which the second summation is for the positive value as well as the negative for the actual m. The value of n_{max} will here be taken as 16, defined as that n for which 99.99 per cent of all power is included.

The dependences of $P_{rad}^{(n)}$ and $P_{rad}^{(m)}$ are shown in Fig. 6.12 for the reference field as well as for the cases of truncation. For these latter cases, it is seen how $P_{rad}^{(m)}$ for small m is unaffected by the truncation but a distinct cut-off occurs at the indicated value of m_{max} given by

$$m_{max} = n_{max} \sin \theta_t. \tag{6.10}$$

With $n_{max} = 16$ we have $m_{max} = 8$ for $\theta_t = 30°$ and $m_{max} = 11$ for $\theta_t = 45°$.

The power spectrum in n, $P_{rad}^{(n)}$, is disturbed for all n and the modes for large n contain much more power than in the reference case. This shows that the mode expansion is not able to account for the nonphysical discontinuity which is introduced in the near field by the truncation in θ.

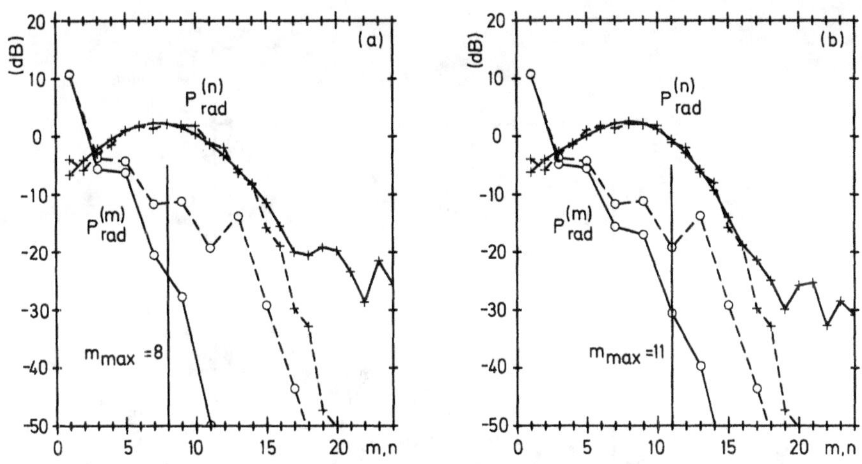

Fig. 6.12 *Mode spectra $P_{rad}^{(n)}$ and $P_{rad}^{(m)}$ of spherical harmonics for near fields truncated at (a) $\theta_t = 30°$ and (b) $\theta_t = 45°$. Same normalization as in Fig. 6.11, i.e. total radiated power of reference field is 1 per steradian (0 dB)*

Spectrum	$P_{rad}^{(n)}$	$P_{rad}^{(m)}$
truncated	—+—+—+—	—O—O—O—
reference	– + – + – + –	– O – O – O –

The relation (6.10) is independent of the measurement distance and the power spectra $P_{rad}^{(n)}$, $P_{rad}^{(m)}$ evaluated from a truncation at $\theta_t = 45°$ at a measurement distance of $A = 1000\lambda$ are indistinguishable from those of Fig. 6.12(b) for $A = 8\lambda$. The derived far-field patterns are, however, different since

$$\theta_{tf} \simeq 31° \qquad \text{for } A = 8\lambda$$

$$\theta_{tf} \simeq \theta_t = 45° \qquad \text{for } A = 1000\lambda$$

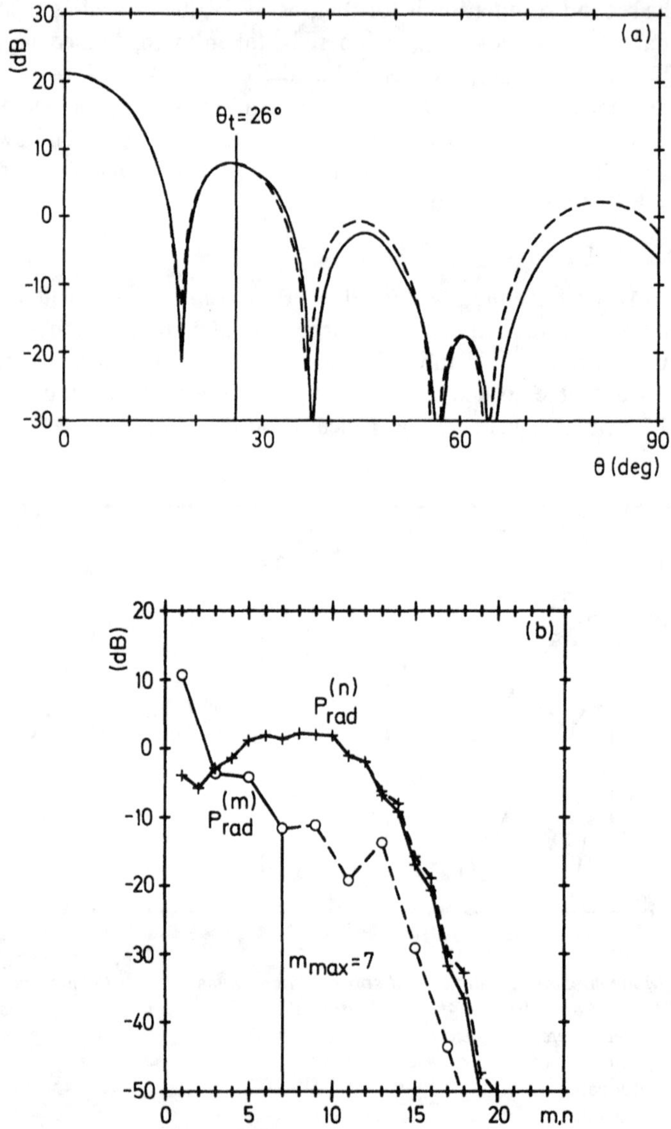

Fig. 6.13 *Mode truncation at $m = m_{max} = 7$ corresponding to $\theta_t = 26°$*
(*a*) Far field patterns, $\phi = 0$
———— mode truncated field
-------- reference far field, dBi
(*b*) Mode spectra

	$P_{rad}^{(n)}$	$P_{rad}^{(m)}$
truncated	—+—+—+—	—○—○—○—
reference	– + – + – + –	– ○ – ○ – ○ –

Table 6.3 *Polar truncation, simulated pattern inaccuracies. (a) Pattern parameters for reference far field. (b) Truncation in polar angle θ at θ_t, far field accurate for $\theta \leqslant \theta_{tf}$ (theoretical limit). (c) Truncation in mode order m at m_{max} ($n_{max} = 16$). Dashes (—) indicate values below the level of numerical accuracy.*

Polar truncation	Power of truncated field	Main beam			First side lobe
		Directivity	3 dB beam width *E*-plane	3 dB beam width *H*-plane	Level
(a) Reference values	0 dB	21.15 dBi	13.50°	15.34°	−13.16 dB
	dB	Change in dB	Change in degrees	Change in degrees	Change in dB
(b) $\theta_t = 30°$, $\theta_{tf} = 16°$	−1.26	0.13	−0.43	−0.10	−2.1
$\theta_t = 37.5°$, $\theta_{tf} = 23.5°$	−1.18	−0.09	0.28	0.29	−0.9
$\theta_t = 45°$, $\theta_{tf} = 31°$	−1.12	−0.02	0.08	0.02	0.14
(c) $m_{max} = 7$, $\theta_t = 26°$	−0.05	—	—	—	0.12

according to eqn (6.7). Thus, the mode expansion for the two resultant far fields are different though the power distribution of the modes is the same.

The cut-off property expressed by eqn (6.10) will further have the consequence that if the field is desired for $\theta < \theta_t$, then the mode expansion may be truncated in m independently of the truncation of the measurement sphere. By the truncation in m, the azimuthal spacing $\Delta\phi$ may be increased to

$$\Delta\phi = \frac{180^\circ}{m_{\max}}. \tag{6.11}$$

As an example, we may consider the case when the far field is desired for $\theta < \theta_t = 26^\circ$. The near field is assumed sampled over the full sphere but the expansion may be truncated at $m = m_{\max} = 7$ (based on $n_{\max} = 16$) as shown in Fig. 6.13(b). The resultant far field is shown in Fig. 6.13(a). It is seen that the agreement with the reference far field is good for $\theta < \theta_t$. Furthermore, since $m_{\max} = 7$, the near-field data may be thinned to $\Delta\phi = 25.7^\circ$ (not shown).

Quantitative evaluation of the inaccuracies introduced by truncation is given in Table 6.3. At the top, (a), the pattern parameters for the reference far field are given. Part (b) illustrates the inaccuracies caused by near-field truncation at $\theta = \theta_t$ with θ_t values from 30° to 45°. The far field will be accurate for $\theta < \theta_{tf}$, θ_{tf} ranging from 16° to 31° according to eqn (6.7). It is seen that the inaccuracies decrease with increasing θ_t, and reasonable accuracies are only obtained when the first side lobe (at $\theta = 25.3^\circ$) is within $\theta = \theta_{tf}$ which is the case only for $\theta_t = 45^\circ$, cf. Fig. 6.11.

The bottom line, (c), of Table 6.3 shows inaccuracies for modal truncation at $m_{\max} = 7$ (cf. Fig. 6.13) corresponding to accurate far-field prediction for $\theta < \theta_t = 26^\circ$ according to eqn (6.10). The inaccuracies in the main beam are below the numerical inaccuracies of the simulations. For the side lobe, the inaccuracy is small though the side-lobe position is close to the truncation limit θ_t.

The conclusion on polar truncation is that the measurement time as well as the amount of measured data may be reduced considerably with only a small increase in the inaccuracy of the predicted far-field parameters. The computer time for the near-field far-field processing may be reduced by the processing of a reduced number of modes when $m_{\max}$ is reduced and for truncated measurements, advantage can be taken by transforming measured data obtained with a sparse sampling in phi. A reduction in the theta transformation of the truncated data can be accomplished if the radiation into the back hemisphere can be neglected [9].

6.6.3 *Equatorial truncation*

When a test antenna is mounted with the main beam pointing at a point on the equator of the measurement grid ($\theta = 90^\circ$), truncation gets a new aspect. First, the measurements need to be carried out with the same spacing in phi and theta since the order m of azimuthal modes will be maximum ($m_{\max} = n_{\max}$). Second, a truncation, e.g. around the aperture, shall be performed in both phi and theta.

The simulations in this section are again carried out for a test antenna which is a front-fed paraboloidal reflector antenna, (see Fig. 6.14), now with diameter $D = 10\lambda$ and $f/D = 0.4$, f being the focal length. The minimum radius is $r_0 = 5.56\lambda$ when centred at the focus. The maximum mode number is then $N = kr_0 + 10 \simeq 45$. The sample spacing has been chosen as $\Delta\theta = \Delta\phi = 3°$ allowing modes up to $N = 60$ to be determined. The applied measurement radius is $A = 8\lambda$ and the near field is determined by GTD.

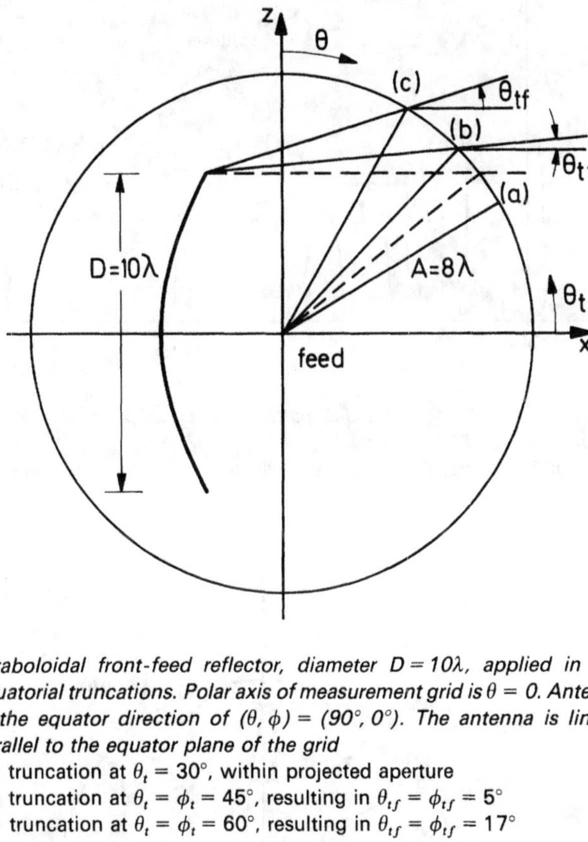

Fig. 6.14 *Paraboloidal front-feed reflector, diameter $D = 10\lambda$, applied in simulations of equatorial truncations. Polar axis of measurement grid is $\theta = 0$. Antenna axis points in the equator direction of $(\theta, \phi) = (90°, 0°)$. The antenna is linearly polarized parallel to the equator plane of the grid*
(*a*) truncation at $\theta_t = 30°$, within projected aperture
(*b*) truncation at $\theta_t = \phi_t = 45°$, resulting in $\theta_{tf} = \phi_{tf} = 5°$
(*c*) truncation at $\theta_t = \phi_t = 60°$, resulting in $\theta_{tf} = \phi_{tf} = 17°$

The antenna near field is shown in the equatorial plane (E-plane) in Fig. 6.15 for $0 \leqslant \phi \leqslant 180°$.

The simulated regions of truncations are indicated by ϕ_t, i.e. the field is sampled only within $|\phi| \leqslant \phi_t$. In the polar angle, the truncation angle θ_t will also be measured from boresight, i.e. the field is sampled only within $|90° - \theta| \leqslant \theta_t$.

First, cases with $\phi_t = \theta_t$ will be discussed. The regions of truncations are shown on the measurement sphere in Fig. 6.16. It is seen that for $\phi_t = \theta_t = 30°$, 86.5 per cent of the radiated energy is within the sample region I. However, the complete aperture field extending to $\theta_t = 39°$ is not sampled, cf. Fig. 6.14, and thus a good far-field prediction cannot be expected.

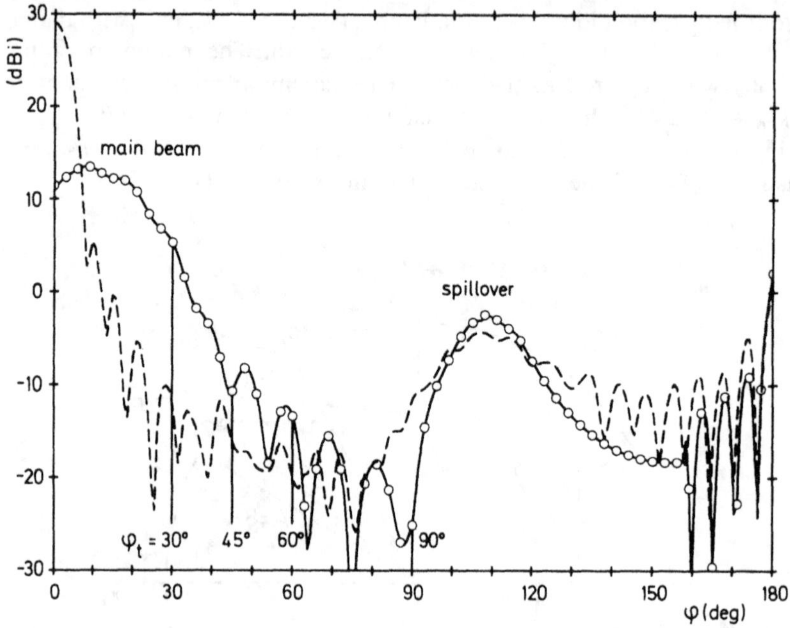

Fig. 6.15 *Near-field and far-field E-plane patterns (equator plane, $\theta = 90°$) for the test antenna of Fig. 6.14*
The near-field pattern is sampled as indicated by the dots. Truncation angles are at ϕ_t

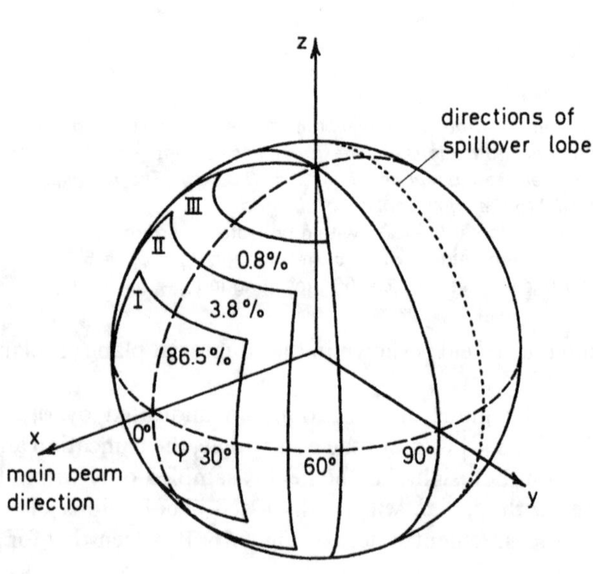

Fig. 6.16 *Power distribution from the test antenna over the near-field sphere, $A = 8\lambda$*

When $\phi_t = \theta_t = 45°$, both regions I and II which contain 90.3 per cent of the radiated energy, are sampled. The far field may be well predicted within the region limited by the ray from the edge of the reflector to the edge of measurement area. This limit is at $\phi_{tf} = \theta_{tf} = 5°$. Application of the general expression (6.7) will lead to a more pessimistic value (1°) of ϕ_{tf} and θ_{tf}. The resulting far field is shown in an equatorial cut in Fig. 6.17(a) in which $\phi_{tf} = 5°$ is also indicated. As in Section 6.5.1, the patterns are normalized to the directivity of the reference field.

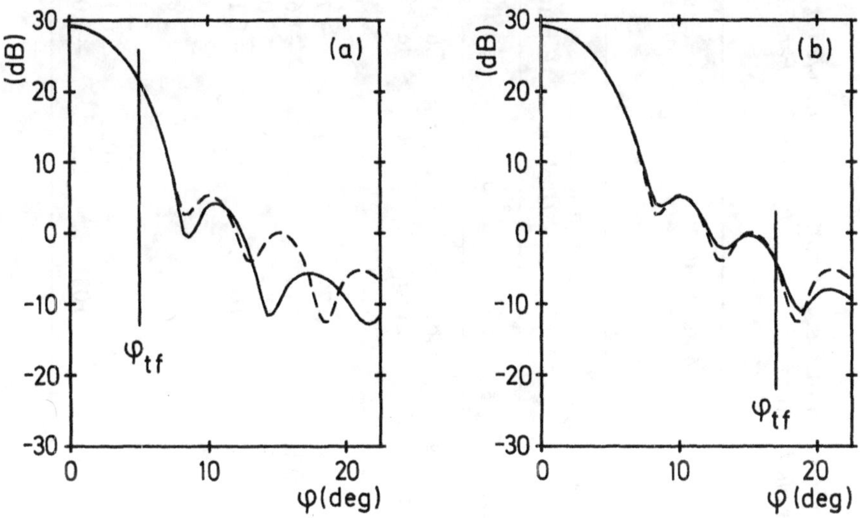

Fig. 6.17 *Near field truncated at (a) $\theta_t = \phi_t = 45°$ and (b) $\theta_t = \phi_t = 60°$, corresponding to $\phi_{tf} = 5°$ and 17°, respectively*
——— far field obtained from truncated near field
-------- reference far field, normalized to dBi

Truncation at $\phi_t = \theta_t = 60°$ results in a far field as shown in Fig. 6.17(b). The measurement surface now includes region III, Fig. 6.16, through which only 0.8 per cent of the total energy is radiated. However, $\theta_{tf} = \phi_{tf}$ is now increased to 17° (Fig. 6.14) and a good agreement with the reference far field is found at this angle. This confirms that it is the angular measurement area, not the percentage of included power, that determines the far-field region with good pattern prediction.

The pattern inaccuracies for various cases of truncation are shown in Table 6.4 in which the first line (a) shows the pattern parameters for the full reference far field. In the next block, (b), are shown the inaccuracies occurring for equal truncation in θ and ϕ ($\theta_t = \phi_t$). In accordance with the increasing measurement area, the accuracies improve considerably for increasing θ_t, ϕ_t. The exception is

Table 6.4 *Equatorial truncation, simulated pattern inaccuracies. (a) Pattern parameters for reference far field; (b) and (c) Truncation at (θ_t, ϕ_t). The near field is sampled within $|90^\circ - \theta| \leqslant \theta_t$, $|\phi| \leqslant \phi_t$. Far field will, approximately, be accurate within $|90^\circ - \theta| \leqslant \theta_{tf}$, $|\phi| \leqslant \phi_{tf}$.*

Equatorial truncation	Power of truncated field	Main beam			First side lobe		Cross-polar lobe
		Directivity	3-dB beamwidth		Level		Level
			E-plane	*H*-plane	*E*-plane	*H*-plane	
(a) Reference values	0 dB	29.12 dBi	6.55°	6.55°	−23.76 dB	−23.79 dB	−35.29 dB
	dB	Change in dB	Change in deg.	Change in deg.	Change in dB	Change in dB	Change in dB
(b) Truncation at $\theta_t = \phi_t$:							
$\theta_t = \phi_t = 30^\circ$, $\theta_{tf} = \phi_{tf} = 0^\circ$	−0.63	−1.2	0.72	0.62	3.7	2.3	−3.2
$\theta_t = \phi_t = 45^\circ$, $\theta_{tf} = \phi_{tf} = 5^\circ$	−0.45	0.13	−0.14	−0.20	−1.6	−1.1	−0.07
$\theta_t = \phi_t = 60^\circ$, $\theta_{tf} = \phi_{tf} = 17^\circ$	−0.41	0.02	−0.04	0.03	−0.05	0.03	0.02
$\theta_t = \phi_t = 90^\circ$, $\theta_{tf} = \phi_{tf} = 51^\circ$	−0.40	0.02	−0.02	−0.01	−0.14	−0.30	0.10
(c) Truncation at $\theta_t = 87^\circ$:							
$\phi_t = 30^\circ$, $\phi_{tf} = 0^\circ$	−0.52	−0.61	0.82	−0.15	3.8	−0.39	−1.1
$\phi_t = 45^\circ$, $\phi_{tf} = 5^\circ$	−0.42	0.07	−0.13	−0.06	−2.2	0.56	−0.13
$\phi_t = 60^\circ$, $\phi_{tf} = 17^\circ$	−0.40	0.02	−0.05	0.03	0.14	0.38	−0.02
$\phi_t = 90^\circ$, $\phi_{tf} = 51^\circ$	−0.39	0.01	−0.03	0.01	−0.01	−0.08	0.09

the case $\theta_t = \phi_t = 60°$ in which the side-lobe and cross-polar lobe levels are well predicted.

For $\theta_t = \phi_t = 30°$, the full aperture is not sampled (Fig. 6.14) which causes a rather poor pattern prediction.

The lower block, (c), of the table is for $\theta_t = 87°$, i.e. the near field is sampled from pole to pole, the poles themselves excluded. ϕ_t increases from 30° to 90° with the same values as above. It is interesting here to see how the large θ_t value gives a considerably improved prediction in the H-plane, the plane with the wide sampling, even for $\phi_t = 30°$. The parameters for the E-plane are roughly unchanged.

Thus, the main beam and first side lobes can be predicted from truncated near-field measurements. The far-field prediction is good within a region determined by the truncation limits of the measurement surface. The truncation shall be determined to a size such that all geometric rays from the antenna to the desired far-field directions pass through the measurement surface. The truncation is not dependent on the level of excluded lobes, e.g. the spillover lobe. The accuracy of a specific far-field parameter will, however, improve with increasing measurement area.

6.6.4 Repetition and overlay of truncated fields [10, 11]

The equatorial truncation considered above gives a reduction in the number of measured near-field values according to the reduced measurement surface. There was no reduction in the near-field far-field processing. Such a reduction is, however, possible by assuming the field to have a shorter period in phi than the physical period of 360°. The reduction is based on the symmetry in ϕ of the measurement grid. If the grid spacing is $\Delta\phi$ and I is an integer, then the grid will cover itself after a rotation of $\phi_r = I\Delta\phi$ in azimuth. Further, if $N_r = 360°/I\Delta\phi$ is an integer, the grid will return to the start position after N_r repeated rotations of each ϕ_r.

By applying the superposition principle, the original truncated measured near-field data may be added at the points of the measurement grid after each of the N_r rotations. In this way an artificial near field is obtained with a period in ϕ of $360°/N_r$. Transformation of this field to its far field will retain this period in ϕ. Since the near-field far-field transformation is linear, this far field is the same as that found by adding the far field obtained by the transformation of the original truncated measured near field N_r times, mutually displaced by ϕ_r. This original far field has negligible amplitude for $|\phi| > \phi_t$ and the superposed far field will then have the same pattern as the original far field for $|\phi| < \phi_t$.

The superposed field is, however, simpler to transform than the original truncated field since the periodicity in ϕ of the former implies that only harmonics with m divisible by N_r will be present. Thus, not only the amount of measured data (i.e. measurement time and data storage requirement) is reduced by the factor N_r but the number of modes is reduced as well (i.e. no additional data storage requirements for the mode storage). This field repetition is utilized to reduce the computer time for the near-field far-field transformation correspondingly.

In the above considerations, the period ϕ_r of the repeated field was limited by the requirement that the full measured near-field region should be included implying that $\phi_r \geqslant 2\phi_t$. A more realistic limitation is, however, that the far-field level shall be sufficiently low outside the possible far-field truncation region given by $|\phi| < \phi_{tf}$. Then, ϕ_r may be reduced to $\phi_{ro} = 2\phi_{tf}$. The grid repetition ϕ_{ro} must be a multiple of the grid spacing ($\phi_{ro} = I_o\Delta\phi$), and the number of periods given by $N_{ro} = 360°/I_o\Delta\phi$ must be an integer. Since $\phi_{tf} < \phi_t$, we will obtain a considerable reduction in the number of modes by a factor N_{ro}. Subsequently, the time for numerical processing for this overlay technique is reduced compared to the field repetition alone, for which the reduction factor is N_r.

The near field should, however, still be measured within $|\phi| \leqslant \phi_t$, since the same amount of information is needed. The measurement time is thus not further reduced compared to the truncated measurements.

The techniques are demonstrated in the following. Repetition without overlay, i.e. field superposition with period $\phi_r \geqslant 2\phi_t$, is illustrated in Fig. 6.18 for the two cases: (a) $\phi_t = 45°$ and (b) $\phi_t = 60°$. ϕ_r is chosen as $2\phi_t$ resulting in a repetition factor $N_r = 360°/\phi_r$ of 4 and 3, respectively. The determined far fields are shown together with the reference far field and good agreement is found for $|\phi| < \phi_{tf}$, $\phi_{tf} = 5°$ and 17°, respectively. Comparison with Fig. 6.17 shows that the errors introduced by the repetition are negligible.

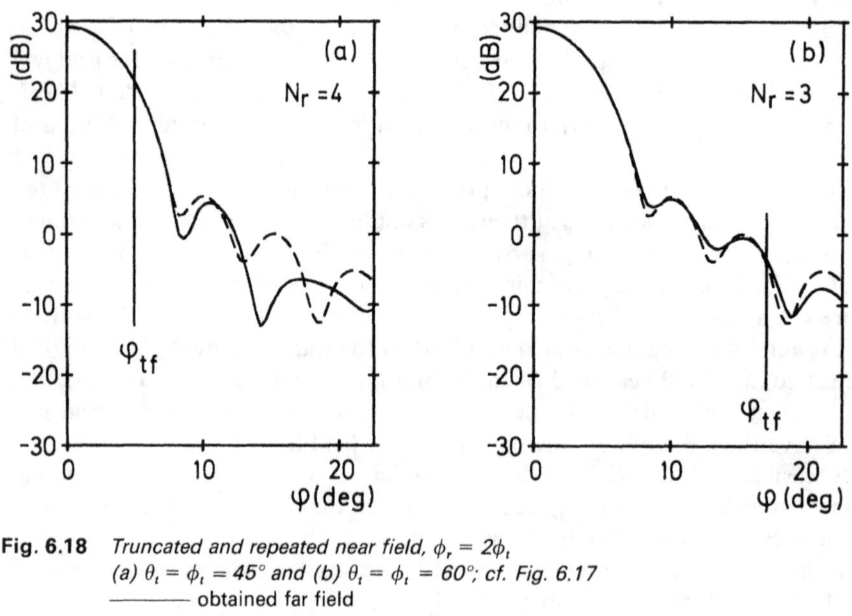

Fig. 6.18 *Truncated and repeated near field, $\phi_r = 2\phi_t$*
(a) $\theta_t = \phi_t = 45°$ and (b) $\theta_t = \phi_t = 60°$; cf. Fig. 6.17
——— obtained far field
-------- reference far field, normalized to dBi

Applying repetition with overlay, i.e. field superposition with period $\phi_r < 2\phi_t$, is illustrated in Fig. 6.19. With $\phi_t = 60°$, three cases are considered: (a) $\phi_{ro} = 30°$, (b) $\phi_{ro} = 45°$ and (c) $\phi_{ro} = 60°$ for which the repetition factor N_{ro} is 12, 8 and 6, respectively. Since the reference field at $\phi = \phi_{tf}$ (=17°) is −28 dB relative to the main beam, reasonably good results can be expected near $\phi = 0$. The level at $\phi = \phi_{ro}/2$ will, however, be wrong by a factor 2 (6 dB) since the field here, from the repeated neighbour far field, is overlaid with the same field value as that of the field itself. This far field periodicity is clearly seen in Fig. 6.19(a). It is obvious that the field is not necessarily well-predicted in the full range $|\phi| < \phi_{tf}$ as was the case for truncated measurements and for repetitions alone. Thus, when the repetition factor is high, as in Fig. 6.19(a), $N_{ro} = 12$, we have $\phi_{ro}/2 = 15°$ which is smaller than $\phi_{tf} = 17°$. Nevertheless, the prediction of the main beam is still good.

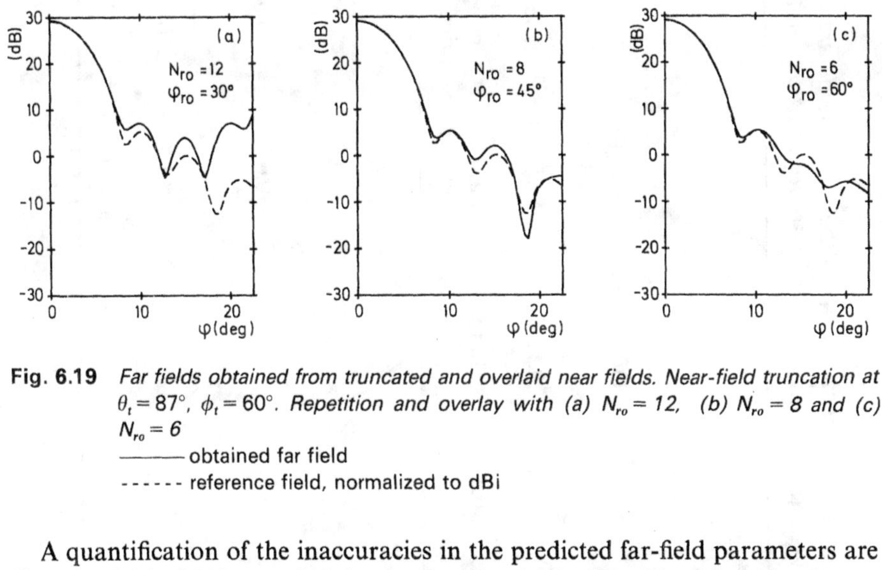

Fig. 6.19 *Far fields obtained from truncated and overlaid near fields. Near-field truncation at $\theta_t = 87°$, $\phi_t = 60°$. Repetition and overlay with (a) $N_{ro} = 12$, (b) $N_{ro} = 8$ and (c) $N_{ro} = 6$*

——— obtained far field

------ reference field, normalized to dBi

A quantification of the inaccuracies in the predicted far-field parameters are given in Table 6.5. As previously, the first line (a) in the table shows the pattern parameters for the nontruncated reference far field. The next two lines, (b), are for truncated repeated near fields (Fig. 6.18). The deviations from the reference field agree well with the results for the truncated near field, cf. Table 6.4. The accuracy in the main-beam region tends to have increased somewhat while the side lobe and especially the cross-polar lobe have decreased in accuracy. Thus, the repetition technique itself changes the level of the inaccuracies only slightly.

In the bottom section, (c), of Table 6.5, results for truncated overlaid near fields are given for the same cases as in Fig. 6.19. As discussed above, the accuracy cannot be good outside the main lobe for $N_{ro} = 12$, and also for $N_{ro} = 8$, some inaccuracies are seen. For $N_{ro} = 6$, the accuracy agrees with that of the corresponding case of truncation (Table 6.4, second-to-last line). Again it is

Table 6.5 *Simulated far-field pattern inaccuracies. (b) When near fields are truncated and repeated, and (c) When near fields are truncated and overlaid. (a) Pattern parameters for the true far-field.*

Truncation with repetition and overlay	Power of repeated field	Main beam			First side lobe		Cross-polar lobe
		Directivity	3-dB beamwidth *E*-plane	3-dB beamwidth *H*-plane	Level *E*-plane	Level *H*-plane	Level
(a) Reference values	0 dB	29.12 dBi	6.55°	6.55°	−23.76 dB	−23.79 dB	−35.29 dB
	dB	Change in dB	Change in degrees	Change in degrees	Change in dB	Change in dB	Change in dB
(b) Repeated near-field							
$\theta_t = \phi_t = 45°$, $\theta_{tf} = \phi_{tf} = 5°$, $N_r = 4$	5.58	0.18	−0.23	−0.18	−1.7	−1.4	−0.99
$\theta_t = \phi_t = 60°$, $\theta_{tf} = \phi_{tf} = 17°$, $N_r = 3$	4.36	0.00	0.01	0.03	−0.01	0.18	0.13
(c) Overlaid near fields							
$\theta_t = 87°$, $\phi_t = 60°$, $\phi_{tf} = 17°$							
$N_{ro} = 12$	10.41	0.06	−0.12	0.07	1.94	−0.94	0.40
$N_{ro} = 8$	8.54	−0.07	−0.03	−0.03	0.07	1.5	0.10
$N_{ro} = 6$	7.35	0.00	−0.06	0.02	0.45	0.52	0.15

found that the main-lobe results tend to improve when overlay is used while the results for the other lobes tend to degrade.

The power of the repeated fields, relative to the reference field, is given in the first column of Table 6.5. Since the truncated fields are repeated N_r (or N_{ro}) times, the power is correspondingly increased with respect to the power of the truncated fields, which explains the high power values. The increase in power by repetition will not necessarily be the value $N_r(N_{ro})$ converted to dB. This depends on the relative phases of the repeated fields when superposed.

Gain values of the test antenna shall be determined in the usual way, e.g. by the substitution method or by measuring the range insertion loss. The antenna gain is not influenced by the truncation which is a numerical, not physical, tool. The antenna loss determined from truncated measurements will include the power radiated in the truncated region. Directivity cannot be predicted from truncated measurements and will always require measurements over the full sphere.

The investigations show that truncation of the near-field measurements is possible irrespective of the direction of the test antenna to the measurement grid. It is important that the measurements are carried out over a sufficiently large part of the measurement sphere, roughly corresponding to a tube surrounding the test antenna and pointing in the direction in which the far field is desired. This tube will include fewer points of the measurement grid the larger the measurement distance is. It is interesting that the spherical wave theory, as based on modes radiating in all directions from the antenna under test as illustrated here, can be used to follow sectoral parts of the energy from the near field to the far field. Thus, the physical ray interpretation of the field can also be used in the interpretation of spherical measurements. This is related to the discussion on weighting functions, cf. Fig. 7.5.

The technique, used in conjunction with a proper probe correction, can improve the accuracy of conventional far-field ranges significantly for measurements of antennas which otherwise would be too large to be measured at an existing far-field range.

The repetition and overlay can only be used for antennas with the main-beam pointing on the equator of the measurement grid. The measurements can, however, be carried out as polar measurements with a certain number of cuts through the main-beam. If the measurement points are placed with the same spacing as would be used for a conventional far-field measurement this will, in connection with Fourier interpolation in theta and phi, give sufficient information for an accurate determination of the field at grid points corresponding to an equatorial region. Then repetition and overlay can be applied. This technique requires the field to be sampled only over the full near-field main-beam or, on intermediate measurement distances, over the main-beam and the first side lobes. Thus, even for large antennas, the number of measurement points need not drastically exceed the number of points for far-field measurements. By the repetition and overlay technique, the time required for numerical processing can also be kept to a reasonable level (for a minicomputer, the process time will be less than the required measurement time).

6.7 Evaluation of measurements

As previously mentioned, linearity exists between the various measurement inaccuracies and the consequential inaccuracies in the predicted far field. This holds in general for mechanical inaccuracies smaller than 0.1λ in linear measure and 0.1° in angular measure. When mechanical and electrical inaccuracies are simulated in combinations, the consequences are found to be additive. Thus, a summation of the individual far-field inaccuracies will give a worst-case estimate of the resulting inaccuracies.

This has been practised for the measurement set-up at TUD for a reflector antenna 30λ in diameter (see Fig. 6.1). The individual inaccuracies have been evaluated and it was found that the performance of the measurements was predicted very accurately. This is summarized in Table 6.6. Further, the inaccuracies as determined from full sphere measurements, Section 5.3.4, are included. The measurements were evaluated by comparing far fields transformed from near fields measured at quite different measurement distances.

The main-beam directivity was predicted accurately to within 0.04 dB and the measurements showed inaccuracies of 0.03 dB. The cross polarization in this direction is disturbed with a signal 59 dB below main beam. The nominal test antenna has no cross polarization in this direction. For the measurements, a cross-polarization level 54 dB below peak was disturbed by 2 dB which corresponds to a disturbing signal at −67 dB. According to the simulations, the inaccuracies are caused by the receiver especially by the amplitude nonlinearities.

The accuracies of the beam width and of the position of the first null were not determined for the measurements.

For the level of the first side lobe, the measurements again show better accuracy than the simulations. For the level of the cross-polar lobe, on the other hand, the simulations are a little too optimistic.

Later studies [12, 13] have detailed these investigations and show that simulations of errors are applicable but approximations in the antenna modelling may limit the region of application. Well-known inaccuracies were introduced in the measurements and the far fields were compared to those of corresponding simulations. Examples are shown in Figs 6.20–6.21 which show errors in directivity for two types of alignment inaccuracy: non-intersecting axes and horizontal depointing, respectively (cf. Table 6.1). The investigated inaccuracies were chosen large compared to realistic alignment inaccuracies in order to dominate over the influence of other non-controlled inaccuracies of the measurements. The simulation model shows reasonable good agreement with the practical measured inaccuracies even though strut scattering and feed blockage were not included in the simulated test antenna.

Table 6.6 *Inaccuracies at the set-up at TUD, simulated (worst case) and measured. (a) The simulations do not include drift of the receiver nor multiple reflections. (b) Pattern parameters for reference far field of simulations. (c) Measured inaccuracies. (d) Far-field pattern parameters of measured field.*

	Main beam			First null	First side lobe	Cross-polar lobe
	Directivity	Cross-polar level	Beam width	Position	Level	Level
	Change in dB	Increased to (dB)	Change in degrees	Change in degrees	Change in dB	Change in dB
(a) TUD facility, *simulated* optimum performance						
mechanical inaccuracies	0.02	−60	—	0.01	0.6	0.1
receiver inaccuracies (S/A 1770)	0.02	−78	0.01	0.03	0.4	2
total inaccuracies	0.04	−59	0.01	0.04	1.0	2
(b) Reference values	38.4 dBi	−∞ dB	2.25°	2.95°	−26.8 dB	−48.2 dB
	Change in dB	Error level (dB)			Change in dB	Change in dB
(c) TUD *measured* inaccuracies						
total inaccuracies	0.03	−67			0.2	2
(d) Measured reference values	36.9 dBi	−54 dB			−18.5 dB	−37 dB

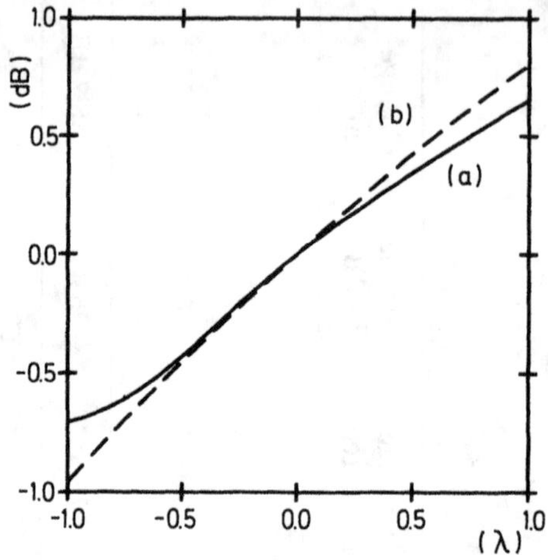

Fig. 6.20 *Error in directivity for non-intersecting axes as determined by (a) simulations and by (b) measurements for various distances between the two axes*

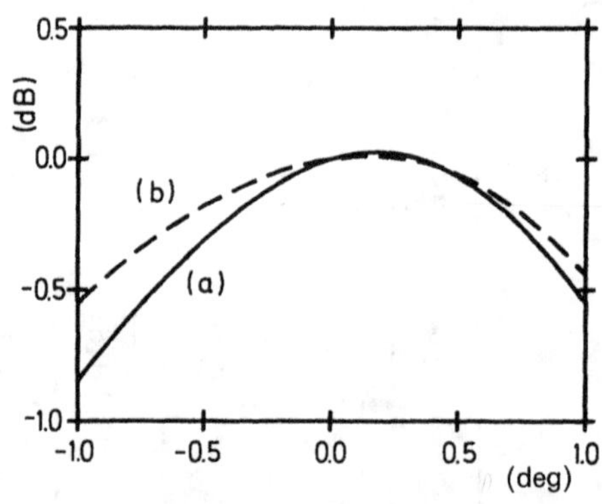

Fig. 6.21 *Error in directivity for horizontal depointing (in degrees) at $\theta = 0$. Determined by (a) simulations and by (b) measurements for various depointing inaccuracies*

6.8 Conclusion

Various mechanical designs for spherical near-field measurements have been presented in the preceding sections. It has been shown that the choice of scanning scheme is important for reducing the influence of alignment inaccuracies.

Numerical simulations of inaccuracies have been shown to be an important tool in evaluation of a measurement set-up, especially to point out which parameters should be addressed for improved accuracy.

Actual simulations have shown that directivity (gain) measurements are mainly disturbed by such inaccuracies, e.g. alignment errors, which apparently reduce (or increase) the aperture area of the test antenna. Side lobes and cross-polar lobes are disturbed mainly by inaccuracies, e.g. drift in the receiver, which cause inconsistent near-field values on the axis.

Probe correction in the near-field far-field transformation is required if precise results are desired. However, the correction may be omitted if the measurement distance is large compared to the test antenna diameter or if the probe is small in terms of wavelengths.

The probe should be calibrated at the applied frequency. However, when the probe pattern is somewhat independent of frequency, the same probe pattern correction can be used over a reasonable frequency band, and a good accuracy can still be obtained.

Probe polarization correction, as for conventional far-field measurements, is required for accurate prediction of the cross polarization, a field parameter for which the tolerances may be tight.

If measurements are carried out on an outdoor range which is too short to measure the far field directly, the spherical near-field technique can be used with advantage to compensate for the finite measurement distance. If ground reflections disturb the measurements, these can be suppressed by applying a directive probe; in such cases it may be necessary to apply a probe correction (see Section 5.2.6).

Truncation of the measurement surface is possible whether the test antenna points on the pole of the measurement grid or on an equator point of the grid. This reduces the measurement time considerably. If the test antenna points on an equator point of the measurement grid, the measured near field can be repeated over the full sphere. This may reduce the requirements to the computer storage and to the transformation time with a factor of about $36/BW$ where BW is the test antenna far-field half-power beam width in degrees. The inaccuracies introduced by such a repetition will be minimal.

The conclusions drawn here are based on simulations for a 30 wavelengths diameter reflector antenna (the truncation simulations are for smaller antennas). The requirements for the measurement set-up will depend on the antenna under test but it is expected that the results and the conclusions are typical for the type of antenna investigated.

Simulations are easy to carry out and give a good general knowledge of the performances of a given measurement set-up. Thus, modifications and improvements in the set-up can be evaluated from examinations of the near field as well as the far field. In this way, the simulations provide an understanding of the spherical technique and its applications.

References

1. JENSEN, F., 1978, 'Computer simulations as a design tool in near-field testing', Conference on Antennas and Propagation, *IEE Conference Publication No.* **169**. Part 1, pp. 111–114, London, England, Nov.

2. JENSEN, F. and LARSEN, FL. HOLM, 1977, 'Spherical near-field technique', *IEEE AP-S Symposium Digest*, pp. 378–381, Stanford, USA, June.
3. BACH, H., FRANDSEN, A. and LARSEN, FL. HOLM, 1981, 'Pattern prediction for a focused reflector antenna using GTD and near-field to far-field transformations', Conference on Antennas and Propagation, *IEE Conference Publication No.* **195**, Part 1, pp. 499–500, York, England, Apr.
4. JENSEN, F., 1981, 'Computer simulations of inaccuracies in spherical near-field testing', Conference on Antennas and Propagation, *IEE Conference Publication No.* **195**, Part 1, pp. 288–292, York, England, Apr.
5. JENSEN, F., 1981, 'Accuracy requirements evaluated from numerical simulations', Experimental spherical near-field antenna test facility, phase 2. Final Report Vol. II, ESA Contract No. 3916/79/NL/DG, *TICRA Report* **S-113-03**, TICRA, Copenhagen, Denmark.
6. FRANDSEN, A., 1983, 'Study of an offset reflector antenna with a sagging feed support', Experimental spherical near-field antenna test facility, phase 3. Final Report Vol. II, ESA Contract 4682, *TICRA Report* **S-179-01**, TICRA, Copenhagen, Denmark.
7. NEWELL, A. C. and REPJAR, A. G., 1977, 'Results of spherical near-field measurements on narrow-beam antennas', *IEEE AP-S Symposium Digest*, pp. 382–385, Stanford, USA, June.
8. JENSEN, F., 1977, 'The spherical near-field far-field technique and its application to satellite antennas', ESA Workshop on Antenna Testing Techniques, **ESA SP 127**, pp. 151–158, ESTEC, Holland, June.
9. LEWIS, R. L., 1981, 'Spherical scanning data processing; An algorithm for halving the data processing effort when the radiation into the back hemisphere is negligible', *IEEE AP-S Symposium Digest*, Vol. **1**, pp. 246–249, Los Angeles, USA, June.
10. SHAPIRA, J., 1979, 'Sampling and processing in antenna pattern measurement', Record of the 11th Convention of IEEE in Israel, Paper No. **B2-5**, Tel Aviv, Israel, Oct.
11. HESS, D. W. and TAVORMINA, J. J., 1982, 'An algorithm with enhanced efficiency for computing the far field from near-field data on a partial spherical surface', URSI National Radio Science Meeting, University of Colorado, p. 185, Boulder, Colorado, USA, Jan.
12. FRANDSEN, A., 1984, 'Comparison between measurements and computer simulations of errors in spherical near-field measurements', Experimental spherical near-field antenna test facility, phase 3, Final Report Vol. IV, ESA Contract No. 4682, *TICRA Report* **S-149-01**, TICRA, Copenhagen, Denmark.
13. FRANDSEN, A., 1985, 'Comparison between measurements and simulations of inaccuracies in spherical near-field measurements', URSI Meeting, Paper No. **A-1-4**, p. 5, Vancouver, June.

Chapter 7

Plane-wave synthesis

7.1 Introduction

The preceding chapters have detailed a method which allows overall prediction of the field from an arbitrary test antenna based on measurements over a spherical surface. The most important application of this method is the prediction of the far field of the test antenna based on measured near-field data.

In the present chapter, we shall consider the general problem of far-field measurements from the point of view of *plane-wave synthesis*. This point of view is straightforward if we consider the measurement of the receiving characteristics of a test antenna. The far-field pattern can here be measured with the receiving test antenna rotated in an incident plane-wave field. Any deviation from plane-wave behaviour of the incident field in the volume occupied by the test antenna—caused, for example, by compromises in the design of the range, reflections from surroundings or multiple interaction—will cause errors in the measured far-field data. Arguments of this kind show that any far-field measurement technique must, in some way, realize or synthesize a quasi-plane-wave field in the test zone. The creation of a plane-wave zone of sufficient quality can be achieved in a number of different ways which may be used to classify the various types of far-field measurement ranges. The quality of the plane-waves produced allows comparison of the different ranges.

If the far-field pattern in the transmitting mode of operation is needed and if the test antenna is not reciprocal, then the test range must be operated in the receiving mode. Even then, however, the plane-wave synthesis property is inherent in the range.

Figure 7.1 depicts some test ranges and the corresponding quasi-plane-wave zones.

In a *conventional far-field range*, the plane-wave zone is generated by a small probe at some sufficiently large distance. The conventional far-field criterion prescribes a measurement distance not smaller than the *Rayleigh distance*

$$R = \frac{2D^2}{\lambda} \tag{7.1}$$

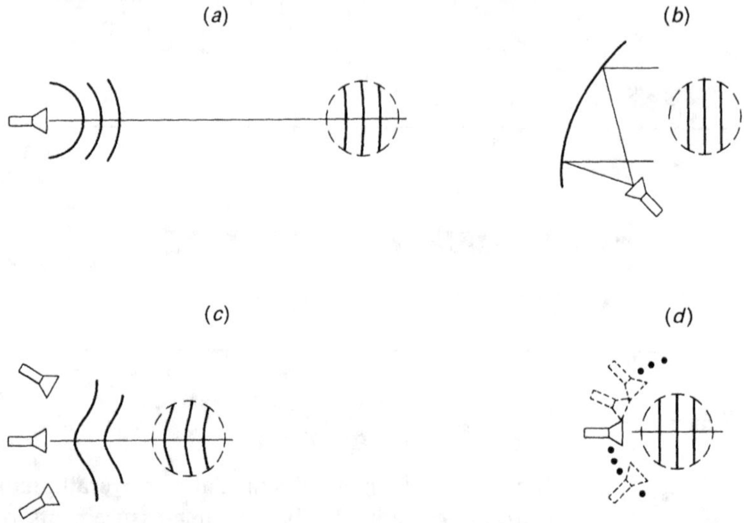

Fig. 7.1 *Different types of test ranges with wave fronts and quasi-plane wave zones*
(*a*) Far-field range
(*b*) Compact range
(*c*) Array of probes
(*d*) Scanning of a single probe

D being the largest cross-section dimension of the test zone. At the Rayleigh distance the spherical form of the wave fronts causes a maximum phase error equal to 22.5°. Thus, for very accurate measurements, a somewhat larger measurement distance is required.

The *compact range* belongs to the class of collimating ranges where the spherical wave fronts from the probe are transformed into a planar form in the test region thereby reducing the distance requirement. The collimator may be a microwave lens or a parabolic reflector. Since the design of collimators relies basically on ray tracing, this class of methods applies only at higher frequencies (usually above 1 GHz).

Instead of using a single probe in connection with a collimator to create a plane-wave field, one can alternatively use an *array of probes* excited in an appropriate manner. A technique of this kind provides a higher degree of control over the test zone field and is applicable at lower frequencies as well. However, depending on the size of the plane-wave zone and the measurement distance, a very large number of probes will often be needed, each with an excitation (a weight) that has been carefully adjusted in amplitude and phase, taking into account mutual coupling.

The need for a large number of probes can be avoided by *scanning a single probe* over the positions of the array elements. For each scan position, the response of the test antenna to a unit excitation on the probe is recorded. Then, provided the test antenna is electrically linear, the response to the entire array illumination (the incident quasi-plane wave) can be computed as the weighted sum of the responses to the single probes. In this procedure, the complex weights of the array elements exist only in the computer. Therefore, based on a single scan, different sets of weights synthesizing plane waves propagating in different directions may be used to obtain far-field information about the test antenna in different directions. Thus, in scan techniques, the plane-wave field never exists physically but only as a concept inherent in the technique.

Physically constructing the array has the advantage that the generated plane wave can be applied for radar cross-section measurements and for electromagnetic susceptibility tests [1].

The problem of establishing a set of weights that will provide a plane-wave zone of sufficient size and quality can be, and has been, approached in a number of different ways. A distinction can be made according to the receiving/transmitting operation of the test antenna assumed in the analysis of the problem.

In the case where the analysis is carried through with the test antenna receiving, it is the array that establishes the field to be measured. This emphasizes the plane-wave synthesis nature of the problem. Therefore, considering this mode of operation, a natural objective in the determination of a set of optimum weights would be the achievement of a plane-wave zone of sufficient size and quality. Methods that are derived from this objective are referred to as plane-wave synthesis methods for far-field estimation. In a plane-wave synthesis method, typically the weighting function will be computed initially and then either used in a hardware implementation of the array or stored in the computer for the processing of subsequent measurements.

Martsafey [2] analyzed the use of a planar array for plane-wave synthesis and considered the significance of the directional properties of the probe as well. However, no general method for the determination of a weighting function for a finite array was proposed. Only homogeneous illumination was investigated, and a smooth tapering of the weights at the boundaries of the array was suggested.

At the University of Sheffield, Bennett and Schoessow [3–4] presented and analyzed a plane-wave synthesis method including an iterative numerical procedure for computation of a weighting function [5–8]. The method of Bennett and Schoessow, and later rather different methods by Hill [1] and Hald [9], are treated in some detail later in this chapter.

Next, let us consider the case where the far-field estimation is analyzed with the test antenna being operated as a transmitter. In this case, the test antenna creates the field to be measured, and the outputs from the array elements are naturally interpreted as samples of the test antenna near field, especially if a scanning probe is used. The problem of determining the far field is therefore naturally treated as that of estimating the distant field with the near field known in amplitude and phase over some measurement surface. This formulation of the problem has led to the development of the near-field to far-field transformation methods in which the near field is probed over some elementary surface (planar, cylindrical or spherical) and the far field is obtained through expansion of the field in a series of modes.

In the near-field to far-field transformation methods based on characteristic mode expansions, the weighting function corresponding to the estimation of a specific far-field component is not explicitly generated. The weights are, however, implicit since each far-field component is obtained effectively as a weighted sum of the single probe responses.

Ludwig and Larsen [10, 11] pointed out that, if these implicit weights were used as excitations for the entire probe array, then, by reciprocity, a quasi-plane wave field would be established in the test zone. Comparison of different test ranges through the size and quality of the synthesized plane-wave zones has the advantage of being independent of a specific test antenna. Ludwig and Larsen suggested that near-field to far-field transformation algorithms can be validated through evaluation of the synthesized plane wave and the algorithm described in Chapter 4 was in fact analyzed that way. Note, that the significance of probe correction as well as the influence of measurement errors can be easily studied in a validation procedure based on computation of the synthesized plane-wave field. The problem of estimating the errors, caused by plane-wave imperfections, in the measurement of parameters such as gain for a given test antenna is left unsolved using a validation of this kind.

Section 7.2 contains a presentation of some of the results of Ludwig and Larsen. In Section 7.3, some basic possibilities and limitations related to generation of a plane-wave field in a finite region are discussed, and Section 7.4 outlines some basic ideas behind the plane-wave synthesis method of Bennett and Schoessow. Section 7.5 presents some figures of merit for the deviation from ideal plane-wave behaviour of a quasi-plane wave. In the numerical synthesis methods by Hill and by Hald, presented in Section 7.6, these figures of merit are used to define least-squares fits of the independent probe fields to the ideal plane-wave.

7.2 Spherical near-field far-field transformation as a plane-wave synthesis method

The first part of this section briefly introduces the algorithm of Ludwig and Larsen [10] for computation of the probe weights which are effectively used by the transformation algorithm of Chapter 4 in computing a specific far-field component. In the second part, a weighting function, as well as a synthesized quasi-plane wave field, is investigated. The results all originate from the work of Ludwig and Larsen [10, 11].

The spherical scan, which provides the input data to the transformation algorithm, consists of measuring equidistantly in theta and phi the test antenna's near field with two polarization orientations of the probe. Figure 7.2 depicts the

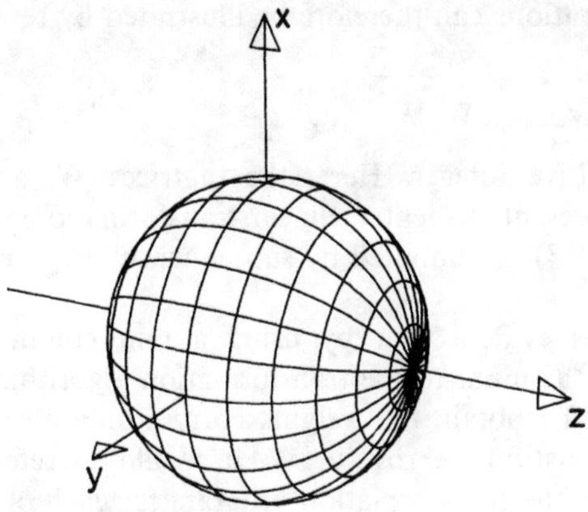

Fig. 7.2 *Spherical measurement grid*

spherical measurement grid. We denote by J_θ and J_ϕ the number of grid points over 2π in theta and phi, respectively.

From the scan data, the coefficients in a spherical wave expansion of the test antenna field are calculated, and the far field is then obtained by evaluation of the mode series. The highest mode degree N, used in the transformation, must satisfy $J_\theta \geqslant 2N + 1$. The highest mode order M must satisfy $J_\phi \geqslant 2M + 1$. Usually, N assumes its maximum value and $J_\phi = J_\theta$, $M = N$. Since the transformation is based on the use of FFTs, the far field is naturally obtained in the set of directions (θ, ϕ) defining the measurement grid.

By this method, the computed far field will be a linear function of the near-field samples. This means that in principle the far field at each point is calculated as a weighted sum of the input near-field data. If the weights associated with one far-field value are used as excitation coefficients of the spherical array, this will create an incident plane-wave field within a sphere surrounding the antenna under test. The direction of propagation will be the opposite of the far-field direction and the polarization will match that of the far-field component to be calculated.

However, the near-field to far-field transformation algorithm outlined in Chapter 4 does not lend itself to an easy calculation of these weighting coefficients. If the transformation is visualized as a multiplication of a column matrix $\mathbf{w}_{\text{near}}$, containing all the near-field data, by a transformation matrix $\mathbf{W}$

$$\mathbf{w}_{\text{far}} = \mathbf{W}\mathbf{w}_{\text{near}} \tag{7.2}$$

then the weighting coefficients corresponding to far-field value number j constitute row number j in $\mathbf{W}$. In the transformation program, the matrix $\mathbf{W}$ is never actually calculated. Instead, the total transformation, visualized as a multiplication by $\mathbf{W}$, is performed as a series of smaller linear operations such as Fourier transforms, ref. Section 4.4.2. These partial linear operations can also be written as matrix multiplications. The implementation of the transform as a series of linear operations can therefore be illustrated by rewriting (7.2) in the following form

$$\mathbf{w}_{\text{far}} = \mathbf{W}_\nu \mathbf{W}_{\nu-1} \ldots \mathbf{W}_2 \mathbf{W}_1 \mathbf{w}_{\text{near}} \tag{7.3}$$

ν being some positive integer. Here, the matrices $\mathbf{W}_i$ are not calculated either; they just represent elementary linear transformations of data. Thus, the right-hand side in (7.3) is computed by subsequently exposing the data to the transforms $\mathbf{W}_i$.

From (7.2), it is evident that by using a unit column matrix $\mathbf{w}_{\text{near}} = (0, \ldots, 0, 1, 0, \ldots, 0)^{\text{T}}$, as input to the transformation algorithm, a column of $\mathbf{W}$ could be obtained. To obtain the weights corresponding to a single far-field direction, which constitute a row in $\mathbf{W}$, it would therefore in general be necessary to execute the transformation program for each of the unit matrices $\mathbf{w}_{\text{near}}$.

To circumvent this problem, Ludwig and Larsen [10] modified the transformation algorithm to provide the weights producing far-field value number j corresponding to the direction (θ_j, ϕ_j) and polarization χ_j. Details of the modified algorithm shall not be given, but the basic idea is to rearrange all the partial matrix multiplications in a way that corresponds to transposing eqn (7.3), i.e. reverse the sequence of matrix multiplications and transpose all the matrices.

Following the discussion in connection with (7.2) and (7.3) we have

$$\mathbf{W}^{\text{T}} = \mathbf{W}_1^{\text{T}} \mathbf{W}_2^{\text{T}} \ldots \mathbf{W}_{\nu-1}^{\text{T}} \mathbf{W}_\nu^{\text{T}} \tag{7.4}$$

and the weights corresponding to the far-field direction (θ_j, ϕ_j) and polarization χ_j constitute column j in $\mathbf{W}^{\text{T}}$. It is easy to verify that if $[\mathbf{W}^{\text{T}}]_j$ denotes column j in $\mathbf{W}^{\text{T}}$, and $[\mathbf{W}_\nu^{\text{T}}]_j$ column j in $\mathbf{W}_\nu^{\text{T}}$, then because of (7.4), we have

$$[\mathbf{W}^{\text{T}}]_j = \mathbf{W}_1^{\text{T}} \mathbf{W}_2^{\text{T}} \ldots \mathbf{W}_{\nu-1}^{\text{T}} [\mathbf{W}_\nu^{\text{T}}]_j \tag{7.5}$$

which is an expression for the set of weights needed. The matrix multiplications by $\mathbf{W}_i^{\text{T}}$, $i < \nu$, correspond to simple linear operations, e.g. Fourier transforms, just like the multiplications by $\mathbf{W}_i$. In (7.3), the matrix $\mathbf{W}_\nu$ performs inverse Fourier transformations in chi, phi and theta, with each row corresponding to a specific value of (χ, θ, ϕ). The other partial linear transformations $\mathbf{W}_i$, $i < \nu$, are independent of theta, phi and chi. Therefore, $[\mathbf{W}_\nu^{\text{T}}]_j$ contains a set of exponentials which define the far-field direction (θ_j, ϕ_j) and polarization χ_j, ref. [10], and the weights corresponding to this direction and polarization are then obtained by exposing $[\mathbf{W}_\nu^{\text{T}}]_j$ to the linear operations represented through $\mathbf{W}_i^{\text{T}}$, $i = \nu - 1, \nu - 2, \ldots, 2, 1$.

The size of the synthesized plane-wave zone can be estimated through the following consideration. In the transformation algorithm, all the spherical wave components of the test antenna field with $n \leqslant N$ are transformed exactly into the far-field region. Conversely, by a reciprocity argument, the quasi-plane wave field generated by the spherical probe array will have a spherical wave expansion which is exactly that of the ideal plane wave for $n \leq N$. This result, in connection with the fact that the spherical wave functions $\vec{F}^{(1)}_{smn}$ with $n > N$ become small for $kr < N$, implies a vanishing plane-wave deviation for $r < N/k$. The radius N/k of the plane-wave zone is seen to be proportional to the highest mode degree N used in the transformation. Using $N \simeq kr_0$, as discussed in Section 2.2.3, the plane-wave zone just covers the test antenna minimum sphere.

The size of the plane-wave zone is limited by the radius A of the measurement sphere. If the spacing $A\Delta\theta$ between the measurement points approaches half a wavelength, and the maximum value of N, $N \simeq J_\theta/2$, is used, then the plane-wave zone radius N/k approaches its upper limit A.

The quality of the synthesized plane-wave will depend on various parameters. First of all, the measurement sphere radius A and the radius N/k of the plane-wave zone have some influence especially when N/k approaches A. But also the

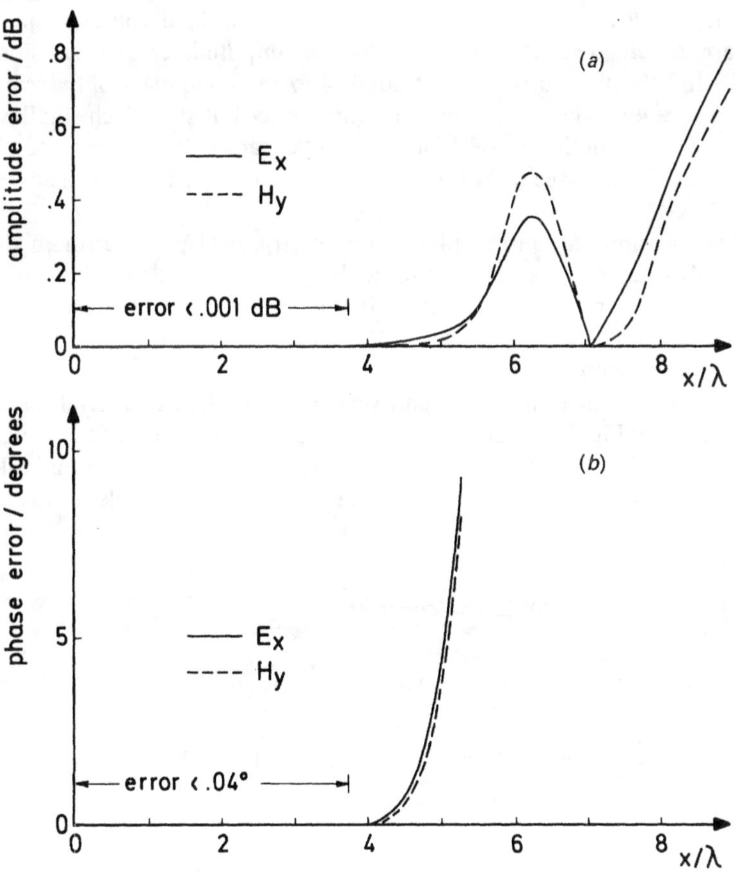

Fig. 7.3 *Deviation from the ideal plane wave along the x-axis*
(*a*) Amplitude deviation
(*b*) Phase deviation

direction (θ, ϕ) of the plane-wave incidence and the polarization of the ideal plane wave may affect the quality of the field. Finally, the field will of course depend slightly on the probe radiation pattern even though probe correction is applied.

We shall consider one nominal case selected as follows: a short dipole probe, plane-wave incidence angle $\theta = 0°$, x-polarized electric field, measurement grid radius $A = 66.7\lambda$ and plane-wave zone radius approximately equal to 4.8λ $(\Delta\theta = 6°, N \simeq J_\theta/2)$. At 10 GHz this corresponds to a 2 m radius measurement sphere and a 30 cm diameter plane-wave zone. The conventional far-field criterion $2D^2/\lambda$, D being the diameter of the plane-wave zone, would be 6 m for this case so the measurement distance is about one third of the normal far-field criterion.

Figure 7.3 shows the amplitude and phase deviations from ideal plane-wave behaviour along the x-axis for the principal polarization components of the $\vec{E}$- and $\vec{H}$-fields. Clearly, there is a 'quiet zone' in which the amplitude and phase errors are extremely small. For $|x| < 3.8\lambda$, the amplitude error is smaller than 0.001 dB and the phase error is less than 0.04°. At the boundary of the test zone, for $r = N/k \simeq 4.8\lambda$, the errors are somewhat larger but still much smaller than the error accepted in the conventional far-field criterion. If the error is too large, the extent of the high-accuracy quiet zone can be increased by increasing N; see eqn (2.32).

Figure 7.4 shows a contour plot of the electric field phase error in the yz-plane. Only the 0.1° and 10° contours have been included. This reveals a high-quality zone (phase error <0.1°) with an approximate spherical shape, and a medium-quality zone (phase error <10°) which is closer to a cylindrical shape over a certain region.

The weighting function will depend very much on the radius A of the measurement sphere. Figure 7.5 depicts the amplitude distribution of the weighting function around the angle of plane-wave incidence for another nominal case described with plane-wave incidence along the x-axis and a plane-wave zone

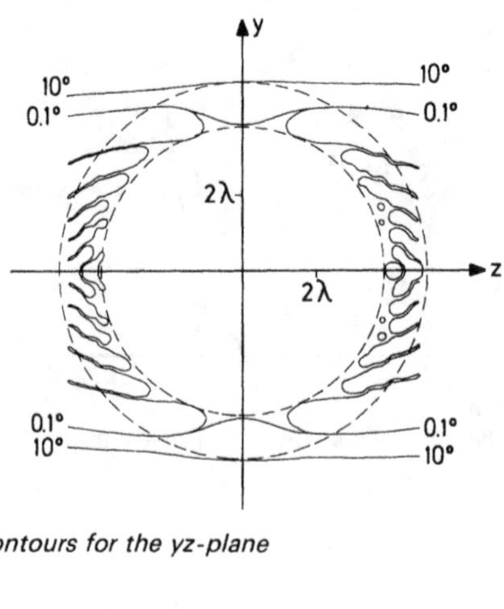

Fig. 7.4 *Phase error contours for the yz-plane*

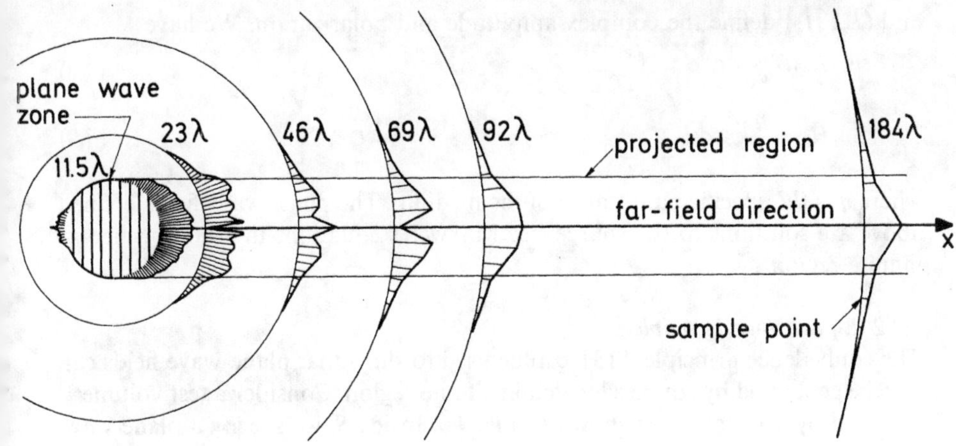

Fig. 7.5 *Amplitude distribution of weights at various distances*

radius approximately equal to 11.5λ ($\Delta\theta = 2.5°$, $N \simeq J_\theta/2$). It is seen that the weights tend to decay outside the area appearing as the projection, in the direction of plane-wave incidence, of the plane-wave zone on the measurement sphere. Outside an area with a radius equal to twice the radius of the test zone, the weights are very small.

The distribution of the weights revealed by Fig. 7.5 has led to the idea of truncating the full spherical array to a smaller array suited for hardware implementation; see Ludwig and Larsen [10, 11]. An array of this kind based on a truncated weighting function will not however be optimal. Better solutions can be obtained more efficiently by application of the methods outlined in the following sections.

7.3 Basic concepts in generation of plane-wave fields

7.3.1 Introduction

The present section provides basic theory related to the dedicated plane-wave synthesis methods. The possibility of generating a plane-wave field by sources in a finite region is discussed briefly and some problems associated with practical implementations are touched upon. For a more extensive treatment of these topics the reader is referred to Hill [1] and Turchin and Tseytlin [12].

The electric and magnetic fields $\vec{E}_{pl}$ and $\vec{H}_{pl}$ of a time-harmonic plane wave coming from a direction (θ_o, ϕ_o) can be written in the following form

$$\vec{E}_{pl} = \vec{E}_o e^{i\vec{k}_o \cdot \vec{r}} \tag{7.6a}$$

$$\vec{H}_{pl} = \vec{H}_o e^{i\vec{k}_o \cdot \vec{r}} \tag{7.6b}$$

where $\vec{k}_o$ is the propagation vector:

$$\vec{k}_o = -k \sin\theta_o \cos\phi_o \hat{x} - k \sin\theta_o \sin\phi_o \hat{y} - k \cos\theta_o \hat{z} \tag{7.7}$$

and $(\vec{E}_o, \vec{H}_o)$ define the complex amplitude and polarization. We have

$$\vec{k}_o \cdot \vec{E}_o = 0 \tag{7.8}$$

$$\vec{H}_o = \frac{\eta}{k}\vec{k}_o \times \vec{E}_o \tag{7.9}$$

where η is the specific admittance of the medium. The plane-wave fields defined above are solutions to the source free Maxwell's equations in a homogeneous, infinite region.

7.3.2 Equivalence principle

The equivalence principle, [13], can be used to show that plane-wave fields can also be generated by sources located in a finite region. Consider a test volume V enclosed by a surface S as shown in Fig. 7.6. Inside S, we specify a plane-wave

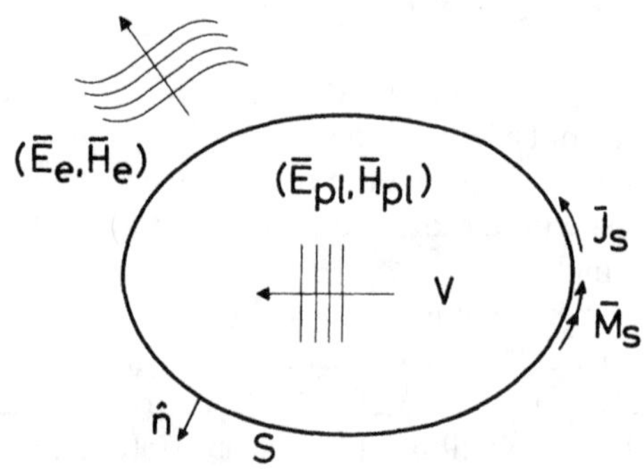

Fig. 7.6 *Electric and magnetic surface currents, $\vec{J}_s$ and $\vec{M}_s$ on a closed surface S. These currents generate an ideal plane wave within S*

field $(\vec{E}_{\rm pl}, \vec{H}_{\rm pl})$ as defined in eqns (7.6–7.9) and outside S we assume an arbitrary outgoing field $(\vec{E}_e, \vec{H}_e)$ satisfying Maxwell's equations. By outgoing fields we mean that they satisfy the radiation condition at infinity. From the equivalence principle, these fields can be generated by the following electric and magnetic surface currents, $\vec{J}_s$ and $\vec{M}_s$, on S:

$$\vec{J}_s = \hat{n} \times (\vec{H}_e - \vec{H}_{\rm pl}) \tag{7.10a}$$

$$\vec{M}_s = (\vec{E}_e - \vec{E}_{\rm pl}) \times \hat{n} \tag{7.10b}$$

where $\hat{n}$ is the outward unit normal to S. Because the field $(\vec{E}_e, \vec{H}_e)$ in (7.10) is arbitrary, the source currents $\vec{J}_s$ and $\vec{M}_s$ are not unique. There is an infinity of current distributions which will generate a perfect plane wave inside S. Non-uniqueness is typical of inverse problems where the field in a region is specified and the sources are unknown. The practical problem of plane-wave synthesis involves that of finding a solution of (7.10) where the currents $\vec{J}_s$ and $\vec{M}_s$ can be well approximated by an array of realistic antenna elements.

The exterior field $(\vec{E}_e, \vec{H}_e)$ is arbitrary in principle because it does not affect the response of a test antenna in V. In practice, however, it is desirable to keep the field outside the test volume small thereby avoiding reflections from surrounding objects that would distort the plane-wave field.

Although the source distributions (7.10) are highly idealized, they show that there is no basic limitation to generating a perfect plane wave by sources close to the plane-wave zone. The practical limitations arise from the use of realistic antennas in attempting to approximate the smooth current distributions (7.10).

The signal w received by a test antenna immersed in the plane-wave field generated by the sources $(\vec{J}_s, \vec{M}_s)$ can be expressed as a reaction integral over the surface S:

$$w = C \int_S (\vec{J}_s \cdot \vec{E}_t - \vec{M}_s \cdot \vec{H}_t)\, \mathrm{d}S \tag{7.11}$$

where $(\vec{E}_t, \vec{H}_t)$ is the field of the test antenna as a radiator. C is a constant for the particular test antenna and $\mathrm{d}S$ is a surface area element. Formula (7.11), which can be derived from the reciprocity theorem (A1.75) (see, for example, [9] and [12]), expresses the far field of the test antenna as an integral transform of the near field. The kernels in the transformation are simply the electric and magnetic surface currents $\vec{J}_s$ and $\vec{M}_s$ producing the incident plane-wave field. This observation is analogous to the fact that the arrays simulated in spherical near-field techniques produce a plane-wave field in the test zone. As a consequence, we need not distinguish between weighting functions and plane-wave source distributions.

7.3.3 Relationship to a scattering problem

In general, the set of surface current solutions given by (7.10) involve both electric and magnetic surface currents. Among these solutions, those requiring electric currents only or magnetic currents only are particularly useful. An electric current might be approximated by an array of dipole antennas, and a magnetic current distribution might be approximated by an array of loops (magnetic dipoles). The desired currents can be found by examining the relevant scattering problems as shown below.

Consider a field $(\vec{E}_i, \vec{H}_i)$, incident on a closed perfect electric conductor as shown in Fig. 7.7(a). Only an electric surface current $\vec{J}_s$ will be set up on the surface S and that current distribution will radiate a scattered field $(\vec{E}_s, \vec{H}_s)$. The total field everywhere is the sum of the incident and scattered fields. Inside S, the total field must be zero in accordance with Waterman's extended boundary condition [14]:

$$\left.\begin{aligned} \vec{E}_i + \vec{E}_s &= \vec{0} \\ \vec{H}_i + \vec{H}_s &= \vec{0} \end{aligned}\right\} \quad \text{Inside } S. \tag{7.12}$$

If we choose the incident field to be the negative of the desired plane wave

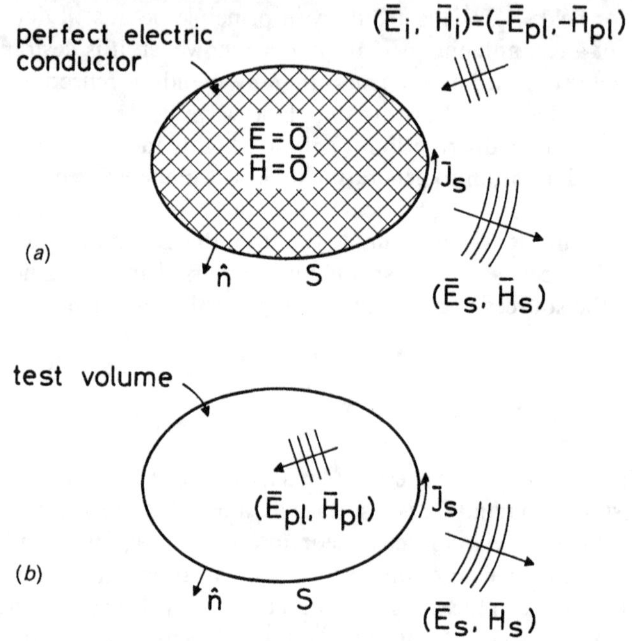

Fig. 7.7 *Illustrating the relation between a scattering problem and a plane-wave synthesis problem*
(*a*) Scattering problem for plane-wave incidence on a perfect conductor
(*b*) Source problem for plane-wave synthesis

$(\vec{E}_i = -\vec{E}_{\text{pl}}, \vec{H}_i = -\vec{H}_{\text{pl}})$, then from (7.12), the scattered field inside S is exactly the desired plane wave

$$\left.\begin{aligned} \vec{E}_s &= \vec{E}_{\text{pl}} \\ \vec{H}_s &= \vec{H}_{\text{pl}} \end{aligned}\right\} \quad \text{Inside } S. \tag{7.13}$$

If we now remove the incident field and let the surface currents $\vec{J}_s$ radiate in free space, the result will be the desired plane wave inside S and the scattered field, $(\vec{E}_s, \vec{H}_s)$, outside S as shown in Fig. 7.7(b). Thus, a solution to the problem of finding an electric surface current distribution which generates a plane wave inside S can be obtained by solving the problem of plane-wave scattering by the same shape. Analytical solutions are available for plane-wave scattering by simple shapes [15] such as spheres and cylinders. Numerical methods are required for general shapes, and integral equation solutions [16, 17] are typically utilized.

Solutions (7.10) involving only magnetic surface currents can be achieved similarly by considering the corresponding scattering problem for a perfect magnetic conductor, or alternatively, by application of duality to the electric current solution [1].

Usually, the scattering problems associated with plane-wave synthesis problems either have very complicated analytical solutions or must be solved by a complex numerical method. One case of practical interest is when the surface S of the perfect electric conductor in Fig. 7.7(a) does not have a large curvature. In this case, the physical optics approximation [13] yields the following simple result

$$\vec{J}_s = \begin{cases} 2\hat{n} \times (-\vec{H}_{pl}), & \text{illuminated area} \\ \vec{0}, & \text{shadow area.} \end{cases} \tag{7.14}$$

The errors caused by the physical optics approximation are mainly due to the coarseness of approximation in the region of the boundary between the 'illuminated' and the 'shadow' portions of S. The effects on the generated plane-wave field are therefore very similar to the effects of a truncation of the exact surface current distribution treated in the following section.

7.3.4 Truncation and sampling of continuous sources of plane waves

In practical implementations of current distributions like the one given in (7.10), often only a truncated surface will be utilized, and a sampling of the continuous current distribution is applied. As long as a source surface is large with only a small curvature and the physical optics current (7.14) is considered, the effects of truncation and sampling will be similar to the planar case. We shall therefore assume the source surface S to be the xy-plane.

Figure 7.8 depicts a sheet of physical optics current defined by a fictitious plane wave incident at an angle θ_o relative to the normal of the conducting source plane. The near field of the planar current sheet will consist of a plane-wave zone in the shadow region of the sheet, a transition zone along the shadow

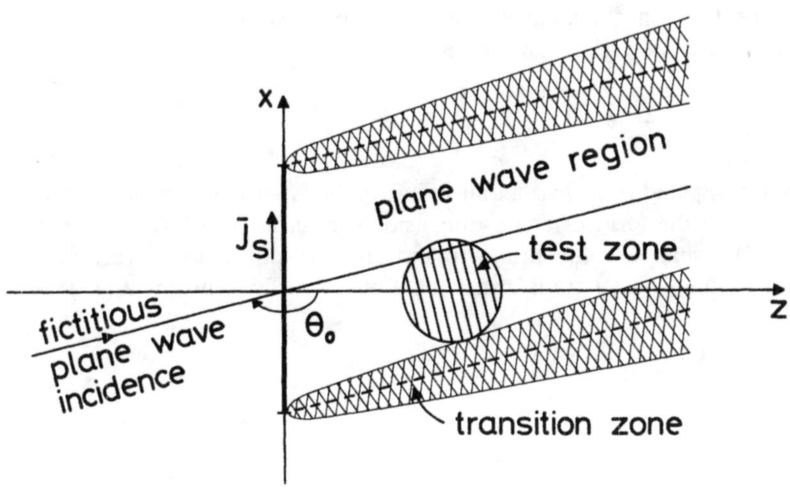

Fig. 7.8 *Near field of truncated physical optics current*

boundary, and a very small field outside the shadow region. Provided the dimensions are large compared to wavelength, diffraction theory [1] applies to the investigation of the transition zone and of the ripples in the plane-wave field caused by the edge diffraction. Following the analysis in [1], the width d of the transition zone will depend on the wavelength λ and the distance ρ to the edge of the current sheet through the relation

$$d = \alpha\sqrt{\lambda\rho} \tag{7.15}$$

where α is a constant depending primarily on the ripple that can be tolerated in the plane-wave zone. Thus, if the current sheet has a width L, the test zone has a diameter D, and the distance between the current sheet and the test zone is A, then for θ_o close to π

$$L \geqslant D + 2A|\tan\theta_o| + \alpha\sqrt{\lambda A} \tag{7.16}$$

in order that the test zone is within the plane-wave region. The relation (7.16) limits the range over which the angle θ_o of plane-wave incidence can be scanned. It has been assumed that the test zone and the current sheet have a common axis, ref. Fig. 7.8.

In the plane-wave region, the interference between the desired plane wave and the edge-diffracted fields produces ripples. At a distance of 30λ from a 30λ by 30λ square current sheet, the amplitude ripple amounts to about 1.6 dB for the case of normal plane-wave incidence ($\theta_o = \pi$), [1]. For $\theta_o \neq \pi$, the error will be larger. A way of reducing the ripple is to use a smooth taper instead of the sharp cut-off at the edges of the sheet. A tapering of this kind will, however, reduce the width of the plane-wave zone [1].

The main effect of an equidistant sampling of the continuous planar surface current can be understood by considering the effect on the far-field pattern. In the far-field region, the magnetic vector potential $\vec{A}(\vec{r})$ is related to the surface current $\vec{J}_s(\vec{r}\,')$ on S, through the relation [18]

$$\vec{A}(\vec{r}) = \frac{\mu}{4\pi r} e^{ikr} \int_S \vec{J}_s(\vec{r}\,') e^{-ik\hat{r}\cdot\vec{r}\,'} dS \tag{7.17}$$

where $r \equiv |\vec{r}|$, and $\hat{r} \equiv \vec{r}/r$ is a unit vector in the far-field direction. The integral in (7.17) has the form of a two-dimensional Fourier transform of the current $\vec{J}_s(\vec{r}\,')$. We shall use the prime to indicate two-dimensional vectors. Introduction of the Fourier transform $\vec{U}_s(\vec{k}\,')$ of the surface current $\vec{J}_s(\vec{r}\,')$

$$\vec{U}_s(\vec{k}') \equiv \int_S \vec{J}_s(\vec{r}\,') e^{-i\vec{k}\cdot\vec{r}\,'} dS \tag{7.18}$$

allows the far-field vector potential to be expressed as

$$\vec{A}(\vec{r}) = \frac{\mu}{4\pi r} e^{ikr} \vec{U}_s(k\hat{r}') \tag{7.19}$$

where $\hat{r}'$ is the projection of $\hat{r}$ onto the xy-plane. Thus, the far-field pattern is determined by that part of the Fourier transformed surface current $\vec{U}_s(\vec{k}')$, which is within the 'visible region' $|\vec{k}'| \leqslant k$.

From the theory of sampling and Fourier transformation, it is known [19] that replacement of the continuous function $\vec{J}_s(\vec{r}')$ by a discrete surface current obtained by sampling with intervals Δx and Δy in x and y leads to the introduction of a periodic replica of $\vec{U}_s$ with intervals $\Delta k_x = 2\pi/\Delta x$ and $\Delta k_y = 2\pi/\Delta y$ in the Fourier domain $\vec{k}'$. Those replicas which fall inside the visible region, $|\vec{k}'| \leqslant k$, give rise to grating lobes in the radiation pattern for the sampled surface current. The main lobe, created by the continuous current, will be centred at $\vec{k}'$ equal to the projection $\vec{k}'_o$ of the ideal plane-wave propagation vector $\vec{k}_o$ onto the xy-plane. Now, assume that the main lobe is narrow. This will be true if the extent of the surface current is large compared to the wavelength. The replicas will then fall outside the visible region independent of $\vec{k}_o$, provided $\Delta k_x \geqslant 2k$ and $\Delta k_y \geqslant 2k$, or equivalently

$$\Delta x \leqslant \frac{\lambda}{2} \quad \text{and} \quad \Delta y \leqslant \frac{\lambda}{2}. \tag{7.20}$$

For the special case of plane-wave propagation orthogonal to the source plane, i.e. $\vec{k}'_o = \vec{0}$, the replicas are outside the visible region if only $\Delta k_x > k$ and $\Delta k_y > k$, or equivalently

$$\Delta x < \lambda \quad \text{and} \quad \Delta y < \lambda. \tag{7.21}$$

If the replicas are completely outside the visible region, there are no grating lobes and the field from the sampled current will equal the field from the continuous current plus some evanescent fields which decay very quickly in directions away from the source plane; see ref. [2].

Milder sampling requirements than (7.20) and (7.21) can be used if allowance is made for grating lobes. The minimum requirement is that the grating lobes do not significantly affect the field within the test zone. If a fixed probe and a scanning test antenna is used, an array of probes is formed as seen from the test antenna but only a single probe is seen from the environment. In that case a high directivity of the probe array cannot be used to reduce the anechoic requirements. This justifies the allowance for grating lobes.

Figure 7.9 illustrates a case where the ideal plane wave propagates orthogonal to the source plane and the first grating lobe just misses the test aperture. The test aperture with diameter D and the physical optics current sheet with diameter L are both assumed to have a common centre normal. The angle v from the z-axis to the direction of propagation of the first grating lobe is the angle between the propagation vector $\vec{k}_o = k\hat{z}$ of the ideal plane wave and the propagation vector $\Delta k_x\hat{x} + \sqrt{k^2 - (\Delta k_x)^2}\,\hat{z}$ of the grating lobe. Therefore v must satisfy $\sin v = \Delta k_x/k$. Combination of this expression with the condition that $\tan v \geqslant (L + D)/(2A)$ for the grating lobe to miss the test aperture, leads to

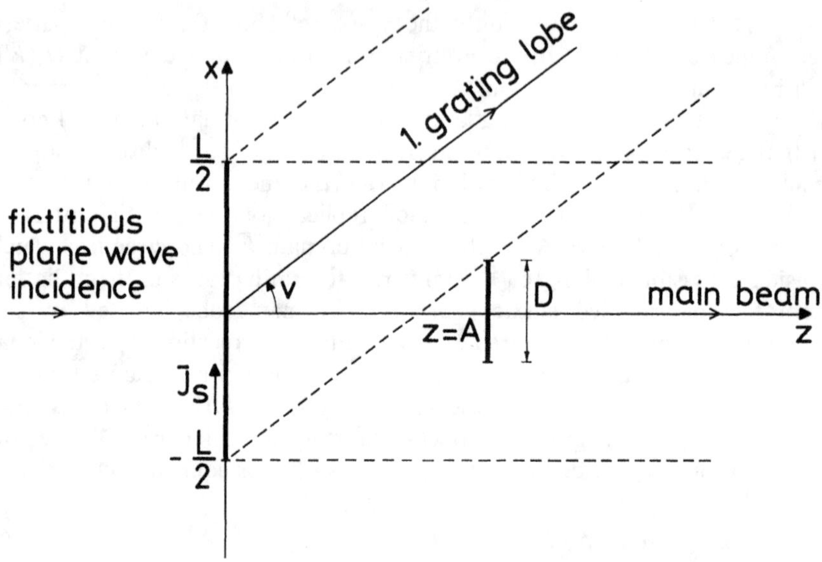

Fig. 7.9 *Illustrating the avoidance of grating lobes in the test zone*

the sampling criterion

$$\Delta x \leqslant \lambda \sqrt{1 + \left(\frac{2A}{L+D}\right)^2} \tag{7.22}$$

where A is the distance between the source plane and the test aperture. In this derivation we have neglected the width of the transition zone, and scanning of the plane wave has not been considered. The treatment of sampling Δy in y is of course completely analogous.

Even larger sampling spacings may be allowed if the continuous surface current is obtained differently. First of all, we need to specify a plane-wave field to exist only in the limited test volume. Secondly, we may choose the surface current in a way that causes the effects of the sampling to be filtered out during the propagation to the test zone (spatial filtering). As an example, the spherical measurement technique allows an angular grid spacing up to $\Delta\theta = \lambda/D$ (ref. Sections 2.2.3 and 4.4.3), which on a linear scale corresponds to

$$\Delta x \leqslant \lambda \frac{A}{D}. \tag{7.23}$$

Here the higher spatial frequencies introduced by the sampling are filtered out through the cut-off properties of the spherical wave functions. A similar spatial filtering may of course take place in connection with a planar source or equivalently, a planar measurement surface. The method, described in the following section, for generation of a surface source distribution, serves as an example.

7.4 The Sheffield method

7.4.1 Introduction

A method, for the synthesis of a surface source distribution (weighting function) specifying a plane wave to exist only in the test zone, has been developed at the University of Sheffield, [3–8]. This section contains some basic theory related to the method. For a detailed description of the theory, implementations and applications, the reader is referred to the original literature.

7.4.2 Measurement without probe correction

7.4.2.1 Introduction: The basic theory not including probe correction is based on the plane-to-plane diffraction theory presented in [20]. With reference to Fig. 7.10, a plane-wave field ($\vec{E}$, $\vec{H}$) with sources for $z < 0$ is specified over the plane

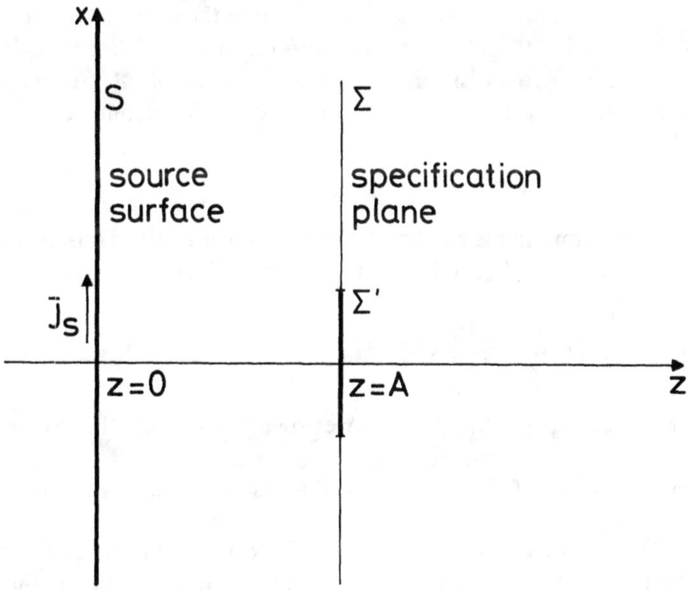

Fig. 7.10 *Geometry in plane-to-plane diffraction theory*

$z = A$, $A > 0$. This specification plane, denoted Σ, contains a central cross-section Σ' of the test zone. The plane-to-plane diffraction theory then enables the field to be computed at another plane, $z = \text{const} \geqslant 0$. By computing the field at $z = 0$, we may obtain a surface source distribution there, producing the specified field.

The source surface is denoted by S. Initially, the problem of specifying the field at $z = A$ will be treated.

7.4.2.2 Specification of an electromagnetic field on a plane: Requiring the specified field to represent a solution to the vector wave equation (2.3) introduces some restrictions which may be formulated in terms of the plane-wave spectrum $\vec{P}(k_x, k_y, A)$ of the field $\vec{E}(x, y, A)$:

$$\vec{P}(k_x, k_y, A) \equiv \iint_{-\infty}^{\infty} \vec{E}(x, y, A)\, \mathrm{e}^{-i(k_x x + k_y y)}\, \mathrm{d}x\, \mathrm{d}y. \tag{7.24}$$

The fact that each rectangular component of the field satisfies the scalar Helmholtz equation implies that [18]

$$\vec{E}(x, y, z) = \frac{1}{(2\pi)^2} \iint_{-\infty}^{\infty} \vec{P}(k_x, k_y, A)\, \mathrm{e}^{i(k_x x + k_y y + k_z(z - A))}\, \mathrm{d}k_x\, \mathrm{d}k_y \tag{7.25}$$

with

$$k_z \equiv \sqrt{k^2 - k_x^2 - k_y^2}. \tag{7.26}$$

Here, the positive imaginary root must be chosen in the evanescent wave region $k_x^2 + k_y^2 > k^2$ in order to satisfy the radiation condition for $z > 0$. In order for the field $\vec{E}(x, y, z)$ to satisfy the vector wave equation, it must satisfy the Helmholtz equation and also be purely solenoidal. The second requirement is fulfilled if [18]

$$k_x P_x + k_y P_y + k_z P_z = 0. \tag{7.27}$$

Hence, only two components can be specified independently. Thus, if E_x and E_y are specified, then P_x and P_y can be determined from (7.24). Equation (7.27) then prescribes

$$P_z = -\frac{1}{k_z}(k_x P_x + k_y P_y) \tag{7.28}$$

which determines $E_z(x, y, z)$. The above specification of the electric field uniquely determines the magnetic field.

Specification instead of the magnetic field leads to an analogous formulation.

7.4.2.3 Plane-to-plane diffraction theory: Once the field has been specified over the plane $z = A$, eqns (7.24–7.26) in principle allow prediction of the field in the entire half space $z \geqslant 0$.

For $z > A$, this prediction procedure is equivalent to the Rayleigh diffraction formula of the first kind, [20],

$$\vec{E}(x, y, z) = \frac{1}{2\pi} \iint_{-\infty}^{\infty} \vec{E}(x', y', A) \frac{\partial}{\partial A} \left\{ \frac{\mathrm{e}^{ikR}}{R} \right\} \mathrm{d}x'\, \mathrm{d}y' \qquad \text{for } z > A \tag{7.29}$$

where

$$R \equiv \sqrt{(x-x')^2+(y-y')^2+(z-A)^2}. \tag{7.30}$$

Formula (7.29) has the form of a convolution integral.

For $z < A$, however, the kernel in the corresponding convolution formulation takes the form of a divergent integral because the transfer function, $e^{ik_z(z-A)}$, in the two-dimensional Fourier domain is unbounded for increasing spatial frequencies k_x and k_y. In connection with physical fields with sources for $z < 0$, the integral (7.25) will be convergent because $|\vec{P}(k_x, k_y, A)|$ will be quickly decaying when the spatial frequencies tend toward infinity. If A exceeds just a couple of wavelengths, the evanescent wave components have been attenuated to a very low level at $z = A$.

Let us first assume that the field specified at $z = A$ contains no evanescent components, i.e.

$$\vec{P}(k_x, k_y, A) = 0 \qquad \text{for } k_x^2 + k_y^2 > k^2. \tag{7.31}$$

Then, we may choose the opposite sign for the imaginary root in (7.26) and still obtain the correct solution. After this change, the prediction of the field in the region $0 \leqslant z < A$ may be expressed through the following convolution [20]:

$$\vec{E}(x, y, z) = -\frac{1}{2\pi}\iint_{-\infty}^{\infty} \vec{E}(x', y', A)\frac{\partial}{\partial A}\left\{\frac{e^{-ikR}}{R}\right\}dx'\,dy' \qquad 0 \leqslant z < A. \tag{7.32}$$

Now, if the specified field $\vec{E}(x, y, A)$ contains evanescent components, then because of the sign change on k_z, these components will be attenuated exponentially by a factor $e^{|k_z|(z-A)}$ for decreasing z. Thus, if A is large, such components have been filtered out at $z = 0$ when (7.32) is used for the prediction.

7.4.2.4 Surface source distributions derived from diffraction theory: Formula (7.32) and the corresponding formula for the magnetic field

$$\vec{H}(x, y, z) = -\frac{1}{2\pi}\iint_{-\infty}^{\infty} \vec{H}(x', y', A)\frac{\partial}{\partial A}\left\{\frac{e^{-ikR}}{R}\right\}dx'\,dy' \qquad 0 \leqslant z < A \tag{7.33}$$

enable the field at $z = 0$ to be calculated. The problem of finding a set of surface sources producing a specified field in a closed region was treated in Sections 7.3.2 and 7.3.3 for the special case of the specified field being a perfect plane wave. To obtain a distribution of electric surface currents at $z = 0$, producing the field (7.32–7.33) for $z > 0$, we apply the method of Section 7.3.3. Let us imagine the entire half space $z \geqslant 0$ is filled with a perfect conductor, and let us imagine a field defined as the negative of the field (7.32–7.33) is incident for

$z < 0$. The surface current at $z = 0$

$$\vec{J}_s(x, y, 0) = 2(-\hat{z}) \times (-\vec{H}(x, y, 0))$$
$$= 2\hat{z} \times \vec{H}(x, y, 0) \tag{7.34}$$

produced in this fictitious set-up will generate the desired field for $z \geqslant 0$. Again, note that evanescent wave components in the field specified at $z = A$ will not be reproduced because they are heavily attenuated in the backward diffraction process (7.32–7.33) and will be further attenuated through the physical propagation of the field generated by the surface current.

So far the source surface S has been the plane $z = 0$. Since (7.32–7.33) apply for backward diffraction to the entire region $0 \leqslant z < A$, we may choose other source surfaces in this region. As an example, we may choose a surface consisting of a plane $z = B$, $0 < B < A$, with a central portion around the z-axis replaced by a sphere with radius A and centred at $(x, y, z) = (0, 0, A)$; see Fig. 7.11.

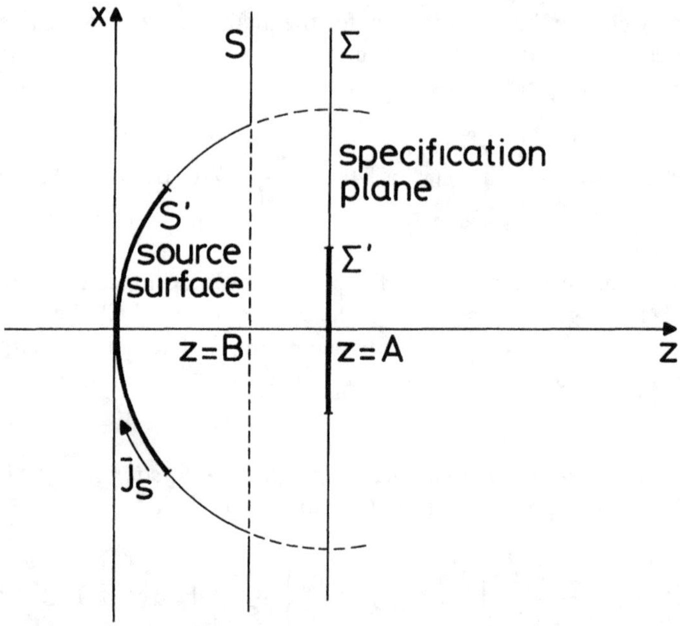

Fig. 7.11 *Spherical source surface*

Provided the current on the planar part is negligible, we obtain a spherical surface source. If the radius A of the sphere is large compared to wavelength, we may use the physical optics approximation for the surface current on the conducting sphere

$$\vec{J}_s(x, y, z) = -2\hat{n} \times \vec{H}(x, y, z). \tag{7.35}$$

Here $\hat{n}$ is an outward normal to the sphere, (x, y, z) is a point on the spherical surface and $\vec{H}(x, y, z)$ is calculated from (7.33).

7.4.2.5 Specification of a plane-wave field: Having described the method for generation of surface source distributions, we return to the problem of specifying a proper field over the plane Σ, ref. Fig. 7.11. On Σ' we specify $(\vec{E}, \vec{H})$ to be a plane-wave field, while outside Σ' the field is in principle arbitrary. However, in order to minimize reflections from surrounding objects, the field outside Σ' should preferably have a small amplitude. A choice has to be made since the field must be specified over the entire plane Σ. In selecting the field outside Σ' the following points should be considered:

1. Evanescent wave components in the field will not be reproduced. With a bad choice of the field outside Σ', a significant part of the specified plane-wave field over Σ' may become part of the evanescent wave field that will not be reproduced. The consequence will be ripples in the plane-wave field produced.
2. The field should preferably be concentrated within a rather small region over the source surface S in order to minimize the extent of the source distribution.
3. The plane-wave properties of the synthesized field must be retained over some range in the z-axis direction.

The depth-of-field requirement in point 3 is approached by restricting the source to cover only a limited area S' on the source surface S. A truncation of the source area will inevitably introduce ripples in the plane-wave zone. However, if over Σ the plane-wave field has been specified in a way to concentrate the backward diffracted field within the area S', the ripples can be kept below an acceptable level.

Bennett and Schoessow [3] introduced an iterative technique which allows an initial specification of a zero-field outside Σ' in the plane Σ. With this choice, serious ripples will be present in the synthesized plane-wave field due to the large content of evanescent waves as well as the truncation of the source distribution. The source distribution must be rather wide to produce the fast cut-off of the plane-wave field at the boundary of Σ'. These ripples are reduced through the following iterative procedure. In each step, an error field is diffracted back to the source surface S. Over Σ', the error field equals the difference between the plane-wave field actually synthesized and the ideal plane wave. Outside Σ' the error field is specified to be zero. The contribution from this field is then subtracted from the source distribution within the area S' and a new synthesized field can be computed. Notice that the field outside Σ' is not propagated backwards. Thus the field outside Σ' is rendered free to grow during the progress of the iteration. Gradually, this field will approach a form which meets the requirements in points 1–3 above, [5].

Pereira, Anderson and Bennett [8] suggested a Butterworth type of amplitude distribution as another initial specification for the field on the plane Σ, e.g.

$$|\vec{E}(x, y, A)| = \frac{E_0}{\sqrt{1 + \left\{\varepsilon^2 \frac{x^2 + y^2}{\rho_0^2}\right\}^N}} \tag{7.36}$$

where ρ_0 defines the boundary of Σ'. By careful selection of the small number ε and the integer N, the requirements formulated in points 1–3 above can be more or less met by the source distribution function calculated from the initially specified field. If the field produced by the first source distribution proves to be unsatisfactory, the iterative procedure described previously can be applied.

7.4.2.6 Sampling of the continuous surface currents: For practical applications, a discrete set of probes must replace the determined continuous current distribution. As outlined in Section 7.3.4, the maximum sampling interval depends very much on the directivity requirements for the probe array.

For the case of a spherical measurement surface, Schoessow [5] derived a sampling criterion ensuring the first grating lobe to be outside the plane-wave zone. He showed that the angular spacing criterion $\Delta\theta \leqslant \lambda/D$, known from the theory of spherical scanning, applies for the Sheffield method. The only requirement is that the measurement distance, A, exceeds about $2D$, D being the diameter of the plane-wave zone.

If the high directivity obtained by the continuous source distribution is to be maintained for the probe array the spacing must be chosen so as to avoid grating lobes. In general, this requires a spacing less than $\lambda/2$, ref. (7.20). In the case where angular scanning of the plane-wave is not applied or is restricted to a small angle, the spacing may be increased to a value between $\lambda/2$ and λ, ref. eqn (7.21). Pereira, Anderson and Bennett [8] used a spacing equal to 0.75λ in a 55×55-element planar array to produce a plane-wave region of size $20\lambda \times 20\lambda$ at a 30λ measurement distance. With $\varepsilon = 0.04$, $N = 9$ (ref. eqn (7.36)) and plane-wave propagation orthogonal to the array plane, a maximum amplitude ripple equal to 0.02 dB and a maximum phase ripple equal to 0.11° was obtained without use of the iterative technique for reduction of such ripples. The radiation pattern did not contain grating lobes.

The directional properties of the array were exploited to reduce the anechoic requirements to the environment. Note that this is possible only if a scanning probe or a complete array is applied, ref. Section 7.3.4.

7.4.2.7 Use in connection with spherical scanning: According to the discussion in connection with formula (7.11), a source distribution for generation of a plane wave is the same as a weighting function for obtaining a component of the far field from a test antenna based on measurement of its near field. The surface source distributions treated in this section may therefore be designated as weighting functions.

We shall consider briefly the use of a weighting function for a spherical surface in connection with a spherical scan where the measurements are taken equidistantly in the two spherical polar coordinates θ and ϕ for two orientations χ of a probe (see Fig. 7.12). For a measurement of this kind, a single weighting

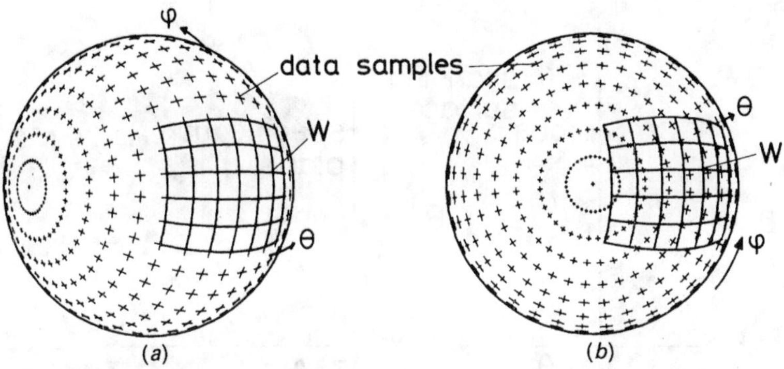

Fig. 7.12 *Application of a weighting function $W(\chi, \theta, \phi)$ in a set of spherical scan data*
(*a*) On principal elevation axis
(*b*) Shifted off axis

function $W(\chi, \theta, \phi)$ defined on the spherical measurement surface can be applied for prediction of the far field in any direction by proper positioning of the weighting function relative to the measurement grid. The main difficulty in this process arises from the fact that for practical reasons the weighting function will be calculated and stored only for a discrete set of points, e.g. for some part of a spherical grid.

Therefore, to obtain measurement data in the weighting function grid, some interpolation procedure must be applied to the measured data, ref. Fig. 7.12. Schoessow [5] investigated a set of interpolation methods and chose a Lagrange four-point approach. The interpolation of measured data is based on the fact that the test antenna field will contain spherical harmonics only up to some upper degree and order, ref. Section 2.2.3.

Both the weighting functions and the necessary set of interpolation parameters can be precalculated and applied in connection with several measurements [3, 5].

7.4.3 Measurement with probe correction

A weighting function taking into account the directional properties of the probe can be derived through application of a reciprocity integral in connection with a symmetrical solution to Maxwell's equations.

In the derivation we consider a physical realization of the entire infinite probe array over the source surface S. The array elements must be arranged in a way that makes mutual interaction between them negligible and at the same time enables the specified field to be established over the specification plane Σ by proper weighting of the array elements.

Considering a set-up of this kind, producing a plane-wave field $(\vec{E}, \vec{H})$ of sufficient quality, we wish to determine the excitation weight W at a specific probe position as shown in Fig. 7.13.

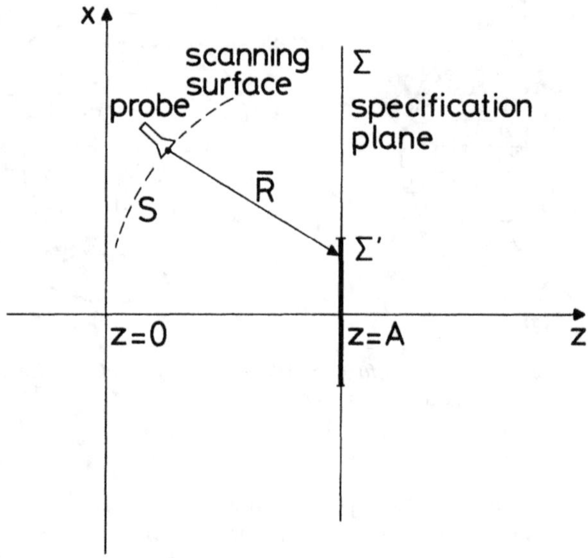

Fig. 7.13 *Geometry used in definition of probe correction*

To this end, we denote by $(\vec{E}_p, \vec{H}_p)$ the field radiated by the selected probe with an impressed unit excitation and with the remaining probes connected to matched loads. Assuming the specification plane Σ at $z = A$ to be entirely in the far-field region relative to the probe, and neglecting scattering from the other probes, means that the probe field can be expressed as

$$\vec{E}_p = \vec{E}_o(\hat{R}) \frac{e^{ikR}}{kR} \quad \text{on } \Sigma \tag{7.37}$$

$$\vec{H}_p = \eta \hat{R} \times \vec{E}_p \quad \text{on } \Sigma \tag{7.38}$$

where $\vec{R}$ is a vector from the probe position to the field point and $R \equiv |\vec{R}|$, $\hat{R} \equiv \vec{R}/|\vec{R}|$. With an excitation W of the probe, the probe field becomes $(W\vec{E}_p, W\vec{H}_p)$.

Further, let $(\vec{E}', \vec{H}')$ be the *symmetrical solution* to Maxwell's equations relative to the plane wave field $(\vec{E}, \vec{H})$. The symmetrical solution is obtained by complex conjugation of all fields and currents, and by changing the sign on the magnetic field and the electric currents,

$$(\vec{E}', \vec{H}') \equiv (\vec{E}^*, -\vec{H}^*). \tag{7.39}$$

For the probe field the symmetrical solution is $(W^*\vec{E}_p^*, -W^*\vec{H}_p^*)$. This applies as well in the waveguide used for exciting the probe which shows that the weight on the inward propagating waveguide field $(\vec{E}_p^*, -\vec{H}_p^*)$ is W^*.

A formula for the complex weight W^* on the received waveguide field can now be obtained by application of the reciprocity theorem (A1.75) to the fields $(\vec{E}', \vec{H}')$ and $(\vec{E}_p, \vec{H}_p)$. Two assumptions must be made: First, the part of the field $(\vec{E}, \vec{H})$ from the probe array which does not pass through the specification plane Σ must be negligible. Secondly, the field on Σ radiated by the particular probe must be unaffected by the scattering from the remaining probes. Under these conditions application of the reciprocity theorem leads to the following expression

$$W^* = C^* \iint_\Sigma (\vec{E}' \times \vec{H}_p - \vec{E}_p \times \vec{H}') \cdot \hat{z}\, \mathrm{d}x'\, \mathrm{d}y'. \tag{7.40}$$

Here, C is a constant for the particular probe, i.e. independent of probe position, ref. e.g. [9] where a similar transmission formula is derived for the case of only one receiving probe. Application of (7.37–7.39) in (7.40) leads to the expression

$$W = C \iint_\Sigma \frac{\mathrm{e}^{-ikR}}{kR} \left\{ \vec{E}_o^* \times \vec{H} + \eta \vec{E} \times (\hat{R} \times \vec{E}_o^*) \right\} \cdot \hat{z}\, \mathrm{d}x'\, \mathrm{d}y'. \tag{7.41}$$

In (7.41), we may use a vector identity in connection with the last term and neglect the second-order quantity $(\hat{z} \cdot \vec{E}_o^*)(\hat{R} \cdot \vec{E})$. We then obtain

$$W \simeq C \iint_\Sigma \frac{\mathrm{e}^{-ikR}}{kR} \left\{ \hat{z} \cdot (\vec{E}_o^* \times \vec{H}) + \eta (\hat{z} \cdot \hat{R})(\vec{E} \cdot \vec{E}_o^*) \right\} \mathrm{d}x'\, \mathrm{d}y'. \tag{7.42}$$

Further simplification is obtained if we assume that $\vec{E}$ and $\vec{H}$ are related as in an ideal plane-wave

$$\vec{H} = \eta \hat{z} \times \vec{E}. \tag{7.43}$$

Inserting (7.43) into (7.42) and neglecting another second-order term $(\hat{z} \cdot \vec{E})(\hat{z} \cdot \vec{E}_o^*)$, leads to

$$W \simeq C\eta \iint_\Sigma \frac{\mathrm{e}^{-ikR}}{kR} (1 + \hat{z} \cdot \hat{R}) \vec{E}_o^* \cdot \vec{E}\, \mathrm{d}x'\, \mathrm{d}y'. \tag{7.44}$$

The relation (7.43) can be used as an additional restriction in the initial specification of the field $(\vec{E}, \vec{H})$ in the plane Σ. However, in possible iterative procedures for suppression of ripples, the relation (7.43) is unlikely to hold for the error field.

Numerical results for a slightly different probe-correction formulation are given in [4] and [5].

7.5 Figures of merit for plane-wave deviation

7.5.1 Introduction

The idea of using the quasi-plane wave generated or synthesized in a given test range, for validation of the range and for comparison with other ranges, was introduced in Section 7.1. Such procedures require the use of some suitable figures of merit to characterize the deviation between a quasi-plane wave and the corresponding ideal plane wave in the test zone.

Traditionally, the imperfect plane wave is evaluated in terms of the amplitude and phase deviation, as described, for example, in [21]. One main reason for this is that these quantities are obtained relatively easily by measurements in the test zone when the quasi-plane wave is physically established. Another reason is the simple interpretation in connection with the conventional far-field ranges. Since here the test zone is in the far-field region relative to the probe, the quasi-plane wave has the form of a spherical wave with an amplitude distribution described through the pattern of the probe. A third reason is that there exists some experimental and theoretical knowledge about the influence on the measured patterns of the amplitude and phase deviations [21].

The amplitude and phase deviations used in the evaluation of test zones are, however, not convenient for use as figures of merit for design of test ranges where some objective function expressing the deviation over the entire test zone is minimized. In a procedure of this kind, it is not important whether or not a quantity describing the plane-wave deviation is difficult to measure since it must be obtained through the application of a mathematical model of the test range. Instead, a figure of merit to be used in connection with a numerical design procedure should have the following properties:

1. Amplitude, phase and polarization errors must be taken into account.
2. The computation of the deviation figure must be simple and efficient.
3. The amount of data necessary to express the plane-wave deviation over the entire test zone should be as small as possible.

Section 7.5.2 introduces some general figures of merit for plane-wave deviation that can be applied independent of the test zone geometry. The *relative plane-wave deviation*, T, defined on the basis of an *error field*, $(\vec{E}_{err}, \vec{H}_{err})$, expresses the total relative deviation from the ideal plane-wave at a point, including amplitude, phase and polarization errors. For use in optimization of plane-wave zones, the *combined field error vector*, $\vec{F}_{err}$, is introduced.

In Section 7.5.3, some figures of merit emphasizing the spherical extent of a plane-wave zone are presented. The *spherical profile* of the relative plane-wave deviation is expressed directly in terms of field quantities whereas the *error spectrum*, $\{u_n\}$, and the *test zone error norm*, U_N, are based on spherical wave expansions.

Evaluation of test ranges based on the figures of merit for plane-wave deviation presented in Sections 7.5.2 and 7.5.3 has the advantage of being independent of a specific test antenna. However, these figures do not predict the measurement errors introduced by the plane-wave deviation. Section 7.5.4 presents two measurement error upper bounds that apply for different classes of test antennas.

7.5.2 *General figures of merit for plane-wave deviation*

Denoting a quasi-plane wave field by $(\vec{E}, \vec{H})$ and the corresponding ideal plane wave by $(\vec{E}_{pl}, \vec{H}_{pl})$, the *error field* $(\vec{E}_{err}, \vec{H}_{err})$ is defined as the difference between these two fields

$$\vec{E}_{err} \equiv \vec{E} - \vec{E}_{pl} \tag{7.45}$$

$$\vec{H}_{err} \equiv \vec{H} - \vec{H}_{pl}. \tag{7.46}$$

Thus, $(\vec{E}_{err}, \vec{H}_{err})$ is a solution to Maxwell's equations with sources where $(\vec{E}, \vec{H})$ has sources.

We now define the *relative plane-wave deviation*, $T(\vec{r})$, through the expression

$$T(\vec{r}) \equiv \frac{|\vec{E}_{err}(\vec{r})|^2 + \eta^{-2}|\vec{H}_{err}(\vec{r})|^2}{|\vec{E}_{pl}(\vec{r})|^2 + \eta^{-2}|\vec{H}_{pl}(\vec{r})|^2} \tag{7.47}$$

where $\vec{r}$ is a position vector and η is the intrinsic admittance of the medium. It is apparent that $T(\vec{r})$ is the ratio between the time-average energy densities of the error field and the ideal plane wave. Clearly, this figure of merit takes into account amplitude, phase and polarization errors.

In relation to point 3 in Section 7.5.1, the function $T(\vec{r})$ has the following important property: if $T(\vec{r}) = 0$ over a closed surface Σ in a source-free region (ref. Fig. 7.14) then $T(\vec{r}) = 0$ in the entire volume V bounded by Σ. To prove this, begin first by noting that $T(\vec{r}) = 0$ if and only if the error field $(\vec{E}_{err}, \vec{H}_{err})$ equals zero. Thus, if $T(\vec{r}) = 0$ on the closed surface Σ, then the tangential components of $\vec{E}_{err}$ and $\vec{H}_{err}$ vanish over Σ which is sufficient to guarantee that $\vec{E}_{err} = \vec{0}$ and $\vec{H}_{err} = \vec{0}$ in V. This is due to the fact that $(\vec{E}_{err}, \vec{H}_{err})$ is a source-free solution to Maxwell's equations in V; see, for example, [22].

On the basis of this property of T, we state the hypothesis that if T is small on a closed surface Σ in the source-free region, then it will be small in the volume V bounded by Σ. Provided this hypothesis holds true, then in order to suppress the plane-wave deviation in a test zone V, we need only suppress the relative plane-

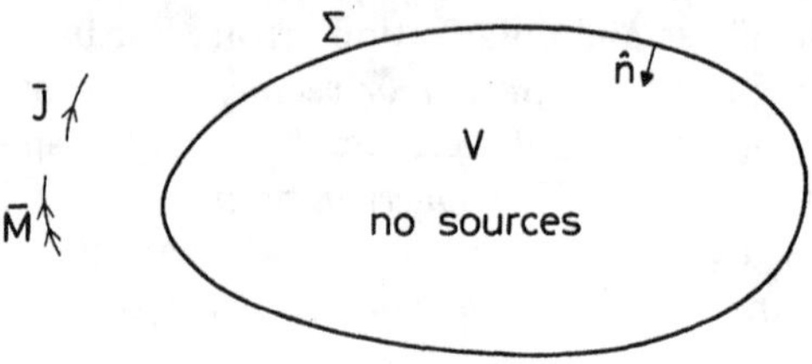

Fig. 7.14 *Test zone V with boundary Σ. Arbitrary test zone geometry*

wave deviation T on the boundary Σ of V. This property of T is very useful in connection with a numerical design procedure.

Since the property is hypothetical, a plane-wave field obtained by minimization of T over Σ should, however, be checked, e.g. by computation of $T(\vec{r})$ over parts of V. Actual use of $T(\vec{r})$ in a numerical design procedure has confirmed the feasibility of considering $T(\vec{r})$ on the boundary of the plane-wave zone only [9].

From the definition (7.47) it is evident that direct minimization of $T(\vec{r})$ involves minimization of the six complex components in $\vec{E}_{err}(\vec{r})$ and $\vec{H}_{err}(\vec{r})$. However, according to the discussion above it is sufficient to minimize the four components of $\vec{E}_{err}$ and $\vec{H}_{err}$ that are tangential to Σ. Hill [1] showed that it is even sufficient to minimize the two components in the *combined field error vector*, $\vec{F}_{err}$:

$$\vec{F}_{err} \equiv \vec{E}_{err,tan} - \eta^{-1}\hat{n} \times \vec{H}_{err} \tag{7.48}$$

which is tangential to Σ, $\vec{E}_{err,tan}$ being the component of $\vec{E}_{err}$ tangential to Σ, and $\hat{n}$ being the inward unit normal to Σ, ref. Fig. 7.14. He proved that if $\vec{F}_{err} = \vec{0}$ over Σ, then $\vec{E}_{err} = \vec{0}$ and $\vec{H}_{err} = \vec{0}$ inside Σ. In general, any source-free field $(\vec{E}, \vec{H})$ in a finite region is uniquely determined by the combined field vector $\vec{F} \equiv \vec{E}_{tan} - \eta^{-1}\hat{n} \times \vec{H}$ over the boundary of the region.

The small number of field components involved in $\vec{F}_{err}$ as compared to $(\vec{E}_{err,tan}, \vec{H}_{err,tan})$ is important in test range design problems where each field component is a linear function of the design parameters. For such problems, an optimal parameter set can be obtained by the approximate solution of an overdetermined set of complex linear equations, the number of equations being equal to the number of sampling points on Σ times the number of field components, ref. Section 7.6.2.

However, as a figure of merit to be used for mapping of the plane-wave deviation over regions in space, the relative plane-wave deviation T must be preferred over the combined field error vector $\vec{F}_{err}$ or the magnitude of this vector. The quantity T simply represents the magnitude of the error field in each point. A map of T is therefore very simple to interpret. As opposed to this, the vector $\vec{F}_{err}$ only represents the error field components tangential to a prescribed surface.

7.5.3 Spherical test zone geometry

Usually, a test range will be used for measurements on a series of test antennas exhibiting different shapes. The single test antenna will be rotated in the synthesized plane-wave field in order to obtain full angular information about its pattern. As a consequence, the spherical extent of the plane-wave zone is an important characteristic of the test range. A long and narrow test zone geometry cannot normally be exploited.

Consideration of the spherical extent of a plane-wave zone requires the definition of a centre. In the following, we shall describe the plane-wave field in the test zone coordinate system with its origin in the centre of the test zone, ref. Fig. 7.15. Let $\vec{r}$ denote a position vector to a point (r, θ, ϕ) in this coordinate system.

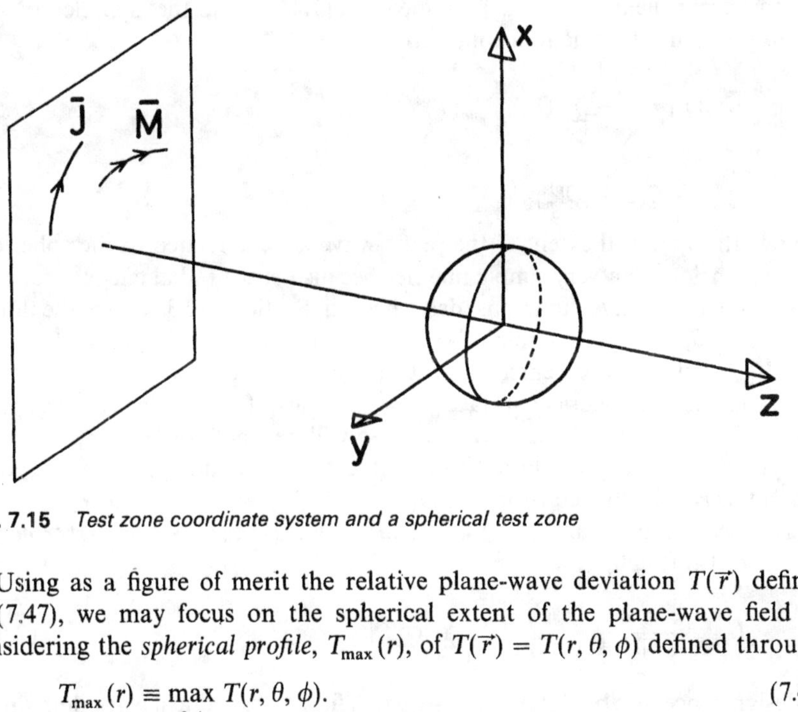

Fig. 7.15 *Test zone coordinate system and a spherical test zone*

Using as a figure of merit the relative plane-wave deviation $T(\vec{r})$ defined in (7.47), we may focus on the spherical extent of the plane-wave field by considering the *spherical profile*, $T_{max}(r)$, of $T(\vec{r}) = T(r, \theta, \phi)$ defined through

$$T_{max}(r) \equiv \max_{\theta,\phi} T(r, \theta, \phi). \tag{7.49}$$

From the preceding section, we know that if $T_{max}(r_0) = 0$ for some positive r_0, then $T_{max}(r) = 0$ for $r < r_0$.

A complete procedure for the design of a test range producing a plane-wave zone with radius r_0 will typically comprise (1) a design step and (2) a verification step:

1. Minimize $T_{max}(r_0)$ by selecting an optimal set of design parameters.
2. Verify the design by computation of $T_{max}(r)$ or $T(\vec{r})$.

Instead of minimizing $T_{max}(r_0)$, we could minimize the combined field error vector $\vec{F}_{err}(\vec{r})$ defined in (7.48) over the sphere $|\vec{r}| = r_0$.

Other figures of merit emphasizing the spherical extent of the plane-wave zone can be achieved by adapting a description of the plane-wave fields in terms of their spherical wave expansions.

The ideal plane wave $(\vec{E}_{pl}, \vec{H}_{pl})$ and the quasi-plane wave $(\vec{E}, \vec{H})$ are expanded in spherical waves of the standing-wave type, e.g. for the electric fields

$$\vec{E}_{pl}(\vec{r}) = \frac{k}{\sqrt{\eta}} \sum_{smn} Q_{smn}^{pl} \vec{F}_{smn}^{(1)}(\vec{r}) \tag{7.50}$$

$$\vec{E}(\vec{r}) = \frac{k}{\sqrt{\eta}} \sum_{smn} Q_{smn} \vec{F}_{smn}^{(1)}(\vec{r}). \tag{7.51}$$

For the error field $(\vec{E}_{err}, \vec{H}_{err})$ defined in (7.45–7.46), the spherical wave expansion then takes the following form

$$\vec{E}_{err}(\vec{r}) = \frac{k}{\sqrt{\eta}} \sum_{smn} Q_{smn}^{err} \vec{F}_{smn}^{(1)}(\vec{r}) \tag{7.52}$$

with

$$Q_{smn}^{err} \equiv Q_{smn} - Q_{smn}^{pl}. \tag{7.53}$$

Now, the spherical extent of the plane-wave zone is related to the spherical wave expansions in a very simple manner because of the radial cut-off property of the spherical wave functions described in Section 2.2.3. Since the function $\vec{F}_{smn}^{(1)}(r, \theta, \phi)$ becomes small for $r < n/k$, the fields $(\vec{E}, \vec{H})$, $(\vec{E}_{pl}, \vec{H}_{pl})$ and $(\vec{E}_{err}, \vec{H}_{err})$ will be represented almost entirely through wave functions with $n \leqslant kr_0$ inside the spherical surface $|\vec{r}| = r_0$. Thus, if the coefficients Q_{smn}^{err} are small for $n \leqslant kr_0$, the plane-wave deviation will be small for $|\vec{r}| \leqslant r_0$.

A possible design procedure with the object of obtaining a plane-wave zone with radius r_0 therefore consists in minimizing $|Q_{smn}^{err}|$ for $n \leqslant N$ with $N \simeq kr_0$. If, as the objective function in the minimization, we use the *test zone error norm*, U_N, defined through

$$U_N \equiv \frac{1}{\sqrt{4\pi}\,|\vec{E}_{pl}|} \frac{k}{\sqrt{\eta}} \sqrt{\sum_{n=1}^{N} \sum_{sm} |Q_{smn}^{err}|^2} \tag{7.54}$$

the design procedure becomes a least-squares fit in the spherical wave domain of the field from the test range to a predefined ideal plane wave $(\vec{E}_{pl}, \vec{H}_{pl})$. The normalization of U_N has been selected such that U_N equals an upper bound on the relative measurement error; this will be discussed later. The normalization also makes U_N dimensionless and independent of the amplitude $|\vec{E}_{pl}|$ of the selected ideal plane-wave field. U_N is an integral figure of merit for the relative plane-wave deviation for $|\vec{r}| \lesssim N/k$.

A more elaborate figure of merit for the relative plane-wave deviation, containing information about the spherical extent of the plane-wave zone, is obtained by splitting U_N into its contributions u_n from different degrees n of spherical waves:

$$u_n \equiv \frac{1}{\sqrt{4\pi}\,|\vec{E}_{pl}|}\frac{k}{\sqrt{\eta}}\sqrt{\sum_{sm}|Q^{err}_{smn}|^2}. \tag{7.55}$$

From (7.55) and (7.54), it is evident that

$$U_N = \sqrt{\sum_{n=1}^{N} u_n^2}. \tag{7.56}$$

The sequence $\{u_n | n = 1, 2, \ldots\}$ is denoted the *error spectrum* of the quasi-plane wave $(\vec{E}, \vec{H})$ with respect to the ideal plane wave $(\vec{E}_{pl}, \vec{H}_{pl})$ in the selected co-ordinate system.

It follows from the preceding discussion that the relative plane-wave deviation is small for $r \leqslant r_0$ if and only if u_n is small for $n \leqslant kr_0$. Then, just like the spherical profile of the relative plane-wave deviation $T_{max}(r)$, the sequence $\{u_n\}$ can be used to check the spherical extent and quality of the plane wave zone. However, $\{u_n\}$ does not directly express the plane-wave deviation. The level of the relative plane-wave deviation can be estimated by comparing the error spectrum $\{u_n\}$ with the error spectrum for a simple known quasi-plane-wave, e.g. the field from a short dipole at the Rayleigh distance $8r_0^2/\lambda$.

Using the figures of merit U_N and $\{u_n\}$, we may define a new procedure for designing a test range producing a plane-wave zone with radius r_0:

1. Minimize U_N with $N \simeq kr_0$ by selecting an optimal set of design parameters.
2. Verify the design by computation of $\{u_n\}$ and/or $T_{max}(r)$.

If a high plane-wave quality is required over the entire test zone, N should be set to approximately $kr_0 + 10$, ref. eqn (2.32).

The description of plane-wave deviation in terms of spherical wave expansions takes a very simple form in connection with the spherical near-field to far-field transformation method of Chapter 4. Following the discussion in Section 7.2, the synthesized and ideal plane-waves have identical plane-wave expansions for $n \leqslant N$. In other words,

$$Q^{err}_{smn} = 0 \qquad \text{for } n \leqslant N \tag{7.57}$$

in the spherical near-field antenna measurement technique implying that

$$u_n = 0 \qquad \text{for } n \leqslant N. \tag{7.58}$$

Therefore, the relative plane-wave deviation T is small for $|\vec{r}| < N/k$ and very small for $|\vec{r}| < (N - 10)/k$, ref. eqn (2.32).

7.5.4 Measurement error upper bounds

It has already been mentioned that the figure of merit U_N defined in (7.54) is closely related to the problem of estimating an upper bound on the measurement error introduced by the plane-wave deviation. To treat this problem in more detail, we introduce a test antenna in the test zone.

The field ($\vec{E}_t$, $\vec{H}_t$) radiated by this antenna is expanded in spherical waves of the outward propagating type in the test zone coordinate system, ref. Fig. 7.15:

$$\vec{E}_t(r, \theta, \phi) = \frac{k}{\sqrt{\eta}} \sum_{smn} Q^t_{smn} \vec{F}^{(3)}_{smn}(r, \theta, \phi). \tag{7.59}$$

Then, according to eqn (2.24), the power P_{rad} radiated by the test antenna can be expressed as

$$P_{\text{rad}} = \frac{1}{2} \sum_{smn} |Q^t_{smn}|^2. \tag{7.60}$$

Since the spherical waves radiate power independently, we may split P_{rad} into the partial power quantities radiated in spherical waves of different degree n:

$$P^{(n)}_{\text{rad}} \equiv \frac{1}{2} \sum_{sm} |Q^t_{smn}|^2 \qquad n = 1, 2, 3, \ldots. \tag{7.61}$$

The set of partial power quantities $\{P^{(n)}_{\text{rad}} | n = 1, 2, \ldots\}$ shall be denoted the *test antenna power spectrum.* It is evident that

$$P_{\text{rad}} = \sum_{n} P^{(n)}_{\text{rad}}. \tag{7.62}$$

An important property of the test antenna power spectrum is that it is invariant under a rotation of the test antenna about the origin of the coordinate system. This can be easily verified from the expansion (A2.2) of a spherical wave in a rotated coordinate system.

To describe the measurement error at a given angular position of the test antenna relative to the plane-wave field, let Δw be the error in the far-field measurement caused by the plane-wave deviation. Further, let w_o be the far field that would be measured by the ideal plane-wave field, if the power P_{rad} from the test antenna was radiated isotropically and with polarization matching to the ideal plane-wave field.

Under the assumption of strict band limitation

$$P^{(n)}_{\text{rad}} = 0 \qquad \text{for } n > N' \tag{7.63}$$

where N' is a given integer, it may be shown [9] that

$$\left| \frac{\Delta w}{w_o} \right| \leqslant U_{N'}. \tag{7.64}$$

The assumption (7.63) will be reasonable for test antennas that are within the spherical surface $|\vec{r}| = N'/k$.

As shown in Chapter 4, spherical near-field scanning is an exact method for test antennas for which the radiated field contains spherical modes of degree only up to the maximum mode degree N processed in the transformation algorithm. This result can also be obtained from the error bound (7.63–7.64) by making use of the fact that the error spectrum $\{u_n\}$ for the synthesized quasi-plane wave field equals zero for $n \leqslant N$. Since u_n equals zero for $n \leqslant N$, we have $U_{N'} = 0$ for $N' \leqslant N$, and the result follows directly from (7.63–7.64).

If, instead of (7.63), we assume the power spectrum $\{P_{\mathrm{rad}}^{(n)}\}$ of the test antenna to be known, it can be shown [9] that

$$\left|\frac{\Delta w}{w_o}\right| \leqslant \sum_n u_n \sqrt{\frac{P_{\mathrm{rad}}^{(n)}}{P_{\mathrm{rad}}}}. \tag{7.65}$$

Note that the upper bound on the relative measurement error formulated in (7.65) is invariant under a rotation of the test antenna about the origin of the coordinate system. This is because the power spectrum $\{P_{\mathrm{rad}}^{(n)}\}$ is invariant under such a rotation.

In both (7.64) and (7.65), the equality may be true so that (7.64) and (7.65) are the lowest bounds that can be given under the assumptions upon which they are based. If (7.63) holds true for some N', and the power spectrum $\{P_{\mathrm{rad}}^{(n)}\}$ of the test antenna is known, eqn (7.65) will give the least pessimistic bound since it relies on the most restrictive assumptions.

Figure 7.16 illustrates the significance of the measurement error upper bounds (7.64) and (7.65) in connection with measurement by a short dipole (at the Rayleigh distance) of the far-field radiation pattern from a maximum directivity antenna (see Section 2.4.4).

7.6 Numerical optimization methods for plane-wave synthesis

7.6.1 Introduction

In the present section we shall treat the following problem: which probe excitations should be applied to obtain the best plane-wave field for a given test zone and probe array geometry? The methods to be presented for solution of this problem apply linear least-squares optimization techniques in connection with some of the figures of merit for plane-wave deviation defined in Section 7.5.

The method to be presented in Section 7.6.2 was introduced by Hill [1]. It is based on the application of the combined field vector $\vec{F}$, ref. eqn (7.48), and is applicable with arbitrary test zones and array geometries.

A more specialized method, introduced by Hald [9], is described in Section 7.6.3. The method is based on the test zone error norm U_N, ref. eqn (7.54), which means that the test zone must be spherical. Further, only uniform ring arrays of probes are used thus excluding the possibility of scanning. These restrictions are exploited to achieve a method requiring a minimum of computation.

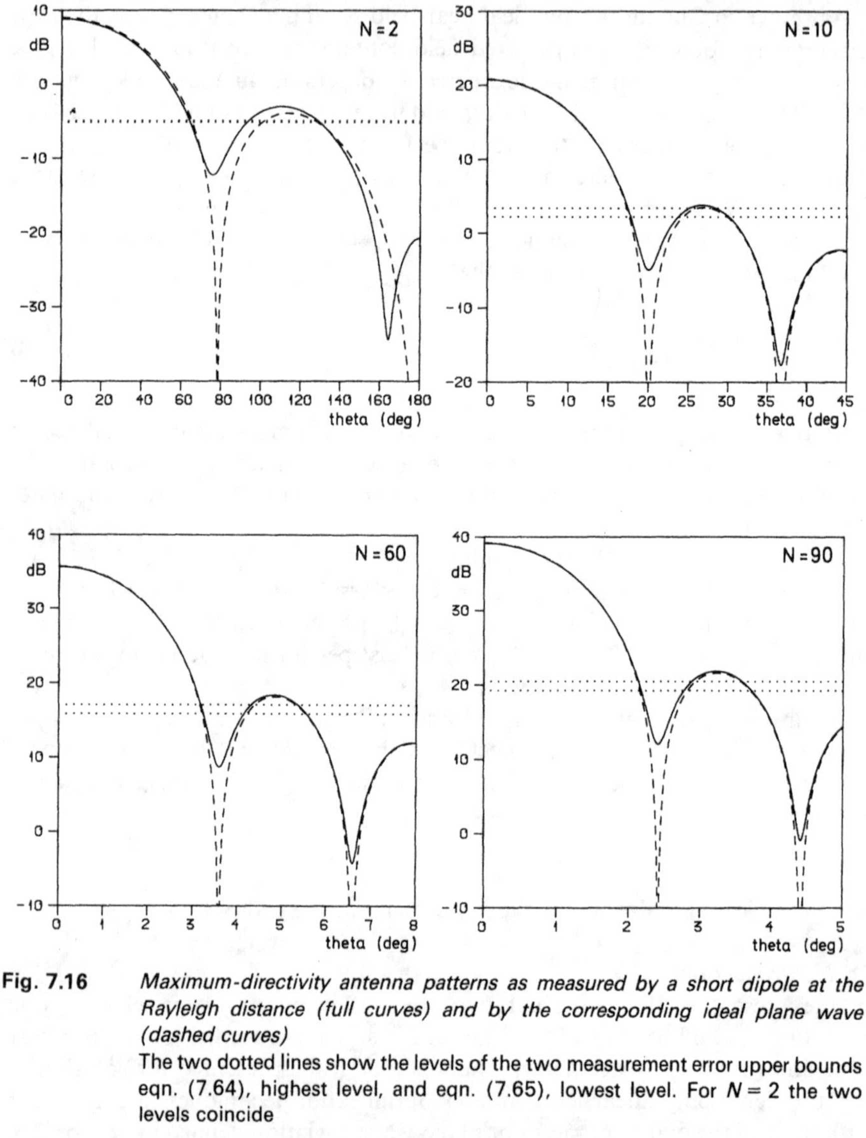

Fig. 7.16 *Maximum-directivity antenna patterns as measured by a short dipole at the Rayleigh distance (full curves) and by the corresponding ideal plane wave (dashed curves)*

The two dotted lines show the levels of the two measurement error upper bounds eqn. (7.64), highest level, and eqn. (7.65), lowest level. For $N = 2$ the two levels coincide

7.6.2 Arbitrary array and test zone geometry

With reference to Fig. 7.17 we assume the quasi-plane-wave field $(\vec{E}, \vec{H})$ to be produced by I probes. Letting the field $(\vec{E}_i, \vec{H}_i)$ be the field from probe i when unit excitation is applied, and letting $\vec{F}_i$ be the corresponding combined field vector on the boundary Σ of the test zone

$$\vec{F}_i \equiv \vec{E}_{i,\tan} - \eta^{-1}\hat{n} \times \vec{H}_i \tag{7.66}$$

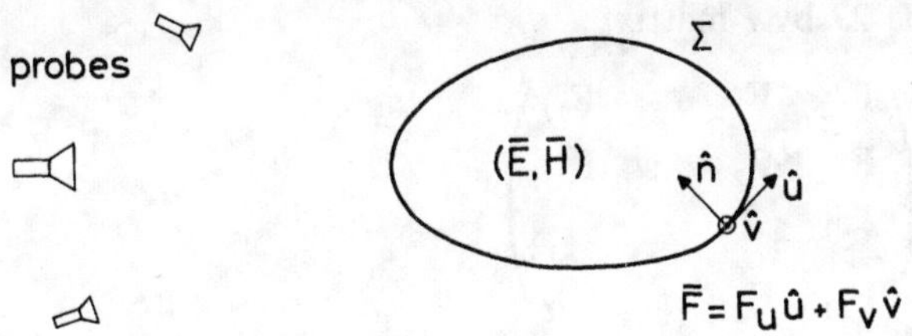

Fig. 7.17 *General geometry for plane-wave synthesis*
A set of probes generates a quasi-plane wave within the closed surface Σ

then the total combined field vector $\vec{F}$ on Σ may be expressed as

$$\vec{F}(\vec{r}) = \sum_{i=1}^{I} c_i \vec{F}_i(\vec{r}). \tag{7.67}$$

Here, $\vec{r}$ is a position vector to a point on Σ, and c_i is the complex excitation of probe number i.

According to the discussion in Section 7.5.2, the resulting field $(\vec{E}, \vec{H})$ will equal the ideal plane wave $(\vec{E}_{pl}, \vec{H}_{pl})$ inside Σ if, and only if, we have

$$\vec{F}(\vec{r}) = \vec{F}_{pl}(\vec{r}) \qquad \text{on } \Sigma \tag{7.68}$$

with

$$\vec{F}_{pl} \equiv \vec{E}_{pl,tan} - \eta^{-1} \hat{n} \times \vec{H}_{pl}. \tag{7.69}$$

Substitution of (7.67) into (7.68) leads to

$$\sum_{i=1}^{I} c_i \vec{F}_i(\vec{r}) = \vec{F}_{pl}(\vec{r}) \qquad \text{on } \Sigma. \tag{7.70}$$

Equation (7.70) can be solved approximately by sampling over a sufficiently dense grid on Σ and subsequently solving the resulting overdetermined system of linear equations by a least-squares procedure.

Let us denote the set of grid points by $\vec{r}_j$, $j = 1, 2, \ldots, J$, and define the column matrices

$$\mathbf{F}_{ij} \equiv \begin{pmatrix} \vec{F}_i(\vec{r}_j) \cdot \hat{u}(\vec{r}_j) \\ \vec{F}_i(\vec{r}_j) \cdot \hat{v}(\vec{r}_j) \end{pmatrix} \tag{7.71}$$

$$\mathbf{F}_{pl,j} \equiv \begin{pmatrix} \vec{F}_{pl}(\vec{r}_j) \cdot \hat{u}(\vec{r}_j) \\ \vec{F}_{pl}(\vec{r}_j) \cdot \hat{v}(\vec{r}_j) \end{pmatrix} \tag{7.72}$$

where the orthogonal unit vectors, $\hat{u}$ and $\hat{v}$, tangential to Σ, are defined in Fig. 7.17. We then obtain the following matrix equation from eqn (7.70)

$$\mathbf{Mc} = \mathbf{P} \tag{7.73}$$

where $\mathbf{M}$ is the $2J$ by I matrix

$$\mathbf{M} \equiv \begin{pmatrix} \mathbf{F}_{11} & \mathbf{F}_{12} & \cdots & \mathbf{F}_{1I} \\ \mathbf{F}_{21} & \mathbf{F}_{22} & \cdots & \mathbf{F}_{2I} \\ \vdots & \vdots & & \vdots \\ \mathbf{F}_{J1} & \mathbf{F}_{J2} & \cdots & \mathbf{F}_{JI} \end{pmatrix} \tag{7.74}$$

$\mathbf{c}$ is the column matrix of complex excitation coefficients

$$\mathbf{c} \equiv \begin{pmatrix} c_1 \\ c_2 \\ \vdots \\ c_I \end{pmatrix} \tag{7.75}$$

and $\mathbf{P}$ contains the combined field components of the ideal plane wave at the grid points:

$$\mathbf{P} \equiv \begin{pmatrix} \mathbf{F}_{\mathrm{pl},1} \\ \mathbf{F}_{\mathrm{pl},2} \\ \vdots \\ \mathbf{F}_{\mathrm{pl},J} \end{pmatrix}. \tag{7.76}$$

Normally, $2J$ is greater than I and eqn (7.73) can only be solved approximately. The best two-norm solution is obtained by solving the linear least-squares problem

$$\underset{\mathbf{c}}{\text{minimize}} \, \|\mathbf{Mc} - \mathbf{P}\|_2 \tag{7.77}$$

where $\|\mathbf{z}\|_2 \equiv \sqrt{\sum_i z_i^* z_i}$, $\mathbf{z}$ being a complex column matrix with elements z_i. The solution to (7.77) is [23]

$$\mathbf{c} = (\mathbf{M}^+\mathbf{M})^{-1}\mathbf{M}^+\mathbf{P} \tag{7.78}$$

where $\mathbf{M}^+$ designates the Hermitian conjugate of $\mathbf{M}$.

However, the solution (7.78) is often unsatisfactory because it produces for the probe array a very low directivity towards the test zone. As an example, two or more probes producing almost the same field in the test zone will cause the corresponding columns in the matrix $\mathbf{M}$ to be almost identical. This means that $\mathbf{M}^+\mathbf{M}$ will be ill-conditioned. In general, ill-conditioning of $\mathbf{M}^+\mathbf{M}$ is very likely to cause $\|\mathbf{c}\|_2$ to be large, $\mathbf{c}$ being obtained from eqn (7.78). A large value of $\|\mathbf{c}\|_2$ means a high value of the power radiated by the probe array to produce the prescribed amplitude of the plane-wave field and therefore, a low near-field directivity.

This problem can be circumvented by the introduction of a constraint on

$\|\mathbf{c}\|_2$ in the least-squares problem (7.77). In other words, $\mathbf{c}$ should be obtained as the solution to the problem

$$\underset{\mathbf{c}}{\text{Minimize}}\ \|\mathbf{Mc} - \mathbf{P}\|_2 \qquad \text{subject to } \|\mathbf{c}\|_2 \leqslant C \tag{7.79}$$

where C is a positive real number. Methods for solving this type of problem will be discussed in Section 7.6.3.5.

The main advantage of the method outlined above is its generality: array geometry and test zone geometry can be arbitrarily chosen and different sets of weights can be computed corresponding to different directions of plane-wave propagation relative to the array.

The major problems associated with the method are the high dimension of the constrained least-squares problem (7.79) for large arrays and the problem of proper sampling of the combined field $\vec{F}$ over the boundary Σ of the test volume. The general sampling criterion requires a distance between the sampling points not exceeding half a wavelength.

For large arrays and large test zones, the method will in general be computationally too expensive unless inherent characteristics, such as symmetries, can be exploited to significantly reduce the number of independent probe weights and/or the number of samples of the combined field vector. The problem of identifying and exploiting these inherent characteristics is then an important task.

7.6.3 Ring sources and spherical test zones

7.6.3.1 Introduction: The computational problem of identifying a set of optimal probe weights through a least-squares fit of the probe fields to the ideal plane-wave field can be largely reduced if the test zone and the probe array have a common axis of rotational symmetry.

In the present section, we shall assume a spherical test zone and a probe array which exhibits rotational symmetry in a certain sense about the z-axis of the test-zone coordinate system, ref. Fig. 7.18. More precisely, the probe array is assumed to consist of a set of uniform ring arrays with all the probes having parallel or almost parallel polarizations.

With these restrictions on the array structure, the direction of propagation of the ideal plane wave must be in either the positive or the negative z-direction.

We shall consider plane-wave propagation in the positive z-direction only; see Fig. 7.18. The array and test zone geometry realized in this way is perfectly suited for a description of the fields in terms of spherical wave functions. The ideal plane wave propagating in the z-direction consists of spherical modes of order $m = \pm 1$ only, ref. Section A1.6. If the probe pattern contains only azimuthal modes $\mu = \pm 1$, and if the probe spacing on each ring is sufficiently small, the test zone field from each probe ring can be shown to consist entirely of spherical modes of order $m = \pm 1$ in the test zone coordinate system, ref. Section 7.6.3.3. Thus, within the test zone with radius r_0, each one of the independent

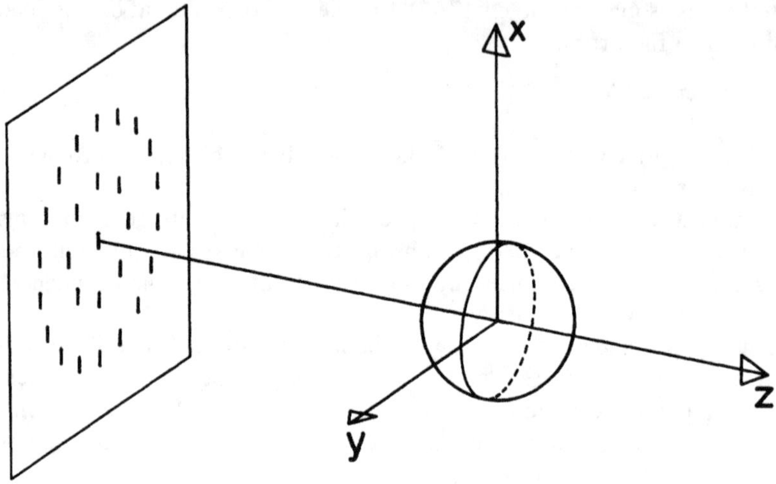

Fig. 7.18 *Ring sources and a spherical test zone*

fields will be completely represented by spherical mode coefficients Q_{smn} with $m = \pm 1$ and $n \leqslant N$, $N \sim kr_0$, because of the radial cut-off properties of the spherical wave functions. The presence of planar symmetries can be easily exploited in the spherical mode representation as described in Section A1.5.2.

The number of independent excitation coefficients to be determined has been reduced to equal the number of probe rings by only considering uniform ring arrays. In reference [9] it was shown for some small ring arrays of short dipoles that virtually no improvement in the plane-wave quality can be gained by allowing different excitations of the elements on each ring or by the use of two orthogonally polarized probes in each probe position.

The main drawback associated with the rotationally symmetric structure is its inability to scan. However, in connection with a hardware implementation or when a set of weights for a spherical array geometry is to be applied in connection with a spherical scan, this will usually be of minor importance. In connection with a spherical scan, a single set of weights applies for prediction in all directions of the far field, ref. Section 7.4.2.7.

7.6.3.2 Probe rings: As described previously, the entire probe array for creation of the plane-wave field is assumed to consist of a set of uniform ring arrays having the z-axis as their common axis. Furthermore, we shall assume the entire composite array to have either a planar or a spherical geometry. In the planar case (ref. Fig. 7.18) each one of the uniform ring arrays is designated as a *planar probe ring*, while in the spherical case (ref. Fig. 7.19), the term *spherical probe ring* is used.

The problem to be treated in Sections 7.6.3.3 and 7.6.3.4 is that of describing the test zone fields from spherical and planar probe rings in terms of expansions in spherical wave functions.

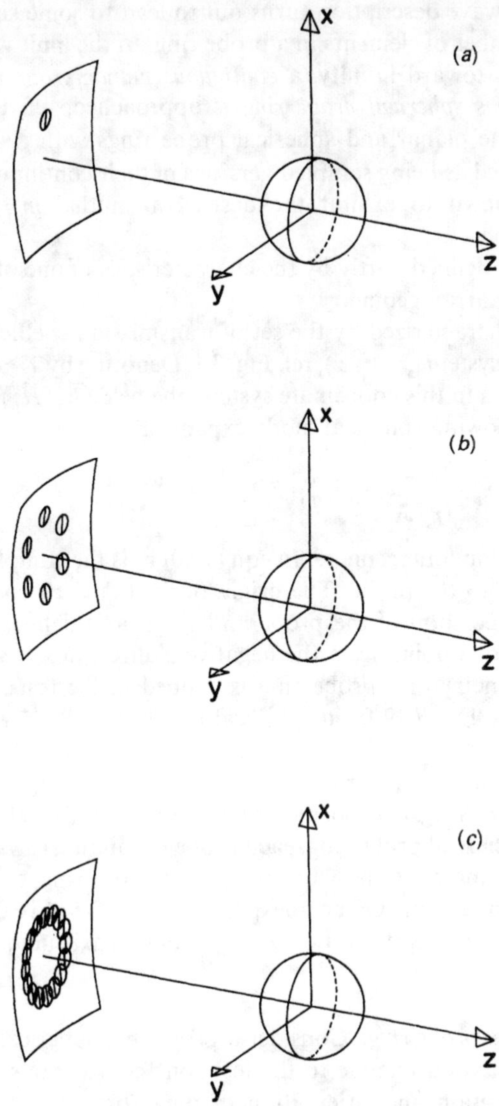

Fig. 7.19 *Definition of a discrete and a continuous (x-polarized) spherical probe ring. Compare also to Fig. 7.20*
(*a*) Single probe at $\theta = \theta_r$, $\phi = 0$ on a spherical measurement surface $r = A$. The probe is rotated to $\chi = \pi$
(*b*) Six-element spherical probe ring at $\theta = \theta_r$. The elements are identical, equally excited and positioned equidistantly along the circle $\theta = \theta_r$ on the measurement sphere. Probe No. *l* is rotated to $\chi_l = \phi_l + \pi$, ϕ_l defining the probe position on the ring
(*c*) Illustration of a continuous spherical probe ring at $\theta = \theta_r$: an infinite number of equally excited probes positioned equidistantly along the circle $\theta = \theta_r$ on the measurement sphere, each rotated to $\chi_l = \phi_l + \pi$

The spherical wave description turns out to lead to some simple criteria for the necessary number of elements in a probe ring. In the limit when the number of elements tends toward infinity, a *continuous planar probe ring* and respectively, a *continuous spherical probe ring* is approached, ref. Fig. 7.19 for the spherical case. The planar and spherical probe rings with a finite number of elements are treated as being sampled versions of their continuous counterparts which will be shown to exhibit the desired azimuthal $m = \pm 1$ rotational symmetry.

A probe ring is defined partly by the characteristics of one of the probes and partly by the ring array geometry.

The probe is characterized by the set of transmission coefficients $T^p_{\sigma\mu\nu}$ in the probe coordinate system (x', y', z'), ref. Fig. 3.1. Denoting by $\vec{r}\,'$ a position vector to a point (r', θ', ϕ') in this coordinate system, the field $(\vec{E}_p, \vec{H}_p)$ radiated by the probe has the following spherical mode expansion

$$\vec{E}_p(\vec{r}\,') = v_p \frac{k}{\sqrt{\eta}} \sum_{\sigma\mu\nu} T^p_{\sigma\mu\nu} \vec{F}^{(3)}_{\sigma\mu\nu}(\vec{r}\,') \tag{7.80}$$

outside the probe minimum sphere. In eqn (7.80) v_p is the complex amplitude of the signal applied to the probe. The origin of the (x', y', z') coordinate system defines the exact position of the probe and the z'-axis defines the axis of the probe with the probe pointing in the negative z'-direction.

The array geometry of a probe ring is defined in the test zone coordinate system (x, y, z); see Fig. 7.19 for a spherical probe ring. By $\vec{r}$ we will denote a position vector to a point (r, θ, ϕ) in this coordinate system. The position of the ring is defined by the distance r to the origin and by the angle $\theta = \theta_r$ from the positive z-axis to the ring. The position of probe number l on the ring is defined by $\phi = \phi_l$. In a spherical probe ring, each probe is pointing toward the origin of the test zone coordinate system while in a planar probe ring, each probe points in the positive z-direction. Given the position and the axis direction of the particular probe, its angle of rotation $\chi = \chi_l$ about its axis uniquely specifies its position.

7.6.3.3 Spherical probe ring: Consider a probe in a spherical probe ring. Its position and orientation relative to the test zone coordinate system is defined exactly as the position and orientation of the probe in spherical scanning. Following the derivations in Sections 3.2.2 and 3.2.3, the probe coordinate system can be obtained from the test zone coordinate system by rotating it through the Euler angles $(\chi_l, \theta_r, \phi_l)$ followed by a translation A in the positive z'-direction. Formula (3.12) expresses the probe field $(\vec{E}_p, \vec{H}_p)$ in the test zone coordinate system in terms of the transmission coefficients $T^p_{\sigma\mu\nu}$

$$\vec{E}_p(\vec{r}) = v_p \frac{k}{\sqrt{\eta}} \sum_{\substack{\sigma\mu\nu \\ smn}} T^p_{\sigma\mu\nu} C^{\sigma\nu(3)}_{s\mu n}(-kA)\, e^{-i\mu\chi_l} d^n_{m\mu}(-\theta_r)\, e^{-im\phi_l} \vec{F}^{(1)}_{smn}(\vec{r}). \tag{7.81}$$

This expression is valid for $r < A - r_p$, r_p being the minimum sphere radius of the probe.

To generate a plane-wave field propagating in the positive z-direction, the set of probe rings must be centred about the negative z-axis. In the following we shall consider a spherical probe ring of this kind with θ_r close to π.

Figure 7.20 depicts the probe ring as seen from the origin of the test zone coordinate system. By choosing

$$\chi_l = \phi_l + \pi \tag{7.82}$$

all the probes on the ring will have almost the same polarization direction. Further, the x'-axis direction will be almost the same as the x-axis direction, while the y'- and y-axes will have opposite directions.

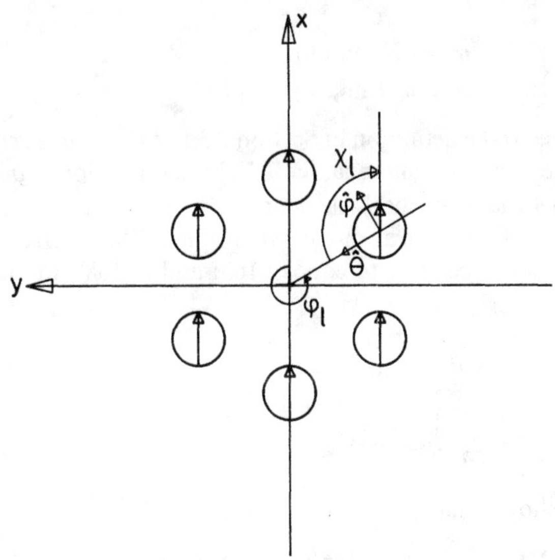

Fig. 7.20 *Spherical probe ring as seen from the origin of the test zone coordinate system. Each probe is rotated to $\chi_l = \phi_l + \pi$, ϕ_l describing the probe position on the ring*

We shall assume the probe ring to consist of L equidistantly spaced probes at the positions $\phi = \phi_l$,

$$\phi_l \equiv (l-1)\frac{2\pi}{L}, \qquad l = 1, 2, \ldots, L \tag{7.83}$$

on the ring, each with excitation

$$v_p = \frac{1}{L} v_r \tag{7.84}$$

where v_r is the complex excitation coefficient of the ring operated as a single antenna. Inserting (7.82–7.84) into eqn (7.81) and summing the contributions from the single probes, leads to the following expression for the field $(\vec{E}_r^L, \vec{H}_r^L)$ from the L-element spherical probe ring

$$\vec{E}_r^L(\vec{r}) = v_r \frac{k}{\sqrt{\eta}} \sum_{\substack{\sigma\mu\nu \\ smn}} \left\{ \frac{1}{L} \sum_{l=1}^{L} \mathrm{e}^{-im\phi_l} \mathrm{e}^{-i\mu(\phi_l+\pi)} \right\}$$

$$d_{m\mu}^n(-\theta_r) C_{s\mu n}^{\sigma\nu(3)}(-kA) T_{\sigma\mu\nu}^p \vec{F}_{smn}^{(1)}(\vec{r}). \tag{7.85}$$

Here, the summation in the curly bracket may be reduced by application of the orthogonality relation (A4.3). The result is

$$\vec{E}_r^L(\vec{r}) = v_r \frac{k}{\sqrt{\eta}} \sum_{\substack{\sigma\mu\nu \\ smn}} (-1)^\mu \delta_{m,-\mu}^L d_{m\mu}^n(-\theta_r) C_{s\mu n}^{\sigma\nu(3)}(-kA) T_{\sigma\mu\nu}^p \vec{F}_{smn}^{(1)}(\vec{r}) \tag{7.86}$$

where

$$\delta_{mm'}^L \equiv \begin{cases} 1 & m = m' \ (\mathrm{mod}\ L) \\ 0 & m \neq m' \ (\mathrm{mod}\ L). \end{cases} \tag{7.87}$$

With reference to the definition in Section 2.3.5 of the *source scattering matrix* for a region interior to an antenna, let $T_{smn}^{\prime rL}$ be the interior region $(r < A - r_p)$ transmission coefficients of the L-element probe ring. Furthermore, let Q_{smn}^{rL} be the coefficients in the spherical mode expansion of the interior field $(\vec{E}_r^L, \vec{H}_r^L)$ from the ring. Then according to eqns (2.163) and (2.168), we have

$$Q_{smn}^{rL} = v_r T_{smn}^{\prime rL} \tag{7.88}$$

which in connection with the mode expansion formula

$$\vec{E}_r^L(\vec{r}) = \frac{k}{\sqrt{\eta}} \sum_{smn} Q_{smn}^{rL} \vec{F}_{smn}^{(1)}(\vec{r}) \tag{7.89}$$

and eqn (7.86) shows that

$$T_{smn}^{\prime rL} = \sum_{\sigma\mu\nu} (-1)^\mu \delta_{m,-\mu}^L d_{m\mu}^n(-\theta_r) C_{s\mu n}^{\sigma\nu(3)}(-kA) T_{\sigma\mu\nu}^p \tag{7.90}$$

$$= \sum_{\mu} (-1)^\mu \delta_{m,-\mu}^L d_{\mu m}^n(\theta_r) \left\{ \sum_{\sigma\nu} C_{\sigma,-\mu,\nu}^{sn(3)}(kA) T_{\sigma\mu\nu}^p \right\}. \tag{7.91}$$

Here, (7.91) is obtained by application of eqns (A2.7) and (A3.15).

In eqn (7.91), the content of the curly bracket has the same form as the expression (3.26) for the probe response constants $P_{s\mu n}^p$. Assuming a reciprocal probe, then by virtue of eqns (3.26) and (2.107),

$$P_{s\mu n}^p(kA) = \tfrac{1}{2}(-1)^\mu \sum_{\sigma\nu} C_{\sigma\mu\nu}^{sn(3)}(kA) T_{\sigma,-\mu,\nu}^p \tag{7.92}$$

which allows eqn (7.91) to be rewritten in the following form

$$T_{smn}^{\prime rL} = 2 \sum_{\mu} \delta_{m,-\mu}^L d_{\mu m}^n(\theta_r) P_{s,-\mu,n}^p(kA). \tag{7.93}$$

The coefficients T'^{rL}_{smn} exhibit a characteristic band structure in m in the nm-plane, the distance between the bands being equal to L due to the form of the expression (7.87) for $\delta^L_{mm'}$. For L increasing towards infinity, the side bands disappear and only the main band constituting the coefficients T'^r_{smn} of the continuous spherical probe ring remain:

$$T'^r_{smn} = 2d^n_{-m,m}(\theta_r)P^p_{smn}(kA). \tag{7.94}$$

The field $(\vec{E}_r, \vec{H}_r)$ created by the continuous ring is therefore

$$\vec{E}_r(\vec{r}) = \frac{k}{\sqrt{\eta}} \sum_{smn} Q^r_{smn} \vec{F}^{(1)}_{smn}(\vec{r}) \tag{7.95}$$

where

$$Q^r_{smn} = v_r T'^r_{smn}. \tag{7.96}$$

For increasing measurement distance A, the coefficients T'^r_{smn} with $m \neq \pm 1$ vanish due to the asymptotic behaviour of the translation coefficients $C^{sn(3)}_{\sigma\mu\nu}(kA)$ stated in eqns (A3.22–3.24). Thus, as should be expected, the field $(\vec{E}_r, \vec{H}_r)$ approaches the azimuthal $m = \pm 1$ rotational symmetry which is inherent in any plane wave propagating in the z-direction. Inspection of eqns (7.92) and (7.94) shows that this $m = \pm 1$ rotational symmetry will be present in the field $(\vec{E}_r, \vec{H}_r)$ from the continuous probe ring independent of the measurement distance A, if the probe has the azimuthal $\mu = \pm 1$ symmetry:

$$T^p_{\sigma\mu\nu} = 0 \qquad \text{for } \mu \neq \pm 1. \tag{7.97}$$

In the following, we shall assume a $\mu = \pm 1$ probe. Then according to eqn (7.92), only the probe response constants $P^p_{s\mu n}$ with $\mu = \pm 1$ differ from zero and eqn (7.94) shows that the transmission coefficients T'^r_{smn} of the continuous probe ring differ from zero for $m = \pm 1$ only.

For the transmission coefficients T'^{rL}_{smn} of the L-element probe ring, eqn (7.93) shows the band structure in m which has already been mentioned. Only the coefficients T'^{rL}_{smn} with $m = \pm 1 + jL$, $j = 0, \pm 1, \pm 2, \ldots$, differ from zero. This band structure in the nm-plane is illustrated in Fig. 7.21. If $L \geqslant 3$, the main band at $m = \pm 1$ (corresponding to $j = 0$) will be unaffected by the side bands corresponding to $j = \pm 1, \pm 2, \ldots$. In this case, the main band coefficients of the L-element probe ring are the same as the coefficients T'^r_{smn} of the continuous ring.

Sampling criteria: The band structure of the mode coefficients in the nm-plane can be exploited to derive a sampling criterion stating how large L should be for the L-element probe ring to reproduce the $m = \pm 1$ rotationally symmetric field of the continuous ring in the test zone.

First, in the mode series representation (7.89) of the field $(\vec{E}^L_r, \vec{H}^L_r)$, only the coefficients Q^{rL}_{smn} with $n \lesssim kr_0$ contribute significantly to the field within the test zone because of the radial cut-off property of the spherical wave functions

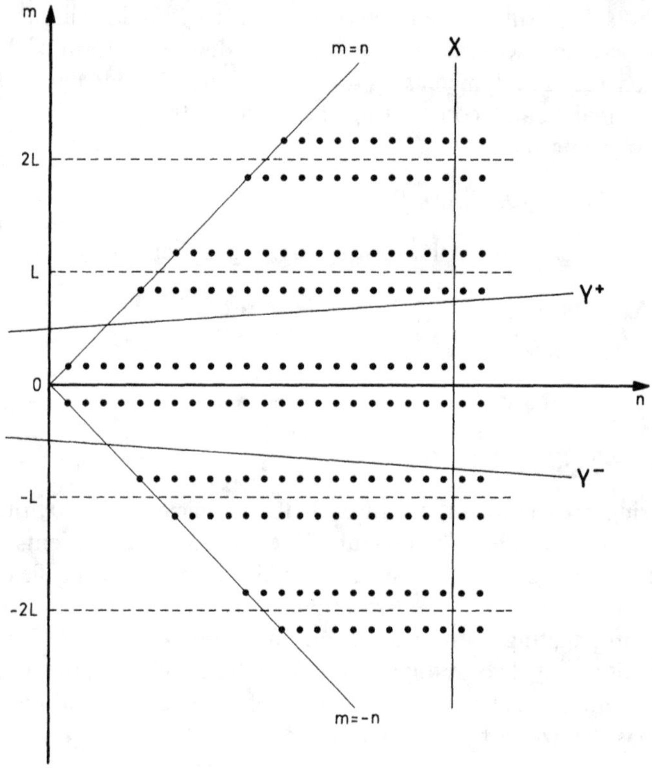

Fig. 7.21 *Band structure in the nm-plane of the transmission coefficients for an L-element ring of $\mu = \pm 1$ probes. The lines X, Y^+ and Y^- represent different cut-offs*

$\vec{F}^{(1)}_{smn}(r, \theta, \phi)$, ref. Section 2.2.3. The implied limitation in the set of significant coefficients T'^{rL}_{smn} is illustrated by the line X in Fig. 7.21.

Secondly, the cut-off of the associated Legendre functions for $\sin\theta \lesssim |m|/n$, ref. Section 2.2.3, implies that only the coefficients T'^{rL}_{smn} with $|m|/n \lesssim \sin\theta_r$ will have a significant amplitude. A verification of this statement can be achieved by considering expression (7.93) for the coefficients T'^{rL}_{smn} in connection with formulas (A2.17–2.19) for the rotation coefficients $d^n_{\pm 1,m}(\theta)$. The cut-off in $|m|$ thus formulated is illustrated by the lines Y^+ and Y^- in Fig. 7.21, although in a modified form which will be explained later.

Now, provided the area in the *nm*-plane bounded by the lines X, Y^+, Y^- and the *m*-axis contains only the main band at $m = \pm 1$ of the coefficients T'^{rL}_{smn}, the test zone fields from the *L*-element array and from the continuous ring will be virtually identical. Using the cut-off criteria $|m| \leqslant n \sin\theta_r$ and $n \leqslant kr_0$, this occurs when

$$L - 1 > kr_0 \sin\theta_r \tag{7.98}$$

which corresponds to a maximum angular spacing approximately equal to λ/r_0

between the probe positions on the ring as seen from the origin of the test zone coordinate system.

The λ/r_0 angular spacing equals twice the angular spacing $\lambda/(2r_0)$ allowed in spherical scanning, the reason being that only plane-wave propagation in the z-direction is considered. If angular scanning were required, the entire region in the nm-plane bounded by the lines X, Y^+, Y^- and the m-axis would contain significant coefficients for the continuous ring. In that case the angular spacing $\lambda/(2r_0)$ would be required to avoid overlapping of the first side band into the main band.

To obtain very good agreement between the test zone fields from the L-element probe ring and from the continuous probe ring, the cut-off criteria formulated above must be revised. First, the radial cut-off criterion should be changed to take into account coefficients T'^{rL}_{smn} with n up to N where $N \simeq kr_0 + 10$, ref. Section 2.2.3. Secondly, the cut-off of the associated Legendre functions in $|m|$ is effective only at $|m| \simeq \frac{4}{3}n \sin\theta + 4$ (see [9]) as illustrated by the lines Y^+ and Y^- in Fig. 7.21. These stronger criteria lead to the following sampling criterion for the probe ring:

$$L - 1 > \tfrac{4}{3}(kr_0 + n_1)\sin\theta_r + 4. \tag{7.99}$$

So far the sampling criteria for the choice of L have taken into account only the field in the test zone with radius r_0. If high directivity is of importance for the ring, then grating lobes introduced by the sampling must be avoided. This can be ensured if we require the $m = \pm 1$ rotational symmetric field of the continuous ring to be reproduced by the discrete ring in the entire inner region $r < A - r_p$. Using the weak cut-off criteria, this leads to the sampling criterion

$$L - 1 > kA\sin\theta_r \tag{7.100}$$

which corresponds to a distance about λ between the probe positions on the ring, in agreement with the criterion (7.21) for sampling of a continuous surface current source.

Finally, it should be mentioned that for the case of identical orientation of the probes on the ring, the use of an odd number of samples can be shown to be advantageous over the use of an even number of samples, [24].

Example: Consider an x'-polarized short dipole probe with the following current distribution function

$$\vec{J}(\vec{r}\,') = d_e \delta(\vec{r}\,')\hat{x}' \tag{7.101}$$

when unit excitation, $v_p = 1$ watts$^{1/2}$, is applied. In eqn (7.101), d_e is a dipole moment, δ is the Dirac delta function and $\hat{x}'$ is a unit vector in the x'-direction.

The spherical mode coefficients $Q^p_{smn} = v_p T^p_{smn}$ of the field from the probe can then be obtained by substituting the current distribution function (7.101) in formula (A1.78). After division by the excitation coefficient v_p, we obtain the following expression for the probe transmission coefficients

$$T^p_{\sigma\mu\nu} = v_p^{-1} \frac{kd_e}{\sqrt{\eta}} (-1)^{\mu+1} \vec{F}^{(1)}_{\sigma,-\mu,\nu}(0, \theta', \phi') \cdot \hat{x}' \tag{7.102}$$

which may be reduced by application of eqns (A1.51–1.53).

Formulas (7.92–7.93) now lead to expressions for the interior region transmission coefficients T''^{rL}_{smn} of the L-element ring. Thus, the mode coefficients $Q^{rL}_{smn} = v_r T''^{rL}_{smn}$ of the field created with excitation v_r of the ring are known. With unit excitation ($v_r = 1$ watts$^{1/2}$) of the ring, the coefficients Q^{rL}_{smn} are identical to the coefficients $Q^{(1)}_{smn}$ of eqns (A1.103). The coefficients $Q^{(1)}_{smn}$ are obtained directly on the basis of the current distribution function of the probe ring which is known in this simple case.

7.6.3.4 Planar probe ring: From the general description of a probe ring in Section 7.6.3.2, it follows that the only difference between a planar and a spherical probe ring is the orientation of the probes. In the planar ring, the probes are pointing in the z-direction. Considering an array of rings, another difference is that a set of planar rings will normally be positioned in a plane, ref. Fig. 7.18, implying that the rings have different distances to the origin of the test zone coordinate system. The planar array of rings corresponds to a scanning technique known as plane polar scanning [25].

Assuming, as for the spherical probe ring, plane-wave propagation in the positive z-direction, we consider a planar probe ring at $z = -A$, $A > 0$, with the probes pointing in the positive z-direction. Thus, the z'-axis in the probe coordinate system is pointing in the negative z-direction of the test zone coordinate system, ref. Fig. 3.1. The rotation angle χ_l of probe number l, at $\phi = \phi_l$ about its axis is chosen in such a way as to make the x- and the x'-axes parallel, just like for the spherical probe ring. The ring itself is characterized by $\theta = \theta_r$ and $r = A/\cos(\pi - \theta_r)$.

It may be verified that starting with the probe coordinate system (r', θ', ϕ') of the probe at $\phi = \phi_l$, the test zone coordinate system can be obtained through the following series of rotations and translations:

1. Rotate ϕ_l in ϕ'.
2. Rotate $-\theta_r$ in θ''.
3. Translate $\dfrac{A}{\cos(\pi - \theta_r)}$ in z'''.
4. Rotate $\theta_r - \pi$ in θ''''.
5. Rotate $\phi_l + \pi$ in ϕ'''''.

Here, an additional prime is added in the coordinate variables after each operation.

Following a procedure similar to the one outlined in Section 3.2 for obtaining the spherical mode expansion of the probe field $(\vec{E}_p, \vec{H}_p)$ in the test zone coordinate system, we obtain from steps 1 to 5 above

$$\vec{E}_p(\vec{r}) = v_p \frac{k}{\sqrt{\eta}} \sum_{\substack{\sigma\mu\nu \\ \gamma \\ smn}} T^p_{\sigma\mu\nu} e^{i\mu\phi_l} d^\nu_{\gamma\mu}(-\theta_r) C^{\sigma\nu(3)}_{s\gamma n}\left(\frac{kA}{\cos(\pi - \theta_r)}\right)$$

$$d^n_{m\gamma}(\theta_r - \pi) e^{im(\phi_l + \pi)} \vec{F}^{(1)}_{smn}(\vec{r}). \tag{7.103}$$

The element positions and excitations in the planar ring being defined as in the spherical probe ring, the following formula for the transmission coefficients T'^r_{smn} of the continuous planar ring can be derived from eqn (7.103)

$$T'^r_{smn} = (-1)^m \sum_{\sigma\gamma\nu} T^p_{\sigma,-m,\nu} d^\nu_{\gamma,-m}(-\theta_r) C^{\sigma\nu(3)}_{s\gamma n}\left(\frac{kA}{\cos(\pi - \theta_r)}\right) d^n_{m\gamma}(\theta_r - \pi). \tag{7.104}$$

The transmission coefficients T'^{rL}_{smn} of the L-element discrete planar probe ring exhibit the same band structure in m as the corresponding transmission coefficients of the L-element spherical probe ring; see Fig. 7.21.

Clearly, the expression (7.104) for the transmission coefficients of the planar ring is more complex and requires more computation than the corresponding set of formulas (7.92), (7.94) for the spherical ring. The planar array geometry is in general less suited for a description of the fields in terms of spherical wave functions. However, a $\mu = \pm 1$ probe will still cause only the coefficients T'^r_{smn} with $m = \pm 1$ to differ from zero.

For cases where the current distribution function of the probe ring is known and simple, it is easier to use this function directly instead of deriving the coefficients T'^r_{smn} of the ring from the coefficients T^p_{smn} of the single probe.

Example: As in the previous section we shall consider an x'-polarized short dipole probe with a current distribution function as defined in eqn (7.101). For this simple probe, only the transmission coefficients T^p_{211} and $T^p_{2,-1,1}$ given in eqn (7.102) are different from zero. The mode coefficients $Q^r_{smn} = v_r T'^r_{smn}$, $v_r = 1$ watts$^{1/2}$, of the field from the continuous probe ring, obtained by application of eqn (7.104), are equal to the mode coefficients $Q^{(1)}_{smn}$ of eqns (A1.97), obtained directly from the current distribution function of the continuous planar probe ring.

7.6.3.5 Least-squares optimization: Consider a set of I spherical or planar probe rings centred on the negative z-axis and intended to produce a plane wave propagating in the positive z-direction within a test zone with radius r_0, ref. Fig. 7.18. The probe rings are uniform ring arrays as described in the preceding sections. We wish to determine the set of complex excitations v_i, $i = 1, 2, \ldots, I$, that should be applied to the set of rings to obtain the best plane-wave quality in the test zone.

As in Section 7.5.3, we denote by Q_{smn}^{pl} the coefficients in a spherical mode expansion of the ideal plane wave in the test zone coordinate system. Since in this case the plane wave propagates in the positive z-direction, we obtain from eqn (A1.106)

$$Q_{smn}^{\mathrm{pl}} = \frac{\sqrt{\eta}}{k}(-1)^m \sqrt{4\pi}\, i\vec{E}_{\mathrm{pl}}(\vec{0}) \cdot \vec{K}_{s,-m,n}(\theta = \pi) \tag{7.105}$$

where $\vec{E}_{\mathrm{pl}}(\vec{0})$ is the electric field of the ideal plane wave at the origin and $\vec{K}_{smn}(\theta = \pi)$ is given by formulas (A1.63–1.64). From the expressions for $\vec{K}_{smn}(\theta = \pi)$, it follows that

$$Q_{smn}^{\mathrm{pl}} = 0 \qquad \text{for } m \neq \pm 1. \tag{7.106}$$

Assuming a $\mu = \pm 1$ probe, then following the derivations in the preceding sections, only modes with $m = \pm 1$ will be present in the test zone field from any one of the probe rings provided the number of elements L in the ring is sufficiently large. We shall assume a sufficient number of elements in each of the probe rings. This means that the test zone field from any one of the probe rings equals the field from the corresponding continuous ring.

By $T_{smn}^{\prime i}$ we shall designate the transmission coefficients $T_{smn}^{\prime r}$ for the continuous ring number i, ref. eqns (7.94) and (7.104). Then, since a $\mu = \pm 1$ probe is applied, we have

$$T_{smn}^{\prime i} = 0 \qquad \text{for } m \neq \pm 1. \tag{7.107}$$

The spherical mode coefficients Q_{smn} of the total field from the set of I continuous rings then equal a linear combination of the coefficients $T_{smn}^{\prime i}$:

$$Q_{smn} = \sum_{i=1}^{I} v_i T_{smn}^{\prime i} \tag{7.108}$$

where v_i is the complex signal applied to ring number i.

Using the mode coefficients Q_{smn}^{pl} and Q_{smn} of the ideal and the generated plane wave, respectively, the test zone error norm U_N defined in eqn (7.54) takes the following form

$$U_N = \frac{1}{\sqrt{4\pi}\,|\vec{E}_{\mathrm{pl}}|}\frac{k}{\sqrt{\eta}}\sqrt{\sum_{n=1}^{N}\sum_{m=\pm 1}\sum_{s=1,2}|Q_{smn} - Q_{smn}^{\mathrm{pl}}|^2} \tag{7.109}$$

where $N \simeq kr_0$.

To formulate the minimization of U_N in the variable excitation coefficients v_i of the rings, we introduce a set of matrices. The matrix $\mathbf{Q}_{\mathrm{pl}}$ is a column matrix containing the coefficients Q_{smn}^{pl} of the ideal plane wave in a specific order

$$\mathbf{Q}_{\mathrm{pl}} = \{Q_{smn}^{\mathrm{pl}}\} \qquad n = 1, 2, \ldots, N, \qquad m = \pm 1, \qquad s = 1, 2 \tag{7.110}$$

and $\mathbf{T}_i'$ is a column matrix containing the transmission coefficients $T_{smn}^{\prime i}$ of the continuous ring i in exactly the same order:

$$\mathbf{T}'_i \equiv \{T'^i_{smn}\} \qquad n = 1, 2, \ldots, N, \qquad m = \pm 1, \qquad s = 1, 2. \tag{7.111}$$

Thus, both $\mathbf{Q}_{\text{pl}}$ and $\mathbf{T}'_i$ are $4N$ by 1 matrices.

As a simple extension of the scattering matrix formulation for a one-port antenna (Section 2.3), we define a column matrix $\mathbf{v}$ containing the variable ring excitations v_i

$$\mathbf{v} \equiv \begin{pmatrix} v_1 \\ v_2 \\ \vdots \\ v_I \end{pmatrix} \tag{7.112}$$

and a matrix $\mathbf{T}'$ containing the column matrices $\mathbf{T}'_i$

$$\mathbf{T}' \equiv (\mathbf{T}'_1 \quad \mathbf{T}'_2 \quad \ldots \quad \mathbf{T}'_I). \tag{7.113}$$

These definitions and eqn (7.108) enable the expression (7.109) for U_N to be rewritten in the following form

$$U_N = \frac{1}{\sqrt{4\pi}\,|\vec{E}_{\text{pl}}|}\frac{k}{\sqrt{\eta}}\|\mathbf{T}'\mathbf{v} - \mathbf{Q}_{\text{pl}}\|_2 \tag{7.114}$$

The two-norm of a complex column matrix was defined in connection with eqn (7.77).

To minimize the test zone error norm U_N as a function of $\mathbf{v}$, we must solve the $4N$ complex equations in I complex variables

$$\mathbf{T}'\mathbf{v} = \mathbf{Q}_{\text{pl}} \tag{7.115}$$

in the least-squares sense. With reference to eqns (7.77–7.78), the solution to this problem is

$$\mathbf{v} = (\mathbf{T}'^{+}\mathbf{T}')^{-1}\mathbf{T}'^{+}\mathbf{Q}_{\text{pl}}. \tag{7.116}$$

Once an optimal set of excitation coefficients $\mathbf{v}$ has been determined, each of the continuous rings must be sampled according to the guidelines given in Section 7.6.3.3. Then, the excitation of the individual probes can be obtained from eqn (7.84).

However, for the same reasons as those given in connection with eqn (7.78), the set of weights $\mathbf{v}$ obtained from eqn (7.116) will for some configurations of rings lead to a low directivity of the array of rings toward the test zone. A consequence will be a high error sensitivity [9].

To avoid these low-directivity solutions, we introduce a constraint on the power radiated by the array of rings in the minimization of $U_N(\mathbf{v})$. For this purpose we need the normalized radiation resistance matrix $\mathbf{R}_{\text{rad}}$ for the set of sampled rings, allowing the total radiated power P_{rad} to be written in the following form

$$P_{\text{rad}} = \mathbf{v}^+ \mathbf{R}_{\text{rad}} \mathbf{v}. \tag{7.117}$$

The radiation resistance matrix $\mathbf{R}_{\text{rad}}$ must be defined for the set of sampled rings because the sampling has a large influence on the radiated power and on the directivity. The element spacing must be less than λ on each ring in order for the directional properties of the discrete ring to approach the directional properties of the continuous ring, ref. eqn (7.100). Consequently, to obtain a set of optimal weights taking into account the near-field directivity, the sampling of the rings must be defined before the weights are computed unless the probe spacings are less than λ.

As a reference level for the power radiated by the array of rings, we introduce the power

$$P_{\text{pl,iso}} \equiv 2\pi A^2 \eta |\vec{E}_{\text{pl}}|^2 \tag{7.118}$$

needed to establish with an isotropic radiation pattern the required amplitude $|\vec{E}_{\text{pl}}(\vec{0})|$. The radiated power P_{rad} normalized with respect to this isotropic reference level shall be denoted the *relative power level* and represented by the symbol P_{rel}:

$$P_{\text{rel}} \equiv \frac{P_{\text{rad}}}{P_{\text{pl,iso}}} \tag{7.119}$$

Now, provided the created plane-wave field $(\vec{E}, \vec{H})$ is a good approximation to the ideal plane wave $(\vec{E}_{\text{pl}}, \vec{H}_{\text{pl}})$ within the test zone, it is apparent that $1/P_{\text{rel}}$ approximates a near-field directivity of the array of rings toward the test zone. Therefore, specifying an upper bound $P_{\text{rel}}^{\max}$ on the relative power level P_{rel} in the minimization of U_N

$$\text{Minimize} \quad U_N(\mathbf{v}) \qquad \text{subject to} \quad P_{\text{rel}}(\mathbf{v}) \leqslant P_{\text{rel}}^{\max} \tag{7.120}$$

is equivalent to introduction of a lower bound on the near-field directivity.

Insertion of eqns (7.119) and (7.117) into (7.120) leads to the following formulation of the constrained least-squares problem

$$\text{Minimize} \quad U_N(\mathbf{v}) \qquad \text{subject to} \quad \mathbf{v}^+ \mathbf{R}_{\text{rad}} \mathbf{v} \leqslant P_{\text{rel}}^{\max} P_{\text{pl,iso}} \tag{7.121}$$

to be solved for the desired set of weights $\mathbf{v}$.

To solve (7.121), we perform a transformation of variables. The radiation resistance matrix $\mathbf{R}_{\text{rad}}$ being Hermitian and positive definite means that there exists an I by I matrix $\mathbf{R}_{\text{rad}}^{\frac{1}{2}}$ satisfying the relation

$$(\mathbf{R}_{\text{rad}}^{\frac{1}{2}})^+ \mathbf{R}_{\text{rad}}^{\frac{1}{2}} = \mathbf{R}_{\text{rad}}. \tag{7.122}$$

The matrix $\mathbf{R}_{\text{rad}}^{\frac{1}{2}}$ can be obtained from $\mathbf{R}_{\text{rad}}$ by Choleski factorization or by eigenvalue/eigenvector factorization of $\mathbf{R}_{\text{rad}}$. Different methods will in general lead to different matrices $\mathbf{R}_{\text{rad}}^{\frac{1}{2}}$ so the same definition has to be applied throughout.

Defining the column matrix **u** of transformed excitation variables by

$$\mathbf{u} \equiv \mathbf{R}_{\text{rad}}^{\frac{1}{2}}\mathbf{v} \tag{7.123}$$

means that the constrained least-squares problem (7.121) can be rewritten in the following form

$$\text{Minimize} \quad \|\tilde{\mathbf{T}}'\mathbf{u} - \mathbf{Q}_{\text{pl}}\|_2 \qquad \text{subject to} \quad \|\mathbf{u}\|_2^2 \leqslant P_{\text{rel}}^{\max} P_{\text{pl,iso}} \tag{7.124}$$

where $\tilde{\mathbf{T}}'$ is a $4N$ by I matrix which is the solution to

$$\mathbf{T}' = \tilde{\mathbf{T}}'\mathbf{R}_{\text{rad}}^{\frac{1}{2}}. \tag{7.125}$$

Once (7.124) has been solved for an optimal column matrix **u**, the set of corresponding optimal excitation coefficients **v** is obtained by solving eqn (7.123), and the excitation of the single probes can be obtained by the application of eqn (7.84) for each ring.

A procedure for the solution of problems like (7.124) has been given by Mautz and Harrington [23]. Their method involves computation of the eigenvalues and eigenvectors of the matrix $\tilde{\mathbf{T}}'^{+}\tilde{\mathbf{T}}'$.

A computationally more stable method, avoiding the computation and use of the matrix $\tilde{\mathbf{T}}'^{+}\tilde{\mathbf{T}}'$, is obtained by application of the *Singular-Value Decomposition* (*SVD*) [26, 27] in connection with the matrix $\tilde{\mathbf{T}}'$. Based on a singular-value decomposition of $\tilde{\mathbf{T}}'$, it is possible to obtain the solution to (7.124) for all values of the maximum relative power level $P_{\text{rel}}^{\max}$ [9, 26].

In an implementation of the SVD method, it is important that explicit calculation of the large unitary matrix containing the left-hand eigenvectors of $\tilde{\mathbf{T}}'$ can be avoided [9]. This unitary matrix is of dimension $4N$ by $4N$ because $\tilde{\mathbf{T}}'$ has $4N$ rows, ref. eqn (7.125). Examples in which the SVD method has been applied are given in the following section.

7.6.3.6 Numerical results: The results to be presented in this section deal with the problem of choosing an appropriate geometry for a planar or a spherical ring array with constant increment between the ring radii: how large should the array of rings be and which ring radius increment should be chosen. Based on our knowledge of the distribution of the probe weights applied in spherical scanning (ref. Section 7.2), it would be reasonable to choose a value for the array radius between r_0 and $2r_0$, r_0 being the test zone radius.

Figure 7.22 depicts a cut along the z-axis of the set-up to be treated. A planar ring array in the plane $z = -A$, $A = 100\lambda$, is used to establish a plane-wave zone with radius $r_0 = 15\lambda$. The selected measurement distance A corresponds to 1/18 of the conventional far-field criterion $8r_0^2/\lambda$. Letting ρ_i be the radius of ring i, we use

$$\rho_i = (i-1)\Delta\rho \qquad i = 1, 2, \ldots, I \tag{7.126}$$

where $\Delta\rho$ is the increment in the ring radii and I is the number of rings. Thus, the

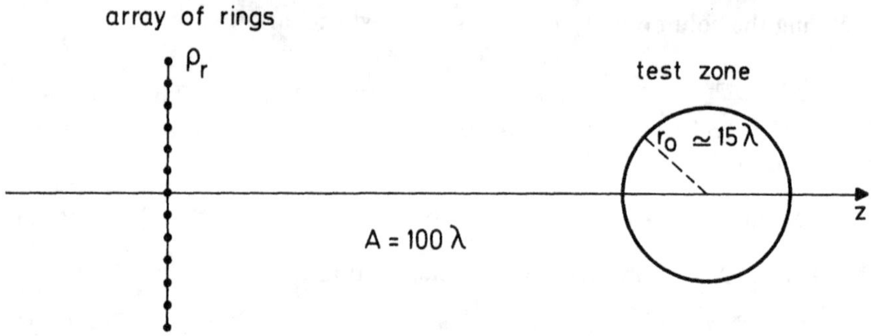

Fig. 7.22 *Cut along the z-axis of the test zone and the planar ring array*

inner ring has a radius equal to zero and the radius ρ_r of the entire array of rings is

$$\rho_r = \rho_I = (I - 1)\Delta\rho. \tag{7.127}$$

The probe is assumed to be an x'-polarized short dipole. This allows the interior region transmission coefficients T'^{i}_{smn} of each ring to be computed directly from the current distribution function of the ring, ref. Sections 7.6.3.4 and A1.5.3. The ideal plane wave to be approximated within the test zone is x-polarized and propagates in the positive z-direction. Its spherical mode expansion coefficients Q^{pl}_{smn} are obtained from eqn (7.105).

The planar geometry of the ring array has been selected for the purpose of simplifying the calculation of the normalized radiation resistance matrix $\mathbf{R}_{\text{rad}}$, ref. eqn (7.117). However, for small ring arrays, i.e. $\rho_r/A \ll 1$, virtually identical results can be obtained for spherical and for planar ring arrays. An algorithm for calculating $\mathbf{R}_{\text{rad}}$ was described in reference [9] which considered the spherical mode expansions of the fields from the rings in a coordinate system with origin at the centre of the ring array.

In all of the examples to be considered, the rings are sampled according to the sampling criterion in eqn (7.99). This means that high-directivity solutions cannot be obtained due to the presence of grating lobes.

For each of the different ring arrays to be considered, the optimal excitation of the rings is obtained by solving the constrained least-squares problem (7.124) with spherical modes of degree up to $N = 90$ ($kr_0 \simeq 90$). From a singular-value decomposition of the matrix $\tilde{\mathbf{T}}'$ the solution for all values of the maximum relative power level $P^{\max}_{\text{rel}}$ is obtained. In particular, we obtain the minimum value of the test zone error norm U_N as a function of the specified maximum relative power level $P^{\max}_{\text{rel}}$. This function contains important information about the significance of using the power (directivity) constraint.

In general, the matrix $\tilde{\mathbf{T}}'$ has $4N$ rows since spherical modes with $s = 1, 2$ and $m = \pm 1$ must be considered for $n = 1, 2, ..., N$. However, for the present problem, where all sources and electric fields have a common symmetry plane containing the z-axis ($\phi = 0$), there is a fixed relation between the mode coefficients with $m = -1$ and those with $m = 1$, ref. Section A1.5.2. Therefore, the coefficients with $m = -1$ need not be considered thus reducing the number of equations in the constrained least-squares problem (7.124) from $4N$ to $2N$.

All the results to be presented below originate from reference [9].

Example 1: In the first example we choose the increment $\Delta\rho$ between the ring radii approximately equal to the spacing required in spherical scanning. From the angular spacing requirement $\Delta\theta_{\text{scan}} = \lambda/(2r_0)$ in spherical scanning we obtain

$$\Delta\rho = A\,\Delta\theta_{\text{scan}} = \frac{\lambda}{2}\frac{A}{r_0}. \tag{7.128}$$

Thus, since $A = 100\lambda$ and $r_0 = 15\lambda$, we use

$$\Delta\rho = 3.33\lambda. \tag{7.129}$$

With $I = 6$ rings, the radius $\rho_r = (I - 1)\Delta\rho$ of the ring array just exceeds the radius $r_0 = 15\lambda$ of the test zone. Using $I = 10$, ρ_r equals twice the test zone radius r_0.

Figure 7.23 shows the minimum value of the test zone error norm U_N as a function of the specified maximum relative power level $P_{\text{rel}}^{\text{max}}$ for the cases when

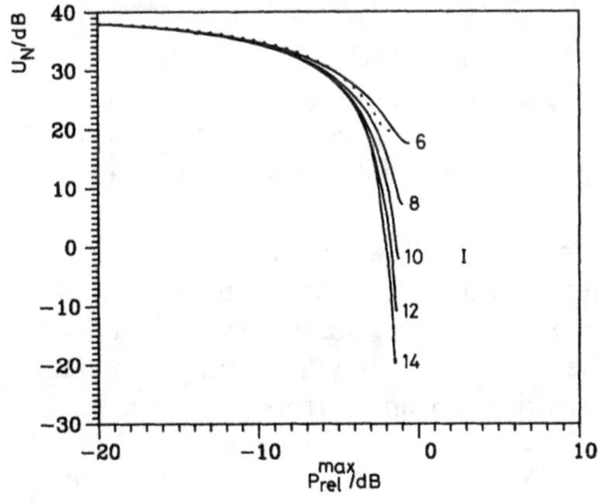

Fig. 7.23 *Minimum test zone error norm U_N as a function of the specified maximum relative power level $P_{\text{rel}}^{\text{max}}$ for the case of 6, 8, 10, 12 and 14 rings with increment $\Delta\rho = 3.33\lambda$ between the ring radii*

6, 8, 10, 12 and 14 rings are used. The dotted curve shows the same relation for the case of a single x-polarized short dipole at the Rayleigh distance $8r_0^2/\lambda$. The parts of the curves where the power constraint is inactive have been omitted. These parts are simply horizontal lines starting at the end points of the curves.

The curves in Fig. 7.23 show that

1. With $\Delta\rho = A\Delta\theta_{scan}$ there is little or no need to use the power (directivity) constraint. There is no low-directivity solution allowing a significant decrease in P_{rel} at the expense of a minor increase in U_N.
2. Using $I = 6$ rings, i.e. for $\rho_r \simeq r_0$, the minimum plane-wave deviation is of the same order of magnitude as the deviation present in the field created by a single short dipole at the Rayleigh distance. The error norm U_N is decreased by about 9 dB each time two rings are added.

Note, that for the parts of the curves where U_N approaches its maximum value, the synthesized test zone field approaches the zero field. For these parts, P_{rel} cannot be interpreted as a reciprocal directivity.

Example 2: This example deals with the situation where the increment $\Delta\rho$ in the ring radii is changed. We use $\Delta\rho = 3.5\lambda$, $\Delta\rho = 3.0\lambda$ and $\Delta\rho = 2.5\lambda$. For each one of these three values of $\Delta\rho$, a series of functions $U_N(P_{rel}^{max})$ corresponding to different numbers I of rings have been calculated and plotted, just as for the case $\Delta\rho = 3.33\lambda$ in Fig. 7.23. For each value of $\Delta\rho$, the values of I have been selected to make the radius ρ_r of the ring array vary from r_0 to $2r_0$.

Figure 7.24 contains the results for the three values of $\Delta\rho$. It appears that

1. For $\Delta\rho < A\Delta\theta_{scan} = 3.33\lambda$, it is important to use a power (directivity) constraint because a low-directivity solution exists and provides the smallest error norm U_N.
2. Using a value of $\Delta\rho$ less than $A\Delta\theta_{scan}$ in connection with a power (directivity) constraint allows a small value of U_N to be achieved with a directivity larger than 0 dB. In connection with a scan technique this will, however, require a grid of scan positions with angular spacing less than $\Delta\theta_{scan}$.

Conclusion: In conclusion, if we choose $\Delta\rho \geqslant A\Delta\theta_{scan}$, we can omit the directivity constraint and calculate the set of optimal excitation coefficients from eqn (7.116). In that case, the fields from the rings are sufficiently different within the test zone that the matrix $\mathbf{T}'^{+}\mathbf{T}'$ is well conditioned and the solution obtained will be relatively insensitive to small errors [9].

Fig. 7.24 *Minimum test zone error norm U_N as a function of specified maximum relative power level P_{rel}^{max}*
(a) $\Delta\rho = 3.5\lambda$
(b) $\Delta\rho = 3.0\lambda$
(c) $\Delta\rho = 2.5\lambda$
For each of the three values of $\Delta\rho$, the range of values of I has been selected to make ρ_r vary from r_o to $2r_o$

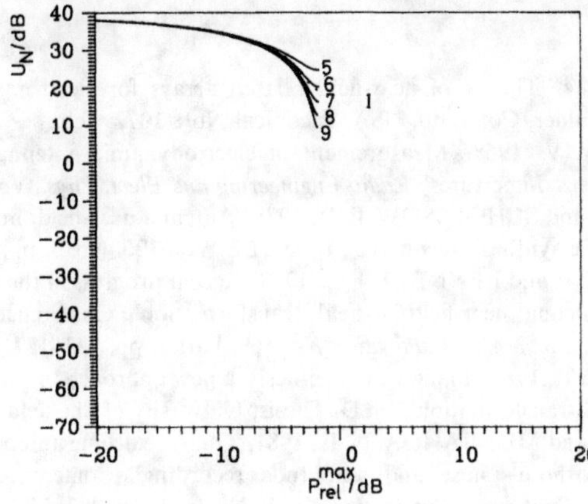
U_N/dB
40
30
20
10
0
−10
−20
−30
−40
−50
−60
−70
−20
−10
0
10
20
P_{rel}^{max}/dB
5
6
7
8
9
I

(*a*)

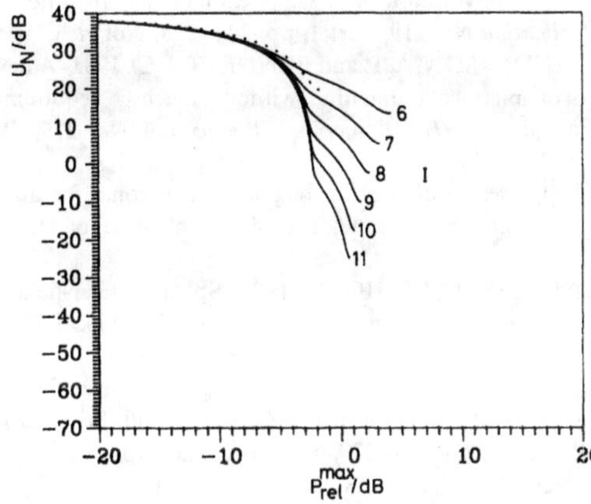
U_N/dB
40
30
20
10
0
−10
−20
−30
−40
−50
−60
−70
−20
−10
0
10
20
P_{rel}^{max}/dB
6
7
8
9
10
11
I

(*b*)

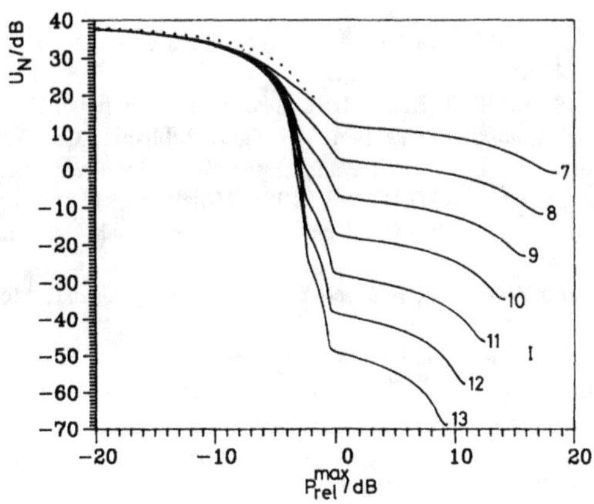
U_N/dB
40
30
20
10
0
−10
−20
−30
−40
−50
−60
−70
−20
−10
0
10
20
P_{rel}^{max}/dB
7
8
9
10
11
12
13
I

(*c*)

References

1. HILL, D. A., 1984, 'Theory of near-field phased arrays for electromagnetic susceptibility testing', NBS, Boulder, Colorado, USA, Technical Note **1072**, Feb.
2. MARTSAFEY, V. V., 1968, 'Measurement of electrodynamic antenna parameters by the method of synthesized apertures', *Radio Engineering and Elect. Phys.*, Vol. **13**, pp. 1869–1873.
3. BENNETT, J. C. and SCHOESSOW, E. P., 1978, 'Antenna near-field/far-field transformation using a plane-wave-synthesis technique', *Proc. IEE*, Vol. **125**, No. 3, pp. 179–184, Mar.
4. SCHOESSOW, E. P. and BENNETT, J. C., 1978, 'Recent progress in the plane-wave synthesis technique for antenna near-field/far-field transformation', Conference on Antennas and Propagation, *IEE Conference Publication No.* **169**, Part 1, pp. 123–127, London, Nov.
5. SCHOESSOW, E. P., 1980, 'Plane wave synthesis: a new approach to the problem of antenna near-field/far-field transformation', Ph.D. Thesis, University of Sheffield, Sep.
6. BENNETT, J. C. and MUNTANGA, N. E., 1981, 'Quasi-real-time antenna near-field/far-field transformation without phase and amplitude recording', Conference on Antennas and Propagation, *IEE Conference Publication No.* **195**, Part 1, pp. 283–287, York, Apr.
7. BENNETT, J. C. and GRAZIOTIS, A., 1983, 'Removal of environmental effects from antenna radiation patterns by deconvolution processing', Conference on Antennas and Propagation, *IEE Conference Publication No.* **219**, Part 1, pp. 224–228, Norwich, Apr.
8. PEREIRA, J. F. R., ANDERSON, A. P. and BENNETT, J. C., 1983, 'A new procedure for near-field measurements of microwave antennas without anechoic requirements', Conference on Antennas and Propagation, *IEE Conference Publication No.* **219**, Part 1, pp. 219–223, Norwich, Apr.
9. HALD, J., 1983, 'Numerical optimization of plane-wave zones for antenna measurements', Ph.D. Thesis, Electromagnetics Institute, Technical University of Denmark, Report **LD 51**, Oct.
10. LUDWIG, A. C. and LARSEN, FL. HOLM, 1979, 'Spherical near-field measurements from a "compact-range" viewpoint', Electromagnetics Institute, Technical University of Denmark, Report **R 213**, Apr.
11. LUDWIG, A. C. and LARSEN, FL. HOLM, 1981, 'Spherical near-field measurements from a "compact-range" viewpoint', Conference on Antennas and Propagation, *IEE Conference Publication No.* **195**, Part 1, pp. 274–277, York, England, Apr.
12. TURCHIN, V. I. and TSEYTLIN, N. M., 1979, 'Antenna testing based on near-field measurements (Review)', *Radio Engineering & Electronic Physics*, Vol. **24**, pp. 1–26, Dec.
13. HARRINGTON, R. F., 1961, *Time-harmonic electromagnetic fields*, McGraw-Hill, Inc., New York.
14. WATERMAN, P. C., 1965, 'Matrix formulation of electromagnetic scattering', *Proc. IEEE*, Vol. **53**, pp. 805–812, Aug.
15. BOWMAN, J. J., SENIOR, T. B. and USLENGHI, P. L. E. (eds), 1969, *Electromagnetic and acoustic scattering by simple shapes*, North-Holland Publishing Co., Amsterdam.
16. MITTRA, R. (ed.), 1973, *Computer techniques for electromagnetics*, Pergamon Press, Oxford.
17. MAUTZ, J. R. and HARRINGTON, R. F., 1978, 'H-field, E-field and combined field solutions for conducting bodies of revolution', *Archiv für Elektronik und Übertragungstechnik*, Vol. **32**, pp. 157–164.
18. COLLIN, R. E. and ZUCKER, F. J. (eds), 1969, *Antenna Theory*, McGraw-Hill, Inc., New York.
19. PAPOULIS, A., 1977, *Signal analysis*, McGraw-Hill, Inc., New York.
20. SHEWELL, J. R. and WOLF, E., 1968, 'Inverse diffraction and a new reciprocity theorem', *Journal of the Optical Society of America*, Vol. **58**, pp. 1596–1603, Dec.
21. IEEE, 1979, 'IEEE Standard test procedures for antennas', The Institute of Electrical and Electronics Engineers, Inc., *IEEE Std.* **149**.
22. MULLER, C., 1969, *Foundations of the mathematical theory of electromagnetic waves*, Springer Verlag, Berlin.

23. MAUTZ, J. R. and HARRINGTON, R. F., 1975, 'Computational methods for antenna pattern synthesis', *IEEE Trans.*, Vol. **AP-23**, pp. 507–512.
24. KNUDSEN, H. L., 1951, 'The necessary number of elements in a directional ring aerial', *Journal of Applied Physics*, Vol. **22**, No. 11, pp. 1299–1306, Nov.
25. RAHMAT-SAMII, Y., GALINDO-ISRAEL, V. and MITTRA, R., 1980, 'A plane-polar approach for far-field construction from near-field measurements', *IEEE Trans.*, Vol. **AP-28**, No. 2, pp. 216–230, Mar.
26. LAWSON, C. L. and HANSON, R. J., 1974, *Solving least squares problems*, Prentice Hall, Inc.
27. HANSEN, P. C., 1984, 'SVD-theory and applications', Institute for Numerical Analysis, Technical University of Denmark, Report **NI-84-05**, Lyngby, Nov.

Appendix A1

Spherical wave functions, notation and properties

A1.1 Notation

A1.1.1 The general spherical wave expansion

Electromagnetic fields may be expanded into spherical waves in source-free regions of space limited by spherical surfaces centred at the origin of a spherical coordinate system (r, θ, ϕ)

$$\vec{E}(r, \theta, \phi) = \frac{k}{\sqrt{\eta}} \sum_{csmn} Q_{smn}^{(c)} \vec{F}_{smn}^{(c)}(r, \theta, \phi) \tag{A1.1}$$

$$\vec{H}(r, \theta, \phi) = -ik\sqrt{\eta} \sum_{csmn} Q_{smn}^{(c)} \vec{F}_{3-s,m,n}^{(c)}(r, \theta, \phi). \tag{A1.2}$$

In these expansions, $\exp(-i\omega t)$ is the (suppressed) time dependence, $k = \omega\sqrt{\mu\varepsilon}$ is the wave number and $\eta = \sqrt{\varepsilon/\mu}$ is the specific admittance of the medium.

The spherical wave functions $\vec{F}_{smn}^{(c)}(r, \theta, \phi)$ are dimensionless. Their spherical components are products of elementary functions of r, θ and ϕ, and are thus separated with respect to the dependence on the radial, polar and azimuthal coordinates, ref. Section A1.3.

The upper index (c) indicates the particular function in the radial dependencies of the spherical wave functions. We have

- $c = 1$ Spherical Bessel function representing a radial standing wave, finite at the origin.
- $c = 2$ Spherical Neumann function representing a radial standing wave, infinite at the origin.
- $c = 3$ Spherical Hankel function of the first kind, representing a radial outgoing wave, infinite at the origin.
- $c = 4$ Spherical Hankel function of the second kind, representing a radial incoming wave, infinite at the origin.

The index s distinguishes with $s = 1$ and $s = 2$ between the two spherical wave functions $\vec{F}^{(c)}_{1mn}$ and $\vec{F}^{(c)}_{2mn}$. The index n indicates the degree of the wave functions and takes the values $n = 1, 2, 3, \ldots, \infty$. The highest degree retained in computations is denoted N. The index m indicates the order of $\vec{F}^{(c)}_{1mn}$ and $\vec{F}^{(c)}_{2mn}$. The values taken by m are $m = -n, -n+1, \ldots, 0, \ldots, n-1, n$.

Of the two spherical wave functions only $\vec{F}^{(c)}_{2mn}$ has a radial component and is analogous to Stratton's $\vec{n}$-function. The other function, $\vec{F}^{(c)}_{1mn}$, is purely transverse with respect to the radial direction. It is analogous to Stratton's $\vec{m}$-function [1].

In the above expansion of an electromagnetic field $Q^{(c)}_{1mn}$ therefore represents the transverse electric (TE) part while $Q^{(c)}_{2mn}$ is representing the transverse magnetic (TM) part. The wave coefficients $Q^{(c)}_{smn}$ are of dimension watts$^{\frac{1}{2}}$.

The spherical wave functions are orthogonal (ref. Section A1.4) and power-normalized (ref. Section 2.2.4). The power radiated by an outgoing wave field ($c = 3$) is given by

$$P = \frac{1}{2}\sum_{smn} |Q^{(3)}_{smn}|^2 \quad \text{watts.} \tag{A1.3}$$

A1.1.2 Single index convention

The summation over the indices (s, m, n) in eqn (A1.1) may be performed in different orders

$$\sum_{smn} Q^{(c)}_{smn}\vec{F}^{(c)}_{smn}$$

$$= \sum_{s=1}^{2}\sum_{m=-N}^{N}\sum_{\substack{n=|m|\\(n\neq 0)}}^{N} Q^{(c)}_{smn}\vec{F}^{(c)}_{smn} = \sum_{s=1}^{2}\sum_{n=1}^{N}\sum_{m=-n}^{n} Q^{(c)}_{smn}\vec{F}^{(c)}_{smn}. \tag{A1.4}$$

On the right-hand side of (A1.4) the inner summation is with respect to m while on the left-hand side the inner summation is with respect to n. Both triple sums have the finite number of terms $J = 2N(N+2)$ where N is the maximum value of n. The terms in each of the sums may be renumbered to form single summations.

A suitable index transformation for the right-hand side is

$$j = 2\{n(n+1) + m - 1\} + s$$

where $j = 1, 2, \ldots, J$. Hereby we may write

$$\sum_{s=1}^{2}\sum_{n=1}^{N}\sum_{m=-n}^{n} Q^{(c)}_{smn}\vec{F}^{(c)}_{smn} = \sum_{j=1}^{J} Q^{(c)}_{j}\vec{F}^{(c)}_{j}$$

Conversely, if the index j is given, the triplet (s, m, n) can be found by the following three-step algorithm:

Step	Operation
1	$s = \begin{cases} 1 & j \text{ odd} \\ 2 & j \text{ even} \end{cases}$
2	$n = \text{integer part of } \sqrt{\frac{(j-s)}{2} + 1}$
3	$m = \frac{(j-s)}{2} + 1 - n(n+1)$

A1.1.3 Representation of far fields

In order to describe far fields, the asymptotic forms of $\vec{F}^{(3)}_{1mn}(r, \theta, \phi)$ and $\vec{F}^{(3)}_{2mn}(r, \theta, \phi)$ as kr tends to infinity may be introduced by substituting for the radial functions the large argument approximations given by eqns (2.13) and (2.15). However, since functions that are dependent on the angular variables only are more convenient to work with, the common r-dependence $\exp(ikr)/(kr)$ of the asymptotic forms of the wave functions may be left out. We therefore introduce the far-field pattern functions defined through

$$\vec{K}_{smn}(\theta, \phi) = \lim_{kr \to \infty} \left\{ \sqrt{4\pi} \frac{kr}{e^{ikr}} \vec{F}^{(3)}_{smn}(r, \theta, \phi) \right\} \tag{A1.5}$$

where an additional factor $\sqrt{4\pi}$ is included in the definitions. This will imply that expressions for gain and directivity become simpler.

A1.2 Elementary functions

A1.2.1 Radial functions

Definitions

The radial dependence of the spherical wave functions may be expressed through the function $R^{(c)}_{sn}(kr)$ defined by

$$R^{(c)}_{sn}(kr) = \begin{cases} z^{(c)}_n(kr), & s = 1 \\ \dfrac{1}{kr} \dfrac{\mathrm{d}}{\mathrm{d}(kr)} \{ kr z^{(c)}_n(kr) \}, & s = 2 \end{cases} \tag{A1.6}$$

where

$z^{(1)}_n(kr) = j_n(kr)$ (spherical Bessel function)

$z^{(2)}_n(kr) = n_n(kr)$ (spherical Neumann function)

$z^{(3)}_n(kr) = h^{(1)}_n(kr) = j_n(kr) + i n_n(kr)$ (spherical Hankel function of the first kind)

$$z_n^{(4)}(kr) = h_n^{(2)}(kr) = j_n(kr) - in_n(kr) \quad \text{(spherical Hankel function of the second kind).}$$

Recurrence relations

$$\frac{z_n^{(c)}(kr)}{kr} = \frac{1}{2n+1}\{z_{n-1}^{(c)}(kr) + z_{n+1}^{(c)}(kr)\} \tag{A1.7}$$

$$\frac{1}{kr}\frac{\mathrm{d}}{\mathrm{d}(kr)}\{krz_n^{(c)}(kr)\} = z_{n-1}^{(c)}(kr) + n\frac{z_n^{(c)}(kr)}{kr} \tag{A1.8}$$

$$= (n+1)\frac{z_n^{(n)}(kr)}{kr} - z_{n+1}^{(c)}(kr) \tag{A1.9}$$

$$= \frac{1}{2n+1}\{(n+1)z_{n-1}^{(c)}(kr) - nz_{n+1}^{(c)}(kr)\}. \tag{A1.10}$$

Wronskian

$$R_{sn}^{(c)}(kr)R_{3-s,n}^{(\gamma)}(kr) - R_{sn}^{(\gamma)}(kr)R_{3-s,n}^{(c)}(kr) = -(-1)^s\frac{A^{(c,\gamma)}}{(kr)^2} \tag{A1.11}$$

where the constant $A^{(c,\gamma)}$ is given by

$A^{(c,\gamma)}$	$\gamma=1$	$\gamma=2$	$\gamma=3$	$\gamma=4$
$c=1$	0	1	i	$-i$
$c=2$	-1	0	-1	-1
$c=3$	$-i$	1	0	$-2i$
$c=4$	i	1	$2i$	0

(A1.12)

Cross product

$$R_{1n}^{(1)}(kr)R_{2n}^{(1)}(kr) + R_{1n}^{(2)}(kr)R_{2n}^{(2)}(kr) = \left\{\frac{1}{kr} + \frac{1}{2}\frac{\mathrm{d}}{\mathrm{d}(kr)}\right\}|h_n^{(1)}(kr)|^2. \tag{A1.13}$$

Asymptotic behaviour

Large-argument approximations valid for $kr \gg n$ are

$$R_{1n}^{(3)}(kr) = z_n^{(3)}(kr) \to (-i)^{n+1}\exp(ikr)/(kr) \qquad kr\to\infty \tag{A1.14}$$

$$R_{1n}^{(4)}(kr) = z_n^{(4)}(kr) \to i^{n+1}\exp(-ikr)/(kr) \qquad kr\to\infty \tag{A1.15}$$

$$R_{2n}^{(3)}(kr) = \frac{1}{kr}\frac{\mathrm{d}}{\mathrm{d}(kr)}\{krz_n^{(3)}(kr)\} \to (-i)^n\exp(ikr)/(kr) \qquad kr\to\infty \tag{A1.16}$$

$$R_{2n}^{(4)}(kr) = \frac{1}{kr}\frac{\mathrm{d}}{\mathrm{d}(kr)}\{krz_n^{(4)}(kr)\} \to i^n\exp(-ikr)/(kr) \qquad kr\to\infty \tag{A1.17}$$

Special values

$$j_n(0) = 0, \quad n \geqslant 1 \tag{A1.18}$$

$$\lim_{x\to 0} \frac{j_n(x)}{x} = \begin{cases} \frac{1}{3}, & n = 1 \\ 0, & n > 1 \end{cases} \tag{A1.19}$$

$$\lim_{x\to 0} \frac{1}{x}\left(\frac{\mathrm{d}}{\mathrm{d}x}(xj_n(x))\right) = \begin{cases} \frac{2}{3}, & n = 1 \\ 0, & n > 1 \end{cases} \tag{A1.20}$$

Low-order functions, s = 1

$R_{1n}^{(1)}(kr) = j_n(kr)$. With $kr = x$:

$$\left.\begin{aligned}
j_0(x) &= x^{-1}\sin x \\
j_1(x) &= x^{-1}\{-\cos x + x^{-1}\sin x\} \\
j_2(x) &= x^{-1}\{-3x^{-1}\cos x + (-1 + 3x^{-2})\sin x\} \\
j_3(x) &= x^{-1}\{(1 - 15x^{-2})\cos x + (-6x^{-1} + 15x^{-3})\sin x\} \\
j_4(x) &= x^{-1}\{(10x^{-1} - 105x^{-3})\cos x + (1 - 45x^{-2} + 105x^{-4})\sin x\} \\
j_5(x) &= x^{-1}\{(-1 + 105x^{-2} - 945x^{-4})\cos x \\
&\qquad + (15x^{-1} - 420x^{-3} + 945x^{-5})\sin x\}
\end{aligned}\right] \tag{A1.21}$$

$R_{1n}^{(3)}(kr) = h_n^{(1)}(kr)$. With $kr = x$:

$$\left.\begin{aligned}
h_0^{(1)}(x) &= x^{-1}\exp(ix)\{-i\} \\
h_1^{(1)}(x) &= x^{-1}\exp(ix)\{-1 - ix^{-1}\} \\
h_2^{(1)}(x) &= x^{-1}\exp(ix)\{-3x^{-1} + i(1 - 3x^{-2})\} \\
h_3^{(1)}(x) &= x^{-1}\exp(ix)\{1 - 15x^{-2} + i(6x^{-1} - 15x^{-3})\} \\
h_4^{(1)}(x) &= x^{-1}\exp(ix)\{10x^{-1} - 105x^{-3} + i(-1 + 45x^{-2} - 105x^{-4})\} \\
h_5^{(1)}(x) &= x^{-1}\exp(ix)\{-1 + 105x^{-2} - 945x^{-4} \\
&\qquad + i(-15x^{-1} + 420x^{-3} - 945x^{-5})\}
\end{aligned}\right] \tag{A1.22}$$

Low-order functions, s = 2

$$R_{2n}^{(1)}(kr) = \frac{1}{kr}\frac{\mathrm{d}}{\mathrm{d}(kr)}\{krj_n(kr)\}. \qquad \text{With } kr = x:$$

$$\left.\begin{aligned}
R_{20}^{(1)}(x) &= x^{-1}\{\cos x\}\\
R_{21}^{(1)}(x) &= x^{-1}\{x^{-1}\cos x + (1 - x^2)\sin x\}\\
R_{22}^{(1)}(x) &= x^{-1}\{(-1 + 6x^{-2})\cos x + (3x^{-1} - 6x^{-3})\sin x\}\\
R_{23}^{(1)}(x) &= x^{-1}\{(-6x^{-1} + 45x^{-3})\cos x\\
&\qquad + (-1 + 21x^{-2} - 45x^{-4})\sin x\}\\
R_{24}^{(1)}(x) &= x^{-1}\{(1 - 55x^{-2} + 420x^{-4})\cos x\\
&\qquad + (-10x^{-1} + 195x^{-3} - 420x^{-5})\sin x\}\\
R_{25}^{(1)}(x) &= x^{-1}\{(15x^{-1} - 630x^{-3} + 4725x^{-5})\cos x\\
&\qquad + (1 - 120x^{-2} + 2205x^{-4} - 4725x^{-6})\sin x\}
\end{aligned}\right] \quad \text{(A1.23)}$$

$$R_{2n}^{(3)}(kr) = \frac{1}{kr}\frac{\mathrm{d}}{\mathrm{d}(kr)}\{krh_n^{(1)}(kr)\}. \qquad \text{With } kr = x:$$

$$\left.\begin{aligned}
R_{20}^{(3)}(x) &= x^{-1}\exp(ix)\\
R_{21}^{(3)}(x) &= x^{-1}\exp(ix)\{x^{-1} + i(-1 + x^{-2})\}\\
R_{22}^{(3)}(x) &= x^{-1}\exp(ix)\{(-1 + 6x^{-2}) + i(-3x^{-1} + 6x^{-3})\}\\
R_{23}^{(3)}(x) &= x^{-1}\exp(ix)\{(-6x^{-1} + 45x^{-3}) + i(1 - 21x^{-2} + 45x^{-4})\}\\
R_{24}^{(3)}(x) &= x^{-1}\exp(ix)\{(1 - 55x^{-2} + 420x^{-4})\\
&\qquad + i(10x^{-1} - 195x^{-3} + 420x^{-5})\}\\
R_{25}^{(3)}(x) &= x^{-1}\exp(ix)\{(15x^{-1} - 630x^{-3} + 4725x^{-5})\\
&\qquad + i(-1 + 120x^{-2} - 2205x^{-4} + 4725x^{-6})\}
\end{aligned}\right] \quad \text{(A1.24)}$$

A1.2.2 Angular functions

Definitions

The theta dependence of the spherical components of the spherical wave functions are expressed through three functions, all defined in the interval $0 \leqslant \theta \leqslant \pi$,

$$\bar{P}_n^{|m|}(\cos\theta), \qquad \frac{m}{\sin\theta}\bar{P}_n^{|m|}(\cos\theta) \qquad \text{and} \qquad \frac{\mathrm{d}}{\mathrm{d}\theta}\bar{P}_n^{|m|}(\cos\theta)$$

where $\bar{P}_n^m(\cos\theta)$ is the normalized associated Legendre function [2]. Note that only functions of non-negative order $|m|$ are employed. $\bar{P}_n^m(\cos\theta)$ is defined by the equation

$$\bar{P}_n^m(\cos\theta) = \sqrt{\frac{2n+1}{2}\frac{(n-m)!}{(n+m)!}}\,P_n^m(\cos\theta) \tag{A1.25}$$

where $P_n^m(\cos\theta)$ is the associated Legendre function, defined for $n > 0$ and $m \geqslant 0$ by

$$P_n^m(\cos\theta) = (\sin\theta)^m \frac{\mathrm{d}^m P_n(\cos\theta)}{\mathrm{d}(\cos\theta)^m}$$

where

$$P_n(\cos\theta) = \frac{1}{2^n n!}\frac{\mathrm{d}^n}{\mathrm{d}(\cos\theta)^n}(\cos^2\theta - 1)^n$$

is the Legendre polynomial. The definition of $P_n^m(\cos\theta)$ employed in the present book is the same as that of Stratton [1]. Another definition which includes a factor $(-1)^m$ relative to the present definition is given in [3]. The latter definition is also in use in electromagnetics [4].

Orthogonality

While the unnormalized associated Legendre functions satisfy

$$\int_{-1}^{1} P_k^m(\cos\theta)P_n^m(\cos\theta)\,\mathrm{d}(\cos\theta) = \frac{2}{2n+1}\frac{(n+m)!}{(n-m)!}\delta_{nk} \tag{A1.26}$$

we have for the normalized associated Legendre function

$$\int_{-1}^{1} \bar{P}_k^m(\cos\theta)\bar{P}_n^m(\cos\theta)\,\mathrm{d}(\cos\theta) = \delta_{nk}. \tag{A1.27}$$

In these formulas $\delta_{nk} = 1$ for $n = k$ while $\delta_{nk} = 0$ for $n \neq k$ (Kronecker's delta).

Other orthogonality integrals for the unnormalized associated Legendre functions are

$$\int_{-1}^{1} P_n^m(\cos\theta) P_n^k(\cos\theta) \sin^{-2}\theta \, \mathrm{d}(\cos\theta) = \frac{1}{m}\frac{(n+m)!}{(n-m)!}\delta_{mk} \qquad (m,k) \neq (0,0) \qquad \text{(A1.28)}$$

$$\int_{-1}^{1} \left\{ \frac{\mathrm{d}P_n^m(\cos\theta)}{\mathrm{d}\theta}\frac{\mathrm{d}P_k^m(\cos\theta)}{\mathrm{d}\theta} + \frac{m^2}{\sin^2\theta} P_n^m(\cos\theta) P_k^m(\cos\theta) \right\} \mathrm{d}(\cos\theta) = \frac{2}{2n+1}\frac{(n+m)!}{(n-m)!} n(n+1)\,\delta_{nk} \qquad \text{(A1.29)}$$

$$\int_{-1}^{1} \left\{ \frac{P_n^m(\cos\theta)}{\sin\theta}\frac{\mathrm{d}P_k^m(\cos\theta)}{\mathrm{d}\theta} + \frac{P_k^m(\cos\theta)}{\sin\theta}\frac{\mathrm{d}P_n^m(\cos\theta)}{\mathrm{d}\theta} \right\} \sin\theta \, \mathrm{d}(\cos\theta) = 0. \qquad \text{(A1.30)}$$

Recurrence relations

$$(n-m+1)P_{n+1}^m(\cos\theta) - (2n+1)\cos\theta\, P_n^m(\cos\theta) + (n+m)P_{n-1}^m(\cos\theta) = 0 \qquad \text{(A1.31)}$$

$$\sin\theta\, P_n^{m+1}(\cos\theta) - 2m\cos\theta\, P_n^m(\cos\theta) + (n+m)(n-m+1)\sin\theta\, P_n^{m-1}(\cos\theta) = 0 \qquad \text{(A1.32)}$$

$$P_n^n(\cos\theta) - (2n-1)\sin\theta\, P_{n-1}^{n-1}(\cos\theta) = 0 \qquad \text{(A1.33)}$$

$$\frac{mP_n^m(\cos\theta)}{\sin\theta} = \begin{cases} 0, & m = 0 \\ \tfrac{1}{2}\cos\theta\{(n-m+1)(n+m)P_n^{m-1}(\cos\theta) + P_n^{m+1}(\cos\theta)\} \\ \quad + m\sin\theta\, P_n^m(\cos\theta), & m > 0 \end{cases} \qquad \text{(A1.34a)}$$

$$\frac{\mathrm{d}P_n^m(\cos\theta)}{\mathrm{d}\theta} = \begin{cases} -P_n^1(\cos\theta), & m = 0 \\ \tfrac{1}{2}\{(n-m+1)(n+m)P_n^{m-1}(\cos\theta) \\ \quad - P_n^{m+1}(\cos\theta)\}, & m > 0 \end{cases} \qquad \text{(A1.34b)}$$

Fourier expansion

In the Fourier expansion

$$P_n^m(\cos\theta) = \sum_{m'=-n}^{n} c_{m'}\, \mathrm{e}^{\mathrm{i}m'\theta} \qquad \text{(A1.35a)}$$

the coefficients vanish for $(m'+n)$ odd. They satisfy the recurrence relation, ref. eqn (A2.37)

$$(n+m'+2)(n-m'-1)c_{m'+2} - 2(n^2 - m'^2 + n - 2m^2)c_{m'} + (n+m'-1)(n-m'+2)c_{m'-2} = 0 \qquad \text{(A1.35b)}$$

with the initial values

$$c_n = \left(-i\frac{m}{|m|}\right)^m \frac{1}{2^{2n}} \frac{(2n)!}{(n-m)!\,n!} \quad \text{and} \quad c_{n-2} = \frac{n-2m^2}{2n-1} c_n.$$

For $m = 0$, eqn (A1.35b) may be shown to take the form

$$(n+m')(n-m'+1)c_{m'} - (n+m'-1)(n-m'+2)c_{m'-2} = 0. \tag{A1.35c}$$

Special values

With the symbol

$$n!! = \begin{cases} n(n-2)\ldots 3\cdot 1, & n \text{ odd} \\ n(n-2)\ldots 4\cdot 2, & n \text{ even} \end{cases}$$

we have

$$P_n^{|m|}(\cos\theta)\bigg|_{\theta=0} = \begin{cases} 1 & |m| = 0 \\ 0 & |m| > 0 \end{cases} \tag{A1.36}$$

$$P_n^{|m|}(\cos\theta)\bigg|_{\theta=\pi/2} = \begin{cases} (-1)^{(n-|m|)/2}\dfrac{(n+|m|-1)!!}{(n-|m|)!!} & (n+|m|) \text{ even} \\ 0 & (n+|m|) \text{ odd} \end{cases} \tag{A1.37}$$

$$P_n^{|m|}(\cos\theta)\bigg|_{\theta=\pi} = \begin{cases} (-1)^n & |m| = 0 \\ 0 & |m| > 0 \end{cases} \tag{A1.38}$$

Further,

$$\frac{mP_n^{|m|}(\cos\theta)}{\sin\theta}\bigg|_{\theta=0} = \begin{cases} 0 & m \neq \pm 1 \\ \pm\dfrac{n(n+1)}{2} & m = \pm 1 \end{cases} \tag{A1.39}$$

$$\frac{mP_n^{|m|}(\cos\theta)}{\sin\theta}\bigg|_{\theta=\pi/2} = \begin{cases} m(-1)^{(n-|m|)/2}\dfrac{(n+|m|-1)!!}{(n-|m|)!!} & (n+|m|) \text{ even} \\ 0 & (n+|m|) \text{ odd} \end{cases} \tag{A1.40}$$

$$\frac{mP_n^{|m|}(\cos\theta)}{\sin\theta}\bigg|_{\theta=\pi} = \begin{cases} 0 & m \neq \pm 1 \\ \pm(-1)^{n+1}\dfrac{n(n+1)}{2} & m = \pm 1 \end{cases} \tag{A1.41}$$

and

$$\left.\frac{dP_n^{|m|}(\cos\theta)}{d\theta}\right|_{\theta=0} = \begin{cases} 0 & |m| \neq 1 \\ \dfrac{n(n+1)}{2} & |m| = 1 \end{cases} \tag{A1.42}$$

$$\left.\frac{dP_n^{|m|}(\cos\theta)}{d\theta}\right|_{\theta=\pi/2} = \begin{cases} 0 & (n+|m|) \text{ even} \\ (-1)^{(n-|m|+1)/2}\dfrac{(n+|m|)!!}{(n-|m|-1)!!} & (n+|m|) \text{ odd} \end{cases} \tag{A1.43}$$

$$\left.\frac{dP_n^{|m|}(\cos\theta)}{d\theta}\right|_{\theta=\pi} = \begin{cases} 0 & |m| \neq 1 \\ (-1)^n\dfrac{n(n+1)}{2} & |m| = 1 \end{cases} \tag{A1.44}$$

Low-order normalized functions

$\bar{P}_n^{|m|}(\cos\theta)$

	$n = 1$	$n = 2$	$n = 3$	$n = 4$	$n = 5$
$\lvert m\rvert = 5$					$\frac{3\sqrt{154}}{512}(\sin 5\theta - 5\sin 3\theta + 10\sin\theta)$
$\lvert m\rvert = 4$				$\frac{3\sqrt{35}}{128}(\cos 4\theta - 4\cos 2\theta + 3)$	$\frac{3\sqrt{385}}{256}(\cos 5\theta - 3\cos 3\theta + 2\cos\theta)$
$\lvert m\rvert = 3$			$-\frac{\sqrt{70}}{32}(\sin 3\theta - 3\sin\theta)$	$-\frac{3\sqrt{70}}{64}(\sin 4\theta - 2\sin 2\theta)$	$-\frac{\sqrt{770}}{512}(9\sin 5\theta - 13\sin 3\theta - 6\sin\theta)$
$\lvert m\rvert = 2$		$-\frac{\sqrt{15}}{8}(\cos 2\theta - 1)$	$-\frac{\sqrt{105}}{16}(\cos 3\theta - \cos\theta)$	$-\frac{3\sqrt{5}}{64}(7\cos 4\theta - 4\cos 2\theta - 3)$	$-\frac{\sqrt{1155}}{128}(3\cos 5\theta - \cos 3\theta - 2\cos\theta)$
$\lvert m\rvert = 1$	$\frac{\sqrt{3}}{2}\sin\theta$	$\frac{\sqrt{15}}{4}\sin 2\theta$	$\frac{\sqrt{42}}{32}(5\sin 3\theta + \sin\theta)$	$\frac{3\sqrt{10}}{64}(7\sin 4\theta + 2\sin 2\theta)$	$\frac{\sqrt{165}}{256}(21\sin 5\theta + 7\sin 3\theta + 2\sin\theta)$
$\lvert m\rvert = 0$	$\frac{\sqrt{6}}{2}\cos\theta$	$\frac{\sqrt{10}}{8}(3\cos 2\theta + 1)$	$\frac{\sqrt{14}}{16}(5\cos 3\theta + 3\cos\theta)$	$\frac{3\sqrt{2}}{128}(35\cos 4\theta + 20\cos 2\theta + 9)$	$\frac{\sqrt{22}}{256}(63\cos 5\theta + 35\cos 3\theta + 30\cos\theta)$

$$\frac{m}{\sin\theta}\overline{P}_n^{|m|}(\cos\theta)$$

	$n = 1$	$n = 2$	$n = 3$	$n = 4$	$n = 5$
$m = 5$					$\frac{15\sqrt{154}}{256}(\cos 4\theta - 4\cos 2\theta + 3)$
$m = 4$				$-\frac{3\sqrt{35}}{16}(\sin 3\theta - 3\sin\theta)$	$-\frac{3\sqrt{385}}{32}(\sin 4\theta - 2\sin 2\theta)$
$m = 3$			$-\frac{3\sqrt{70}}{16}(\cos 2\theta - 1)$	$-\frac{9\sqrt{70}}{32}(\cos 3\theta - \cos\theta)$	$-\frac{3\sqrt{770}}{256}(9\cos 4\theta - 4\cos 2\theta - 5)$
$m = 2$		$\frac{\sqrt{15}}{2}\sin\theta$	$\frac{\sqrt{105}}{4}\sin 2\theta$	$\frac{3\sqrt{5}}{16}(7\sin 3\theta + 3\sin\theta)$	$\frac{\sqrt{1155}}{32}(3\sin 4\theta + 2\sin 2\theta)$
$m = 1$	$\frac{\sqrt{3}}{2}$	$\frac{\sqrt{15}}{2}\cos\theta$	$\frac{\sqrt{42}}{16}(5\cos 2\theta + 3)$	$\frac{3\sqrt{10}}{32}(7\cos 3\theta + 9\cos\theta)$	$\frac{\sqrt{165}}{128}(21\cos 4\theta + 28\cos 2\theta + 15)$
$m = 0$	0	0	0	0	0

$$\frac{d}{d\theta}\bar{P}_n^{|m|}(\cos\theta)$$

	$n=1$	$n=2$	$n=3$	$n=4$	$n=5$
$\lvert m\rvert=5$					$\frac{15\sqrt{154}}{512}(\cos 5\theta - 3\cos 3\theta + 2\cos\theta)$
$\lvert m\rvert=4$				$-\frac{3\sqrt{35}}{32}(\sin 4\theta - 2\sin 2\theta)$	$-\frac{3\sqrt{385}}{256}(5\sin 5\theta - 9\sin 3\theta + 2\sin\theta)$
$\lvert m\rvert=3$			$-\frac{3\sqrt{70}}{32}(\cos 3\theta - \cos\theta)$	$-\frac{3\sqrt{70}}{16}(\cos 4\theta - \cos 2\theta)$	$-\frac{3\sqrt{770}}{512}(15\cos 5\theta - 13\cos 3\theta - 2\cos\theta)$
$\lvert m\rvert=2$		$\frac{\sqrt{15}}{4}\sin 2\theta$	$\frac{\sqrt{105}}{16}(3\sin 3\theta - \sin\theta)$	$\frac{3\sqrt{5}}{16}(7\sin 4\theta - 2\sin 2\theta)$	$\frac{\sqrt{1155}}{128}(15\sin 5\theta - 3\sin 3\theta - 2\sin\theta)$
$\lvert m\rvert=1$	$\frac{\sqrt{3}}{2}\cos\theta$	$\frac{\sqrt{15}}{2}\cos 2\theta$	$\frac{\sqrt{42}}{32}(15\cos 3\theta + \cos\theta)$	$\frac{3\sqrt{10}}{16}(7\cos 4\theta + \cos 2\theta)$	$\frac{\sqrt{165}}{256}(105\cos 5\theta + 21\cos 3\theta + 2\cos\theta)$
$\lvert m\rvert=0$	$-\frac{\sqrt{6}}{2}\sin\theta$	$-\frac{3\sqrt{10}}{4}\sin 2\theta$	$-\frac{3\sqrt{14}}{16}(5\sin 3\theta + \sin\theta)$	$-\frac{15\sqrt{2}}{32}(7\sin 4\theta + 2\sin 2\theta)$	$-\frac{15\sqrt{22}}{256}(21\sin 5\theta + 7\sin 3\theta + 2\sin\theta)$

A1.3 Wave functions

A1.3.1 The spherical wave functions

General expressions

$$\vec{F}^{(c)}_{1mn}(r,\theta,\phi) = \frac{1}{\sqrt{2\pi}}\frac{1}{\sqrt{n(n+1)}}\left(-\frac{m}{|m|}\right)^m$$

$$\left\{z_n^{(c)}(kr)\frac{im\bar{P}_n^{|m|}(\cos\theta)}{\sin\theta}e^{im\phi}\hat{\theta} - z_n^{(c)}(kr)\frac{d\bar{P}_n^{|m|}(\cos\theta)}{d\theta}e^{im\phi}\hat{\phi}\right\} \qquad \text{(A1.45)}$$

$$\vec{F}^{(c)}_{2mn}(r,\theta,\phi) = \frac{1}{\sqrt{2\pi}}\frac{1}{\sqrt{n(n+1)}}\left(-\frac{m}{|m|}\right)^m$$

$$\left\{\frac{n(n+1)}{kr}z_n^{(c)}(kr)\bar{P}_n^{|m|}(\cos\theta)\,e^{im\phi}\hat{r}\right.$$

$$+\frac{1}{kr}\frac{d}{d(kr)}\{krz_n^{(c)}(kr)\}\frac{d\bar{P}_n^{|m|}(\cos\theta)}{d\theta}e^{im\phi}\hat{\theta}$$

$$\left.+\frac{1}{kr}\frac{d}{d(kr)}\{krz_n^{(c)}(kr)\}\frac{im\bar{P}_n^{|m|}(\cos\theta)}{\sin\theta}e^{im\phi}\hat{\phi}\right\} \qquad \text{(A1.46)}$$

Special values

Note that the wave functions in (A1.47–1.50) may be expressed in terms of the rectangular unit vectors $\hat{x}$, $\hat{y}$ and $\hat{z}$, ref. Section A1.7.3.

$$\vec{F}^{(c)}_{1mn}(r,0,\phi) = 0, \qquad m > 1$$

$$\vec{F}^{(c)}_{11n}(r,0,\phi) = -\frac{1}{4}\sqrt{\frac{2n+1}{\pi}}\,z_n^{(c)}(kr)i\,e^{i\phi}(\hat{\theta}+i\hat{\phi})$$

$$\vec{F}^{(c)}_{10n}(r,0,\phi) = 0$$

$$\vec{F}^{(c)}_{1,-1,n}(r,0,\phi) = -\frac{1}{4}\sqrt{\frac{2n+1}{\pi}}\,z_n^{(c)}(kr)i\,e^{-i\phi}(\hat{\theta}-i\hat{\phi})$$

$$\vec{F}^{(c)}_{1mn}(r,0,\phi) = 0, \qquad m < -1 \qquad \text{(A1.47)}$$

$$\left.\begin{aligned}
&\vec{F}^{(c)}_{2mn}(r, 0, \phi) = 0, \qquad m > 1\\
&\vec{F}^{(c)}_{21n}(r, 0, \phi) = -\frac{1}{4}\sqrt{\frac{2n+1}{\pi}}\frac{1}{kr}\frac{\mathrm{d}}{\mathrm{d}(kr)}\{krz^{(c)}_n(kr)\}\,\mathrm{e}^{i\phi}(\hat{\theta} + i\hat{\phi})\\
&\vec{F}^{(c)}_{20n}(r, 0, \phi) = \sqrt{\frac{n(n+1)(2n+1)}{4\pi}}\frac{z^{(c)}_n(kr)}{kr}\hat{r}\\
&\vec{F}^{(c)}_{2,-1,n}(r, 0, \phi) = \frac{1}{4}\sqrt{\frac{2n+1}{\pi}}\frac{1}{kr}\frac{\mathrm{d}}{\mathrm{d}(kr)}\{krz^{(c)}_n(kr)\}\,\mathrm{e}^{-i\phi}(\hat{\theta} - i\hat{\phi})\\
&\vec{F}^{(c)}_{2mn}(r, 0, \phi) = 0, \qquad m < -1
\end{aligned}\right] \tag{A1.48}$$

$$\left.\begin{aligned}
&\vec{F}^{(c)}_{1mn}(r, \pi, \phi) = 0, \qquad m > 1\\
&\vec{F}^{(c)}_{11n}(r, \pi, \phi) = (-1)^n\frac{1}{4}\sqrt{\frac{2n+1}{\pi}}\,z^{(c)}_n(kr)i\,\mathrm{e}^{i\phi}(\hat{\theta} - i\hat{\phi})\\
&\vec{F}^{(c)}_{10n}(r, \pi, \phi) = 0\\
&\vec{F}^{(c)}_{1,-1,n}(r, \pi, \phi) = (-1)^n\frac{1}{4}\sqrt{\frac{2n+1}{\pi}}\,z^{(c)}_n(kr)i\,\mathrm{e}^{-i\phi}(\hat{\theta} + i\hat{\phi})\\
&\vec{F}^{(c)}_{1mn}(r, \pi, \phi) = 0, \qquad m < -1
\end{aligned}\right] \tag{A1.49}$$

$$\left.\begin{aligned}
&\vec{F}^{(c)}_{2mn}(r, \pi, \phi) = 0, \qquad m > 1\\
&\vec{F}^{(c)}_{21n}(r, \pi, \phi) = -(-1)^n\frac{1}{4}\sqrt{\frac{2n+1}{\pi}}\frac{1}{kr}\frac{\mathrm{d}}{\mathrm{d}(kr)}\{krz^{(c)}_n(kr)\}\mathrm{e}^{i\phi}(\hat{\theta} - i\hat{\phi})\\
&\vec{F}^{(c)}_{20n}(r, \pi, \phi) = (-1)^n\sqrt{\frac{n(n+1)(2n+1)}{4\pi}}\frac{z^{(c)}_n(kr)}{kr}\hat{r}\\
&\vec{F}^{(c)}_{2,-1,n}(r, \pi, \phi) = (-1)^n\frac{1}{4}\sqrt{\frac{2n+1}{\pi}}\frac{1}{kr}\frac{\mathrm{d}}{\mathrm{d}(kr)}\{krz^{(c)}_n(kr)\}\mathrm{e}^{-i\phi}(\hat{\theta} + i\hat{\phi})\\
&\vec{F}^{(c)}_{2mn}(r, \pi, \phi) = 0, \qquad m < -1
\end{aligned}\right] \tag{A1.50}$$

$$\left.\begin{aligned}
&\vec{F}^{(1)}_{111}(0, \theta, \phi) = 0\\
&\vec{F}^{(1)}_{101}(0, \theta, \phi) = 0\\
&\vec{F}^{(1)}_{1,-1,1}(0, \theta, \phi) = 0
\end{aligned}\right] \tag{A1.51}$$

$$\vec{F}^{(1)}_{211}(0,\theta,\phi) = -\frac{\sqrt{3}}{6\sqrt{\pi}} e^{i\phi}(\sin\theta\,\hat{r} + \cos\theta\,\hat{\theta} + i\hat{\phi}) = -\frac{\sqrt{3}}{6\sqrt{\pi}}(\hat{x} + i\hat{y})$$

$$\vec{F}^{(1)}_{201}(0,\theta,\phi) = \frac{\sqrt{6}}{6\sqrt{\pi}}(\cos\theta\,\hat{r} - \sin\theta\,\hat{\theta}) = \frac{\sqrt{6}}{6\sqrt{\pi}}\hat{z}$$

$$\vec{F}^{(1)}_{2,-1,1}(0,\theta,\phi) = \frac{\sqrt{3}}{6\sqrt{\pi}} e^{-i\phi}(\sin\theta\,\hat{r} + \cos\theta\,\hat{\theta} - i\hat{\phi}) = \frac{\sqrt{3}}{6\sqrt{\pi}}(\hat{x} - i\hat{y}) \qquad \text{(A1.52)}$$

$$\vec{F}^{(1)}_{smn}(0,\theta,\phi) = 0, \qquad n \neq 1. \qquad \text{(A1.53)}$$

ecial relation

$$\vec{F}^{(3)*}_{smn}(r,\theta,\phi) = (-1)^m \vec{F}^{(4)}_{s,-m,n}(r,\theta,\phi). \qquad \text{(A1.54)}$$

e functions with $n = 1$

$$\vec{F}^{(1)}_{111}(r,\theta,\phi) = \frac{-\sqrt{3}}{4\sqrt{\pi}} e^{i\phi}\frac{1}{kr}\left\{-\cos(kr) + \frac{\sin(kr)}{kr}\right\}(i\hat{\theta} - \cos\theta\hat{\phi})$$

$$\vec{F}^{(1)}_{101}(r,\theta,\phi) = \frac{\sqrt{6}}{4\sqrt{\pi}}\frac{1}{kr}\left\{-\cos(kr) + \frac{\sin(kr)}{kr}\right\}\sin\theta\hat{\phi}$$

$$\vec{F}^{(1)}_{1,-1,1}(r,\theta,\phi) = \frac{-\sqrt{3}}{4\sqrt{\pi}} e^{-i\phi}\frac{1}{kr}\left\{-\cos(kr) + \frac{\sin(kr)}{kr}\right\}(i\hat{\theta} + \cos\theta\hat{\phi}) \qquad \text{(A1.55)}$$

$$\begin{aligned}
\vec{F}^{(1)}_{211}(r,\theta,\phi) &= \frac{-\sqrt{3}}{2\sqrt{\pi}} e^{i\phi}\frac{1}{(kr)^2}\left\{-\cos(kr) + \frac{\sin(kr)}{kr}\right\}\sin\theta\hat{r} \\
&\quad - \frac{\sqrt{3}}{4\sqrt{\pi}} e^{i\phi}\frac{1}{kr}\left\{\sin(kr) + \frac{\cos(kr)}{kr} - \frac{\sin(kr)}{(kr)^2}\right\}(\cos\theta\hat{\theta} + i\hat{\phi}) \\
\vec{F}^{(1)}_{201}(r,\theta,\phi) &= \frac{\sqrt{6}}{2\sqrt{\pi}}\frac{1}{(kr)^2}\left\{-\cos(kr) + \frac{\sin(kr)}{kr}\right\}\cos\theta\hat{r} \\
&\quad - \frac{\sqrt{6}}{4\sqrt{\pi}}\frac{1}{kr}\left\{\sin(kr) + \frac{\cos(kr)}{kr} - \frac{\sin(kr)}{(kr)^2}\right\}\sin\theta\hat{\theta} \\
\vec{F}^{(1)}_{2,-1,1}(r,\theta,\phi) &= \frac{\sqrt{3}}{2\sqrt{\pi}} e^{-i\phi}\frac{1}{(kr)^2}\left\{-\cos(kr) + \frac{\sin(kr)}{kr}\right\}\sin\theta\hat{r} \\
&\quad + \frac{\sqrt{3}}{4\sqrt{\pi}} e^{-i\phi}\frac{1}{kr}\left\{\sin(kr) + \frac{\cos(kr)}{kr} - \frac{\sin(kr)}{(kr)^2}\right\}(\cos\theta\hat{\theta} - i\hat{\phi})
\end{aligned} \qquad \text{(A1.56)}$$

$$\vec{F}^{(3)}_{111}(r,\theta,\phi) = \frac{\sqrt{3}}{4\sqrt{\pi}} e^{i\phi} \frac{e^{ikr}}{kr}\left(1+\frac{i}{kr}\right)(i\hat{\theta} - \cos\theta\hat{\phi})$$

$$\vec{F}^{(3)}_{101}(r,\theta,\phi) = -\frac{\sqrt{6}}{4\sqrt{\pi}} \frac{e^{ikr}}{kr}\left(1+\frac{i}{kr}\right)\sin\theta\hat{\phi}$$

$$\vec{F}^{(3)}_{1,-1,1}(r,\theta,\phi) = \frac{\sqrt{3}}{4\sqrt{\pi}} e^{-i\phi} \frac{e^{ikr}}{kr}\left(1+\frac{i}{kr}\right)(i\hat{\theta} + \cos\theta\hat{\phi}) \tag{A1.57}$$

$$\begin{aligned}\vec{F}^{(3)}_{211}(r,\theta,\phi) &= \frac{\sqrt{3}}{2\sqrt{\pi}} e^{i\phi} \frac{e^{ikr}}{(kr)^2}\left(1+\frac{i}{kr}\right)\sin\theta\hat{r} \\ &\quad - \frac{\sqrt{3}}{4\sqrt{\pi}} e^{i\phi} \frac{e^{ikr}}{kr}\left(-i+\frac{1}{kr}+\frac{i}{(kr)^2}\right)(\cos\theta\hat{\theta} + i\hat{\phi}) \\ \vec{F}^{(3)}_{201}(r,\theta,\phi) &= -\frac{\sqrt{6}}{2\sqrt{\pi}} \frac{e^{ikr}}{(kr)^2}\left(1+\frac{i}{kr}\right)\cos\theta\hat{r} \\ &\quad - \frac{\sqrt{6}}{4\sqrt{\pi}} \frac{e^{ikr}}{kr}\left(-i+\frac{1}{kr}+\frac{i}{(kr)^2}\right)\sin\theta\hat{\theta} \\ \vec{F}^{(3)}_{2,-1,1}(r,\theta,\phi) &= -\frac{\sqrt{3}}{2\sqrt{\pi}} e^{-i\phi} \frac{e^{ikr}}{(kr)^2}\left(1+\frac{i}{kr}\right)\sin\theta\hat{r} \\ &\quad + \frac{\sqrt{3}}{4\sqrt{\pi}} e^{-i\phi} \frac{e^{ikr}}{kr}\left(-i+\frac{1}{kr}+\frac{i}{(kr)^2}\right)(\cos\theta\hat{\theta} - i\hat{\phi})\end{aligned} \tag{A1.58}$$

A1.3.2 The far-field pattern functions

Definition: See eqn (A1.5).

General expressions

$$\begin{aligned}\vec{K}_{1mn}(\theta,\phi) &= \sqrt{\frac{2}{n(n+1)}}\left(-\frac{m}{|m|}\right)^m e^{im\phi}(-i)^{n+1} \\ &\quad \left\{\frac{im\bar{P}_n^{|m|}(\cos\theta)}{\sin\theta}\hat{\theta} - \frac{d\bar{P}_n^{|m|}(\cos\theta)}{d\theta}\hat{\phi}\right\}\end{aligned} \tag{A1.59}$$

$$\begin{aligned}\vec{K}_{2mn}(\theta,\phi) &= \sqrt{\frac{2}{n(n+1)}}\left(-\frac{m}{|m|}\right)^m e^{im\phi}(-i)^{n} \\ &\quad \left\{\frac{d\bar{P}_n^{|m|}(\cos\theta)}{d\theta}\hat{\theta} + \frac{im\bar{P}_n^{|m|}(\cos\theta)}{\sin\theta}\hat{\phi}\right\}\end{aligned} \tag{A1.60}$$

Special values

Note that the wave functions in (A1.61–1.64) may be expressed in terms of the rectangular unit vectors $\hat{x}$, $\hat{y}$ and $\hat{z}$, ref. Section A1.7.3.

$$\left.\begin{aligned}
\vec{K}_{1mn}(0,\phi) &= 0, \qquad m > 1\\
\vec{K}_{11n}(0,\phi) &= -(-i)^n \tfrac{1}{2}\sqrt{2n+1}\,\mathrm{e}^{i\phi}(\hat{\theta} + i\hat{\phi})\\
\vec{K}_{10n}(0,\phi) &= 0\\
\vec{K}_{1,-1,n}(0,\phi) &= -(-i)^n \tfrac{1}{2}\sqrt{2n+1}\,\mathrm{e}^{-i\phi}(\hat{\theta} - i\hat{\phi})\\
\vec{K}_{1mn}(0,\phi) &= 0, \qquad m < -1
\end{aligned}\right] \tag{A1.61}$$

$$\left.\begin{aligned}
\vec{K}_{2mn}(0,\phi) &= 0, \qquad m > 1\\
\vec{K}_{21n}(0,\phi) &= -(-i)^n \tfrac{1}{2}\sqrt{2n+1}\,\mathrm{e}^{i\phi}(\hat{\theta} + i\hat{\phi})\\
\vec{K}_{20n}(0,\phi) &= 0\\
\vec{K}_{2,-1,n}(0,\phi) &= (-i)^n \tfrac{1}{2}\sqrt{2n+1}\,\mathrm{e}^{-i\phi}(\hat{\theta} - i\hat{\phi})\\
\vec{K}_{2mn}(0,\phi) &= 0, \qquad m < -1
\end{aligned}\right] \tag{A1.62}$$

$$\left.\begin{aligned}
\vec{K}_{1mn}(\pi,\phi) &= 0, \qquad m > 1\\
\vec{K}_{11n}(\pi,\phi) &= i^n \tfrac{1}{2}\sqrt{2n+1}\,\mathrm{e}^{i\phi}(\hat{\theta} - i\hat{\phi})\\
\vec{K}_{10n}(\pi,\phi) &= 0\\
\vec{K}_{1,-1,n}(\pi,\phi) &= i^n \tfrac{1}{2}\sqrt{2n+1}\,\mathrm{e}^{-i\phi}(\hat{\theta} + i\hat{\phi})\\
\vec{K}_{1mn}(\pi,\phi) &= 0, \qquad m < -1
\end{aligned}\right] \tag{A1.63}$$

$$\left.\begin{aligned}
\vec{K}_{2mn}(\pi,\phi) &= 0, \qquad m > 1\\
\vec{K}_{21n}(\pi,\phi) &= -i^n \tfrac{1}{2}\sqrt{2n+1}\,\mathrm{e}^{i\phi}(\hat{\theta} - i\hat{\phi})\\
\vec{K}_{20n}(\pi,\phi) &= 0\\
\vec{K}_{2,-1,n}(\pi,\phi) &= i^n \tfrac{1}{2}\sqrt{2n+1}\,\mathrm{e}^{-i\phi}(\hat{\theta} + i\hat{\phi})\\
\vec{K}_{2mn}(\pi,\phi) &= 0, \qquad m < -1
\end{aligned}\right] \tag{A1.64}$$

$$\begin{aligned}
|\vec{K}_{smn}(0,\phi)|^2 &= |\vec{K}_{smn}(0,\phi)\cdot\hat{\theta}|^2 + |\vec{K}_{smn}(0,\phi)\cdot\hat{\phi}|^2\\
&= \begin{cases} \dfrac{2n+1}{2} & m = \pm 1\\ 0 & m \neq \pm 1. \end{cases}
\end{aligned} \tag{A1.65}$$

Far-field pattern functions with n = 1

$$\vec{K}_{111}(\theta,\phi)=\frac{\sqrt{3}}{2}e^{i\phi}(i\hat{\theta}-\cos\theta\hat{\phi})$$

$$\vec{K}_{101}(\theta,\phi)=-\frac{\sqrt{6}}{2}\sin\theta\hat{\phi}$$

$$\vec{K}_{1,-1,1}(\theta,\phi)=\frac{\sqrt{3}}{2}e^{-i\phi}(i\hat{\theta}+\cos\theta\hat{\phi}) \quad \text{(A1.66)}$$

$$\vec{K}_{211}(\theta,\phi)=i\frac{\sqrt{3}}{2}e^{i\phi}(\cos\theta\hat{\theta}+i\hat{\phi})$$

$$\vec{K}_{201}(\theta,\phi)=i\frac{\sqrt{6}}{2}\sin\theta\hat{\theta}$$

$$\vec{K}_{2,-1,1}(\theta,\phi)=i\frac{\sqrt{3}}{2}e^{-i\phi}(-\cos\theta\hat{\theta}+i\hat{\phi}) \quad \text{(A1.67)}$$

A1.4 Orthogonality of spherical wave functions

In the formulas below, $\delta_{ij}=1$ for $i=j$ while $\delta_{ij}=0$ for $i\neq j$ (Kronecker's delta).

A1.4.1 Product of radial components

$$\int_{\phi=0}^{2\pi}\int_{\theta=0}^{\pi}\{\vec{F}^{(c)}_{smn}(r,\theta,\phi)\cdot\hat{r}\}\{\vec{F}^{(\gamma)}_{\sigma\mu\nu}(r,\theta,\phi)\cdot\hat{r}\}\sin\theta\,d\theta\,d\phi$$

$$=\delta_{s\sigma}\delta_{s2}\delta_{m,-\mu}\delta_{n\nu}(-1)^m n(n+1)\frac{z_n^{(c)}(kr)}{kr}\frac{z_n^{(\gamma)}(kr)}{kr}. \quad \text{(A1.68)}$$

A1.4.2 Products of tangential components

$$\int_{\phi=0}^{2\pi}\int_{\theta=0}^{\pi}\Big[\{\vec{F}^{(c)}_{smn}(r,\theta,\phi)\cdot\hat{\theta}\}\{\vec{F}^{(\gamma)}_{\sigma\mu\nu}(r,\theta,\phi)\cdot\hat{\theta}\}$$

$$+\{\vec{F}^{(c)}_{smn}(r,\theta,\phi)\cdot\hat{\phi}\}\{\vec{F}^{(\gamma)}_{\sigma\mu\nu}(r,\theta,\phi)\cdot\hat{\phi}\}\Big]\sin\theta\,d\theta\,d\phi$$

$$=\delta_{s\sigma}\delta_{m,-\mu}\delta_{n\nu}(-1)^m R^{(c)}_{sn}(kr)R^{(\gamma)}_{sn}(kr). \quad \text{(A1.69)}$$

A1.4.3 Scalar product

$$\int_{\phi=0}^{2\pi}\int_{\theta=0}^{\pi} \vec{F}^{(c)}_{smn}(r,\theta,\phi)\cdot\vec{F}^{(\gamma)}_{\sigma\mu\nu}(r,\theta,\phi)\sin\theta\,\mathrm{d}\theta\,\mathrm{d}\phi$$

$$= \delta_{s\sigma}\delta_{m,-\mu}\delta_{n\nu}(-1)^m\left\{R^{(c)}_{sn}(kr)\,R^{(\gamma)}_{sn}(kr) + \delta_{s2}n(n+1)\frac{z^{(c)}_n(kr)}{kr}\frac{z^{(\gamma)}_n(kr)}{kr}\right\}. \tag{A1.70}$$

A1.4.4 Vector product

$$\int_{\phi=0}^{2\pi}\int_{\theta=0}^{\pi} \{\vec{F}^{(c)}_{smn}(r,\theta,\phi)\times\vec{F}^{(\gamma)}_{\sigma\mu\nu}(r,\theta,\phi)\}\cdot\hat{r}\sin\theta\,\mathrm{d}\theta\,\mathrm{d}\phi$$

$$= -\delta_{s,3-\sigma}\delta_{m,-\mu}\delta_{n\nu}(-1)^{m+s}R^{(c)}_{sn}(kr)\,R^{(\gamma)}_{3-s,n}(kr). \tag{A1.71}$$

A1.4.5 Reciprocity integral [5]

For the two fields, both representing a single mode with wave coefficients equal to 1 watts$^{1/2}$

$$(\vec{E}^{(c)}_{smn},\vec{H}^{(c)}_{smn}) = \left(\frac{k}{\sqrt{\eta}}\vec{F}^{(c)}_{smn}(r,\theta,\phi),\ -ik\sqrt{\eta}\,\vec{F}^{(c)}_{3-s,m,n}(r,\theta,\phi)\right) \tag{A1.72}$$

$$(\vec{E}^{(\gamma)}_{\sigma\mu\nu},\vec{H}^{(\gamma)}_{\sigma\mu\nu}) = \left(\frac{k}{\sqrt{\eta}}\vec{F}^{(\gamma)}_{\sigma\mu\nu}(r,\theta,\phi),\ -ik\sqrt{\eta}\,\vec{F}^{(\gamma)}_{3-\sigma,\mu,\nu}(r,\theta,\phi)\right) \tag{A1.73}$$

we have

$$\int_S \{\vec{E}^{(c)}_{smn}\times\vec{H}^{(\gamma)}_{\sigma\mu\nu} - \vec{E}^{(\gamma)}_{\sigma\mu\nu}\times\vec{H}^{(c)}_{smn}\}\cdot\mathrm{d}\vec{S}$$

$$= \delta_{s\sigma}\delta_{m,-\mu}\delta_{n\nu}(-1)^m(-i)A^{(c,\gamma)} \tag{A1.74}$$

where the constant $A^{(c,\gamma)}$ is given by (A1.12). The right-hand side contains the product of two wave coefficients, both equal to 1 watts$^{1/2}$.

The surface of integration in (A1.74) is a spherical surface S of radius r with outward area element $\mathrm{d}\vec{S} = \hat{r}r^2\sin\theta\,\mathrm{d}\theta\,\mathrm{d}\phi$. Consider another closed surface S' with outward area element $\mathrm{d}\vec{S}'$ such that there are no sources between S and S'. Then, according to the reciprocity theorem (ref. Section A1.5.1) S and $\mathrm{d}\vec{S}$ may be replaced by S' and $\mathrm{d}\vec{S}'$, respectively, in eqn (A1.74) without consequences for the value of the integral.

A1.5 Calculation of spherical wave coefficients from current distributions

A1.5.1 General formulation

The reciprocity theorem

Consider a volume V bounded by a surface S and assume two possible sets of sources $(\vec{J}_1, \vec{M}_1)$ and $(\vec{J}_2, \vec{M}_2)$ producing the fields $(\vec{E}_1, \vec{H}_1)$ and $(\vec{E}_2, \vec{H}_2)$, respectively, in V and on S. The general form of the reciprocity theorem is then [6]

$$\int_S (\vec{E}_1 \times \vec{H}_2 - \vec{E}_2 \times \vec{H}_1) \cdot \mathrm{d}\vec{S} = \int_V (\vec{E}_2 \cdot \vec{J}_1 - \vec{E}_1 \cdot \vec{J}_2 - \vec{H}_2 \cdot \vec{M}_1 + \vec{H}_1 \cdot \vec{M}_2)\,\mathrm{d}V \quad \text{(A1.75)}$$

where $\mathrm{d}V$ is the volume element and $\mathrm{d}\vec{S}$ the outward-oriented area element.

The reciprocity theorem will be applied below for the determination of the spherical wave coefficients of the field $(\vec{E}, \vec{H})$ radiated by a given distribution of electric and magnetic currents $(\vec{J}, \vec{M})$ in a linear, isotropic and homogeneous medium. Two distinct cases will be considered, the exterior field and the interior field.

The exterior field

With reference to Fig. A1.1, we consider the field $(\vec{E}, \vec{H})$ in the spherical waveguide section outside a spherical surface S enclosing the sources $(\vec{J}, \vec{M})$. In this region $(\vec{E}, \vec{H})$ has a spherical mode expansion involving only spherical wave functions of the outward-travelling type, e.g. for the electric field

$$\vec{E}(r, \theta, \phi) = \frac{k}{\sqrt{\eta}} \sum_{smn} Q^{(3)}_{smn} \vec{F}^{(3)}_{smn}(r, \theta, \phi). \quad \text{(A1.76)}$$

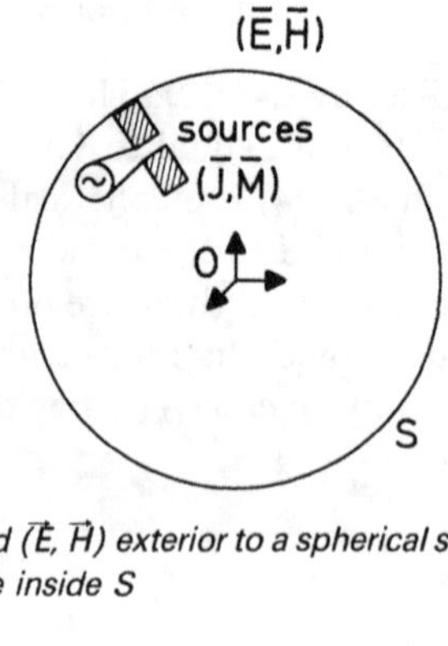

Fig. A1.1 *Illustrating the field $(\vec{E}, \vec{H})$ exterior to a spherical surface S centred at the origin. The sources $(\vec{J}, \vec{M})$ are inside S*

A formula expressing the coefficients $Q^{(3)}_{smn}$ in terms of the known current distribution $(\vec{J}, \vec{M})$ can be achieved through application of the reciprocity theorem. Substitution of $(\vec{E}, \vec{H})$ for one of the fields in this theorem and the single mode $(\vec{E}^{(1)}_{smn}, \vec{H}^{(1)}_{smn})$ of (A1.72) for the other field leads to

$$Q^{(3)}_{smn} = (-1)^{m+1} \int_V (\vec{E}^{(1)}_{s,-m,n} \cdot \vec{J} - \vec{H}^{(1)}_{s,-m,n} \cdot \vec{M})\,\mathrm{d}V \tag{A1.77}$$

$$= (-1)^{m+1} \int_V \left(\frac{k}{\sqrt{\eta}} \vec{F}^{(1)}_{s,-m,n} \cdot \vec{J} + ik\sqrt{\eta}\,\vec{F}^{(1)}_{3-s,-m,n} \cdot \vec{M} \right) dV \tag{A1.78}$$

where V is the interior region containing the sources $(\vec{J}, \vec{M})$. We have made use of the fact that V contains no sources of the mode field $(\vec{E}^{(1)}_{smn}, \vec{H}^{(1)}_{smn})$, and we have used the integral (A1.74).

The interior field

With reference to Fig. A1.2, we now consider the field $(\vec{E}, \vec{H})$ in the spherical waveguide section inside a spherical surface S having the sources $(\vec{J}, \vec{M})$ on its

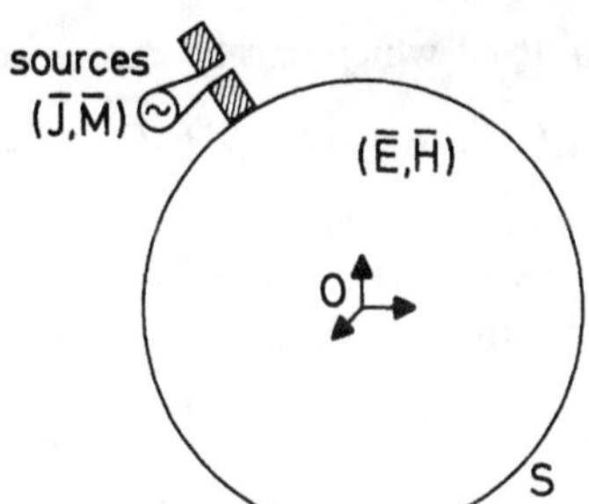

Fig. A1.2 *Illustrating the field $(\vec{E}, \vec{H})$ inside a spherical surface S centred at the origin. The sources $(\vec{J}, \vec{M})$ are outside S*

outside. In this interior region, $(\vec{E}, \vec{H})$ has a spherical mode expansion involving only spherical wave functions of the standing wave type, e.g. for the electric field

$$\vec{E}(r, \theta, \phi) = \frac{k}{\sqrt{\eta}} \sum_{smn} Q^{(1)}_{smn} \vec{F}^{(1)}_{smn}(r, \theta, \phi). \tag{A1.79}$$

Again we apply the reciprocity theorem to obtain a formula for the expansion coefficients $Q^{(1)}_{smn}$. In this case however, the volume V is the region between S and a spherical surface S_∞ with radius tending towards infinity. For one of the fields in the reciprocity theorem we substitute $(\vec{E}^{(3)}_{smn}, \vec{H}^{(3)}_{smn})$ of (A1.72), and for the other field we substitute $(\vec{E}, \vec{H})$. Application of the reciprocity theorem and the integral (A1.74) then leads to

$$Q^{(1)}_{smn} = (-1)^{m+1} \int_V (\vec{E}^{(3)}_{s,-m,n} \cdot \vec{J} - \vec{H}^{(3)}_{s,-m,n} \cdot \vec{M})\,\mathrm{d}V \tag{A1.80}$$

$$= (-1)^{m+1} \int_V \left(\frac{k}{\sqrt{\eta}} \vec{F}^{(3)}_{s,-m,n} \cdot \vec{J} + ik\sqrt{\eta}\,\vec{F}^{(3)}_{3-s,-m,n} \cdot \vec{M} \right) \mathrm{d}V \tag{A1.81}$$

since $(\vec{E}^{(3)}_{smn}, \vec{H}^{(3)}_{smn})$ has no sources in V. The integral over S_∞ vanishes because both $(\vec{E}, \vec{H})$ and $(\vec{E}^{(3)}_{smn}, \vec{H}^{(3)}_{smn})$ satisfy the radiation condition at infinity.

A1.5.2 Source symmetries

Planar and rotational source symmetries related to the z-axis have simple mathematical formulations in spherical coordinates. This fact is easily exploited to derive corresponding symmetries in the spherical mode expansion coefficients for the radiated fields. Without loss of generality we shall restrict the treatment in the remainder of Section A1.5 to comprise electric sources $\vec{J}$ only.

Rotational symmetry

A current distribution $\vec{J}(r, \theta, \phi)$ which can be expressed in the following form

$$\vec{J}(r, \theta, \phi) = \mathrm{e}^{im\phi}\{J_r(r, \theta)\hat{r} + J_\theta(r, \theta)\hat{\theta} + J_\phi(r, \theta)\hat{\phi}\} \tag{A1.82}$$

shall be said to possess rotational symmetry of order m. If this definition is inserted into formulas (A1.78) and (A1.81) for the mode coefficients of the radiated field, it is evident that

$$Q^{(c)}_{sm'n} = 0 \qquad \text{for } m' \neq m. \tag{A1.83}$$

Planar symmetries

Let $\vec{J}(r, \theta, \phi)$ be any current distribution and let $\vec{J}^w(r, \theta, \phi)$ be the geometrical image of $\vec{J}(r, \theta, \phi)$ in the plane $\phi = w$, $\phi = w + \pi$. If $w = u$ is a symmetry plane for $\vec{J}$, then

$$\vec{J}^u(r, \theta, \phi) = \vec{J}(r, \theta, \phi) \qquad \text{(symmetry)} \tag{A1.84}$$

and if $w = v$ is a plane of anti-symmetry, then

$$\vec{J}^v(r, \theta, \phi) = -\vec{J}(r, \theta, \phi) \qquad \text{(anti-symmetry).} \tag{A1.85}$$

The electric field radiated by a current distribution with one or both of the planar symmetries (A1.84–85) will possess the same planar symmetries as the current. Writing

$$\vec{J}(r, \theta, \phi) = J_r(r, \theta, \phi)\hat{r} + J_\theta(r, \theta, \phi)\hat{\theta} + J_\phi(r, \theta, \phi)\hat{\phi} \tag{A1.86}$$

one easily verifies that the imaged current distribution $\vec{J}^w$ has the following form

$$\vec{J}^w(r,\theta,\phi) = J_r(r,\theta,2w-\phi)\hat{r} + J_\theta(r,\theta,2w-\phi)\hat{\theta} - J_\phi(r,\theta,2w-\phi)\hat{\phi}. \tag{A1.87}$$

Let $Q^{(c)}_{smn}$ be spherical mode coefficients of the field radiated by $\vec{J}(r,\theta,\phi)$ and let $Q^{(c)w}_{smn}$ be related to $\vec{J}^w$ the same way. These mode coefficients can be expressed by insertion in formulas (A1.78) and (A1.81) of $\vec{J}$ and $\vec{J}^w$. Comparison of the expressions shows that

$$Q^{(c)w}_{smn} = (-1)^{m+s}\mathrm{e}^{-im2w}Q^{(c)}_{s,-m,n}. \tag{A1.88}$$

Thus, if $\vec{J}(r,\theta,\phi)$ is symmetric with respect to the plane $\phi = u$, $\phi = u + \pi$, then from (A1.84) we have $Q^{(c)u}_{smn} = Q^{(c)}_{smn}$, and therefore

$$Q^{(c)}_{s,-m,n} = (-1)^{m+s}\mathrm{e}^{im2u}Q^{(c)}_{smn} \qquad \text{(symmetry)}. \tag{A1.89}$$

If $\vec{J}$ is anti-symmetric with respect to the plane $\phi = v$, $\phi = v + \pi$, then because of (A1.85) we have $Q^{(c)v}_{smn} = -Q^{(c)}_{smn}$, and (A1.88) shows that

$$Q^{(c)}_{s,-m,n} = (-1)^{m+s+1}\mathrm{e}^{im2v}Q^{(c)}_{smn} \qquad \text{(anti-symmetry)}. \tag{A1.90}$$

The two symmetries (A1.89–1.90) can be present at the same time only if $v = u + (\pi/2 + p\pi)$ for some integer p, in which case (A1.89) and (A1.90) combine to

$$Q^{(c)}_{smn} = 0 \qquad \text{for } m \text{ even} \tag{A1.91a}$$

$$Q^{(c)}_{s,-m,n} = (-1)^{s+1}\mathrm{e}^{im2u}Q^{(c)}_{smn} \qquad \text{for } m \text{ odd}. \tag{A1.91b}$$

This double planar symmetry exists in any linearly polarized plane wave propagating in the direction of the z-axis.

Notice that (A1.91a) is independent of u. Thus, if two current components with different values of u are present, then for the composite current distribution, (A1.91a) will still hold true for the radiated field.

Rotation of current distributions

Let us assume an arbitrary current distribution

$$\vec{J}(r,\theta,\phi) = J_r(r,\theta,\phi)\hat{r} + J_\theta(r,\theta,\phi)\hat{\theta} + J_\phi(r,\theta,\phi)\hat{\phi} \tag{A1.92}$$

with coefficients $Q^{(c)}_{smn}$ for the radiated field. For the same current distribution rotated by an angle ϕ_o about the z-axis

$$\vec{J}_{\phi_o}(r,\theta,\phi) = J_r(r,\theta,\phi-\phi_o)\hat{r} + J_\theta(r,\theta,\phi-\phi_o)\hat{\theta} + J_\phi(r,\theta,\phi-\phi_o)\hat{\phi} \tag{A1.93}$$

the expansion coefficients are given by

$$Q^{(c)}_{smn\phi_o} = \mathrm{e}^{-im\phi_o}Q^{(c)}_{smn} \tag{A1.94}$$

ref. (A1.78) and (A1.81).

A1.5.3 The continuous x-polarized planar current ring

This elementary current distribution serves as an illustration of the symmetries described in Section A1.5.2, and it is a useful component in plane-wave synthesis. Other polarizations are easily obtained by application of the rotation formula (A1.94).

Current distribution function

$$\begin{aligned}\vec{J}(r,\theta,\phi) &\equiv J_o(r,\theta)\hat{x}\\ &= e^{i\phi}\tfrac{1}{2}J_o(r,\theta)\{\sin\theta\hat{r}+\cos\theta\hat{\theta}+i\hat{\phi}\}\\ &\quad+e^{-i\phi}\tfrac{1}{2}J_o(r,\theta)\{\sin\theta\hat{r}+\cos\theta\hat{\theta}-i\hat{\phi}\}\end{aligned} \tag{A1.95a}$$

with

$$J_o(r,\theta)=\frac{\delta(r-r_0)\delta(\theta-\theta_o)}{2\pi r_0^2\sin\theta_o}d_e \tag{A1.95b}$$

where d_e is a dipole moment and δ is the Dirac delta function.

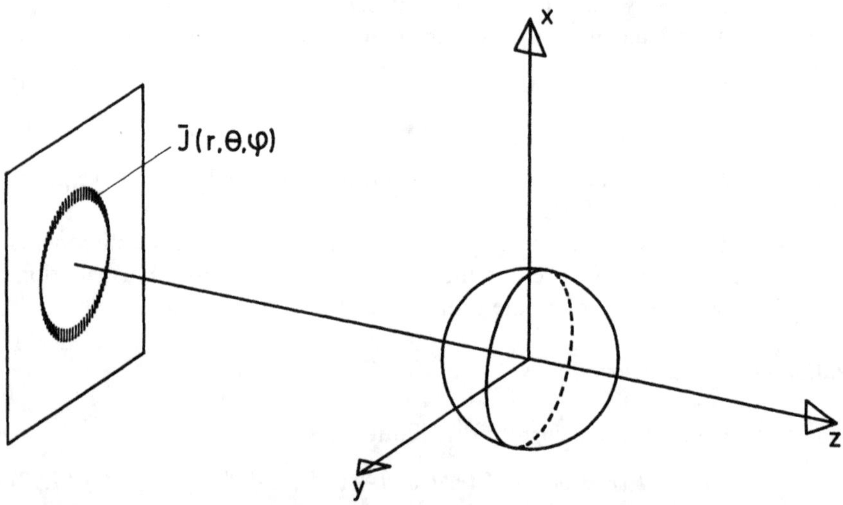

Fig. A1.3 *Continuous x-polarized planar current ring; ref. eqn (A1.95)*

Normalization

The function $J_o(r,\theta)$ has been normalized so as to make the far-field amplitude on the z-axis independent of r_0 and θ_o, which follows from the fact that

$$\int_V \vec{J}(r,\theta,\phi)\,dV=d_e\hat{x} \tag{A1.96}$$

independent of r_0 and θ_o, as long as V encloses the ring. For $\theta_o \to 0$ or $\theta_o \to \pi$ the ring approaches a short dipole on the z-axis.

Symmetries

Comparison with (A1.82) shows that $\vec{J}$ contains components with rotational symmetry of order $m = \pm 1$ only. Also, it is evident that there is symmetry in the plane $\phi = 0$ combined with anti-symmetry in $\phi = \pi/2$ so (A1.91) will hold true with $u = 0$.

Spherical mode expansion

The spherical mode coefficients of the radiated field in the interior region and in the exterior region are found by inserting (A1.95) in (A1.81) and (A1.78), respectively. The result is

$$Q^{(c)}_{smn} = 0 \qquad \text{for } m \neq \pm 1 \tag{A1.97a}$$

and

$$Q^{(c)}_{1mn} = \frac{-ikd_e}{\sqrt{8\pi n(n+1)\eta}} R^{(4-c)}_{1n}(kr_0) \left\{ \cos\theta_o \frac{\bar{P}^1_n(\cos\theta_o)}{\sin\theta_o} + \left. \frac{\mathrm{d}\bar{P}^1_n(\cos\theta)}{\mathrm{d}\theta} \right|_{\theta=\theta_o} \right\} \qquad \text{for } m = \pm 1 \tag{A1.97b}$$

$$Q^{(c)}_{2mn} = \frac{mkd_e}{\sqrt{8\pi n(n+1)\eta}} \left[\frac{n(n+1)}{kr_0} R^{(4-c)}_{1n}(kr_0) \bar{P}^1_n(\cos\theta_o)\sin\theta_o + R^{(4-c)}_{2n}(kr_0) \left\{ \cos\theta_o \left. \frac{\mathrm{d}\bar{P}^1_n(\cos\theta)}{\mathrm{d}\theta} \right|_{\theta=\theta_o} + \frac{\bar{P}^1_n(\cos\theta_o)}{\sin\theta_o} \right\} \right] \qquad \text{for } m = \pm 1 \tag{A1.97c}$$

where $c = 1$ for the interior field and $c = 3$ for the exterior field.

A1.5.4 The sampled x-polarized planar current ring

This elementary current distribution serves as an illustration of the symmetries described in Section A1.5.2, and is a useful component in plane wave synthesis.

Current distribution function

The equidistantly sampled current ring has the following current distribution function

$$\vec{J}^L(r,\theta,\phi) \equiv \frac{2\pi}{L} \sum_{l=1}^{L} \delta(\phi - \phi_l)\vec{J}(r,\theta,\phi) \tag{A1.98a}$$

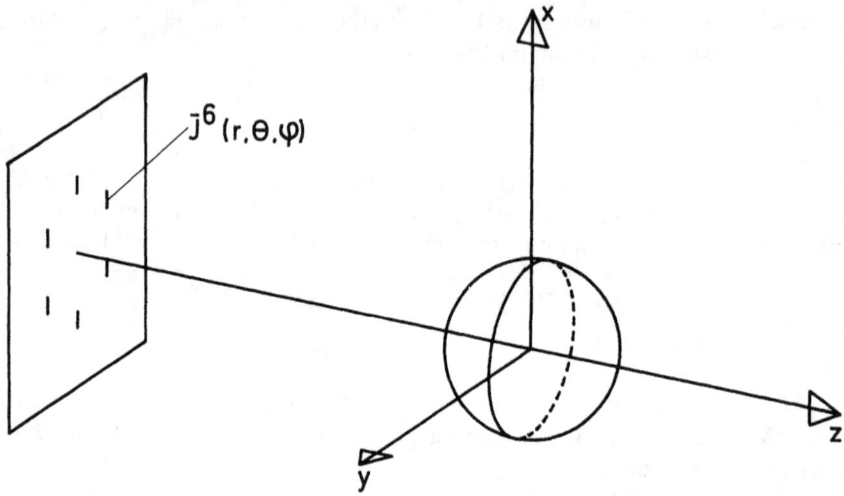

Fig. A1.4 *Sampled x-polarized planar current ring; ref. eqn (A1.98), shown for $L=6$*

where L is the number of samples of the continuous distribution $\vec{J}(r,\theta,\phi)$, ref. eqn (A1.95) and where

$$\phi_l = \phi_o + (l-1)\frac{2\pi}{L}. \tag{A1.98b}$$

The sampled ring is an x-polarized uniform ring array.

Normalization

It follows from the definition (A1.98) that

$$\int_V \vec{J}^L(r,\theta,\phi)\,\mathrm{d}V = d_e\hat{x} \tag{A1.99}$$

as long as V encloses the ring. Thus, each sample is a short dipole with the dipole moment d_e/L.

Symmetries

For $\phi_o = p\pi/L$, p being an integer, there is symmetry about $\phi = 0$. If, further, L is even, there is anti-symmetry about $\phi = \pi/2$.

Spherical mode expansion

Inserting (A1.98) with $\phi_o = 0$ in expressions (A1.78) and (A1.81) for the mode coefficients of the radiated field leads to

$$Q_{1mn}^{(c)} = \frac{ikd_e}{\sqrt{8\pi n(n+1)\eta}}\left(-\frac{m}{|m|}\right)^m R_{1n}^{(4-c)}(kr_0)$$

$$\left\{\frac{m\bar{P}_n^{|m|}(\cos\theta_o)}{\sin\theta_o}\cos\theta_o(\delta_{m1}^L + \delta_{m,-1}^L)\right.$$

$$\left.+\left.\frac{\mathrm{d}\bar{P}_n^{|m|}(\cos\theta)}{\mathrm{d}\theta}\right|_{\theta=\theta_o}(\delta_{m1}^L - \delta_{m,-1}^L)\right\} \tag{A1.100a}$$

$$Q_{2mn}^{(c)} = \frac{-kd_e}{\sqrt{8\pi n(n+1)\eta}}\left(-\frac{m}{|m|}\right)^m$$

$$\left[\frac{n(n+1)}{kr_0}R_{1n}^{(4-c)}(kr_0)\bar{P}_n^{|m|}(\cos\theta_o)\sin\theta_o(\delta_{m1}^L + \delta_{m,-1}^L)\right.$$

$$+R_{2n}^{(4-c)}(kr_0)\left\{\left.\frac{\mathrm{d}\bar{P}_n^{|m|}(\cos\theta)}{\mathrm{d}\theta}\right|_{\theta=\theta_o}\cos\theta_o(\delta_{m1}^L + \delta_{m,-1}^L)\right.$$

$$\left.\left.+\frac{m\bar{P}_n^{|m|}(\cos\theta_o)}{\sin\theta_o}(\delta_{m1}^L - \delta_{m,-1}^L)\right\}\right] \tag{A1.100b}$$

where

$$\delta_{m\mu}^L \equiv \begin{cases}1 & m = \mu \pmod L\\ 0 & m \neq \mu \pmod L\end{cases} \tag{A1.101}$$

The sampling is seen to introduce sidebands for $m = \pm 1 + jL$, $j = \pm 1$, $\pm 2, \ldots$. These are shown in the nm-plane in Fig. 7.21. If $L \geqslant 3$ these sidebands will not interfere with the main band at $m = \pm 1$. In this case, the main band coefficients ($m = \pm 1$) of the sampled ring are identical with the coefficients of the continuous ring.

In the case $\phi_o \neq 0$, $\delta_{m\mu}^L$ in (A1.100) should be replaced by $\delta_{m\mu}^L\,\mathrm{e}^{i(m-\mu)\phi_o}$.

A1.5.5 Ring of currents tangential to a sphere

Current distribution function

The current distribution function of the continuous ring with approximate x-polarization for θ_o close to π is defined by

$$\vec{J}_t(r,\theta,\phi) \equiv J_o(r,\theta)\hat{p}_t \tag{A1.102a}$$

where $J_o(r,\theta)$ is defined in (A1.95b) and

$$\begin{aligned}\hat{p}_t &\equiv -\cos\phi\hat{\theta} - \sin\phi\hat{\phi}\\ &= \mathrm{e}^{i\phi}\tfrac{1}{2}(-\hat{\theta} + i\hat{\phi}) + \mathrm{e}^{-i\phi}\tfrac{1}{2}(-\hat{\theta} - i\hat{\phi}).\end{aligned} \tag{A1.102b}$$

For $\theta \to \pi$ the ring approaches an x-polarized short dipole on the negative z-axis.

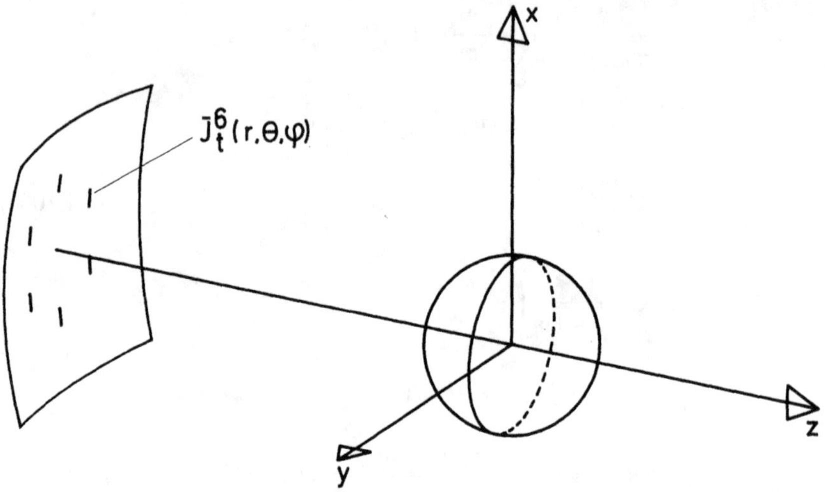

Fig. A1.5 *Sampled ring of currents tangential to a sphere, ref eqns (A1.102) and (A1.98), shown for $L=6$*

The current distribution function of the sampled ring is defined, as in the case of the planar ring, through (A1.98a).

Spherical mode expansion

Only the expansion coefficients of the sampled ring shall be given, since again the coefficients of the continuous ring are just the main band ($m = \pm 1$) for a sampled ring with $L \geqslant 3$. The result is for $\phi_o = 0$,

$$Q_{1mn}^{(c)} = \frac{ikd_e}{\sqrt{8\pi n(n+1)\eta}} \left(-\frac{m}{|m|}\right)^m R_{1n}^{(4-c)}(kr_0)$$

$$\left\{\frac{-m\bar{P}_n^{|m|}(\cos\theta_o)}{\sin\theta_o}(\delta_{m1}^L + \delta_{m,-1}^L) + \left.\frac{\mathrm{d}\bar{P}_n^{|m|}(\cos\theta)}{\mathrm{d}\theta}\right|_{\theta=\theta_o}(\delta_{m1}^L - \delta_{m,-1}^L)\right\} \tag{A1.103a}$$

$$Q_{2mn}^{(c)} = \frac{-kd_e}{\sqrt{8\pi n(n+1)\eta}} \left(-\frac{m}{|m|}\right)^m R_{2n}^{(4-c)}(kr_0)$$

$$\left\{-\left.\frac{\mathrm{d}\bar{P}_n^{|m|}(\cos\theta)}{\mathrm{d}\theta}\right|_{\theta=\theta_o}(\delta_{m1}^L + \delta_{m,-1}^L) + \frac{m\bar{P}_n^{|m|}(\cos\theta_o)}{\sin\theta_o}(\delta_{m1}^L - \delta_{m,-1}^L)\right\} \tag{A1.103b}$$

A1.6 The plane wave

Consider a plane wave $\vec{E}_o \exp(i\vec{k}_o \cdot \vec{r})$ coming from a direction (θ_o, ϕ_o) in free space. Here, $\vec{E}_o$ describes the amplitude and complex polarization of the wave while $\vec{k}_o$ is the propagation vector given by

$$\vec{k}_o = -k\sin\theta_o\cos\phi_o\hat{x} - k\sin\theta_o\sin\phi_o\hat{y} - k\cos\theta_o\hat{z}. \tag{A1.104}$$

The wave has finite amplitude everywhere and may be represented by the spherical wave expansion

$$\vec{E}_o\exp(i\vec{k}_o\cdot\vec{r}) = \frac{k}{\sqrt{\eta}}\sum_{s=1}^{2}\sum_{n=1}^{N}\sum_{m=-n}^{n} Q_{smn}\vec{F}^{(1)}_{smn}(r,\theta,\phi) \tag{A1.105}$$

valid for $r < R \approx N/k$. The expansion coefficients are then given by

$$Q_{smn} = \frac{\sqrt{\eta}}{k}(-1)^m\sqrt{4\pi}\, i\vec{E}_o\cdot\vec{K}_{s,-m,n}(\theta_o,\phi_o) \tag{A1.106}$$

where $\vec{K}_{smn}$ is the far-field pattern function defined in Section A1.3.2.

Example

For an $\hat{x}$-polarized wave travelling in the direction of the positive z-axis, we have

$$\hat{x}E_o\exp(ikz) = \frac{k}{\sqrt{\eta}}\sum_{s=1}^{2}\sum_{n=1}^{N}\sum_{m=-n}^{n} Q_{smn}\vec{F}^{(1)}_{smn}(r,\theta,\phi) \tag{A1.107}$$

where E_o is an amplitude constant. The expansion coefficients all vanish except

$$\begin{aligned} Q_{11n} &= Q_{1,-1,n} = Q_{21n} = -Q_{2,-1,n} \\ &= \frac{\sqrt{\eta}}{k}E_o\sqrt{4\pi}\, i^{n+1}\tfrac{1}{2}\sqrt{2n+1}. \end{aligned} \tag{A1.108}$$

A1.7 Relations between spherical and rectangular coordinates

A1.7.1 Coordinate transformations

$x = r\sin\theta\cos\phi$	$r = \sqrt{x^2+y^2+z^2}$
$y = r\sin\theta\sin\phi$	$\theta = \mathrm{ATAN2}(\sqrt{x^2+y^2}, z)$
$z = r\cos\theta$	$\phi = \mathrm{ATAN2}(y, x)$

The ATAN2 function is standard FORTRAN.

A1.7.2 Scalar products of unit vectors

	$\hat{x}$	$\hat{y}$	$\hat{z}$
$\hat{r}$	$\sin\theta\cos\phi$	$\sin\theta\sin\phi$	$\cos\theta$
$\hat{\theta}$	$\cos\theta\cos\phi$	$\cos\theta\sin\phi$	$-\sin\theta$
$\hat{\phi}$	$-\sin\phi$	$\cos\phi$	0

A1.7.3 Behaviour of certain vector distributions at $\theta = 0$ and $\theta = \pi$

Vector distribution on spherical surface	Value at $\theta = 0$	Value at $\theta = \pi$
$\hat{r}$	$\hat{z}$	$-\hat{z}$
$\hat{\theta}$	discontinuous	discontinuous
$\hat{\phi}$	discontinuous	discontinuous
$\hat{\theta}\cos\phi - \hat{\phi}\sin\phi$	$\hat{x}$	discontinuous
$\hat{\theta}\sin\phi + \hat{\phi}\cos\phi$	$\hat{y}$	discontinuous
$\hat{\theta}\cos\phi + \hat{\phi}\sin\phi$	discontinuous	$-\hat{x}$
$\hat{\theta}\sin\phi - \hat{\phi}\cos\phi$	discontinuous	$-\hat{y}$
$\exp(i\phi)(\hat{\theta} + i\hat{\phi})$	$\hat{x} + i\hat{y}$	discontinuous
$\exp(-i\phi)(\hat{\theta} - i\hat{\phi})$	$\hat{x} - i\hat{y}$	discontinuous
$\exp(i\phi)(\hat{\theta} - i\hat{\phi})$	discontinuous	$-\hat{x} - i\hat{y}$
$\exp(-i\phi)(\hat{\theta} + i\hat{\phi})$	discontinuous	$-\hat{x} + i\hat{y}$

References

1. STRATTON, J. A., 1941, *Electromagnetic Theory*, Chapter VII, McGraw-Hill, Inc., New York.
2. BELOUSOV, S. L., 1962, *Tables of normalized associated Legendre polynomials*, Pergamon Press, Oxford.
3. ABRAMOWITZ, M. and STEGUN, I. A. (eds), 1965, *Handbook of mathematical functions*, Dover Publications, Inc., New York.
4. HARRINGTON, R. F., 1961, *Time-harmonic electromagnetic fields*, Appendix E, McGraw-Hill, Inc., New York.
5. WOOD, P. J., 1980, *Reflector antenna analysis and design*, Peter Peregrinus, London.
6. COLLIN, R. E. and ZUCKER, F. J. (eds), 1969, *Antenna Theory*, McGraw-Hill, Inc., New York.
7. COLLIN, R. E., 1986, 'Dyadic Green's function expansions in spherical coordinates', *Electromagnetics*, Vol. **6**, pp. 183–207.

Appendix A2

Rotation of spherical waves

A2.1 Euler angles

We consider two right-handed rectangular coordinate systems, initially coincident. Let the first, the unprimed system (x, y, z), remain fixed in space while the other, the primed system (x', y', z'), is rotated about the origin from its initial orientation to some arbitrary orientation in space and kept there.

Any such final orientation of the primed coordinate system may be arrived at through a succession of three rotations about its own axes†.

We indicate the orientation after the first rotation by a fixed coordinate system (x_1, y_1, z_1). The orientation after the second rotation is indicated as another fixed coordinate system (x_2, y_2, z_2). The three rotations of the primed coordinate system (x', y', z') are:

1. A rotation about the z-axis (through an angle ϕ_o).
2. A rotation about the y_1-axis (through an angle θ_o).
3. A rotation about the z_2-axis (through an angle χ_o).

The angles $(\chi_o, \theta_o, \phi_o)$, in that sequence, are the three *Euler angles* describing the rotation from the unprimed to the primed coordinate system, ref. Fig. A2.1. Note that the order of the Euler angles is specified as reversed with respect to the sequence of the above three rotations.

In the following sections, we shall consider spherical coordinate systems defined in the usual manner with respect to the rectangular systems (x, y, z) and (x', y', z').

A2.2 Notation

A spherical vector wave function $\vec{F}^{(c)}_{smn}(r, \theta, \phi)$ defined in the unprimed coordinate system (r, θ, ϕ) may be expressed as a combination of spherical waves

† The positive direction of rotation about an axis is defined such that a positive rotation of $\pi/2$ about the z-axis carries the x-axis into the initial position of the y-axis (right-hand rule).

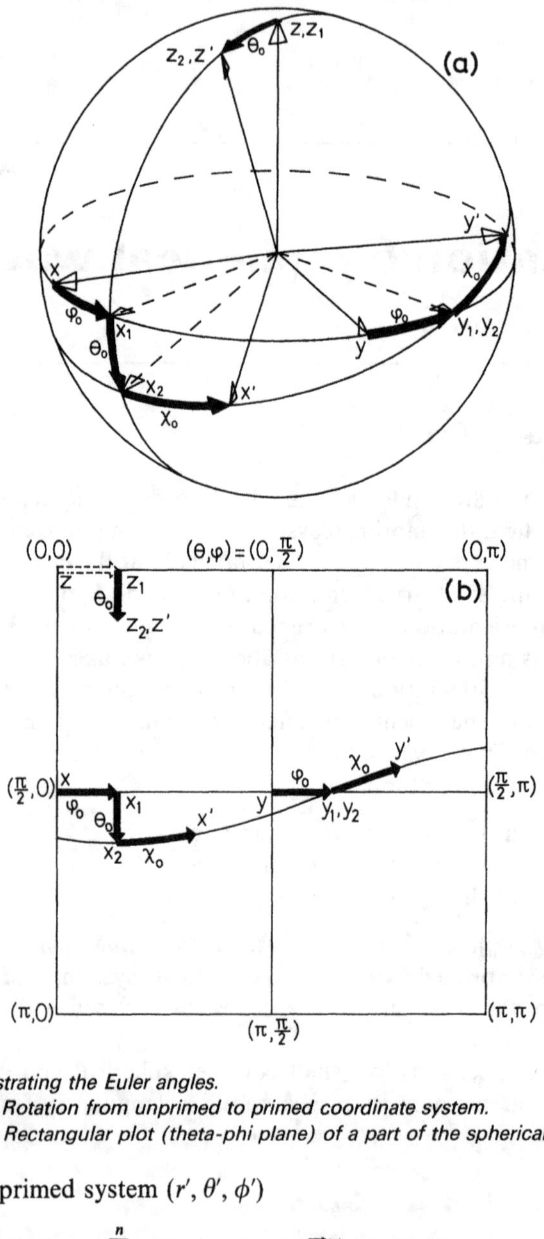

Fig. A2.1 *Illustrating the Euler angles.*
(a) Rotation from unprimed to primed coordinate system.
(b) Rectangular plot (theta-phi plane) of a part of the spherical surface in (a)

defined in the primed system (r', θ', ϕ')

$$\vec{F}^{(c)}_{smn}(r, \theta, \phi) = \sum_{\mu=-n}^{n} D^n_{\mu m}(\chi_o, \theta_o, \phi_o)\vec{F}^{(c)}_{s\mu n}(r', \theta', \phi') \tag{A2.1}$$

where the arguments to the function $D^n_{\mu m}$ are the Euler angles defined in Section A2.1. Note that the index, s, on the spherical wave function is invariant under rotation. This is in accordance with the requirement that waves which are transverse (electric or magnetic with respect to r) must remain so under rotation.

Also, it is seen that the spherical wave functions on the right-hand side all contain the *same index n* as the left-hand side. Independent of the rotation, all terms in (A2.1) have the same radial dependence. Explicitly introducing the dependence on the three Euler angles, eqn (A2.1) may be rewritten

$$\vec{F}^{(c)}_{smn}(r,\theta,\phi) = \sum_{\mu=-n}^{n} e^{im\phi_o} d^n_{\mu m}(\theta_o)\, e^{i\mu\chi_o} \vec{F}^{(c)}_{s\mu n}(r',\theta',\phi'). \tag{A2.2}$$

From (A2.2) it is seen that the first and the last of the three elementary rotations into which a general rotation may be decomposed give rise to phase shifts only. The rotation through θ_o is more complicated and is described by the rotation coefficient $d^n_{\mu m}(\theta_o)$ dealt with in Section A2.3.

A2.3 The rotation coefficient

For brevity, the subscript on θ_o is omitted below.

General expressions

From Edmonds [1, eqn (4.1.15)]

$$d^n_{\mu m}(\theta) = \left\{\frac{(n+\mu)!(n-\mu)!}{(n+m)!(n-m)!}\right\}^{1/2} \sum_\sigma \binom{n+m}{n-\mu-\sigma}\binom{n-m}{\sigma}(-1)^{n-\mu-\sigma} \left(\cos\frac{\theta}{2}\right)^{2\sigma+\mu+m}\left(\sin\frac{\theta}{2}\right)^{2n-2\sigma-\mu-m} \tag{A2.3}$$

where the symbol

$$\binom{i}{j} = \frac{i!}{(i-j)!j!} \tag{A2.4}$$

is the binomial coefficient.

As may be seen, $d^n_{\mu m}(\theta)$ is a real function of θ. The summation over σ involves all terms in which the binomial coefficients do not lead to negative arguments for the factorials. From Edmonds [1, eqn (4.1.23)] we have the alternative formula

$$d^n_{\mu m}(\theta) = \left\{\frac{(n+\mu)!(n-\mu)!}{(n+m)!(n-m)!}\right\}^{1/2} \left(\cos\frac{\theta}{2}\right)^{\mu+m}\left(\sin\frac{\theta}{2}\right)^{\mu-m} P^{(\mu-m,\mu+m)}_{n-\mu}(\cos\theta). \tag{A2.5}$$

The function $P^{(\alpha,\beta)}_n(x)$ is the Jacobi polynomial [2].

Normalization property

$$\sum_{\mu=-n}^{n} (d^n_{\mu m}(\theta))^2 = 1, \qquad \text{all } (m,n) \tag{A2.6}$$

Symmetries

$$d^n_{\mu m}(\theta) = d^n_{m\mu}(-\theta) \tag{A2.7}$$

$$d^n_{\mu m}(\theta) = (-1)^{\mu+m} d^n_{m\mu}(\theta) \tag{A2.8}$$

$$d^n_{\mu m}(\theta) = (-1)^{\mu+m} d^n_{-\mu,-m}(\theta). \tag{A2.9}$$

By a geometrical consideration, $d^n_{\mu m}(\theta)$ is a periodic function of θ with a period of 2π. Equation (A2.7) and (A2.8) show that it has the same parity as $(\mu + m)$.

Orthogonality

Edmonds [1, eqn (4.6.1)]:

$$\int_0^\pi d^n_{\mu m}(\theta) d^{n'}_{\mu m}(\theta) \sin\theta \, \mathrm{d}\theta = \frac{2}{2n+1}\delta_{nn'} \tag{A2.10}$$

where $\delta_{nn'}$ is the Kronecker delta.

Fourier expansion

The finite Fourier series

$$d^n_{\mu m}(\theta) = i^{\mu-m} \sum_{m'=-n}^{n} \Delta^n_{m'\mu} \Delta^n_{m'm} \mathrm{e}^{-im'\theta} \tag{A2.11}$$

$$= i^{m-\mu} \sum_{m'=-n}^{n} \Delta^n_{m'\mu} \Delta^n_{m'm} \mathrm{e}^{im'\theta} \tag{A2.12}$$

is a key point in the Wacker algorithm discussed in Chapter 4. The Fourier coefficients have the form of the product of two rotation coefficients of argument $\pi/2$, ref. Edmonds [1, eqn (4.5.2)],

$$\Delta^n_{m'm} = d^n_{m'm}\left(\frac{\pi}{2}\right) \tag{A2.13}$$

Formulas involving $\Delta^n_{m'm}$ are given in Section A2.4.

Recurrence relation [3]

$$\sqrt{(n+\mu+1)(n-\mu)} \sin\theta d^n_{\mu+1,m}(\theta)$$
$$+ \sqrt{(n+\mu)(n-\mu+1)} \sin\theta d^n_{\mu-1,m}(\theta)$$
$$+ (2m - 2\mu\cos\theta) d^n_{\mu m}(\theta) = 0 \tag{A2.14}$$

Special values

$$d^n_{\mu m}(0) = \delta_{\mu m}, \qquad \text{for all } n \tag{A2.15}$$

$$d^n_{\mu m}(\pi) = (-1)^{n+m} \delta_{\mu,-m}. \tag{A2.16}$$

Special cases [4]

$$d^n_{0m}(\theta) = \left(-\frac{m}{|m|}\right)^m \sqrt{\frac{2}{2n+1}}\,\bar{P}^{|m|}_n(\cos\theta) \tag{A2.17}$$

$$d^n_{1m}(\theta) + d^n_{-1m}(\theta) = -\frac{2}{\sqrt{n(n+1)}}\frac{m d^n_{0m}(\theta)}{\sin\theta} \tag{A2.18}$$

$$d^n_{1m}(\theta) - d^n_{-1m}(\theta) = -\frac{2}{\sqrt{n(n+1)}}\frac{\mathrm{d}}{\mathrm{d}\theta}\{d^n_{0m}(\theta)\}. \tag{A2.19}$$

Insertion of (A2.17) into (A2.18) and (A2.19) and subsequent use of the definition of the far-field pattern function in eqns (A1.59–60) yields, respectively

$$d^n_{1m}(\theta) + d^n_{-1m}(\theta) = \frac{-2i^n}{\sqrt{2n+1}}\mathrm{e}^{-im\phi}\{\vec{K}_{1mn}(\theta,\phi)\}_\theta \tag{A2.20}$$

$$= \frac{2i^{n+1}}{\sqrt{2n+1}}\mathrm{e}^{-im\phi}\{\vec{K}_{2mn}(\theta,\phi)\}_\phi \tag{A2.21}$$

$$d^n_{1m}(\theta) - d^n_{-1m}(\theta) = \frac{2i^{n+1}}{\sqrt{2n+1}}\mathrm{e}^{-im\phi}\{\vec{K}_{1mn}(\theta,\phi)\}_\phi \tag{A2.22}$$

$$= \frac{-2i^n}{\sqrt{2n+1}}\mathrm{e}^{-im\phi}\{\vec{K}_{2mn}(\theta,\phi)\}_\theta. \tag{A2.23}$$

A2.4 The deltas

Since $-n \leqslant m' \leqslant n$ and $-n \leqslant m \leqslant n$ the constants $\Delta^n_{m'm}$ may be visualized as ordered in the so-called delta pyramid [5]. In Fig. A2.2, a delta pyramid for $0 \leqslant n \leqslant N$ is shown.

General expressions

From (A2.3) and (A2.13)

$$\Delta^n_{m'm} = \left\{\frac{(n+m')!(n-m')!}{(n+m)!(n-m)!}\right\}^{1/2} \frac{1}{2^n}\sum_\sigma \binom{n+m}{n-m'-\sigma}\binom{n-m}{\sigma}(-1)^{n-m'-\sigma} \tag{A2.24}$$

From (A2.5) and (A2.13)

$$\Delta^n_{m'm} = \left\{\frac{(n+m')!(n-m')!}{(n+m)!(n-m)!}\right\}^{1/2} \frac{1}{2^{m'}} P^{(m'-m,m'+m)}_{n-m'}(0) \tag{A2.25}$$

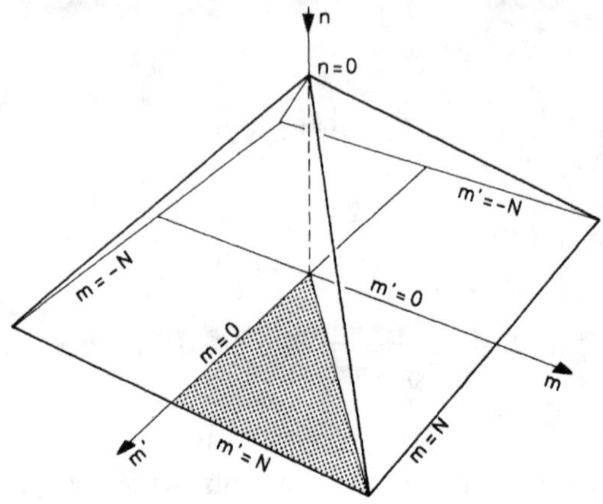

Fig. A2.2 *The delta pyramid*

Symmetries

Symmetries exist [1], from which a total horizontal layer in the pyramid of $\Delta^n_{m'm}$ terms can be calculated from the values in the shaded triangle, ref. Fig. A2.2.

$$\Delta^n_{m'm} = (-1)^{m'+m}\Delta^n_{mm'} \tag{A2.26}$$

$$\Delta^n_{m'm} = (-1)^{n+m'}\Delta^n_{m,-m'} \tag{A2.27}$$

$$\Delta^n_{m'm} = (-1)^{n+m}\Delta^n_{-m'm} \tag{A2.28}$$

$$\Delta^n_{m'm} = (-1)^{m'+m}\Delta^n_{-m',-m} \tag{A2.29}$$

$$\Delta^n_{m'm} = \Delta^n_{-m,-m'} \tag{A2.30}$$

$$\Delta^n_{m'm} = (-1)^{n+m}\Delta^n_{-mm'} \tag{A2.31}$$

$$\Delta^n_{m'm} = (-1)^{n+m'}\Delta^n_{m',-m} \tag{A2.32}$$

Recurrence relations
From (A2.14) with $\theta = \pi/2$:

$$\sqrt{(n+m'+1)(n-m')}\,\Delta^n_{m'+1,m} + \sqrt{(n+m')(n-m'+1)}\,\Delta^n_{m'-1,m} + 2m\Delta^n_{m'm} = 0. \tag{A2.33}$$

From (A2.33) and the symmetry relations:

$$\sqrt{(n+m+1)(n-m)}\,\Delta^n_{m',m+1} + \sqrt{(n+m)(n-m+1)}\,\Delta^n_{m',m-1} - 2m'\Delta^n_{m'm} = 0. \tag{A2.34}$$

Lewis' recurrence [6]:

$$\frac{\sqrt{(n+m'+1)(n-m')}\sqrt{(n+m+1)(n-m)}}{m'+m+1}\Delta^n_{m'+1,m+1}$$
$$+\frac{\sqrt{(n+m')(n-m'+1)}\sqrt{(n+m)(n-m+1)}}{m'+m-1}\Delta^n_{m'-1,m-1}$$
$$=\frac{2(m'+m)}{(m'+m)^2-1}\{n(n+1)-(m'+m)^2+m'm+1\}\Delta^n_{m'm}. \qquad \text{(A2.35)}$$

Vertical recurrence relation obtained from recurrences for Jacobi polynomials, ref. Abramowitz and Stegun [2, p. 782]:

$$\sqrt{(n+m'+1)(n-m'+1)(n+m+1)(n-m+1)}\, n\Delta^{n+1}_{m'm}$$
$$+\sqrt{(n+m')(n-m')(n+m)(n-m)}\,(n+1)\Delta^{n-1}_{m'm}$$
$$+(2n+1)m'm\Delta^n_{m'm}=0. \qquad \text{(A2.36)}$$

From (A2.33) in general and (A2.33) for $m=0$:

$$(n+m'+2)(n-m'-1)\Delta^n_{m'+2,m}\Delta^n_{m'+2,0}$$
$$+(n+m'-1)(n-m'+2)\Delta^n_{m'-2,m}\Delta^n_{m'-2,0}$$
$$-2(n^2-m'^2+n-2m^2)\Delta^n_{m'm}\Delta^n_{m'0}=0 \qquad \text{(A2.37)}$$

This expansion is a recurrence formula for the coefficients in the Fourier expansions of $d^n_{0m}(\theta)$ as well as for the Legendre functions $\bar{P}^m_n(\cos\theta)$ and $P^m_n(\cos\theta)$, because of eqns (A2.12), (A2.17) and (A1.25), respectively.

Special cases

From (A2.17), the symmetry relations and [2, eqn 8.6.1]

$$\Delta^n_{m'0}=\begin{cases}0 & \text{for } (n+m') \text{ odd}\\ (-1)^{(n-m')/2}\dfrac{1}{2^n}\left\{\dbinom{n+m'}{\frac{n+m'}{2}}\dbinom{n-m'}{\frac{n-m'}{2}}\right\}^{1/2} & \text{for } (n+m') \text{ even}\end{cases} \qquad \text{(A2.38)}$$

$$\Delta^n_{0m}=\begin{cases}0 & \text{for } (n+m) \text{ odd}\\ (-1)^{(n+m)/2}\dfrac{1}{2^n}\left\{\dbinom{n+m}{\frac{n+m}{2}}\dbinom{n-m}{\frac{n-m}{2}}\right\}^{1/2} & \text{for } (n+m) \text{ even}\end{cases} \qquad \text{(A2.39)}$$

From (A2.24) with $m = n$

$$\Delta^n_{m'n} = (-1)^{n+m'} \frac{1}{2^n} \sqrt{\binom{2n}{n-m'}} \tag{A2.40}$$

and by the symmetry relations

$$\Delta^n_{nm} = \frac{1}{2^n} \sqrt{\binom{2n}{n-m}} \tag{A2.41}$$

with the special case

$$\Delta^n_{nn} = \frac{1}{2^n}. \tag{A2.42}$$

A2.5 On the computation of rotation coefficients

A2.5.1 Stability of three-term recurrences

A three-term linear recurrence equation has a double infinity of solutions. In principle the two starting values determine which solution is found by the recursion. However, if one seeks to numerically compute a solution that decreases relative to some other solution, rounding errors will introduce the second solution which sooner or later will dominate over the wanted solution. In such cases the recursion should be performed in the opposite direction. The possibility for such a condition can be tested by means of the theory for linear difference equations with constant and real coefficients a and b

$$y_{n+2} + ay_{n+1} + by_n = 0. \tag{A2.43}$$

Seeking a solution of the form $y_n = K^n$, where K is a constant, and inserting into (A2.43) we obtain

$$K^n(K^2 + aK + b) = 0. \tag{A2.44}$$

Hence, K^n is a solution when K is a root of the characteristic polynomial. If there are two different real roots K_1 and K_2 with $|K_1| < |K_2|$, we have a situation where the absolute value of one solution K_1^n decreases relative to the other solution K_2^n.

If the polynomial has two complex conjugate roots, the real part and the imaginary part of K^n will constitute two real and independent solutions to the difference equation (A2.43). These solutions will be oscillating but have the same envelope. Thus, the recurrence will be stable in both directions.

A2.5.2 Stability of delta recurrences

Rotation coefficients $d^n_{\mu m}(\theta)$ for a single value of θ can be calculated by the recurrence formula (A2.14) starting at $\mu = n$ with $d^n_{nm}(\theta)$ found by setting $\mu = n$ in eqn (A2.3). However, coefficients for many equidistant values of theta are often needed. In such cases it is efficient to first compute $\Delta^n_{m'm}$ and $\Delta^n_{m'\mu}$ by parallel recursion in m' using formula (A2.33). The rotation coefficients $d^n_{\mu m}(\theta_i)$ are then found in a single operation by carrying out the discrete Fourier transformation in eqns (A2.11–2.12); see Appendix 4.

Assuming $(n + m')$ and $(n - m')$ in eqn (A2.33) to be large enough that we can approximate the two square roots by $\sqrt{n^2 - m'^2}$, we find that the characteristic polynomial has the following two roots

$$K_{\frac{1}{2}} = -\sqrt{\frac{m^2}{n^2 - m'^2}} \pm \sqrt{\frac{m^2}{n^2 - m'^2} - 1}. \tag{A2.45}$$

Thus for $m'^2 + m^2 > n^2$ we have two negative real roots, K_1 and K_2, where $|K_1| < 1 < |K_2|$. In other words, we have two solutions, each with alternating signs but with decreasing and increasing amplitude respectively, for $|m'|$ increasing. Figure A2.3 shows the absolute values of the deltas in a horizontal layer, i.e. n constant, of the delta pyramid, ref. Fig. A2.2. The change in

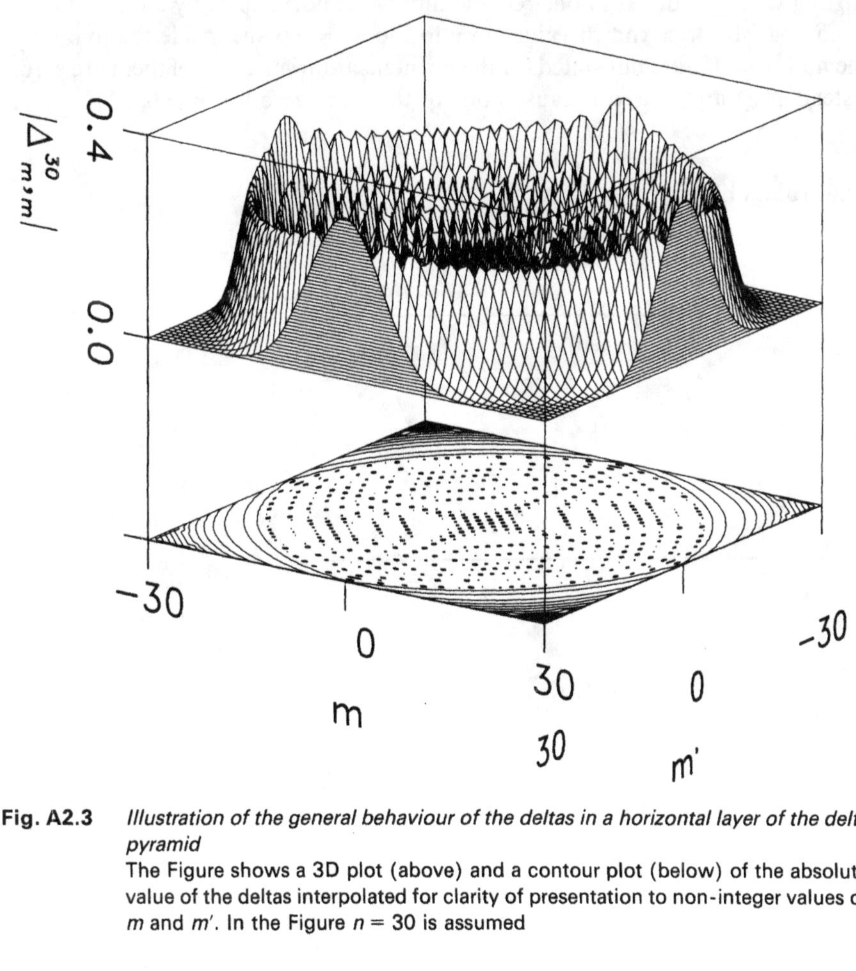

Fig. A2.3 *Illustration of the general behaviour of the deltas in a horizontal layer of the delta pyramid*

The Figure shows a 3D plot (above) and a contour plot (below) of the absolute value of the deltas interpolated for clarity of presentation to non-integer values of m and m'. In the Figure $n = 30$ is assumed behaviour at the circle $m'^2 + m^2 = n^2$ is clearly seen and it can also be noted that the deltas form a decreasing solution outside the circle. In order to assure stability the recursion (A2.33) must therefore be carried out from the face of the pyramid and inwards. Similarly, it can be shown that eqns (A2.34) and (A2.35) are stable going inwards, and eqn (A2.36) is stable going from the face of the pyramid and vertically downwards.

The small values of the deltas outside the circle lead to underflow in a computer for large m and n. The corner element given in eqn (A2.42) is the smallest and can be used to estimate for which n this problem arises. In case of underflow, the recursion can be started inside the pyramid but outside the circle with starting values $\Delta^n_{m'+1,n}$ equal to zero and $\Delta^n_{m'n}$ equal to a small number larger than zero. Going inwards, the deltas constitute a dominant solution outside the circle. Here, the recursion converges rapidly provided it is not started too close to the circle. However, if it is started too close to the face $m' = n$, overflow might happen at the other end of the recursion. Due to the arbitrary starting value, all deltas in one recursion must be normalized by a factor found by comparison to a known value close to the $m' = 0$ plane inside the pyramid. The numbers Δ^n_{0m} are not suited for the normalization since half of them are zero. Instead Δ^n_{1m} can be used because none of these are zero for $m \neq 0$, [4].

A2.6 Tables of low-order deltas

$\Delta^0_{00} = 1$

$\Delta^1_{m'm}$	$m = 0$	$m = 1$
$m' = 0$	0	$-\frac{\sqrt{2}}{2}$
$m' = 1$	$\frac{\sqrt{2}}{2}$	$\frac{1}{2}$

$\Delta^2_{m'm}$	$m=0$	$m=1$	$m=2$
$m'=0$	$-\frac{1}{2}$	0	$\frac{\sqrt{6}}{4}$
$m'=1$	0	$-\frac{1}{2}$	$-\frac{1}{2}$
$m'=2$	$\frac{\sqrt{6}}{4}$	$\frac{1}{2}$	$\frac{1}{4}$

$\Delta^3_{m'm}$	$m=0$	$m=1$	$m=2$	$m=3$
$m'=0$	0	$\frac{\sqrt{3}}{4}$	0	$-\frac{\sqrt{5}}{4}$
$m'=1$	$-\frac{\sqrt{3}}{4}$	$-\frac{1}{8}$	$\frac{\sqrt{10}}{8}$	$\frac{\sqrt{15}}{8}$
$m'=2$	0	$-\frac{\sqrt{10}}{8}$	$-\frac{1}{2}$	$-\frac{\sqrt{6}}{8}$
$m'=3$	$\frac{\sqrt{5}}{4}$	$\frac{\sqrt{15}}{8}$	$\frac{\sqrt{6}}{8}$	$\frac{1}{8}$

$\Delta^4_{m'm}$	$m=0$	$m=1$	$m=2$	$m=3$	$m=4$
$m'=0$	$\frac{3}{8}$	0	$-\frac{\sqrt{10}}{8}$	0	$\frac{\sqrt{70}}{16}$
$m'=1$	0	$\frac{3}{8}$	$\frac{\sqrt{2}}{8}$	$-\frac{\sqrt{7}}{8}$	$-\frac{\sqrt{14}}{8}$
$m'=2$	$-\frac{\sqrt{10}}{8}$	$-\frac{\sqrt{2}}{8}$	$\frac{1}{4}$	$\frac{\sqrt{14}}{8}$	$\frac{\sqrt{7}}{8}$
$m'=3$	0	$-\frac{\sqrt{7}}{8}$	$-\frac{\sqrt{14}}{8}$	$-\frac{3}{8}$	$-\frac{\sqrt{2}}{8}$
$m'=4$	$\frac{\sqrt{70}}{16}$	$\frac{\sqrt{14}}{8}$	$\frac{\sqrt{7}}{8}$	$\frac{\sqrt{2}}{8}$	$\frac{1}{16}$

$\Delta^5_{m'm}$	$m = 0$	$m = 1$	$m = 2$	$m = 3$	$m = 4$	$m = 5$
$m' = 0$	0	$-\frac{\sqrt{30}}{16}$	0	$\frac{\sqrt{35}}{16}$	0	$-\frac{3\sqrt{7}}{16}$
$m' = 1$	$\frac{\sqrt{30}}{16}$	$\frac{1}{16}$	$-\frac{\sqrt{7}}{8}$	$-\frac{\sqrt{42}}{32}$	$\frac{\sqrt{21}}{16}$	$\frac{\sqrt{210}}{32}$
$m' = 2$	0	$\frac{\sqrt{7}}{8}$	$\frac{1}{4}$	$-\frac{\sqrt{6}}{16}$	$-\frac{\sqrt{3}}{4}$	$-\frac{\sqrt{30}}{16}$
$m' = 3$	$-\frac{\sqrt{35}}{16}$	$-\frac{\sqrt{42}}{32}$	$\frac{\sqrt{6}}{16}$	$\frac{13}{32}$	$\frac{9\sqrt{2}}{32}$	$\frac{3\sqrt{5}}{32}$
$m' = 4$	0	$-\frac{\sqrt{21}}{16}$	$-\frac{\sqrt{3}}{4}$	$-\frac{9\sqrt{2}}{32}$	$-\frac{1}{4}$	$-\frac{\sqrt{10}}{32}$
$m' = 5$	$\frac{3\sqrt{7}}{16}$	$\frac{\sqrt{210}}{32}$	$\frac{\sqrt{30}}{16}$	$\frac{3\sqrt{5}}{32}$	$\frac{\sqrt{10}}{32}$	$\frac{1}{32}$

References

1. EDMONDS, A. R., 1974, *Angular momentum in quantum mechanics*, 2nd edition, Princeton University Press, Princeton, New Jersey.
2. ABRAMOWITZ, M. and STEGUN, I. A. (eds), 1965, *Handbook of mathematical functions*, Dover Publications, Inc., New York.
3. FANO, U. and RACAH, G., 1959, *Irreducible tensorial sets*, Academic Press.
4. LARSEN, FL. HOLM, 1980, 'Probe-corrected spherical near-field antenna measurements', Ph.D. Thesis, Electromagnetics Institute, Technical University of Denmark, Report **LD 36,** Dec.
5. WACKER, P. F., 1975, 'Non-planar near-field measurements: Spherical scanning', Report **NBSIR 75-809,** Electromagnetics Division, Inst. for Basic Standards, Nat. Bur. Stds., Boulder, Colorado, June.
6. LEWIS, R. L., 1977, 'The use of three-term recursion relations for numerical computations as applied to near-field scanning', URSI Symposium on Electromagnetic Wave Theory, Proceedings, pp. 224–226, Stanford, USA, June.

Appendix A3

Translation of spherical waves

A3.1 Notation

We consider two right-handed rectangular coordinate systems, initially coincident. Let the first, the unprimed system (x, y, z), remain fixed in space while the other, the primed system, is translated a distance A in the positive direction of the z-axis; see Fig. A3.1. Here, we shall consider the spherical coordinate systems defined in the usual manner with respect to the rectangular systems.

A spherical vector wave function $\vec{F}^{(c)}_{smn}$ defined in the unprimed coordinate system (r, θ, ϕ) may be expressed as a combination of spherical waves defined in the primed system (r', θ', ϕ')

$$\vec{F}^{(c)}_{s\mu n}(r, \theta, \phi) = \sum_{\sigma=1}^{2} \sum_{\substack{\nu=|\mu| \\ \nu\neq 0}}^{\infty} C^{sn(c)}_{\sigma\mu\nu}(kA)\vec{F}^{(1)}_{\sigma\mu\nu}(r', \theta', \phi') \qquad r' < |A| \tag{A3.1}$$

and

$$\vec{F}^{(c)}_{s\mu n}(r, \theta, \phi) = \sum_{\sigma=1}^{2} \sum_{\substack{\nu=|\mu| \\ \nu\neq 0}}^{\infty} C^{sn(1)}_{\sigma\mu\nu}(kA)\vec{F}^{(c)}_{\sigma\mu\nu}(r', \theta', \phi') \qquad r' > |A| \tag{A3.2}$$

where the translation A is contained in the argument of the translation coefficient $C^{sn(c)}_{\sigma\mu\nu}(kA)$. Note that in (A3.1) the spherical wave functions in the right-hand side are of the type $c = 1$ as it must be since $\vec{F}^{(c)}_{s\mu n}(r, \theta, \phi)$ is finite at the origin of (r', θ', ϕ'). Note also that only one azimuthal index μ is involved on both sides of the equations. This reflects the fact that the ϕ-dependence of the wave functions must be preserved under z-translation. The z-directed (axial) translation is sufficient for the purpose here; it is a special case of a general translation of the primed coordinate system in an arbitrary direction [1, 2]. Note, that arbitrary translations may be accomplished by a succession of three operations: rotation, axial translation, and inverse rotation.

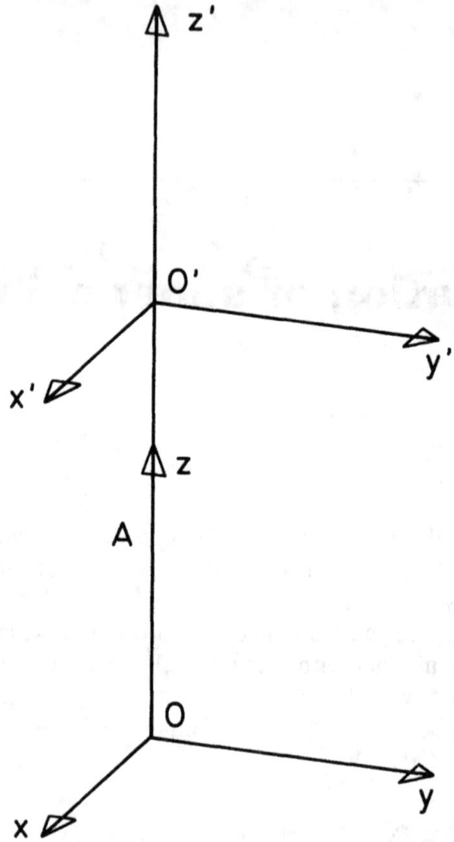

Fig. A3.1 *Translation $A = |\overrightarrow{OO'}|$ of the primed coordinate system relative to the unprimed system*

A3.2 The translation coefficient

General expression [3, 4]

Translation coefficients with positive arguments ($kA > 0$) may be computed from

$$C^{sn(c)}_{\sigma\mu\nu}(kA) = \sqrt{\frac{(2n+1)(2\nu+1)}{n(n+1)\nu(\nu+1)}}\sqrt{\frac{(\nu+\mu)!(n-\mu)!}{(\nu-\mu)!(n+\mu)!}}(-1)^{\mu}\tfrac{1}{2}i^{n-\nu}$$
$$\sum_{p=|n-\nu|}^{n+\nu}\Big[i^{-p}(\delta_{s\sigma}\{n(n+1)+\nu(\nu+1)-p(p+1)\}$$
$$+\delta_{3-s,\sigma}\{2i\mu kA\})a(\mu, n, -\mu, \nu, p)z^{(c)}_{p}(kA)\Big] \quad \text{(A3.3)}$$

where $a(\mu, n, -\mu, \nu, p)$ is a so-called linearization coefficient defined by the expansion of the product of two *unnormalized* associated Legendre functions

$$P_n^m(x)P_\nu^\mu(x) = \sum_{p=|n-\nu|}^{n+\nu} a(m, n, \mu, \nu, p)P_p^{m+\mu}(x). \tag{A3.4}$$

The specific form of (A3.4) needed in (A3.3) is

$$P_n^\mu(x)P_\nu^{-\mu}(x) = \sum_{p=|n-\nu|}^{n+\nu} a(\mu, n, -\mu, \nu, p)P_p(x). \tag{A3.5}$$

The linearization coefficients may be written in terms of the Wigner $3-j$ symbols

$$a(\mu, n, -\mu, \nu, p) = (2p+1)\left\{\frac{(n+\mu)!(\nu-\mu)!}{(n-\mu)!(\nu+\mu)!}\right\}^{1/2} \begin{pmatrix} n & \nu & p \\ 0 & 0 & 0 \end{pmatrix}\begin{pmatrix} n & \nu & p \\ \mu & -\mu & 0 \end{pmatrix}. \tag{A3.6}$$

It may be shown from the formulas for $3-j$ symbols given by Edmonds [5, p. 47] that

$$\begin{pmatrix} n & \nu & p \\ 0 & 0 & 0 \end{pmatrix} = 0 \qquad \text{for } (n+\nu+p) \text{ odd.} \tag{A3.7}$$

Hence, the summation index p in (A3.3), (A3.4) and (A3.5) only takes the values $p = |n-\nu|, |n-\nu|+2, \ldots, n+\nu-2, n+\nu$.

Recurrence relations for efficient calculation of the linearization coefficients have been given by Bruning and Lo [6].

Translation coefficients with negative arguments ($kA < 0$) may be computed using the symmetry relation (A3.13).

Symmetries

Elementary relations are

$$C_{1\mu\nu}^{1n(c)}(kA) = C_{2\mu\nu}^{2n(c)}(kA) \tag{A3.8}$$

$$C_{2\mu\nu}^{1n(c)}(kA) = C_{1\mu\nu}^{2n(c)}(kA) \tag{A3.9}$$

Symmetry relations exist for the $3-j$ symbols [5] and therefore, in virtue of (A3.6) and (A3.3), also for the linearization coefficients and the translation coefficients. Due to the power normalization of our spherical wave functions, the symmetries of the translation coefficients are simple and convenient [4]

$$C_{\sigma\mu\nu}^{sn(c)}(kA) = (-1)^{n+\nu} C_{s\mu n}^{\sigma\nu(c)}(kA) \tag{A3.10}$$

$$= (-1)^{s+\sigma}(-1)^{n+\nu} C_{s,-\mu,n}^{\sigma\nu(c)}(kA) \tag{A3.11}$$

$$= (-1)^{s+\sigma} C_{\sigma,-\mu,\nu}^{sn(c)}(kA). \tag{A3.12}$$

Translation coefficients with negative arguments may be obtained from [3]

$$C_{\sigma\mu\nu}^{sn(c)}(-kA) = (-1)^{s+\sigma}(-1)^{n+\nu} C_{\sigma\mu\nu}^{sn(c)}(kA) \tag{A3.13}$$

$$= (-1)^{s+\sigma} C_{s\mu n}^{\sigma\nu(c)}(kA) \tag{A3.14}$$

$$= C_{s,-\mu,n}^{\sigma\nu(c)}(kA). \tag{A3.15}$$

Complex conjugation is done as follows

$$[C_{\sigma\mu\nu}^{sn(3)}(kA)]^* = (-1)^{s+\sigma} C_{\sigma\mu\nu}^{sn(4)}(kA) \tag{A3.16}$$

Special cases [3, 4]

1. $\mu = 0,\ \nu \geqslant 1$

$$C_{\sigma 0\nu}^{sn(3)}(kA) = \delta_{s\sigma}\tfrac{1}{2}i^n \sqrt{\frac{2n+1}{n(n+1)}}\, i^{-\nu} \sqrt{\frac{2\nu+1}{\nu(\nu+1)}}$$

$$\sum_{p=|n-\nu|,2}^{n+\nu} \left[\frac{\{n(n+1)+\nu(\nu+1)-p(p+1)\}}{n+\nu+p+1} \frac{\begin{pmatrix} -n+\nu+p \\ \frac{-n+\nu+p}{2} \end{pmatrix} \begin{pmatrix} n-\nu+p \\ \frac{n-\nu+p}{2} \end{pmatrix} \begin{pmatrix} n+\nu-p \\ \frac{n+\nu-p}{2} \end{pmatrix}}{\begin{pmatrix} n+\nu+p \\ \frac{n+\nu+p}{2} \end{pmatrix}} i^{-p}(2p+1)h_p^{(1)}(kA) \right] \tag{A3.17}$$

where in the above summation index p takes the values $p = |n-\nu|, |n-\nu|+2, |n-\nu|+4, \ldots, n+\nu-2, n+\nu$.

2. $\mu = 0,\ \nu = 1$

$$C_{\sigma 01}^{sn(3)}(kA) = \delta_{s\sigma} \sqrt{\frac{3}{2}} \sqrt{n(n+1)(2n+1)}\, \frac{h_n^{(1)}(kA)}{kA}. \tag{A3.18}$$

3. $\mu = 1, \nu \geqslant 1$

$$C^{sn(3)}_{\sigma 1 \nu}(kA) = \tfrac{1}{4} i^n \frac{\sqrt{2n+1}}{n(n+1)} i^{-\nu} \frac{\sqrt{2\nu+1}}{\nu(\nu+1)}$$

$$\left[\sum_{p=|n-\nu|,2}^{n+\nu} \left\{ \delta_{s\sigma} \frac{(n(n+1)+\nu(\nu+1)-p(p+1))^2}{n+\nu+p+1} \right. \right.$$

$$\left. + \delta_{3-s,\sigma} 2ikA \frac{(n(n+1)+\nu(\nu+1)-p(p+1))}{n+\nu+p+1} \right\}$$

$$\left. \frac{\begin{pmatrix} -n+\nu+p \\ \frac{-n+\nu+p}{2} \end{pmatrix} \begin{pmatrix} n-\nu+p \\ \frac{n-\nu+p}{2} \end{pmatrix} \begin{pmatrix} n+\nu-p \\ \frac{n+\nu-p}{2} \end{pmatrix}}{\begin{pmatrix} n+\nu+p \\ \frac{n+\nu+p}{2} \end{pmatrix}} i^{-p}(2p+1) h_p^{(1)}(kA) \right]$$

(A3.19)

where in the above summation index p takes the values $p = |n-\nu|, |n-\nu|+2, |n-\nu|+4, \ldots, n+\nu-2, n+\nu$.

4. $\mu = 1, \nu = 1$

$$C^{sn(3)}_{\sigma 11}(kA) = \frac{\sqrt{3}}{2} \sqrt{2n+1} \{\delta_{s\sigma} R^{(3)}_{2n}(kA) + \delta_{3-s,\sigma} i R^{(3)}_{1n}(kA)\} \qquad \text{(A3.20)}$$

where $R^{(3)}_{sn}(kA)$ is defined in eqn (A1.6).

5. $\mu = -1, \nu \geqslant 1$

$$C^{sn(3)}_{\sigma,-1,\nu}(kA) = (-1)^{s+\sigma} C^{sn(3)}_{\sigma 1 \nu}(kA). \qquad \text{(A3.21)}$$

Asymptotic behaviour [6]

For $kA \to \infty$

$$C^{sn(3)}_{\sigma\mu\nu}(kA) = o\left(\frac{1}{kA}\right) \qquad \text{for } \mu \neq \pm 1 \qquad \text{(A3.22)}$$

$$C^{sn(3)}_{\sigma 1 \nu}(kA) = \frac{\sqrt{(2n+1)(2\nu+1)}}{2} i^{\nu-n-1} \frac{e^{ikA}}{kA} + o\left(\frac{1}{kA}\right) \qquad \text{(A3.23)}$$

$$C^{sn(3)}_{\sigma,-1,\nu}(kA) = \frac{\sqrt{(2n+1)(2\nu+1)}}{2} i^{\nu-n-1}(-1)^{s+\sigma}\frac{e^{ikA}}{kA} + o\left(\frac{1}{kA}\right). \tag{A3.24}$$

A note on the computation of translation coefficients

In (A3.17) and (A3.19) the binomial coefficients are of the form

$$B(J) = \binom{J}{J/2} \tag{A3.25}$$

where J is an even integer. Direct computation of $B(J)$ may lead to overflow. However, if each of the binomial coefficients is replaced by

$$B'(J) = \binom{J}{J/2}2^{-J} \tag{A3.26}$$

the numerical value of the particular combination of four binomials in (A3.17) and (A3.19) will not change. The numbers $B'(J)$ are more convenient for computation since $0.5 > B'(J) > 0.025$ for $J = 2, 4, \ldots, 1000$. $B'(J)$ may be computed from the recurrence relation

$$B'(J+2) = \frac{J+1}{J+2}B'(J) \tag{A3.27}$$

with the initial value $B'(0) = 1$.

References

1. STEIN, S., 1961, 'Addition theorems for spherical wave functions', *J. Quart. Appl. Math.*, Vol. **19,** No. 1, pp. 15–24.
2. CRUZAN, O. R., 1962, 'Translation addition theorems for spherical vector wave functions', *J. Quart. Appl. Math.*, Vol. **20,** No. 1, pp. 33–40.
3. BRUNING, J. H. and LO, Y. T., 1971, 'Multiple scattering of EM Waves by spheres, Part 1—Multipole expansion and ray-optical solutions', *IEEE Trans.*, Vol. **AP-19,** No. 3, pp. 378–390, May.
4. LARSEN, FL. HOLM, 1980, 'Probe-corrected spherical near-field antenna measurements', Ph.D. Thesis, Electromagnetics Institute, Technical University of Denmark, Report **LD 36,** Dec.
5. EDMONDS, A. R., 1974, *Angular momentum in quantum mechanics*, 2nd edition, Princeton University Press, Princeton, New Jersey.
6. BRUNING, J. H. and LO, Y. T., 1969, 'Multiple scattering by spheres', Technical Report No. **69-5,** Antenna Laboratory, Department of Electrical Engineering, University of Illinois, Urbana, Illinois.

Appendix A4

Data processing in antenna measurements

A4.1 Introduction

In antenna pattern measurements in general [1, 2, 3] and in spherical near-field testing in particular [4, 5, 6] extensive use is made of various forms of sampling and reconstruction schemes for band-limited, periodic functions [7]. The purpose of this appendix is to provide an overview of these techniques.

As shown in Chapter 2, the radiation from any antenna can be described by a finite series of spherical wave functions. The complex signal $f(\psi)$ sampled by a probe on an arbitrary great circle of the measurement sphere is therefore a band-limited periodic function (with the period 2π). Since the upper limit N of the polar index in the spherical wave representation of the antenna field is invariant with respect to rotation of the antenna coordinate system, the highest harmonics present in $f(\psi)$ are $e^{\pm iN\psi}$. Without loss of generality we may therefore make the simplifying assumption that the great circle coincides with the equator. We let, therefore, $\psi = \phi$ and $f(\psi) = f(\phi)$ in the following.

The number of complex samples for the complete reconstruction of $f(\phi)$ must equal (or exceed) the number of harmonics $2N + 1$. This corresponds, perhaps surprisingly, to a considerably larger sampling interval (of the order of one half-power beam width) than would normally be required for plotting purposes.

The question of sampling and reconstruction of power patterns is covered by the general theory and needs no special attention. This is due to the fact that when the complex function f is band-limited, the real function $g(\phi) = |f(\phi)|^2$ is band-limited as well. The highest harmonics contained in $g(\phi)$ are $e^{i2N\phi}$ and $e^{-i2N\phi}$ and at least $4N + 1$ real samples are needed.

The reconstruction schemes described are all based on the discrete Fourier transform and its inverse transform. It is hoped that the notation adopted here for sequences of complex numbers will facilitate the use of these transforms. As is well known, an efficient computer algorithm, the Fast Fourier Transform algorithm (FFT), is available for practical applications. After definitions and basic theoretical concepts in Sections A4.2–4.3, noise reduction and various forms of interpolations are dealt with in Section A4.4. For other sampling and reconstruction schemes, e.g. for such purposes as reduction of range reflections, error analysis, error reduction or scanning in sections smaller than 2π, the reader is referred to the literature [2, 5, 8].

Section A4.5 deals with numerical integration of periodic band-limited functions.

A4.2 The discrete Fourier transform

A4.2.1 Definitions

Let us denote an ordered sequence of complex numbers g(m), $m = 0, 1, \ldots$ by a parenthesis { } enclosing the elements arranged in the given specific order. We shall use a notation where the elements of the sequence are listed explicitly, e.g. as in {g(0), g(1), g(2), g(3)} or, alternatively, an indirect notation, e.g. $\{g(m)\,|\,m = 0, 1, \ldots, J-1\}$.

The discrete Fourier transform (DFT) of a sequence of J complex numbers is defined by

$$\mathrm{DFT}\{g(m)\,|\,m = 0, 1, \ldots, J-1\} = \sum_{m=0}^{J-1} g(m)(\varepsilon_J)^{ml}, \qquad l = 0, 1, \ldots, J-1$$

$$= \{G(l)\,|\,l = 0, 1, \ldots, J-1\} \qquad \text{(A4.1)}$$

and is in itself a sequence of J complex numbers. In (A4.1) ε_J is defined by

$$\varepsilon_J = e^{2\pi i/J}. \qquad \text{(A4.2)}$$

By applying the orthogonality with respect to summation of the exponential function

$$\sum_{l=0}^{J-1} (\varepsilon_J)^{l(s-m)} = \begin{cases} J & s = m \pmod{J} \\ 0 & s \neq m \pmod{J} \end{cases} \qquad \text{(A4.3)}$$

eqn (A4.1) may be solved for g(m). The inverse discrete Fourier transform (IDFT) is thereby derived

$$\{g(m)\,|\,m = 0, 1, \ldots, J-1\} = \frac{1}{J}\sum_{l=0}^{J-1} G(l)(\varepsilon_J)^{-lm}, \qquad m = 0, 1, \ldots, J-1$$

$$= \mathrm{IDFT}\{G(l)\,|\,l = 0, 1, \ldots, J-1\}. \qquad \text{(A4.4)}$$

We have $(\varepsilon_J)^{lm} = (\varepsilon_J)^{l(m+J)} = (\varepsilon_J)^{(l+J)m}$. Therefore, G($l$), as defined by eqn (A4.1), forms a periodic sequence with the period J

$$G(l) = G(kJ + l), \qquad k = 0, \pm 1, \pm 2, \ldots. \qquad \text{(A4.5)}$$

Likewise, g(m), if defined by eqn (A4.4), forms a periodic sequence with the period J

$$g(m) = g(kJ + m), \qquad k = 0, \pm 1, \pm 2, \ldots. \qquad \text{(A4.6)}$$

A sequence $\{g(m) \mid m = 0, 1, \ldots, J-1\}$ is *even* if

$$g(m) = g(J-m), \qquad m = 1, 2, \ldots, J-1. \tag{A4.7}$$

It is *odd* if

$$g(0) = 0 \quad \text{and} \quad g(m) = -g(J-m), \qquad m = 1, 2, \ldots, J-1. \tag{A4.8}$$

As a consequence of (A4.8) we have, for odd sequences where the number of elements J is even, that $g(J/2) = 0$.

Any given sequence $\{g(m) \mid m = 0, 1, \ldots, J-1\}$ may be written as the sum of an even part with elements $g^e(m)$ and an odd part with elements $g^o(m)$, where

$$g^e(m) = \begin{cases} g(0), & m = 0 \\ \tfrac{1}{2}(g(m) + g(J-m)), & m = 1, 2, \ldots, J-1 \end{cases} \tag{A4.9}$$

and

$$g^o(m) = \begin{cases} 0, & m = 0 \\ \tfrac{1}{2}(g(m) - g(J-m)), & m = 1, 2, \ldots, J-1 \end{cases} \tag{A4.10}$$

Two examples of this decomposition is shown in Fig. A4.1.

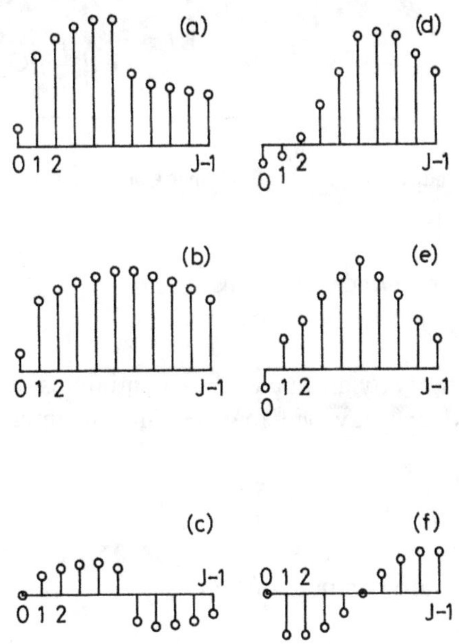

Fig. A4.1 *(a) A sequence having an odd number of elements with (b) its even part and (c) its odd part. (d) A sequence having an even number of elements with (e) its even part and (f) its odd part*
For simplicity, the sequences are shown as real

A4.2.2 Some discrete Fourier transform pairs

Periodic sequence ($m = 0, 1, \ldots, J-1$)	Discrete Fourier Transform ($l = 0, 1, \ldots, J-1$)
$g(m)$	$G(l)$
$g_1(m)$	$G_1(l)$
$g_2(m)$	$G_2(l)$
$ag_1(m) + bg_2(m)$	$aG_1(l) + bG_2(l)$ (linearity)
$g(-m)$	$G(-l)$
$g^*(m)$	$G^*(-l)$
$g^*(-m)$	$G^*(l)$
$g^e(m)$	$G^e(l)$
$g^o(m)$	$G^o(l)$
$g(m-n)$ (shifting)	$(\varepsilon_J)^{ln} G(l)$
$(\varepsilon_J)^{-km} g(m)$	$G(l-k)$ (shifting)
$\sum_{j=0}^{J-1} g_1(m-j) g_2(j)$ (convolution)	$G_1(l)G_2(l)$ (term-by-term product)
$g_1(m)g_2(m)$ (term-by-term product)	$\frac{1}{J}\sum_{j=0}^{J-1} G_1(l-j)G_2(j)$ (convolution)

The asterisk denotes complex conjugation.
The superscripts *e* and *o* indicate the even and odd parts of the sequences.

A4.3 Reconstruction of periodic functions

A4.3.1 Introduction

It is well known [7] that any given piecewise continuous function $f(\phi)$, periodic in ϕ with the period 2π, may be expanded into an infinite series of Fourier harmonics

$$f(\phi) = \sum_{k=-\infty}^{\infty} c_k e^{ik\phi} \tag{A4.11}$$

where the coefficients are computed from

$$c_k = \frac{1}{2\pi}\int_0^{2\pi} f(\phi) e^{-ik\phi} d\phi, \qquad k = \ldots, -1, 0, 1, \ldots. \tag{A4.12}$$

The problem addressed here is that of an alternative determination of the coefficients c_k without reference to the continuous function $f(\phi)$ as in eqn

(A4.12). Instead, only samples $f(l\Delta\phi)$, $l = 0, 1, \ldots, J-1$, are required where $\Delta\phi = 2\pi/J$, J being the number of samples in the interval $0 \leqslant \phi < 2\pi$.

In Section A4.3.2 $f(\phi)$ is assumed to be quasi-band-limited, i.e. the terms in (A4.11) with $|k| \leqslant N$ are dominant. In Section A4.3.3 the case where $f(\phi)$ is strictly band-limited is considered.

The determination of coefficients on the basis of samples is of importance in connection with measured data and numerical computations.

From the samples, the complete continuous function may be reconstructed (interpolated) as will be shown.

A4.3.2 Quasi-band-limited case

We consider first the case where the expansion (A4.11) contains an infinite number of terms of which those with $|k| \leqslant N$ are dominant. In other words,

$$f(\phi) = \sum_{k=-\infty}^{\infty} c_k e^{ik\phi} \tag{A4.13}$$

$$\cong \sum_{k=-N}^{N} c_k e^{ik\phi}. \tag{A4.14}$$

The samples of $f(\phi)$ are given by

$$f(l\Delta\phi) = \sum_{k=-\infty}^{\infty} c_k e^{ik(l\Delta\phi)} \tag{A4.15}$$

$$= \sum_{k=-\infty}^{\infty} c_k(\varepsilon_J)^{kl}, \quad l = 0, 1, \ldots, J-1 \tag{A4.16}$$

where ε_J is given by eqn (A4.2). Let us write the integer k as the sum

$$k = n + rJ \tag{A4.17}$$

where $n = 0, 1, \ldots, J-1$ and $r = \ldots, -1, 0, 1, \ldots$. Equation (A4.16) yields

$$f(l\Delta\phi) = \sum_{n=0}^{J-1} \sum_{r=-\infty}^{\infty} c_{n+rJ}(\varepsilon_J)^{(n+rJ)l} \tag{A4.18}$$

$$= \sum_{n=0}^{J-1} (\varepsilon_J)^{nl} \sum_{r=-\infty}^{\infty} c_{n+rJ}, \quad l = 0, 1, \ldots, J-1 \tag{A4.19}$$

since $(\varepsilon_J)^J = 1$. Defining now the *aliased coefficients* $\bar{c}_n$ by

$$\bar{c}_n = \sum_{r=-\infty}^{\infty} c_{n+rJ}, \qquad n = 0, 1, \ldots, J-1 \tag{A4.20}$$

we obtain from eqn (A4.19)

$$f(l\Delta\phi) = \sum_{n=0}^{J-1} \bar{c}_n(\varepsilon_J)^{nl}, \qquad l = 0, 1, \ldots, J-1. \tag{A4.21}$$

Note, that the aliased coefficients, ref. Fig. A4.2, form a periodic sequence (with

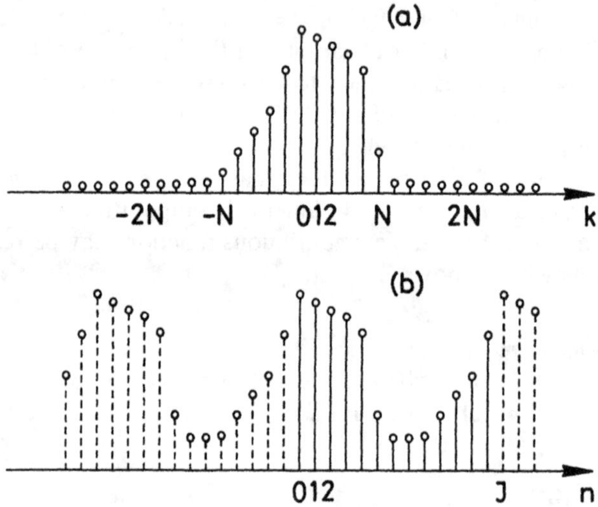

Fig. A4.2 *Illustrating (a) an infinite sequence c_k in which $2N+1$ elements are dominant and (b) its aliased sequence $\bar{c}_n$ of period $J \geqslant 2N+1$*
For simplicity, the sequences are shown as real

the period J) of complex numbers. Equation (A4.21) constitutes J linear equations in the J unknown coefficients $\bar{c}_n$. These may be solved immediately by recognizing that eqn (A4.21) has the form of a discrete Fourier transform (DFT), ref. Section A4.2, between the samples $f(l\Delta\phi)$, $l = 0, 1, \ldots, J-1$ of $f(\phi)$ and the coefficients $\bar{c}_n$, $n = 0, 1, \ldots, J-1$. Taking the inverse discrete Fourier transform (IDFT) on both sides yields

$$\bar{c}_n = \frac{1}{J}\sum_{l=0}^{J-1} f(l\Delta\phi)(\varepsilon_J)^{-ln}, \qquad n = 0, 1, \ldots, J-1. \tag{A4.22}$$

Of course, the original problem was to determine c_k, not $\bar{c}_n$. However, since c_k is small for $|k|$ larger than some integer N, and J (the number of samples in $0 \leqslant \phi < 2\pi$) is chosen larger than $2N$ we obtain from eqn (A4.20) the following approximate relationship

$$c_k \cong \begin{cases} \bar{c}_k & \text{for } 0 \leqslant k \leqslant N \\ \bar{c}_{k+J} & \text{for } -N \leqslant k < 0 \\ 0 & \text{for } |k| > N \end{cases} \qquad (J \geqslant 2N+1) \tag{A4.23}$$

ref. Fig. A4.2. The difference between the left- and right-hand sides of (A4.23) is the aliasing error. From eqns (A4.14), (A4.22) and (A4.23), follows the *approximate reconstruction* of $f(\phi)$ from its samples

$$f(\phi) \cong \sum_{k=-N}^{N} c'_k e^{ik\phi} \tag{A4.24}$$

where the coefficients c'_k are given by

$$\{c'_0, c'_1, \ldots, c'_N, 0, \ldots, 0, c'_{-N}, \ldots, c'_{-1}\} = \mathrm{IDFT}\{f(l\Delta\phi)\,|\,l = 0, 1, \ldots, J-1\}. \tag{A4.25}$$

In eqn (A4.25) both sequences are of length J.

A4.3.3 Band-limited case with 2N + 1 coefficients
An exact reconstruction is possible when $f(\phi)$ is band limited, i.e. $c_k = 0$ for $|k| > N$. We have

$$f(\phi) = \sum_{k=-N}^{N} c_k e^{ik\phi}. \tag{A4.26}$$

Defining the aliased coefficients $\bar{c}_n$ as before

$$\bar{c}_n = \sum_{r=-\infty}^{\infty} c_{n+rJ}, \qquad n = 0, 1, \ldots, J-1 \tag{A4.27}$$

the analysis proceeds as in A4.3.2. Two distinct cases may be considered:

A4.3.3.1 Sufficient number of samples: Provided the number J of samples in $0 \leqslant \phi < 2\pi$ is chosen larger than $2N$, we have the exact relationship

$$c_k = \begin{cases} \bar{c}_k & \text{for } 0 \leqslant k \leqslant N \\ \bar{c}_{k+J} & \text{for } -N \leqslant k < 0 \qquad (J \geqslant 2N+1) \\ 0 & \text{for } |k| > N \end{cases} \tag{A4.28}$$

ref. Fig. A4.3. The aliasing error is zero and we have the *exact reconstruction* of

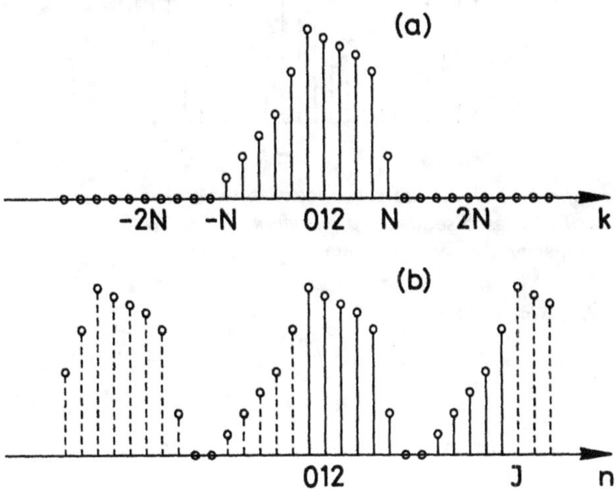

Fig. A4.3 *Illustrating (a) an infinite sequence c_k in which the elements vanish for $|k| > N$ and (b) its aliased sequence $\bar{c}_n$ of period $J \geqslant 2N+1$*
For simplicity, the sequences are shown as real

$f(\phi)$ from its samples

$$f(\phi) = \sum_{k=-N}^{N} c_k e^{ik\phi} \tag{A4.29}$$

where the coefficients are given by

$$\{c_0, c_1, \ldots, c_N, 0, \ldots, 0, c_{-N}, \ldots, c_{-1}\} = \mathrm{IDFT}\{f(l\Delta\phi) \,|\, l = 0, 1, \ldots, J-1\}. \tag{A4.30}$$

In eqn (A4.30) both sequences are of length J.

A4.3.3.2 Undersampling: If, on the other hand, the number J of samples in $0 \leqslant \phi < 2\pi$ is chosen equal to $2N$ or smaller, some overlap (aliasing) will take place when the sequence $\bar{c}_n$ is formed, ref. Fig. A4.4. In such cases it will not be possible to deduce all coefficients c_k, $-N \leqslant k \leqslant N$, from the knowledge of $\bar{c}_k$.

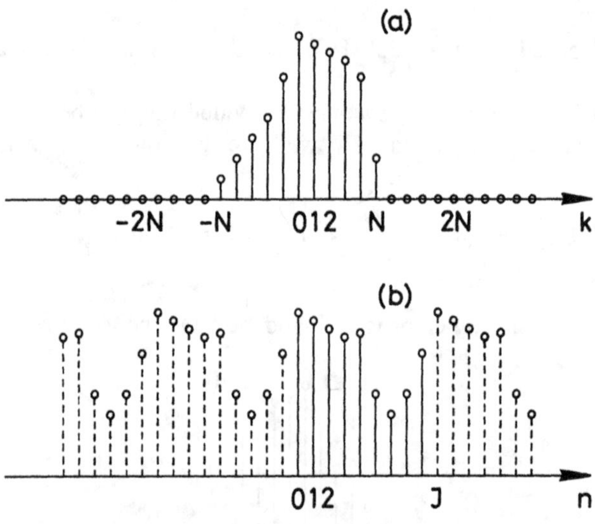

Fig. A4.4 *Illustrating (a) an infinite sequence c_k in which the elements vanish for $|k| > N$, and (b) its aliased sequence $\bar{c}_n$ of period J, $J \leqslant 2N$*
For simplicity, the sequences are shown as real

If $N < J \leqslant 2N$, we have

$$c_k = \begin{cases} \bar{c}_k & \text{for } k = 0, 1, \ldots, J-N-1 \\ \bar{c}_{k+J} & \text{for } k = -(J-N-1), \ldots, -2, -1 \\ \text{undetermined} & \text{for } |k| = J-N, J-N+1, \ldots, N. \end{cases} \qquad (N < J \leqslant 2N) \tag{A4.31}$$

If $J \leqslant N$ then all c_k coefficients are undetermined.

An exception to eqn (A4.31) occurs for $J = 2N$ (undersampling by one) if for the function $f(\phi)$ it is known that $c_N = c_{-N}$. In this particular case all $2N + 1$ coefficients are determined

$$c_k = \begin{cases} \bar{c}_k & \text{for } k = 0, 1, \ldots, N-1 \\ \frac{1}{2}\bar{c}_k & \text{for } k = N \\ \frac{1}{2}\bar{c}_{k+J} & \text{for } k = -N \\ \bar{c}_{k+J} & \text{for } k = -(N-1), \ldots, -2, -1 \end{cases}$$

$$(J = 2N \text{ and } c_N = c_{-N}). \qquad \text{(A4.32)}$$

A4.4 Special cases of reconstruction

A4.4.1 Introduction

In Section A4.3.3 we have considered a band-limited function $f(\phi)$, periodic with the period 2π, and having $2N + 1$ Fourier coefficients c_k

$$f(\phi) = \sum_{k=-N}^{N} c_k \, e^{ik\phi}. \qquad \text{(A4.33)}$$

As we have seen, the values of c_k can be determined from a sequence of equispaced samples of $f(\phi)$. Firstly, we computed the aliased coefficients $\bar{c}_k$ by taking the inverse discrete Fourier transform of the sequence of samples

$$\{\bar{c}_k \,|\, k = 0, 1, \ldots, J-1\} = \text{IDFT}\{f(l\Delta\phi) \,|\, l = 0, 1, \ldots, J-1\} \qquad \text{(A4.34)}$$

where $\Delta\phi = J/2\pi$ is the sampling increment. Secondly, we found the original coefficients c_k from $\bar{c}_k$ in a simple manner. If $J \geqslant 2N + 1$ all coefficients c_k could be obtained and are given by

$$\{c_0, c_1, \ldots, c_N, 0, \ldots, 0, c_{-N}, \ldots, c_{-1}\} = \{\bar{c}_k \,|\, k = 0, 1, \ldots, J-1\} \qquad \text{(A4.35)}$$

where both sequences are of the length J.

With the coefficients known, the finite series in eqn (A4.33) can be evaluated. From its samples, the function $f(\phi)$ is hereby reconstructed for all values of ϕ.

In the following sections some variations and modifications of the above simple case of reconstruction are dealt with.

A4.4.2 Shifting of sample points

Given the sequence of samples $\{f(l\Delta\phi) \,|\, l = 0, 1, \ldots, J-1\}$ of the function $f(\phi)$, ref. Section A4.4.1, we seek the sequence of shifted samples $\{f(l\Delta\phi + \phi_o) \,|\, l = 0, 1, \ldots, J-1\}$. It is assumed that $J \geqslant 2N + 1$.

We have

$$f(l\Delta\phi + \phi_o) = \sum_{k=-N}^{N} c_k \, e^{ik(l\Delta\phi + \phi_o)} = \sum_{k=-N}^{N} e_k c_k \, e^{ikl\Delta\phi} \qquad \text{(A4.36)}$$

where $e_k = e^{ik\phi_o}$, $k = -N, \ldots, 0, \ldots, N$. Introducing the aliased coefficients $\bar{c}_k$, ref. eqn (A4.34) and $\bar{e}_k$, both of period J, where

$$\{\bar{e}_k \,|\, k = 0, 1, \ldots, J-1\} = \{e_0, e_1, \ldots, e_N, 0, \ldots, 0, e_{-N}, \ldots, e_{-1}\} \quad \text{(A4.37)}$$

the sequence sought for is given by

$$\{f(l\Delta\phi + \phi_o) \,|\, l = 0, 1, \ldots, J-1\} = \mathrm{DFT}\{\bar{e}_k \bar{c}_k \,|\, k = 0, 1, \ldots, J-1\} \quad \text{(A4.38)}$$

where the sequence $\{\bar{e}_k \bar{c}_k \,|\, k = 0, 1, \ldots, J-1\}$ on the right-hand side is the term-by-term product sequence of $\bar{e}_k$ and $\bar{c}_k$.

If ϕ_o is a multiple of $\Delta\phi$, eqn (A4.38) corresponds to the shift theorem in Section A4.2.2. Note, that if $|\phi_o| \gg \Delta\phi$ it may be numerically advantageous to reduce $|\phi_o|$ by cyclic shifts of the samples before using the above scheme.

A4.4.3 Increased resolution

Given the sequence of samples $\{f(l\Delta\phi) \,|\, l = 0, 1, \ldots, J-1\}$ of the function $f(\phi)$, ref. Section A4.4.1, we seek the sequence $\{f(l\Delta'\phi) \,|\, l = 0, 1, \ldots, J'-1\}$ corresponding to an increased sampling density, i.e. $J' > J \geqslant 2N+1$. Further, $\Delta'\phi = 2\pi/J'$.

Firstly, the sequence of aliased coefficients $\bar{c}_k$ of period J is found, ref. eqn (A4.34). This sequence contains the original coefficients c_k, possibly interspaced with zeros, ref. eqn (A4.35). Next, from the $\bar{c}_k$ sequence a new, longer sequence $\bar{\bar{c}}_k$ is formed by filling in $(J'-J)$ additional zeros† between the elements c_N and c_{-N}.

$$\{\bar{\bar{c}}_k \,|\, k = 0, 1, \ldots, J'-1\} = \{c_0, c_1, \ldots, c_N, 0, 0, \ldots, 0, 0, c_{-N}, \ldots, c_{-1}\}. \quad \text{(A4.39)}$$

In this equation both sequences are of the length J'. The sequence sought for now follows from

$$\{f(l\Delta'\phi) \,|\, l = 0, \ldots, J'-1\} = \mathrm{DFT}\{\bar{\bar{c}}_k \,|\, k = 0, 1, \ldots, J'-1\} \quad \text{(A4.40)}$$

The above scheme is Whittaker interpolation [9] applied to periodic functions. It is useful in plotting routines for, for example, antenna radiation patterns. If extended to two dimensions, the scheme can also be used for the generation of high-resolution contour plots.

† In cases where N is unknown, or a sharp band limitation of $f(\phi)$ cannot be assumed, the elements shown as zeros in (A4.39) may not be exact zeros. If, furthermore, J is even the sequence (A4.39) contains a middle element $c_{J/2}$ and it becomes a question where to fill in the zeros. It may, however, be argued that a correct procedure is to first split the middle element into two elements (of each $\frac{1}{2}c_{J/2}$) and then fill in $(J'-J-1)$ zeros in the middle of the sequence.

A4.4.4 Suppression of noise

As a primitive scheme of noise suppression when reconstructing $f(\phi)$, ref. Section A4.4.1, from samples with noise, the following method based on oversampling is useful. For simplicity in notation, the primed quantities below denote the corresponding unprimed quantities with superimposed noise contributions.

The number J of samples is chosen such that $J \gg 2N + 1$. Assuming, for a moment, noise-free conditions, the aliased coefficients $\bar{c}_k$ are given by eqn (A4.34). The sequence $\bar{c}_k$ contains $\{J - (2N + 1)\}$ zeros interspaced between the elements of significance, ref. eqn (A4.35).

With noise present, J samples of $f'(\phi)$ are taken and we first compute

$$\{\bar{c}'_k \,|\, k = 0, 1, \ldots, J - 1\} = \mathrm{IDFT}\{f'(l\Delta\phi) \,|\, l = 0, 1, \ldots, J - 1\}. \quad \text{(A4.41)}$$

In this sequence all elements are different from zero, and the $\{J - (2N + 1)\}$ elements interspaced between the elements of significance now contain noise only. As the next step, the mentioned $\{J - (2N + 1)\}$ elements in $\bar{c}'_k$ are all set equal to zero, and the resulting sequence denoted by $\bar{c}''_k$. A reconstruction with reduced noise (indicated by the double prime) can now be based on $\bar{c}''_k$

$$\{f''(l\Delta\phi) \,|\, l = 0, 1, \ldots, J - 1\} = \mathrm{DFT}\{\bar{c}''_k \,|\, k = 0, 1, \ldots, J - 1\}. \quad \text{(A4.42)}$$

If the fine resolution arising from using $J \gg 2N + 1$ samples is not needed, the resolution may be decreased according to the scheme in Section A4.4.5. The above scheme is well suited to antenna measurements, when the value of N is given beforehand by the size of a well-defined minimum sphere. The amount of suppression of noise depends on the noise distribution and the degree of oversampling.

A4.4.5 Decreased resolution

Given the sequence of samples $\{f(l\Delta\phi) \,|\, l = 0, 1, \ldots, J - 1\}$ of the function $f(\phi)$, ref. Section A4.4.1, we seek the sequence $\{f(l\Delta'\phi) \,|\, l = 0, 1, \ldots, J' - 1\}$, where $\Delta'\phi = 2\pi/J'$, corresponding to a decreased sampling density, i.e. $J' < J$. It is assumed that $J \geqslant 2N + 1$.

Firstly, the sequence of aliased coefficients $\bar{c}_k$ of period J is found, ref. eqn (A4.34). This sequence contains the original coefficients c_k, possibly interspaced with zeros, ref. eqn (A4.35). The sequence of original coefficients

$$\{c_k \,|\, k = -N, \ldots, 0, \ldots, N\} \quad \text{(A4.43)}$$

can therefore easily be formed. Next, on the basis of the sequence (A4.43) we generate, ref. eqn (A4.27) a new aliased sequence $\bar{c}'_k$ of period $J' < J$. Of course, since $J' < J$, it is likely that some overlap will take place in this process.

The sequence sought for is now given by

$$\{f(l\Delta'\phi) \,|\, l = 0, 1, \ldots, J' - 1\} = \mathrm{DFT}\{\bar{c}'_k \,|\, k = 0, 1, \ldots, J' - 1\} \quad \text{(A4.44)}$$

valid for all J' in the interval $1 \leqslant J' < J$. Note, that even the trivial case of just one sample, $J' = 1$ is covered by eqn (A4.44).

A4.5 Numerical integration of periodic, band-limited functions

A4.5.1 General case

We consider the numerical evaluation of the integral

$$I_0 = \frac{1}{2\pi}\int_0^{2\pi} \mathrm{f}(\phi)\,\mathrm{d}\phi \tag{A4.45}$$

where $\mathrm{f}(\phi)$is a periodic band-limited function having $2N + 1$ Fourier coefficients c_k, ref. eqn (A4.33). Insertion of (A4.33) into (A4.45) and term-by-term integration yields $I_0 = c_0$. The coefficient c_0 can now be determined from as few as $J = N + 1$ samples of $\mathrm{f}(\phi)$ in the interval $0 \leqslant \phi < 2\pi$, ref. eqns (A4.22) and (A4.31) and we obtain

$$I_0 = \frac{1}{J}\sum_{l=0}^{J-1} \mathrm{f}(l\Delta\phi). \tag{A4.46}$$

Hence, the integral over the period 2π of a periodic function that is band-limited to $k = \pm N$, can be *evaluated exactly by the trapezoidal rule* of numerical integration (equal weights on all samples) with $N + 1$ sample points in the interval $0 \leqslant \phi < 2\pi$. Integration rules of higher order, e.g. Simpson's rule, would not be exact for this number of sample points.

A4.5.2 A special case

As an application of eqns (A4.45–4.46) we shall investigate the numerical evaluation of the often encountered integrals

$$I_m = \frac{1}{2\pi}\int_0^{2\pi} \mathrm{f}(\phi)\,\mathrm{e}^{-\mathrm{i}m\phi}\,\mathrm{d}\phi, \qquad m = -N, \ldots, 0, \ldots, N \tag{A4.47}$$

where $\mathrm{f}(\phi)$ is as specified in Section A4.5.1. The integrands in eqn (A4.47) are again band-limited, however with the highest harmonics now being $\mathrm{e}^{\mathrm{i}2N\phi}$ and $\mathrm{e}^{-\mathrm{i}2N\phi}$. From the result in the general case (Section A4.5.1) it therefore follows that the integral (A4.47) can be evaluated *exactly* by the trapezoidal rule for all values of m, from just $J = 2N + 1$ samples of the integrand

$$I_m = \frac{1}{J}\sum_{l=0}^{J-1} \mathrm{f}(l\Delta\phi)\,\mathrm{e}^{-\mathrm{i}ml\Delta\phi}, \qquad m = -N, \ldots, 0, \ldots, N \tag{A4.48}$$

where $J = 2\pi/\Delta\phi$.

But the sums in (A4.48) are nothing but the elements in the IDFT of a sequence of J samples of $\mathrm{f}(\phi)$, ref. Section A4.3.3

$$\{I_0, I_1, \ldots, I_N, I_{-N}, \ldots, I_{-1}\} = \mathrm{IDFT}\{\mathrm{f}(l\Delta\phi)\,|\,l = 0, 1, \ldots, J-1\}. \tag{A4.49}$$

The IDFT can thus be viewed as a powerful tool by which the trapezoidal rule can be applied to all $2N + 1$ integrals in eqn (A4.47) simultaneously.

References

1. DRABOWITCH, S. and ANCONA, C., 1978, *Antennes*, Tome **II** (Applications), Masson, Paris.
2. SHAPIRA, J., 1979, 'Sampling and processing in antenna pattern measurement', Record of the 11th Convention of IEEE in Israel, Paper No. **B2-5,** Tel Aviv, Israel, Oct.
3. JOY, E. B. and PARIS, D. T., 1972, 'Spatial sampling and filtering in near-field measurements', *IEEE Trans.*, Vol. **AP-20,** No. 3, pp. 253–261, May.
4. LEWIS, R. L., 1981, 'Spherical scanning data processing; An algorithm for halving the data processing effort when the radiation into the back hemisphere is negligible', *IEEE AP-S Symposium Digest*, Vol. **1**, pp. 246–249, Los Angeles, USA, June.
5. HESS, D. W. and TAVORMINA, J. J., 1982, 'An algorithm with enhanced efficiency for computing the far field from near-field data on a partial spherical surface', URSI National Radio Science Meeting, University of Colorado, Boulder, Colorado, USA, Jan.
6. MOSTAFAVI, M. and BOLOMEY, J. C., 1985, 'Correction scheme for continuous rotation of the test antenna using fast near-field measurement techniques', *IEEE Trans.*, Vol. **AP-33,** No. 8, pp. 922–924.
7. PAPOULIS, A., 1977, *Signal analysis*, McGraw-Hill, Inc., New York.
8. HARRIS, F. J., 1978, 'On the use of windows for harmonic analysis with the discrete Fourier transform', *Proc. IEEE*, Vol. **66,** No. 1, pp. 51–83, Jan.
9. WHITTAKER, E. T., 1915, 'On the functions which are represented by the expansions of the interpolation theory', *Proc. Royal Soc. Edinburgh*, Vol. **35,** pp. 181–194.

Appendix A5

List of principal symbols and uses

a_{smn} = incoming wave coefficient page 27
a_j = incoming wave coefficient (single index convention) 27
$\mathbf{a}$ = column matrix with elements $a_1, a_2, \ldots$ 28
b_{smn} = outgoing wave coefficient 27
b_j = outgoing wave coefficient (single index convention) 27
$\mathbf{b}$ = column matrix with elements $b_1, b_2, \ldots$ 28
$d^n_{\mu m}(\theta)$ = rotation coefficient 38, 65, 345–347
$\exp(-i\omega t)$ = time factor (suppressed) 9, 159, 312
$h_n^{(1)}(kr)$ = spherical Hankel function of the first kind 11, 18, 22, 314–317
i = imaginary unit 9
$\hat{\imath}_{co}(\theta, \phi)$ = copolar unit vector 50–52
$\hat{\imath}_{cross}(\theta, \phi)$ = cross-polar unit vector 50–52
i, j = indices (single index convention) 15, 313
$j_n(kr)$ = spherical Bessel function 11, 314–317
k = free-space propagation constant, $2\pi/\lambda$ 10, 312
$\vec{k}_o$ = plane-wave propagation vector 263
$\hat{n}$ = surface unit normal vector 35, 264
r_0 = radius of minimum sphere 12, 17
r = radial vector 10
s, m, n = indices 13–16, 64, 312, 313
t = time 9
u_n = element of spherical wave error spectrum 285
v = complex amplitude of a wave, incoming in a waveguide or applied to a transmitting test antenna 27, 63
v_p = complex amplitude of a wave applied to a transmitting probe 67
w = complex amplitude of a wave, outgoing in a waveguide or received by a receiving probe 27, 67, 77
w_t = complex amplitude of a wave received by a receiving test antenna 68

x, y, z = rectangular coordinates 10
$\hat{x}, \hat{y}, \hat{z}$ = rectangular unit vectors 10
r, θ, ϕ = spherical coordinates 10
$\hat{r}, \hat{\theta}, \hat{\phi}$ = spherical unit vectors 10
A = measurement distance 62, 216
$A_{\theta\phi}$ = amplitude phase factor 154, 161–163
$C_{\sigma\mu\nu}^{sn(3)}(kA)$ = translation coefficient 66, 355–360
$D, D(\theta, \phi)$ = directivity 54–58, 123
$\vec{E}, \vec{H}$ = electric and magnetic field vectors 9
$\vec{F}$ = combined field vector 282
$\vec{F}_{smn}^{(c)}(r, \theta, \phi)$ = spherical vector wave function 13, 312, 325–328
$G, G(\theta, \phi)$ = gain 54–55, 123, 207
$\mathbf{I}$ = unit matrix 29
J = upper limits on indices i and j 15, 313
$\vec{J}, \vec{M}$ = electric and magnetic current densities 9, 332
$\vec{J}_s, \vec{M}_s$ = electric and magnetic surface current densities 264
$\vec{J}^u, \vec{J}^v$ = symmetrical and antisymmetrical current densities 334
$\vec{J}^w$ = geometrical image current density 334
$\vec{J}^L$ = sampled current density 337
$\vec{K}_{smn}(\theta, \phi)$ = far-field pattern function 49, 314, 328–330
$\vec{K}(\theta, \phi)$ = far-field pattern 50, 124
L = number of probes or samples 295, 338
N, M = upper limits on indices n and m 15, 17, 128, 129
N_r = number of repetitions 245
$P_n^m(\cos\theta)$ = associated Legendre function 11, 318–323
$\bar{P}_n^m(\cos\theta)$ = normalized associated Legendre function 13, 318, 322–323
$P_{s\mu n}^{p}$ = probe response constant 74–77, 119
P_{rad} = radiated power 286, 303, 304
$\{P_{\text{rad}}^n | n = 1, 2, \ldots\}$ = power spectrum 236, 286
$\vec{P}(k_x, k_y, z)$ = plane-wave spectrum 272
$Q_{smn}^{(c)}$ = general spherical wave coefficient 14, 312, 313
$Q_{smn}^{(4)}$ = incoming wave coefficient ($= a_{smn}$) 27, 97
$Q_{smn}^{(3)}$ = outgoing wave coefficient ($= Q_{smn} = b_{smn}$) 25, 27, 63, 97
$R_{sn}^{(c)}(kr)$ = radial function 25, 314–317
R_{smn} = receiving coefficient of an antenna 36
R_j = receiving coefficient (single index convention) 27
$\mathbf{R}$ = row matrix with elements $R_1, R_2, \ldots$ 27
S = surface 33, 264
$\vec{S}$ = Poynting's complex vector 23
$S_{smn}^{\sigma\mu\nu}$ = scattering matrix coefficient of an antenna 36
S_{ij} = scattering matrix coefficient (single index convention) 28
$\mathbf{S}$ = matrix with elements $S_{11}, S_{12}, \ldots, S_{21}, S_{22}, \ldots$ 28, 42–45, 46
$T(\vec{r})$ = relative plane-wave deviation 280, 281

$T_{\max}(r)$ = spherical profile of $T(\vec{r})$ 283
T_{smn} = transmitting coefficient of an antenna 36, 63, 101
T_i = transmitting coefficient (single index convention) 28
$\mathbf{T}$ = column matrix with elements $T_1, T_2, \ldots$ 28
V = volume 33, 264
$W(\chi, \theta, \phi)$ = normalized far-field input probe signal 78, 120
$W'(\chi, \theta, \phi)$ = normalized far-field output probe signal 120
$\mathbf{W}$ = matrix of transformation weights 259
$\delta_{ij} = \begin{cases} 1, & i = j \\ 0, & i \neq j \end{cases}$ (Kronecker's delta) 26, 318
$\delta^L_{mn} = \begin{cases} 1, & m = n \pmod L \\ 0, & m \neq n \pmod L \end{cases}$ 296, 339
$\delta(x)$ = Dirac's delta 299, 336
$\Delta^n_{\mu m}$ = delta $(= d^n_{\mu m}(\pi/2))$ 346–354
λ = wavelength 10
ζ, η = specific impedance and admittance 10, 312
ε, μ = permittivity and permeability 9
σ, μ, ν = indices 25, 64
ϕ_r = repetition angle 245
$(\chi_o, \theta_o, \phi_o)$ = Euler angles 343, 344
ω = angular frequency 9
Γ = reflection coefficient 27
∇ = del; differential vector operator 9
∇^2 = Laplacian operator 10
$*$ = superscript, denoting complex conjugate 23, 29
T = superscript, denoting transposed 29
$+$ = superscript, denoting Hermitian conjugate 29
$\hat{}$ = denotes unit vector 10
$\vec{}$ = indicates a vector quantity 9
$\mathbf{A}, \mathbf{a}$ = boldface letters denote matrices 27
$\{g(m) \mid m = 1, 2, \ldots\}$ = sequence 362

Index

www.ingramcontent.com/pod-product-compliance
Lightning Source LLC
LaVergne TN
LVHW011905200926
848484LV00002B/535